ANIMALS WITHOUT BACKBONES

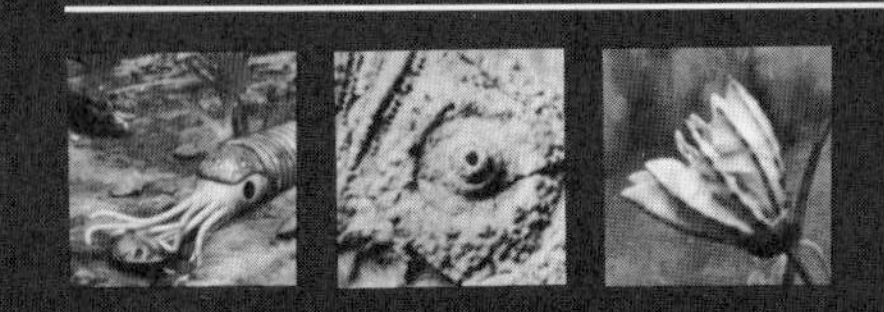

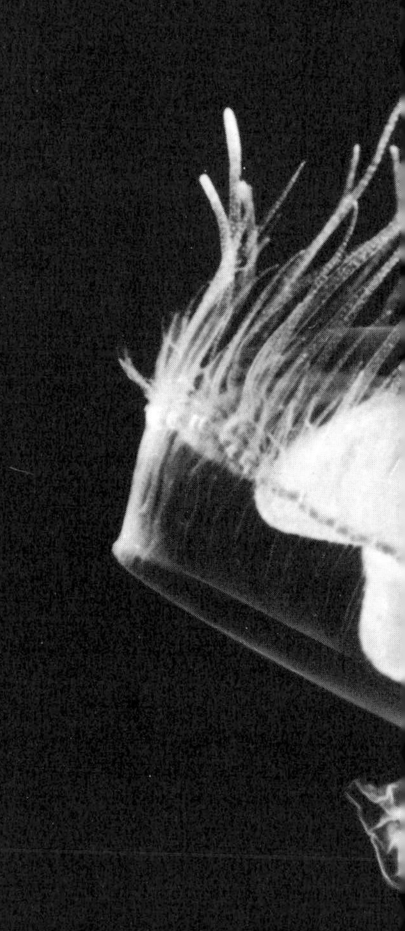

The University of Chicago Press
Chicago and London

Animals Without Backbones

THIRD EDITION

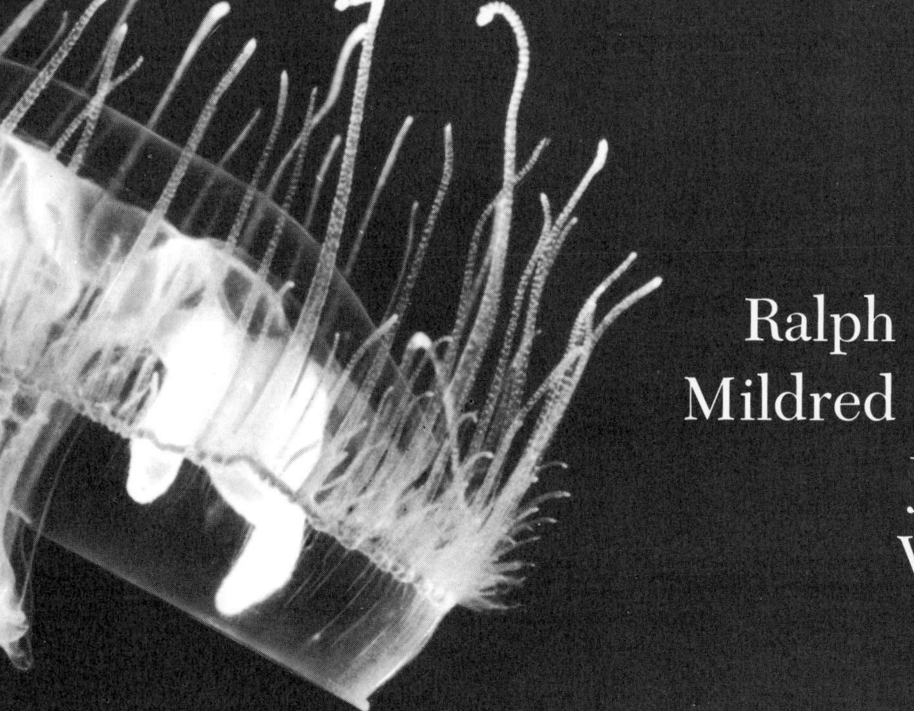

Ralph Buchsbaum
Mildred Buchsbaum
John Pearse
Vicki Pearse

The University of Chicago Press, Chicago 60637
The University of Chicago Press, Ltd., London

Printed in the United States of America

94 93 92 91 90 543

Library of Congress Cataloging-in-Publication Data

Buchsbaum, Ralph Morris, 1907-
Animals without backbones.

Bibliography: p.
Includes index.
1. Invertebrates. I. Title.
QL362.B93 1987 592 86-7046
ISBN 0-226-07873-6
ISBN 0-226-07874-4 (pbk.)

To

the animals without backbones

who have inspired
our continuing interest and affection

and to

the students and teachers
whose enthusiasm has reinforced us.

Contents

Preface

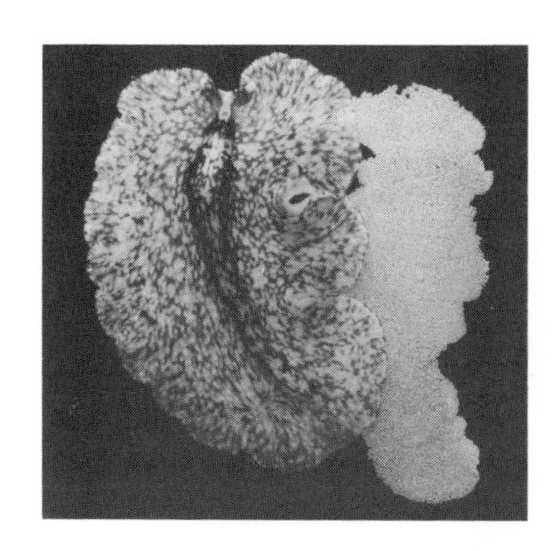

The goal of the Third Edition of *Animals Without Backbones* remains the same as that of previous editions: to present the major groups of invertebrate animals in simple, nontechnical language.

During the nearly 50 years that this book has enjoyed wide readership, it has brought us feedback from a broad variety of users with different levels of biological background and with different needs. On the basis of this experience, we have preserved in this edition most features of the original version. It remains primarily an introduction to the invertebrates for students in advanced high school courses, junior colleges, and colleges, but is designed to be useful to anyone who wants to know about the major kinds of animals.*

In length and level, the present edition of *Animals Without Backbones* remains essentially unchanged, but the contents of the text material have been updated throughout, as have the bibliography and classification. Some of the chapters have been extensively reorganized. In selecting new material from the overwhelming literature now available, we have tried to choose examples of observations and experiments that illuminate the principles of biology and of scientific investigation, and we have aimed at maintaining a balance among natural history, behavior, physiology, evolutionary relationships, and the many other aspects that together make up the biology of each group. The basic morphology is presented with a minimum of burdensome terminology.

The numerous drawings and photographs retain their important role as a complement and supplement to the text. Many of the original drawings have been revised to incorporate recent information. The photographs, which before were restricted to special sections, are now distributed and fully integrated with the text material.

Most of the original drawings, made for the first edition by

*For a more comprehensive and advanced treatment, see our *Living Invertebrates* (Blackwell/Boxwood, 1987).

Elizabeth Buchsbaum Newhall, have been retained, some with minor revisions by Mildred K. Waltrip, as necessitated by new information. Many of the new drawings in this edition were prepared by Mildred K. Waltrip for our book, *Living Invertebrates*; they conform to the style of the drawings in earlier editions of *Animals Without Backbones* and they use the same symbols, making for the ready interpretation that has been a much appreciated feature of this book. The large diagram of a paramecium is by Katherine P. O'Brien. The new substitution for the life history of *Aurelia* is by Jonathan Dimes. About 250 of the photographs in this third edition are new.

As before, we welcome and appreciate corrections, suggestions, or other comments from our readers.

Ralph Buchsbaum
Mildred Buchsbaum
John Pearse
Vicki Pearse

Pacific Grove, California
and
Santa Cruz, California

ANIMALS WITHOUT BACKBONES

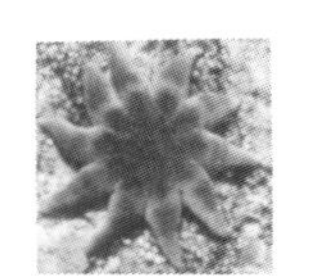

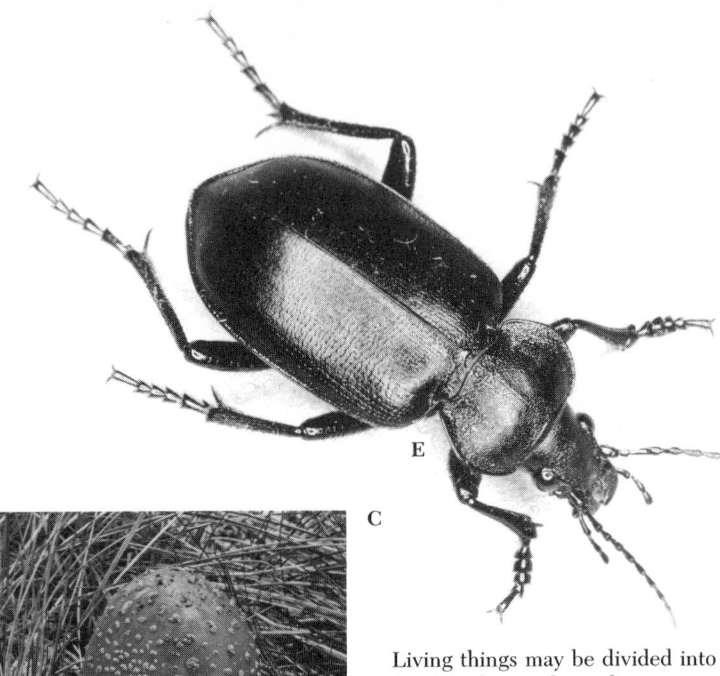

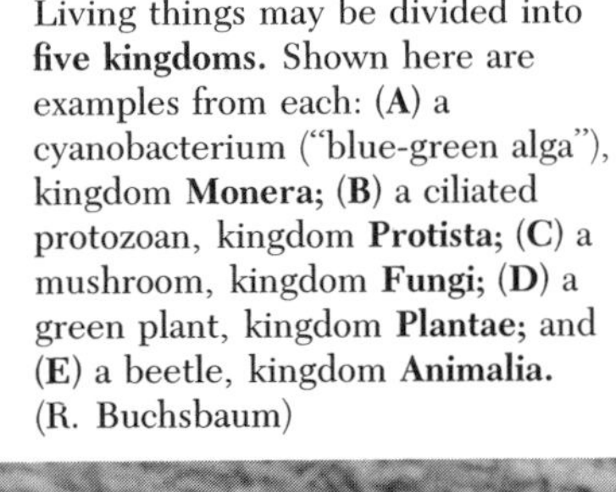

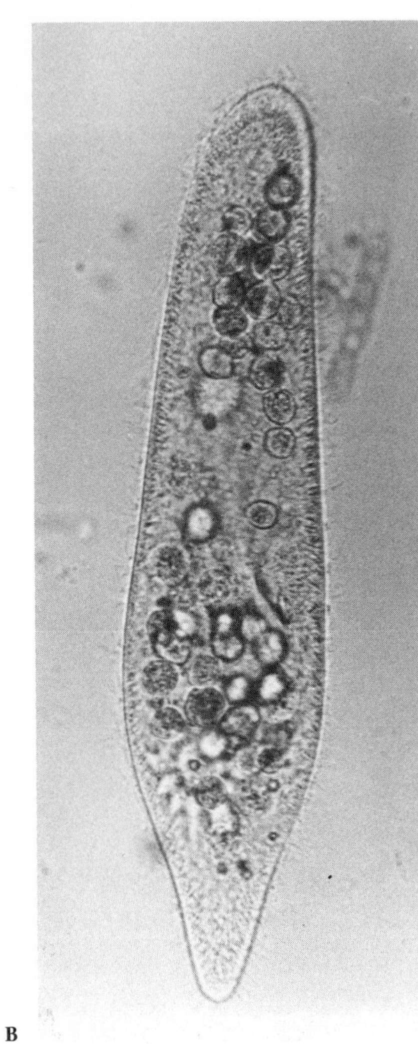

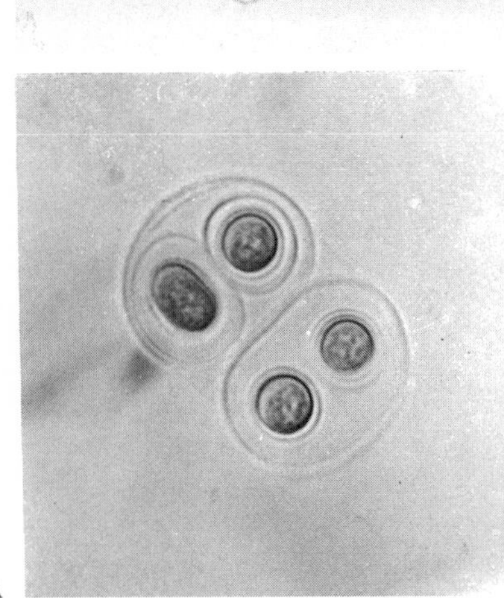

Living things may be divided into **five kingdoms.** Shown here are examples from each: (**A**) a cyanobacterium ("blue-green alga"), kingdom **Monera;** (**B**) a ciliated protozoan, kingdom **Protista;** (**C**) a mushroom, kingdom **Fungi;** (**D**) a green plant, kingdom **Plantae;** and (**E**) a beetle, kingdom **Animalia.** (R. Buchsbaum)

Chapter 1

Introduction: Sorting Out Living Things

Anyone can tell the difference between a tree and a cow. The tree stands still and shows no signs of perceiving your presence or your hand upon its trunk. The cow moves about and appears to notice your approach. This striking difference in the behavior of plants and animals is related to the fundamental difference in plant and animal nutrition.

Plants make their own food from simple constituents in air and soil, or in water. By means of a green pigment, chlorophyll, the leaves of a tree capture solar energy and use it to combine carbon dioxide and water into sugar—a process known as photosynthesis ("putting things together with the aid of light"). The energy stored in the sugar can later be released and used by the tree to combine simple substances into the complex organic substances of which all living things are made.

Animals cannot stand in the sun, soak up energy, and store it in chemicals such as sugar. A cow must get its energy by eating plants. To find a constant supply of energy-giving food, a cow must be able to move from place to place and must react to other animals and to changes in the environment. A pasture offers few threats to cows, except those posed by humans, but any small animals such as rabbits or mice must be alert to bigger animals such as dogs or coyotes that might view them as a tasty source of energy.

Not all animals move about. Corals, for example, grow firmly attached to the sea bottom and depend on water currents to bring a steady supply of small animals within reach of their tentacles. The feeding activities of corals and many other stationary animals were not apparent to early naturalists, who classified them as plants. And in some ways stationary or sedentary animals do resemble plants. Instead of fleeing from predators, such animals protect themselves by developing hard, often spiny skeletons and distasteful or toxic substances in their tissues. Comparable defenses in plants may be tough bark, thorny stems, or tissue toxins.

There are many strategies for obtaining energy, and some stationary organisms neither capture solar energy through photosynthesis nor directly ingest the bodies of other organisms. Instead they absorb energy-rich organic substances. Mushrooms, for example, must live on or near the decaying organic matter that provides their nourishment. Mushrooms and other fungi thus differ sharply from both plants and animals in their mode of nutrition.

Plants and other organisms that supply themselves from simple, inorganic chemicals are called **autotrophs** ("self-nourished) or *producers* (because they produce food). Animals and fungi are called **heterotrophs** ("nourished by others") or *consumers* (because they take in food). Animals and fungi are further distinguished from each other by the mode in which they consume their food: animals are typically *ingestors* while fungi are *absorbers.*

As we examine more and more kinds of life, distinctions of behavior and nutrition grow less and less obvious. Eventually we find microscopic forms that exhibit characteristics possessed by plants, fungi, and animals. Most of these organisms are single cells, or **unicells,** with bodies that are not divided up by cell membranes or walls. Some unicells carry on photosynthesis like plants, but they move about and show the same sensitivity and rapidity of response as do typical animals. Some feed like animals, actively ingesting other small unicells, but also photosynthesize. And as extra insurance against hard times, most unicells can also absorb energy-rich organic matter as fungi do.

Thus, the characteristics of nutrition and behavior commonly used to distinguish between the familiar many-celled organisms, the multicellular plants and animals, are of little use for sorting out the unicellular organisms. Rather, unicells are characterized in part by their *internal cellular organization.* Many are organized like the cells of multicellular organisms: each unit has distinctive little structures called *organelles,* one of them a **nucleus** (pleural: *nuclei*), a discrete body containing most of the hereditary material (genes) of the cell. Whether unicellular or multicellular, organisms in which each cellular unit contains one or more nuclei, and other organelles, are said to be **eukaryotic** ("with a true nucleus"). In contrast, many other unicells, including the bacteria and cyanobacteria ("blue-green algas"), contain few or no organelles and the hereditary material is dispersed in the cell; these are said to be **prokaryotic** ("before a nucleus").

Mostly on the basis of cellular organization and also, for multicellular forms, mode of nutrition, organisms can be divided into five **kingdoms,** as shown in the table below. One kingdom, the **Monera,** includes all prokaryotic organisms, mainly *bacteria,* which may be autotrophs or may absorb nutrients from other organisms, and photosynthetic *cyanobacteria,* which are important producers, especially in stagnant water. The second kingdom, the **Protista,** includes eukaryotic organisms such as *protozoans* that are unicellular or form only simple colonies of cells. The members of the remaining three kingdoms are all eukaryotic multicellular organisms that are distinguished from each other mainly on the basis of nutrition: **Plantae** are primarily photosynthetic autotrophs, **Fungi** are absorbing heterotrophs, and **Animalia** are predominantly ingestive heterotrophs.

Table 1.1. KINGDOMS OF LIVING THINGS

Kingdom	Cellular Organization	Mode of Nutrition	Examples
Monera ("single")	Prokaryotic; Unicellular or simple colonies	Autotrophs & Heterotrophs (absorbers)	Bacteria, Cyanobacteria
Protista ("very first")	Eukaryotic; Unicellular or simple colonies	Autotrophs & Heterotrophs (absorbers & ingestors)	Flagellates, amebas, ciliates
Plantae ("plants")	Eukaryotic; Multicellular	Autotrophs	Trees, grasses, rosebushes
Fungi ("sponge-like")	Eukaryotic; Multicellular	Heterotrophs (absorbers)	Mushrooms, mildews, yeasts
Animalia ("animals")	Eukaryotic; Multicellular	Heterotrophs (ingestors)	Earthworms, lobsters, humans

One of the most plausible hypotheses advanced to account for the origin of life states that early in the earth's history, nearly 4 billion years ago, organic compounds formed and accumulated in shallow pools of water. Energy from the sun, lightning, and volcanic activity produced changes in these substances that made them combine into increasingly complex compounds. Some eventually developed the capability of self-propagation, using the accumulated organic substances around them as energy sources and building blocks.

By analogy, *viruses* today are complex organic compounds (nucleic acids) that occur inside cells of organisms and self-propagate using the organic compounds within the cells as energy sources and building blocks. Sometimes this activity causes severe diseases in the host organisms, including smallpox, herpes, mumps, polio, and AIDS in humans. Before leaving a cell, the viral nucleic acid is enclosed with a protein coat that protects it and enables it to identify and contact more host cells. Because they cannot propagate outside their host cells, viruses are commonly thought to be nuclear remnants and fragments of more complex cells; however, it is possible that some are relicts of the first self-propagating particles in the earliest seas. They appear to be on the borderline between the nonliving and living, and sometimes they are placed in a separate kingdom, the Archetista.

If we imagine that the earliest self-propagating substances were something like viruses that depended on organic material, it is not difficult to suppose that further development of structures such as cell membranes and complex metabolic pathways could lead to larger, more complex organisms at the moneran level, some of which could obtain energy wholly or partly from *in*organic substances. Some bacteria today, for example, which live only in habitats that lack oxygen as did the primordial surface of the earth, obtain energy by combining hydrogen and carbon dioxide to form methane, or capture the energy of the sun using pigments similar to chlorophyll but not producing oxygen. Cyanobacteria, which have a fossil record in excess of 3 billion years, use chlorophyll to capture the sun's energy and produce sugars from carbon dioxide and water, leaving oxygen as a by-product. Oxygen production paved the way for bacteria that can obtain energy by oxidizing inorganic compounds of nitrogen, sulfur, and iron—and for the evolution of more complex forms that depend on oxygen and organic food.

The development of larger, more highly organized eukaryotic protists from prokaryotic monerans is a relatively short jump in complexity, and although no one knows how it happened, or how many times it happened, fossils of protistlike cells some 2 billion years old are known. By forming complex colonies of cells, various types of protists could have given rise to various kinds of multicellular organisms, including plants, fungi, and animals; the oldest fossils of multicellular organisms are about 0.7 billion years old.

When animals are carefully studied and compared, it is found that many of them have a row of bones (vertebras) along the middle of the back, as well as bones inside the limbs and head. Animals having such internal bones, including a vertebral column, or backbone, are known as **vertebrates** and comprise all the fishes, the frogs, toads, and salamanders, the turtles, lizards, snakes, crocodiles, and birds, and the hairy animals known as mammals, such as elephants, lions, dogs, whales, bats, and mice. These more or less familiar animals have a highly exaggerated importance in our minds because they are mostly of large size, because they are similar to us in structure and habits, and because, like us, they often manage to make themselves conspicuous. We are members of this group and share with the others a common body plan; most of the organs and structures in the various kinds of vertebrates, including ourselves, are similar in form and function. Actually, the vertebrate body plan is only one of more than 30 in the animal kingdom. And, in terms of the number of living species, vertebrates comprise only about 3% of the animal kingdom.

The remaining 97% consists of animals without backbones. We are all aware of the difference between these two groups of animals when we indulge in fish and lobster dinners. In the fish, the exterior is relatively soft and inviting, but the interior presents numerous hard bones. In the lobster, on the contrary, the exterior consists of a formidable hard covering, but within this covering is a soft edible interior. A similar situation exists in the oyster, lying soft and defenseless within its hard outer shell. Lobsters and oysters are but samples of the tremendous array of animals that lack internal bones and that are, from their lack of the vertebral column in particular, called **invertebrates.**

A distinction between vertebrates and invertebrates was first recognized by Aristotle, although he did not use these terms but divided animals into those with blood (vertebrates) and those without blood (invertebrates). Unfortunately, Aristotle's neat distinction had little to do with the facts, because many invertebrates possess red blood and most of the others have colorless blood, which he did not recognize as blood at all. Aristotle did remarkably well from the scant knowledge of his time, and the limited time he had for the wealth of observations that he made himself. But the authority of his writings became so entrenched that people stopped looking for themselves, and his error was perpetuated for over two thousand years. With the rebirth of scientific inquiry in eighteenth-century Europe, people began to be skeptical of authority and they set out to investigate nature directly. By the beginning of the nineteenth century, Jean-Baptiste Lamarck and Georges Cuvier of France finally recognized a more accurate distinction between vertebrates and invertebrates, based on fundamental body plans, and Lamarck published his treatise entitled *Histoire naturelle des animaux sans vertèbres.* As scientific inquiry continued, the enormous variety of animal body plans became more fully appreciated, until today it is recognized that the vertebrates comprise only part of a group sharing a body plan that is only one of many.

There is a popular but vague recognition of the difference between vertebrate and invertebrate animals in the expression "spineless as a jellyfish." In this book we shall be concerned not only with jellyfishes, which are seldom seen by most people, but also with many familiar animals without backbones, such as clams, earthworms, lobsters, and fleas. However, there are many other invertebrate animals that generally pass unnoticed because they are too small to be seen without a microscope, because they live in water or in the ground, because they inhabit remote parts of the world, or simply because they escape the unobservant eye. We wish to introduce these animals as well.

Classified Knowledge

Although most vertebrates can be conveniently distinguished by the presence of a backbone, this is only one of many things they have in common. Having determined that an animal is a vertebrate, a zoologist can, without further examination, predict that it has both striated and smooth muscles, a heart and a circulatory system with closed blood vessels, an anterior mouth, a posterior anus, and a digestive tract that includes a large liver; that its eyes and nervous system follow a certain general pattern; that its excretory organs are kidneys; and many other characteristics, including even details of the way in which it developed from its egg.

On the other hand, identifying an animal as an *in*vertebrate, by the absence of a backbone, tells us only that it *lacks* a few uniquely vertebrate characters and gives us no way to predict what characteristics it has. Among the various kinds of invertebrates are dozens of body plans, each distinct from all the others.

Biologists are continually challenged to sort out our facts about both vertebrates and invertebrates in such a way as to increase our powers of prediction. This useful *ability to generalize and predict* with accuracy, even about animals not fully examined, depends upon a system for organizing a vast amount of knowledge. Miscellaneous facts about animal structure, physiology, biochemistry, and behavior are almost entirely useless. They lead us to useful generalizations only when we can relate these facts to each other within the context of a **system of classification.** Such a system would gain us nothing, and could even mislead us, if its categories were based on superficial similarities. For example, if all blue animals were classified together, they would have virtually nothing else in common as a group. Therefore, the fundamental ground rule of the system currently used by biologists is that it should reflect, as closely as possible, *how different animals are related to each other.* And this is judged by the number of basic similarities they share. The system is a *hierarchy.* At the top are a relatively few broad categories whose members share a small number of basic similarities. Each of these categories is divided and redivided into narrower categories whose members are increasingly more closely related and more similar.

The highest level in biological classification, the kingdom, is not based on relationship but, as discussed earlier, on how cells are organized and how energy is obtained. It is likely that many groups of organisms that are included within each kingdom have independent origins and that there is no single ancestral founder of any kingdom.

Table 1.2. Scheme of biological classification, exemplified by the classification of *Anopheles gambiae*, a widespread, malaria-transmitting species of mosquito, and *Homo sapiens*, a widespread species of mammal upon which mosquitoes feed.

Kingdom	Animalia	Animalia
Phylum	Arthropoda	Chordata
Class	Hexapoda	Mammalia
Order	Diptera	Primates
Family	Culicidae	Hominidae
Genus	*Anopheles*	*Homo*
Species	*gambiae*	*sapiens*

The highest level of classification within the animal kingdom is the **phylum** (plural: *phyla*). Animals that share a *common body plan*, and whose body plan is fundamentally different from those of other animals, are grouped in the same phylum. The members of a phylum may live in every kind of habitat, may vary in size and body form, and may differ in their methods of locomotion and feeding—but the distinctiveness of their body plan shows that they are all related and have descended from some common ancestor. Most of the 34 or so phyla of animals living today are well recognized and agreed upon by zoologists, although debate continues over the status of some groups. Any system of classification contains a certain amount of ambiguity, and, depending upon differences in criteria of what makes a body plan "fundamentally different" from all others, some people tend to "lump" similar groups into a single phylum, while other people "split" the same groups into separate phyla. This problem applies to groups at lower levels of classification as well. It does not mean, however, that the groups are arbitrary, artificial, or simply a human invention; rather, the problem reflects the real complexities of animal evolution.

Within each phylum the members are further divided into **classes** on the basis of a *major variation* in the fundamental body plan, usually in adaptation to a particular *way of life*. To take a crude analogy from everyday experience: if we regard all vehicles powered by gasoline engines as belonging to the same "phylum," major variations, for different modes of travel, would be automobiles, airplanes, and motorboats. Similarly, the familiar types of molluscs represent classes of animals that live in quite different ways: snails and slugs (class Gastropoda) mainly crawl about on solid surfaces; clams (class Bivalvia) mainly burrow in soft sediments; and squids (class Cephalopoda) mainly swim in the water.

Each class consists of a number of **orders.** In terms of the analogy given above, the "class" automobiles can be considered to include the "orders" passenger cars, trucks, and racing cars. Among animals, differences between orders are usually great enough to be easily recognized. For instance, the orders of insects include such familiar groups as termites, beetles, flies, butterflies and moths, fleas, and others.

Each order usually contains a number of **families.** The anatomical distinctions between families are not as great as those between orders, but are still great enough to include considerable variety, and they often reflect adaptation to particular habitats or ways of feeding. The insect order of beetles, for example, is divided into some 80 families, such as diving beetles, ground beetles, leaf beetles, wood-boring beetles, dung beetles, and ladybird beetles.

Each family is a grouping of similar **genera** (singular: *genus*). The anatomical criteria used to distinguish one genus from another are usually rather subtle and go unnoticed by most people, until there is reason to notice. Mosquitoes, for example, all belong to one family of flies (order Diptera, family Culicidae). The 30 or so genera of mosquitoes all look very much alike and behave in much the same way, and members of many of the genera suck human blood; but the consequences of giving blood to some mosquitoes can be very different from giving it to others. Malaria, which is transmitted to people when the mosquito is taking a small blood meal, is carried only by mosquitoes of the genus *Anopheles.* One may therefore wish to take a special interest in distinguishing this genus from other genera of mosquitoes.

When we divide a genus into **species,** we are at last dealing with a category that has a precise definition and that represents what scientists usually mean by a "kind of animal." In concept, a species is a population of animals that can interbreed and produce fertile offspring. In practice, all the individuals in a population cannot be tested to make sure that they can and do interbreed, but if they are sufficiently similar, they are assumed to belong to the same species. When some individuals consistently differ from others in certain specific characters, biologists may assume that there are two or more different species. Anatomical

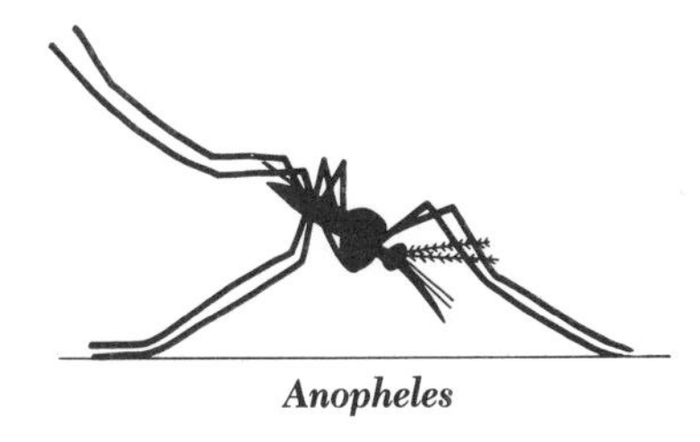

Anopheles

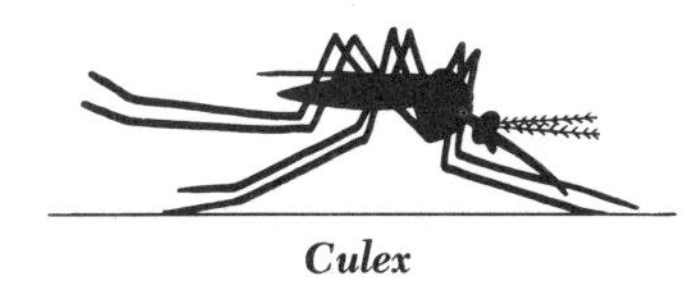

Culex

The members of two genera of mosquitoes, *Anopheles* and *Culex*, can be recognized by the stance they assume while biting. Identifying them correctly is not merely a matter of curiosity, but may be a serious health matter. Only species of *Anopheles* transmit malaria to humans, by injecting sporozoan parasites when they bite. Species of *Culex* do not transmit human malaria but do transmit viruses that cause encephalitis. (Only female mosquitoes suck blood; males drink nectar from flowers.)

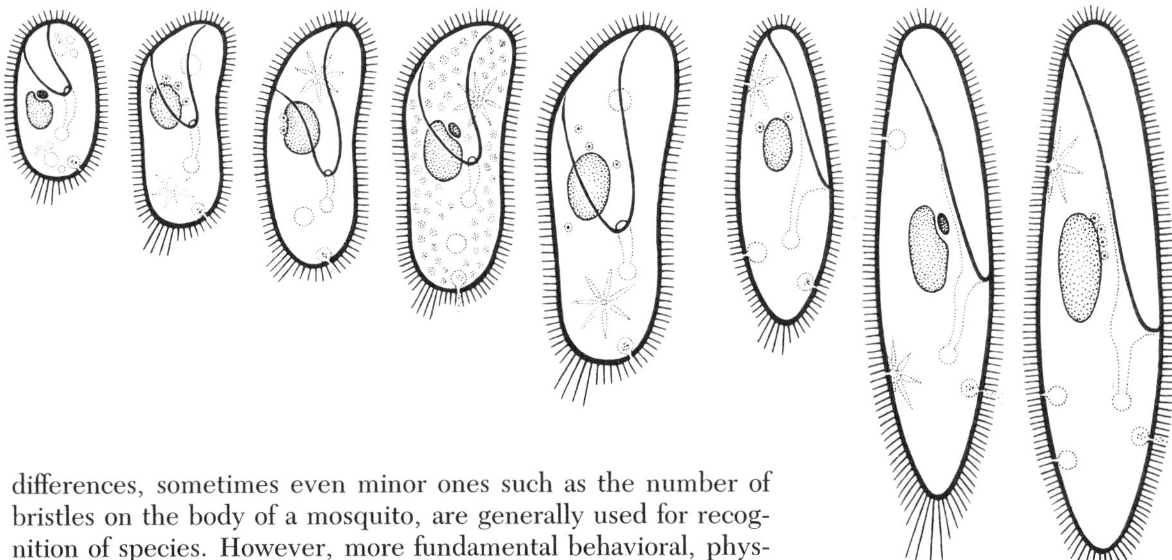

differences, sometimes even minor ones such as the number of bristles on the body of a mosquito, are generally used for recognition of species. However, more fundamental behavioral, physiological, and biochemical differences, which are difficult to determine (and impossible if your specimen is a preserved one in a museum collection), may be more important in preventing interbreeding between closely related species.

Species of the protozoan genus *Paramecium* may be distinguished by the different shapes and sizes of the body, varying numbers of small nuclei, and other features. (After D. H. Wenrich.)

In some cases, closely related species, called *sibling species*, are extremely difficult to distinguish by clearly visible anatomical differences, and physiological or biochemical characteristics must be used to distinguish them. One common method of doing this is to compare various proteins by a technique known as *electrophoresis*. Tissues of different individuals are ground and placed in an electric field. The proteins in the tissues migrate at different rates in the electric field, depending on their specific composition. Species differ in the kinds of proteins they contain, as revealed by electrophoresis, and therefore can be recognized. In most cases, once different species are recognized by electrophoresis, further examination reveals anatomical differences as well, and these can be used for field recognition. However, as more electrophoretic studies are done, more and more species are being recognized that appear to have no anatomically distinctive characters.

Many species consist of separate populations of individuals living in isolated localities, for instance, snails in different ponds or spiders on different mountains. Breeding between individuals in these separate populations may be so uncommon that differences begin to accumulate between them. When these differences are consistent and relatively easy to detect, such as those of color pattern or size, the different populations are often designated as **subspecies.** With time, subspecies that remain isolated from one another may become separate species.

By convention, the **scientific name** of an organism is always written in italics and consists of two parts: first, the genus name (written with a capitalized initial letter) and, second, the species name (not capitalized). The scientific name of the human species, for example, is *Homo sapiens*. (There is only one living species in the genus, although there are several extinct species known only from fossils.) Over a million species of animals have been named in this manner, and there are undoubtedly many more to be discovered and described. Recognizing, describing, and naming new species requires much training, skill, patience, and judgment. Although many new species continue to be described by the dedicated group of biologists trained to do so, the *systematists*, many other species continue to go extinct, sometimes without ever being noticed, collected, or described, especially now as human activities are radically changing the environment around the world.

The scientific name is properly written with the name of the scientist who first adequately described and named the species, and the date when the name was first published. For example, the common edible mussel, *Mytilus edulis* Linnaeus, 1758, was first described in 1758 by Linnaeus (Carl von Linné), a Swedish naturalist who established the system of giving two names to organisms. In his time (1707–78), Latin was the accepted language for scientific writings, and Linnaeus described the plants and animals in Latin and gave them Latin names. This system became universally adopted, and scientists must now produce a latinized name for a newly described species, although the description can be in a modern language. Often a species is named after a person or place so the name must be converted to Latin; for example, the full scientific name for the California mussel is *Mytilus californianus* Conrad, 1837.

The assigning of scientific names to animals is governed by a definite set of regulations as specified in the *International Code of Zoological Nomenclature*. According to these rules, the valid name of a species is the name that was given to it by the first person who published a binomial latinized name for it, together with an adequate description of the species. This seems clear enough. A scientific name cannot be changed simply because the original name given to it seems inappropriate, or is even misspelled or incorrectly latinized; once published, the name remains, with only a few well-regulated exceptions. For example, some descriptions are so incomplete or unclear that it is impossible to determine exactly to which animal the description applies. Or, further study of the different species within a genus may reveal several different groups of species, each of which is distinctive enough to be considered a separate genus.

When such difficulties arise, either the generic or the specific name, or both, may need to be changed, even though they have been in long and familiar use. The changes normally can be made only for reasons clearly defined in the established rules of nomenclature. Cases in which a proposed change is controversial, or for which a motion is made to suspend the rules, are decided by the International Commission on Zoological Nomenclature, made up of internationally established systematists.

When zoologists mention an animal in a scientific paper, they often do not use its common name, because this varies from place to place and from time to time, and there is no established set of rules to regulate common names. What is loosely called a "crayfish" in one locality may be an animal of an entirely different genus, or even family, in another locality. The scientific name is international, recognized the world over as referring only to one clearly defined species. Sometimes a species is so distinctive that it can be recognized on sight. But, more commonly, the number of similar species is so great and the differences between them so small, that they can be accurately identified only by a specialist who knows the particular characteristics of the group. There is no virtue in giving the full name of an animal if the name is not correct. Rather than commit this "scientific crime," one refers to an unidentified animal by using its generic, or even family or order name.

Many animals are not well known to most people so they have not been graced with a common, everyday name, and the only name they carry is their scientific name. In such cases, the generic name, not capitalized, italicized, or latinized, is often used as a common name; for example, "paramecium" is the common name for an individual of any species of the protozoan genus *Paramecium*. Sometimes, after a well-established genus is divided into several genera, all the species in the original genus retain that generic name as their common name; for example, hydras once all belonged to the genus *Hydra* but now there are several genera of hydras.

This book is mainly about members of the largest of the five kingdoms of living things—animals. The kingdom Animalia was named from a Latin word for "breath" or "soul," and is also called Metazoa, a Greek word meaning "later animals." Before going on to these, however, we shall consider a few members of the kingdom Protista: certain amebas, ciliates, and flagellates. Although not members of the animal kingdom, these protists are animal-like in their nutrition and behavior. They are briefly treated in two chapters of this book as an introduction to some modern descendants of the sort of protistan stock from which multicellular animals must have evolved.

Chapter 2

Life Activities

To keep alive and healthy, all organisms must carry on certain **life activities.** Because these activities center about the use of energy, they have often been compared to the functioning of a combustion engine. But the point at which this analogy breaks down is reached quickly. For not only is the living machine a self-feeding, self-tending, and self-perpetuating one, but its form, requirements, and activities may change dramatically in the course of its life. In addition it is a machine that must operate virtually at all times. A stalled motor may easily be repaired. Not so a "stalled" organism, in which the failure of certain functions automatically brings on the disintegration of the machinery itself. A machine can be oiled, covered, and put away on a shelf until ready for use, while an organism is kept running—sometimes rapidly, sometimes very slowly—but continuously from the start until the finish.

Opposite: Some of the **life activities** of termites may be observed by maintaining individuals in a hole in a piece of plywood between two sheets of glass, as shown here. Termites are social insects, and in this genus (*Kalotermes*), the young termites, or nymphs, do the work of the colony. The nymphs are here busily *moving* about and *eating* wood, which they *digest* with the aid of intestinal protozoans that they harbor. The chamber contains many fecal pellets, evidence of *elimination*. And among these is visible evidence of *reproduction*, several elongate white eggs cared for by the workers. In this observation colony, the workers are protected against drying and mechanical injury, as they would be in their nest in nature, but the glass is fitted loosely enough to admit air for *respiration*. Below is a millimeter scale. (R. Buchsbaum)

This continuity of living processes appears, on first thought, to be contradicted by our ordinary observations of animals. For example, we know that crayfishes may lie dormant in the mud of a dried-up pond and then resume their activities when spring rains fill the pond. What we see here is only a temporary cessation in the easily observable activities of the crayfish. Going on all of the time at a greatly slowed but measurable rate is a much more fundamental activity, the liberation of energy. The living "machine" is idling. The energy liberated comes from food stored by the crayfish during the months preceding dormancy. Some other animals are able to enter a dormant state which may last not for a season but for many years; during this time no obvious food stores are present and no measureable energy is liberated, but when the dormancy is broken, life activities are resumed.

As we saw in the first chapter, the means of obtaining energy is one big difference that separates living organisms in the several kingdoms. Most of the processes to be discussed in this chapter are common to members of all the kingdoms, but the fact that animals must eat adds to their list of essential life activities. The same list of tasks must be carried out, although often in a different manner, by animal-like protists.

Locomotion or motility of some type enables most animals to rove about in search of food or to escape from predators that are also on the move hunting their prey. Animals differ strikingly in their modes of locomotion, however; and these differences are related partly to the size and body plan of the animal and partly to the medium through which it moves. *Sessile animals,* which cannot move about but live firmly attached to the substrate, often have moving parts that propel food their way. These include sponges, oysters, barnacles, and others. In addition, there are *sedentary animals,* such as clams or web-building spiders, which feed while remaining for some time in one spot but which can and do move about to escape danger or to take up a new and more profitable feeding station.

Food capture and ingestion is another major activity of animals. Again, the differences in the ways animals eat are related partly to the diversity of animal body plans and partly to the diversity of the food itself. Mouth parts that tear flesh will not do for chewing wood; sucking sap is not the same as sucking blood. Both locomotion and feeding, however, involve certain kinds of common tools: a set of *sensory receptors* to get information about the external environment; *physical mechanisms* for carrying out the necessary actions; and *means of coordinating* the locomotory and feeding mechanisms with the information received from the surroundings so that the net result will be the getting of something to eat and the avoidance of being eaten.

Digestion is the chemical alteration of ingested food into a form in which it is usable as a source of *energy* for the life activities and as a source of *materials* for growth and replacement of worn-out or damaged parts. The food consists of water, carbohydrates, fats, proteins, inorganic salts, and some other substances. The water, some of the salts, and certain of the simpler substances need not be digested; they are immediately available for incorporation into the living animal body. The other substances must be broken down into simpler units because they are too large to pass through cell membranes and because they are too complex to be used directly in growth or in other living processes. Digestion, then, is the breakdown of ingested food into smaller units. To facilitate this breakdown, almost all animals have special internal cavities in which to carry out digestive chemistry.

Secretion is the manufacture and release of specific chemical substances, which may be used on the spot, or may be carried elsewhere in the body, or may be expelled to the outside. Digestive fluids and mucus are well-known animal secretions. Structural materials such as beeswax, silk, sponge fibers, and mammalian hair are also secreted substances. Of the categories of secretions that enter into the chemistry of the basic life activities, among the most important are *enzymes*—complex proteins, manufactured only by living organisms, that speed up chemical reactions. By controlling the times and places at which enzymes are secreted, and the quantities that are manufactured and released (and, when present in excess, destroyed), an organism regulates the multitude of chemical reactions occurring throughout its body.

> Enzymes make possible many kinds of chemical activities that in the nonliving world take place so slowly that they could not possibly serve the needs of a constantly changing living organism. Enzymes are notably specific; most enzymes accelerate only one particular reaction. For example, some digestive enzymes act only on carbohydrates, others only on fats, and still others only on proteins—and each works only on a particular step in the chemical breakdown of these sorts of substances, such as breaking the linkage between certain kinds of amino acids in a protein.
>
> Other important categories of secretions are *hormones*, which circulate among the various parts of the body to coordinate their activities, and *transmitters* secreted by nerves to activate or inhibit other nerves or muscles.

Elimination is the ejection from the body of indigestible food or other accumulated solid wastes. Most plant eaters do not have the enzymes needed to digest completely the woody tissues of the plants they feed upon, and they must eliminate considerable quantities of indigestible matter as solid wastes, or *feces*. Meat eaters tend to expel smaller amounts of feces, but even fluid feeders such as nectar-sipping butterflies are left with a small residue of solid wastes to expel from their digestive machinery.

Assimilation is a constructive chemical process by which materials derived from digestion are incorporated into the living cells or body fluids. After the food is converted from its condition as part of one kind of plant or animal into smaller, simpler units, it is suitable for building the kinds of carbohydrates, fats, proteins and other substances peculiar to the structure and chemistry of the animal concerned. Just as innumerable useful objects can be fashioned by people shaping and combining only a few dozen building materials (stone, wood, steel, etc.), so a few dozen kinds of food units can be built into an almost infinite variety of organisms—each with its own specific composition.

Respiration is a destructive chemical process by which the energy stored in food (originally captured by the photosynthetic processes of green plants) is released. When we burn a log of wood, the chemical process of burning is an *oxidation;* oxygen in the air is combined with the organic substances that compose the wood, and energy is rapidly released in the form of heat and light. Most organisms "burn" sugar or other food substances in somewhat the same way. The food is oxidized, or combined with oxygen from the air, and energy is released; but in this case the process is slow and controlled, and takes place at far lower temperatures. Only some of the energy is released as heat; the rest is captured in stable, high-energy substances from which it may be readily released when and where it is needed to provide power for the chemical work (metabolism) of the organism. Both kinds of oxidation (burning and aerobic respiration) require oxygen, and both produce carbon dioxide and water as by-products that must be removed from the body.

Respiratory exchange of oxygen and carbon dioxide is therefore another essential life activity. These respiratory gases are dissolved in the blood or other body fluids, which carry oxygen from where it enters at the surface of the organism to all the respiring cells of the body; carbon dioxide is removed by the reverse route. In air-breathing vertebrates these exchanges take place in the lungs. Many invertebrates also have specialized structures that provide increased surface for respiratory exchange with the surrounding air or water; in others exchange occurs over the body surface.

Excess water that is produced in respiration, contained in food, or acquired by other means must also be removed from organisms. Animals commonly find it economical to use this water as a solvent for the removal of other nongaseous materials, such as certain salts and especially toxic compounds of nitrogen that are by-products of oxidizing proteins. Thus the life activities of **water regulation** and **excretion** of nongaseous by-products are often performed by the same structures. In humans the kidneys do this work; various devices serve the same functions in other animals.

Reproduction, the production of offspring, represents two distinct types of life activities: *asexual* and *sexual.* These are different processes and not merely alternate means of doing the same thing. Asexual reproduction is a type of **growth** whereby an individual, instead of continuing to get bigger, replicates itself by budding, fission, fragmentation, or other means. All the replicates (except for mutants) are genetically identical to the "parent." Sexual reproduction, on the other hand, involves the **recombination of genes.** Half of the genes of one parent are combined with half of the genes of another parent to produce a new combination. The result is a unique individual, different not only from both of the parents but from its siblings and from all other members of its species.

When eukaryotic cells undergo **cell division** and increase in number, the genetic material they contain is first duplicated, then divided equally by an asexual replicative process called **mitosis;** each of the two new cells has the same genetic material as the original cell. Mitotic cell divisions are involved when a single-celled organism replicates or when a multicellular organism grows by adding new cells. In sexual recombination the sex cells undergo a special division process called **meiosis** whereby each receives *half* the genetic material of the parental cell. The resulting eggs or sperms each carry different combinations of the parental genes, and when an egg and sperm from different parents unite, they begin the life of a new individual uniquely distinct from both parents.

Beginning with one of the least complex of the animal-like protists, we shall see, in the chapters that follow, the spectrum of complexity shown by the living machinery with which the various kinds of invertebrates carry on their life activities. The variety of structure and the diversity of ways in which different animals lead their lives are enlightening in themselves, but they become most meaningful when viewed in relation to their functions and in the context of animal evolution from simple to complex forms.

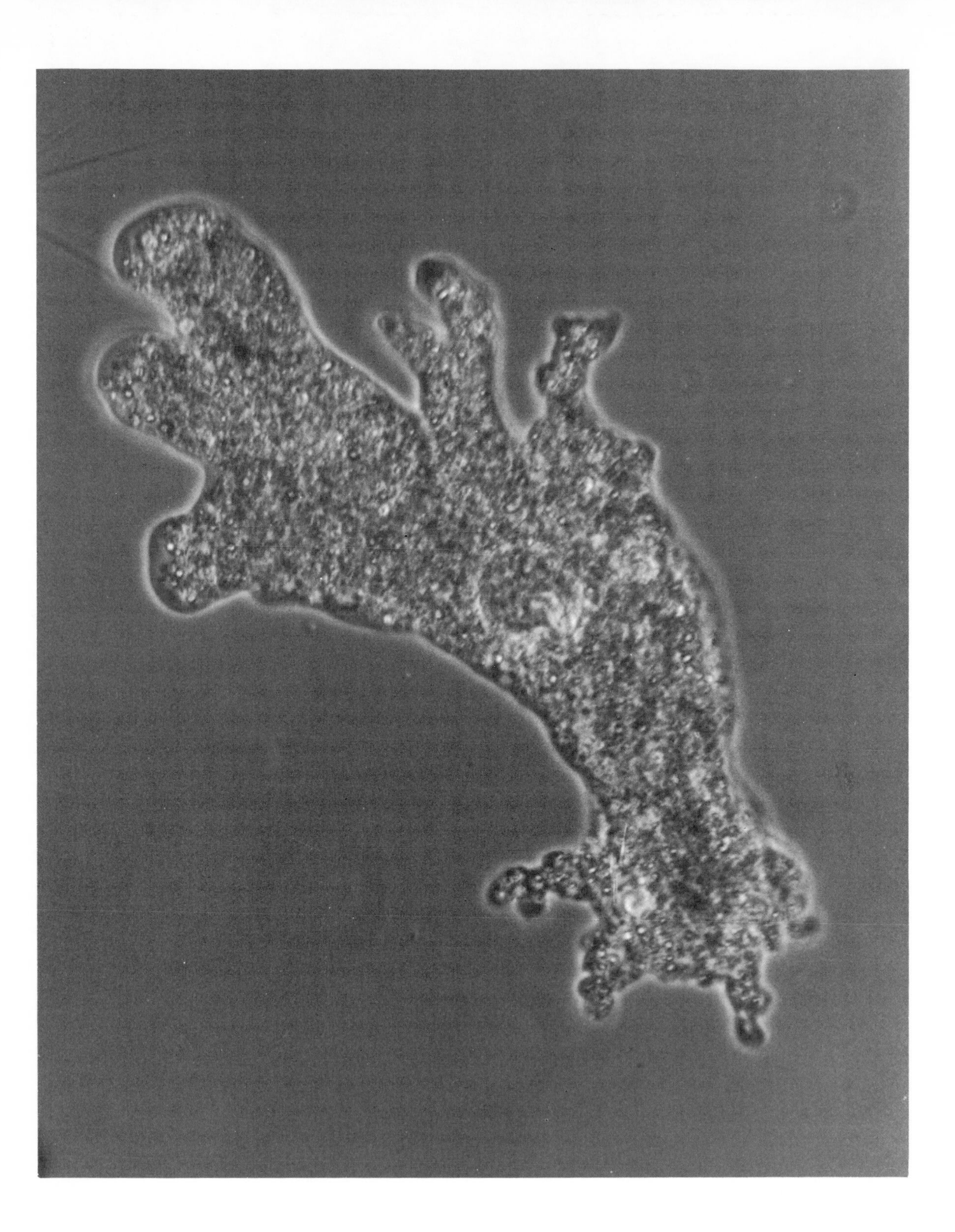

Chapter 3

Simple and Complex Protozoans

Opposite: **Ameba** of freshwater ponds, *Amoeba proteus.* Photomicrograph. (R. Buchsbaum)

To look through a microscope focused on a drop of water from the bottom of a pond rich in organic matter is to enter a microworld unlike any we meet in ordinary life. All five kingdoms of living organisms are represented. But this busy world is inhabited mostly by **unicellular organisms,** the monerans and protists. They form a community as complexly interrelated as any in the macroworld. Some are green and make their own food from materials in the water, with the energy of sunlight, providing the major sources of nutrients and energy for the others. Some absorb organic materials released into the water by decay and are important agents of recycling. And others are ingestors of food, some quietly gathering bacteria, green cells, and organic particles suspended in the water, some swimming rapidly through the water and tackling prey as large as themselves. It is these ingestors, the animal-like protists, that are the subject of this and the next chapter. They are called **protozoans,** a name that means "first animals." Despite this suggestive name, we do not know whether any of the protozoans living today are descendants of those that gave rise to many-celled animals over a half-billion years ago. But protozoans provide models of what the ancestors of animals might have been like; and they are important parts of our contemporary world.

The first person to see this miniature world of creatures was Antony van Leeuwenhoek, a Dutch naturalist who designed early microscopes. His description in 1675 of the numerous "animalcules" that he found in drops of rainwater began a period of lively discovery that continues today as an important branch of biology with a vast scientific literature, much of it published in specialized journals of protozoology.

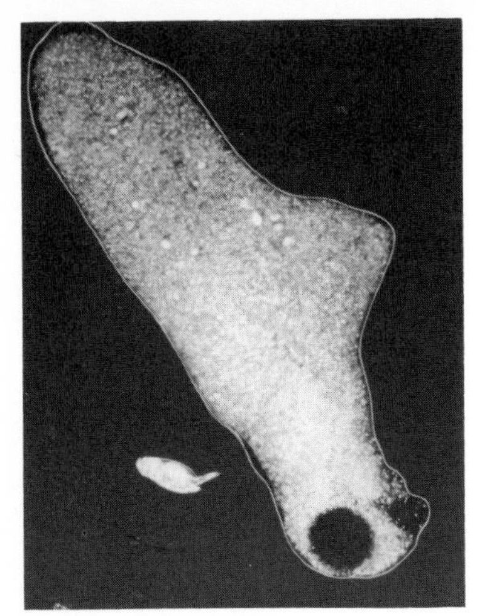

A giant among amebas is *Chaos carolinense,* which can reach a length of 5 mm. This and related species are unusual in having numerous nuclei (large white dots among smaller granules) and many contractile vacuoles, presumably associated with large size. Although one-celled, this large ameba dwarfs the many-celled rotifer (see chapter 14) beside it. Photographed under dark-field illumination, the cell membrane and cytoplasm appear white on a black background. Photomicrograph. (P. S. Tice)

Most protozoans consist of a single cell, organized much like one cell in a many-celled animal. Yet a protozoan behaves much as a *whole* animal does and carries out essentially the same life activities. The internal organization of some protozoans is more complex than that of any single cell in a many-celled animal. Such protozoans have elaborate mechanisms for moving, feeding, defending themselves, excreting, coordinating activities, and performing sexual reproduction—all within the limits of a single cell. We shall examine one of these later in this chapter. We will begin, however, with some simpler protozoans that are perhaps more surprising, in that they seem to have so little permanent structure or differentiated organization. Yet they manage to go on about their activities without any of the fancy equipment that most other organisms require to survive.

A Simple Protozoan: Ameba

The typical **ameba** of freshwater ponds is usually of microscopic size, but large individuals may reach half a millimeter in diameter, being visible to the naked eye as white specks. Each ameba is a little mass of clear gelatinous material containing many granules and droplets. The surface is bounded by a delicate **cell membrane** through which materials pass in and out of the ameba. The membrane possesses the important property of *selective permeability,* admitting certain substances and excluding others, and so enabling the cell to maintain a constant chemical composition which may be different from that of its environment and to some extent independent of changes in the environment.

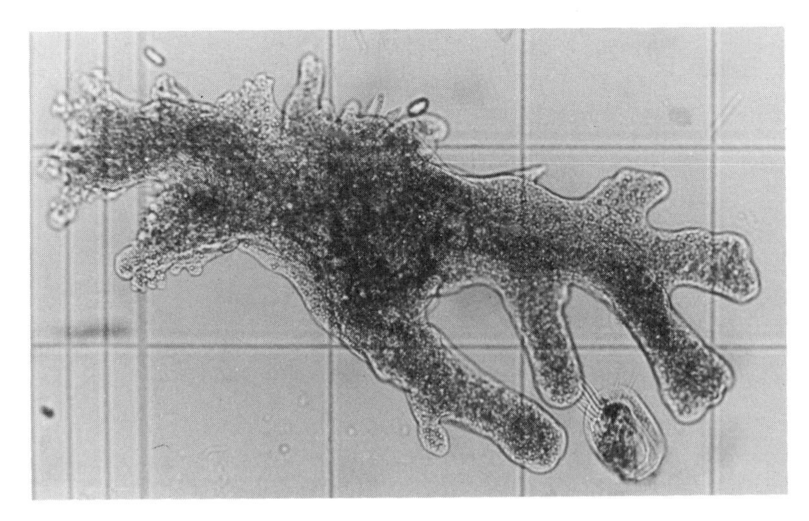

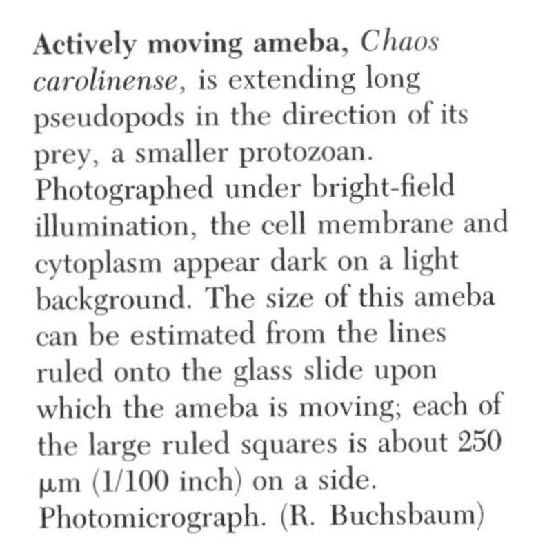

Actively moving ameba, *Chaos carolinense,* is extending long pseudopods in the direction of its prey, a smaller protozoan. Photographed under bright-field illumination, the cell membrane and cytoplasm appear dark on a light background. The size of this ameba can be estimated from the lines ruled onto the glass slide upon which the ameba is moving; each of the large ruled squares is about 250 μm (1/100 inch) on a side. Photomicrograph. (R. Buchsbaum)

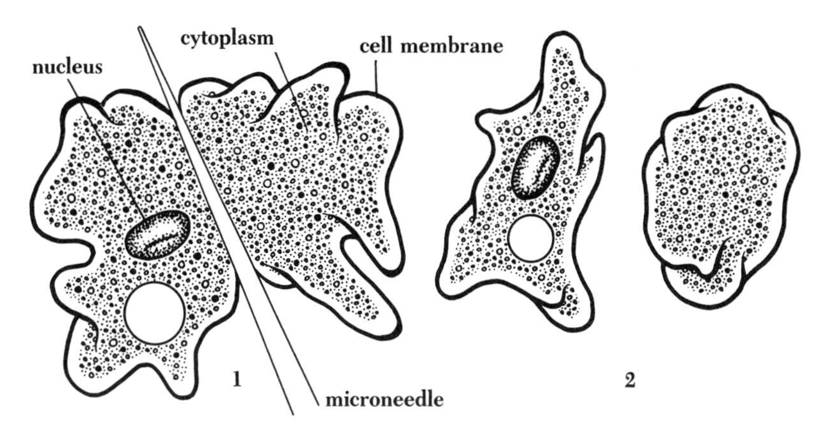

1: An ameba is cut in two with a very fine glass needle. **2:** One piece contains the nucleus and the other lacks a nucleus.

Water passes freely through the cell membrane; but proteins, carbohydrates, fats, and some salts inside are prevented from escaping, while other salts and materials are kept outside. If an ameba is carefully cut in two, each piece rounds up and immediately produces a complete membrane, thereby preventing the loss of the interior contents.

The interior of an ameba, as in almost all eukaryotic cells, is differentiated into nucleus and cytoplasm. The **nucleus** occupies no fixed position, but is generally located away from the cell membrane. Amebas are excellent cells for studying the function of the nucleus, because they can be cut into pieces, some with and others without a nucleus. The pieces with a nucleus behave like normal amebas; they feed and soon grow to full size, and then divide. The pieces without a nucleus move about in more or less normal fashion for a time and even feed, but they are unable to digest food, to grow, or to divide, and they eventually die. From such experiments it can be concluded that the nucleus is largely concerned with the synthesis of new material, growth, and the production of new cells through division.

The **cytoplasm** of an ameba is differentiated into a relatively stiff, jellylike outer layer, the *ectoplasm,* and a more fluid interior, the *endoplasm.* The cytoplasm contains a variety of granules, fat droplets, and food bodies in various stages of digestion, as well as droplets of watery fluid, and these have no fixed positions but continually move around within the ameba.

Motility is one of the striking characteristics of protozoans. The type of movement exhibited by amebas is called, naturally enough, **ameboid movement,** and is found in many kinds of cells, including many of those in our own bodies. It appears to be totally different from the muscular movements of animals, but the chemical reactions involved are basically the same, and the same kinds of *contractile proteins* are involved in ameboid and muscular movements.

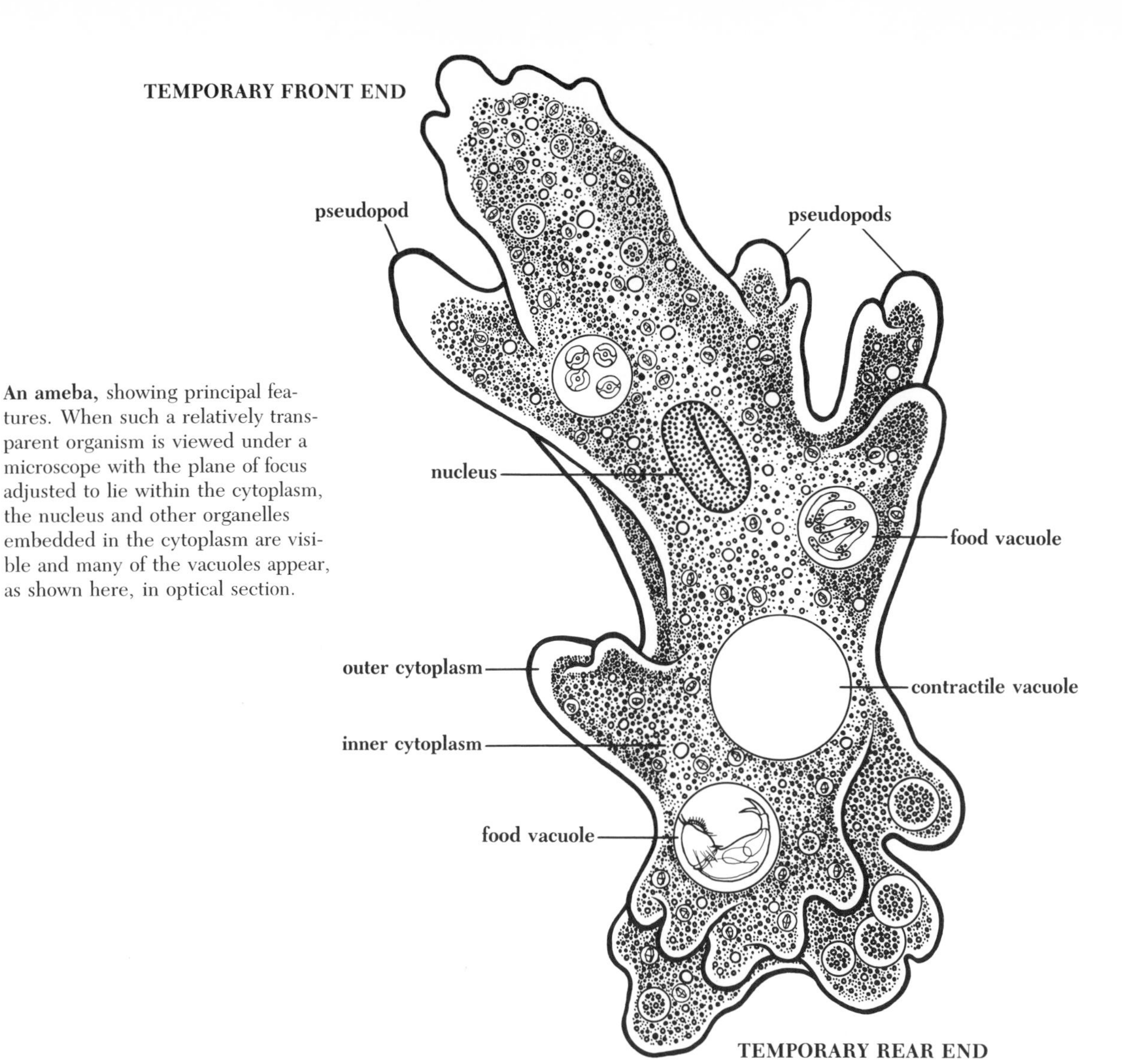

An ameba, showing principal features. When such a relatively transparent organism is viewed under a microscope with the plane of focus adjusted to lie within the cytoplasm, the nucleus and other organelles embedded in the cytoplasm are visible and many of the vacuoles appear, as shown here, in optical section.

Amebas have no distinct head or tail ends but have a surface that is everywhere the same, and any one point on this surface may flow out as a projection, or **pseudopod** ("false foot"). This pseudopod continues to increase in size for some time through the passage into it of some of the material of the ameba, but sooner or later a new projection forms at an adjacent point, and then the cytoplasm flows into the new pseudopod. In this manner the ameba progresses in an irregular fashion—the cytoplasm flowing first into one pseudopod, then into another. The ameba often alters its course by putting out pseudopods on the side away from the previous advance. As new pseudopods form, the

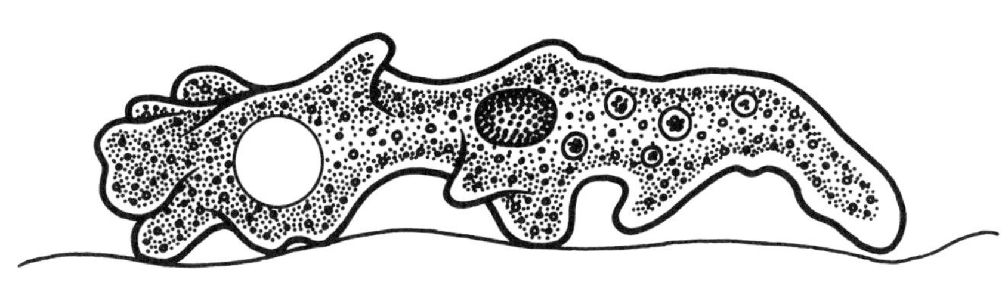

An ameba in profile. When the microscope is arranged so that the organism is viewed from the side, it can be seen that only the tips of the pseudopods are in contact with objects, most of the body being free in the water. The pseudopods act like little legs put out one after another, but they are temporary and each soon flows back into the general cytoplasm. (Based on O. Dellinger.)

old ones flow back into the general mass. Ameboid motion is slow, and the protozoan proceeds along, first in one direction, then another.

Although it has been the subject of considerable research for over a century, ameboid motion remains unexplained. Observations show that it involves a change in the cytoplasm from a relatively stiff **gel** state to a fluid **sol** state, and back again. A pseudopod forms when a portion of the ectoplasm transforms from a gel state to a sol state. Other parts of the ectoplasm contract, using the same contractile proteins as are found in animal muscle, and push some of the sol-state endoplasm forward into the new pseudopod. More endoplasm forms as inner layers of ectoplasm change from a gel to a sol state and join the flow forward. As the pseudopod moves outward, the cytoplasm along the sides changes back from a sol to a gel state and the pseudopod becomes a tube. Any change at the end of the tube that causes the fluid cytoplasm to solidify leads to a change in direction, or even a reversal of movement of the entire ameba as the contents of the pseudopod flow back into the body of the ameba.

Pseudopods participate not only in locomotion but also in the **capture and ingestion** of food. Although they have no special cell structures of taste or smell, amebas can distinguish inert particles from the bacteria, protists, and minute animals upon which they feed. Pseudopods are thrown out around the sides and over the top of the prey objects. In this way the food is held down, then completely surrounded, and finally incorporated into the ameba. The behavior of the ameba varies somewhat with the kind of food it encounters. If the ameba is stalking an active food organism, such as another motile protozoan or minute multicellular animal, the pseudopods are thrown out widely and do not touch or irritate the prey until chemicals released by the ameba immobilize it; then it is completely surrounded and engulfed. When the ameba is ingesting a quiescent object, such as a nonmotile bacterial or algal cell, the pseudopods surround it very closely.

The food usually lies in the ameba's cytoplasm within a drop of water that was taken in when the food was enclosed by the pseudopods. This drop of water containing the food is bounded

An ameba captures and ingests another protist. **1:** The ameba moves toward the prey. **2:** The ameba begins to extend pseudopods. **3:** Pseudopods are thrown out around the sides, and a thin sheet of cytoplasm extends over top of the prey. **4:** A sheet of cytoplasm extends beneath the prey, which is now almost completely enclosed. **5:** The prey lies in the ameba's cytoplasm in a food vacuole. **6:** The ameba stalks another meal. (Based on A. A. Schaeffer.)

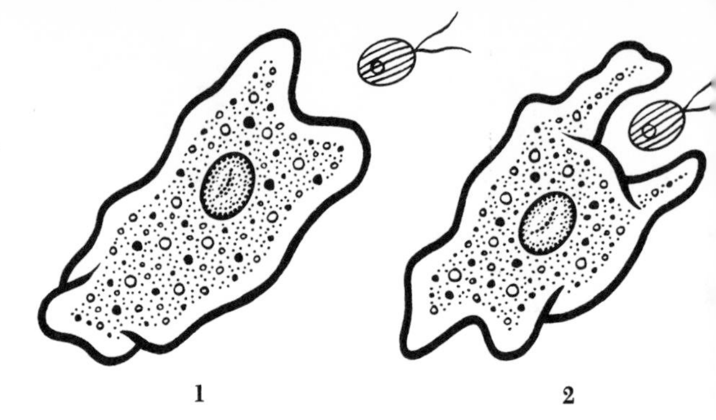

on all sides by pinched-off cell membrane and is called a **food vacuole.** The food soon begins to undergo **digestion** as enzymes are released into the food vacuole through the surrounding cell membrane.

> Once within the newly formed vacuole the ingested prey can sometimes be seen to "wake up" and struggle to escape, but it is too late and the prey soon ceases all activity. When indicator dyes are added to the water, so that they are incorporated into the food vacuoles during feeding, they show that the water in the food vacuoles first becomes acid, then alkaline. The prey begins to undergo perceptible dissolution only after the water becomes alkaline, suggesting that the digestive enzymes act in an alkaline medium as they do in our own intestines.

As the food gradually disintegrates, the dissolved substances pass into the cytoplasm, where they are **assimilated** to be used as energy sources and raw materials for synthesizing new cell parts. The indigestible fragments remaining in the food vacuole are **eliminated** in the simplest fashion possible. They are gradually shifted to the temporary rear end of the animal and then left behind as the ameba flows away.

Exchange of respiratory gases requires no special breathing mechanism in a minute creature such as an ameba. There is no "breathing" in the same sense as this expression is used for people, and its sides do not heave in and out. Rather, the oxygen dissolved in the surrounding water passes into the ameba by *diffusion.* Diffusion is the tendency of matter to disperse in any liquid or gas until its concentration is the same throughout. Since oxygen inside the ameba is constantly being used to oxidize foods and release energy, the oxygen concentration inside the ameba is always lower than that outside. Thus oxygen continually diffuses from the region of higher concentration outside, through the cell membrane, to the region of lower concentration inside.

When oxygen is used to oxidize sugars and fats, only water

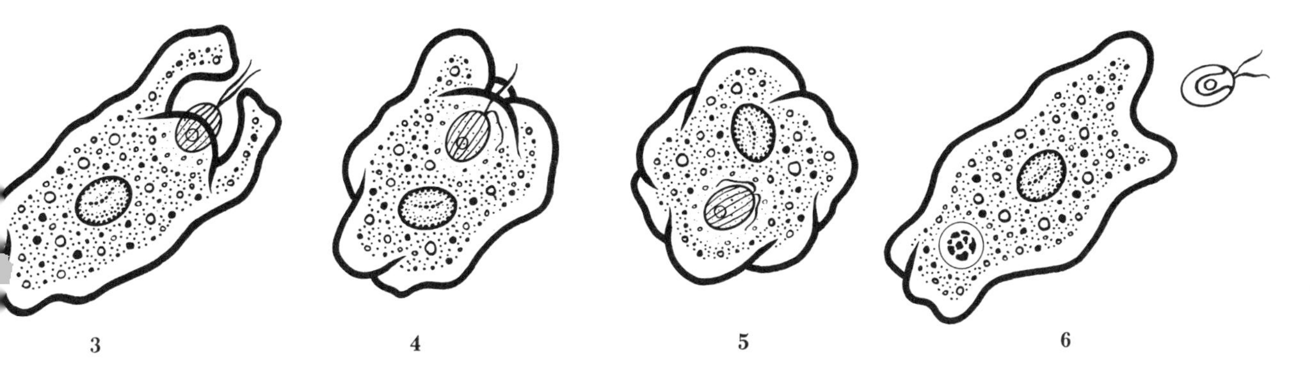

and carbon dioxide remain. The carbon dioxide concentration inside the ameba increases, relative to that outside the ameba, and carbon dioxide diffuses out. This method of respiratory exchange (inward diffusion of oxygen and outward diffusion of carbon dioxide) will work only for minute organisms such as protozoans and for others with an exposed surface that is large in proportion to their bulk or mass. As organisms increase in size they must develop special, greatly enlarged respiratory surfaces, such as gills or lungs.

Oxidation of proteins yields not only carbon dioxide and water, but also nitrogenous substances that are poisonous and must be rapidly **excreted.** Ammonia, the main nitrogenous product produced by aquatic organisms such as amebas, is very soluble in water, and readily diffuses through the cell membrane into the surrounding water; no special excretory mechanism is present.

Near the rear of a moving ameba is a large spherical water vacuole, called the **contractile vacuole,** which contracts at regular intervals, discharging its contents to the exterior. It then forms again from one or more minute droplets and gradually swells to a maximum size, whereupon it again contracts, ejecting its contents through a temporary pore to the outside. The contractile vacuole may best be likened to a pump on a leaking ship, constantly filling with water. Not only is water produced during the oxidation of foods and taken into the ameba when food particles are engulfed, but quantities of water constantly enter through the cell membrane by *osmotic diffusion.* Because the cell membrane retains salts and other materials within the ameba at higher concentrations than outside, the concentration of water is always lower inside than outside. Thus water is constantly diffusing into the ameba and must be expelled if the ameba is to keep from swelling and eventually bursting. Experimentally adding salt or sugar to the water surrounding the ameba lowers the difference in concentration of water inside and outside the ameba; the vacuole swells and contracts less and less frequently,

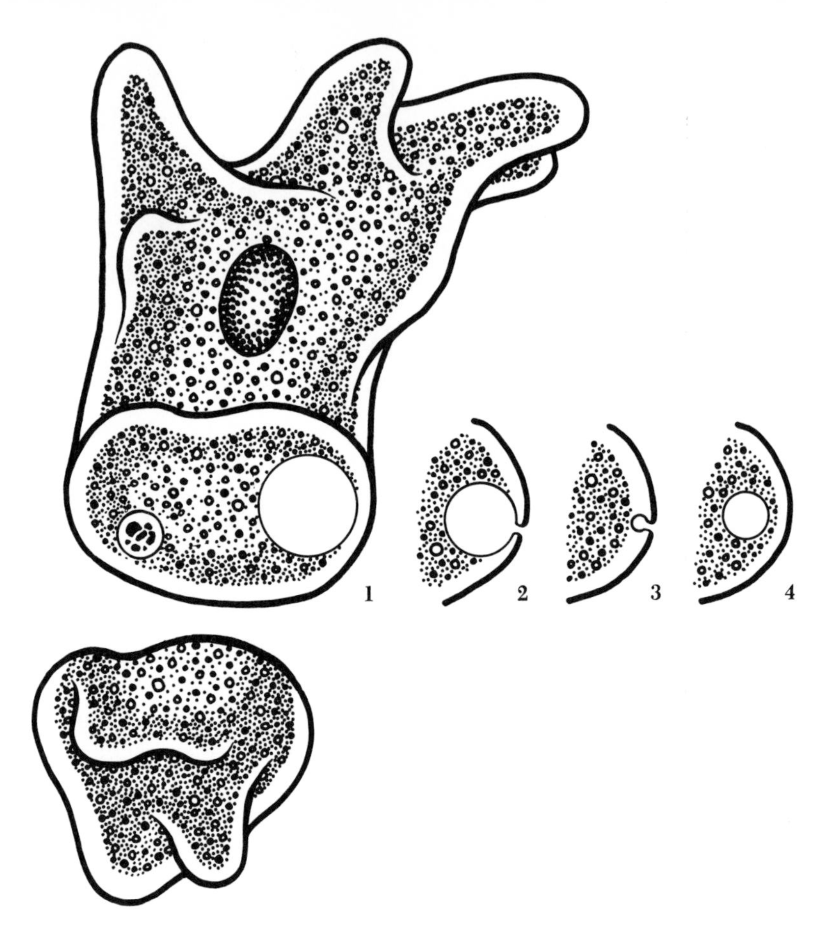

A contractile vacuole as it would appear if we could make a crosswise cut through an ameba at the level of the vacuole and move the two pieces apart without the contents of the ameba spilling out. **1:** The vacuole at maximum size. **2:** Ejecting its contents. **3:** Almost completely emptied. **4:** The vacuole forms again and increases in size. The pore is a temporary structure, formed anew each time the vacuole empties.

and finally it vanishes altogether. Conversely, some marine amebas, which live in seawater with a high salt concentration and normally lack contractile vacuoles, develop them when placed in freshwater. Thus, the main function of the contractile vacuole is to *regulate the water content* of the ameba. The pumping activity requires energy, and if poisons are introduced that prevent energy-producing oxidation in the cell, the contractile vacuole stops pumping and the ameba swells and bursts.

By means of a fine glass micropipet some of the fluid in the contractile vacuole can be withdrawn and analyzed. The fluid contains almost no nitrogenous wastes, so it appears that these vacuoles serve little or no excretory function. There are, however, salts in the fluid of newly forming vacuoles that are at the same concentration as in the cytoplasm, and at much higher concentrations than in the surrounding water. Before the vacuole releases its contents to the outside, the salts are actively absorbed back into the cytoplasm so the cell does not lose large

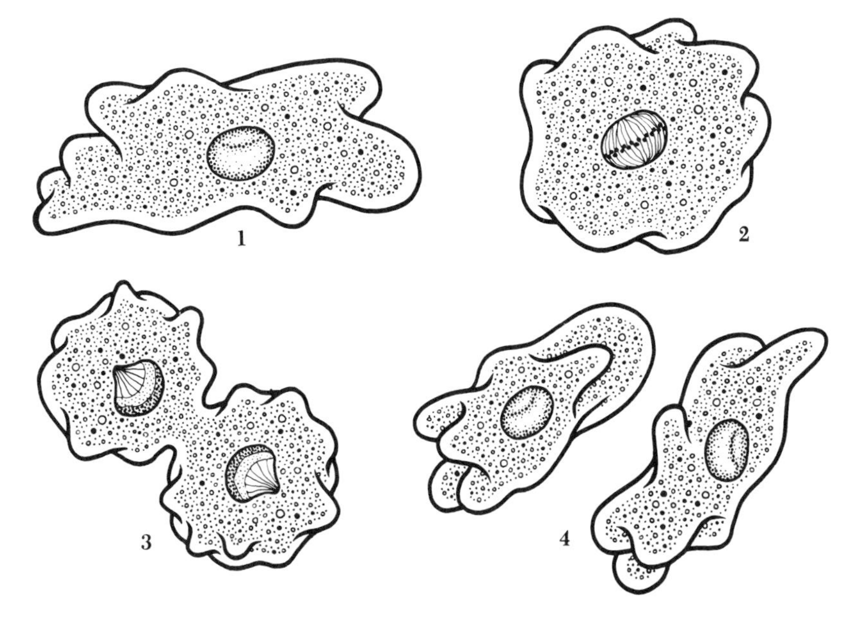

An ameba divides. 1: A large, well-fed ameba. **2:** The ameba rounds up, and the nucleus enters upon the first stages of mitosis. **3:** Both nucleus and cytoplasm divide. **4:** Two small amebas result, each with a nucleus and half the cytoplasm of the original ameba. (Based on Dawson, Kessler, and Silberstein.)

amounts of salts. The kidneys of many freshwater animals, including fishes, also function mainly to retain salts and other small molecules while expelling excess water from the body.

Amebas **grow** by increasing in size and also by increasing in number through **asexual reproduction,** a process of replication whereby each ameba divides into two. After feeding and increasing in size for some time, an ameba rounds up into a spherical mass, the nucleus divides by mitosis (see chapter 2), the cytoplasm constricts, and finally the slender strand that connects the two halves breaks. The entire process requires less than an hour. Each half behaves just like the original ameba and soon increases to maximum size to divide again. Under favorable conditions, a single ameba can soon number in the thousands of nearly identical replicates. *No sexual recombination* has ever been seen in amebas. However the mutations in individual cells may lead to new varieties, and those varieties that survive and replicate may replace earlier forms; we can be nearly certain that no modern strain of amebas is exactly like the ancestral forms.

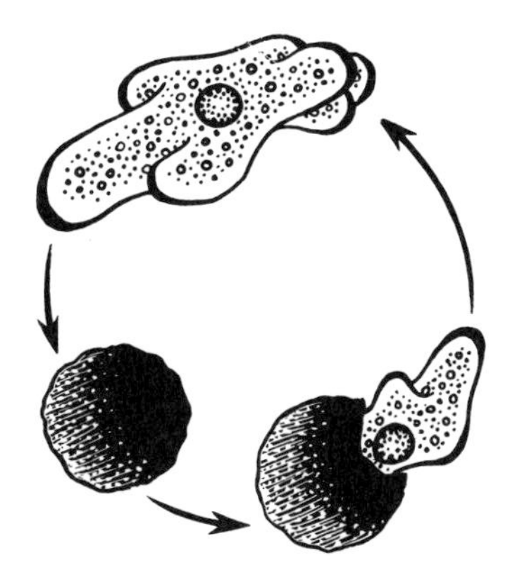

An encysted ameba survives unfavorable conditions. With the return of favorable conditions, the ameba emerges from its cyst.

Amebas carry on their routine activities only when immersed in water that contains a plentiful supply of food. If the water dries up or the food supply runs low, some amebas can round up and secrete around themselves a hard and impervious protective shell called a **cyst.** Within the cyst the ameba's respiration rate falls to nearly zero so that very little energy is used. Living cysts can survive for years in the soil, or be blown about in the wind before returning to a suitable environment where they break open and the enclosed amebas resume their usual activities.

Although an ameba has nothing comparable to our sense organs, its **behavior** leaves the impression that it is sensitive to its surroundings: it reacts to outside events and coordinates its activities. It distinguishes food from particles of no food value, and it approaches moving prey with apparent caution in comparison to stationary food items; probably, moving organisms create disturbances in the water that an ameba can detect. Amebas flow away from bright lights (although they have no eyes) and from injurious chemicals. When poked with a glass needle, an ameba will contract, reverse its direction of flow, and move away. Vigorous disturbance causes it to take on a spherical shape and remain motionless for some time.

Although amebas behave as though they were "conscious," **physicochemical models** duplicate many of their activities. Ameboid movements can be produced by simply injecting a little alcohol into a droplet of clove oil in water. The alcohol changes the surface film of the oil droplet, causing it to send out "pseudopods," and it flows about like an ameba. A drop of chloroform in water appears to be quite as "finicky" in its "eating habits" as an ameba. When offered small pieces of various substances, such a drop will "refuse" sand, wood, and glass, even ejecting them when they are forcibly pushed into the drop. On the other hand, bits of shellac or paraffin are "eagerly" enveloped. If we play a trick on the chloroform drop by "feeding" it a piece of glass coated with shellac, it will engulf this "delicacy," dissolve the shellac, and then "eliminate" the glass. Other mechanical models simulate growth and replication. Although the resemblances between such models and living amebas are usually quite superficial, they do suggest that much of the behavior of living amebas might be explained if we knew more about the purely physical and chemical phenomena involved.

An ameba differs from the physicochemical models in that several models are required to demonstrate the activities that are displayed by a single ameba, a fact that only begins to reveal the complexity of this "simple" protozoan. A more important difference is that the behavior of the ameba is usually *adaptive*, that is, it is of a type likely to result in the survival of the organism.

In amebas we have emphasized apparent simplicity, and a flexible organization of the cell body. There are few highly organized structures in the cytoplasm, and those present move about in a nearly random manner within the cell. Moreover, almost any part of the cell can function alone even when separated from the rest of the ameba. We shall see now in another protozoan that a wholly different kind of organization with many specializations is possible within the limits of a single cell. Such protozoans represent a level of complexity comparable to that of some small many-celled animals.

A Complex Protozoan: Paramecium

Parameciums are found around the world in freshwater ponds. Like an ameba, a paramecium consists of a microscopic mass of living matter which has a relatively fluid interior and a more

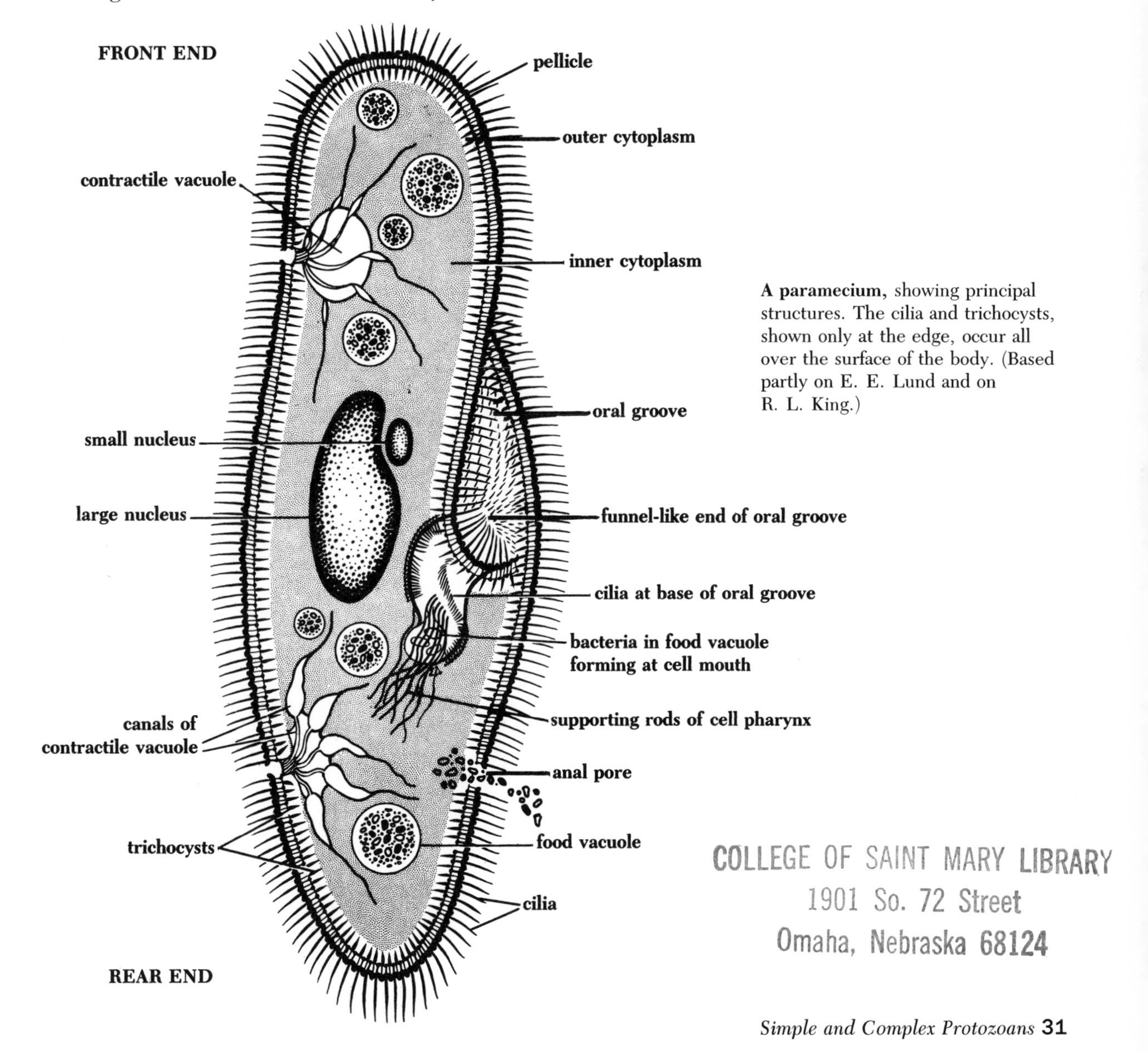

A paramecium, showing principal structures. The cilia and trichocysts, shown only at the edge, occur all over the surface of the body. (Based partly on E. E. Lund and on R. L. King.)

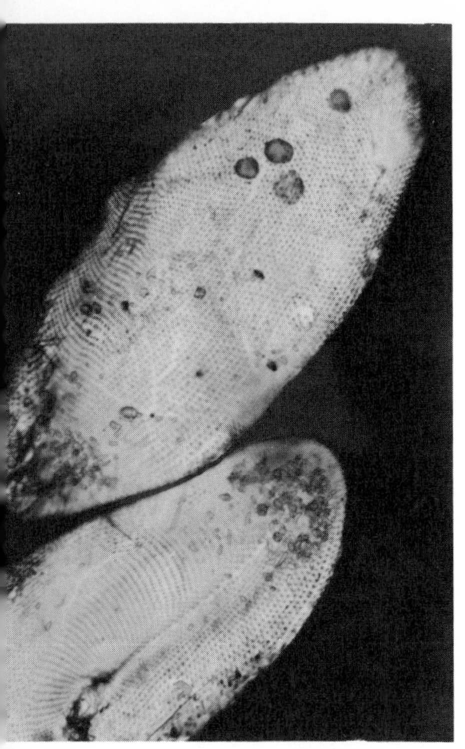

Pellicles of two parameciums, stained to show the surface markings that correspond to the positions of the cilia in the living organisms. The lower pellicle shows the pattern of cilia around the oral groove. (P. S. Tice)

dense outer layer. But many differences between the two protozoans are at once apparent. Instead of a delicate outer membrane, a paramecium is covered by a stiff but flexible outer covering, a *pellicle.* The pellicle gives the paramecium a definite **permanent shape,** somewhat like that of a slipper. Also, a paramecium has **distinct front and rear ends,** the front rounded, the rear pointed—a good example of streamlined form. And most striking of all is the rapid rate at which a paramecium swims about, as compared to the slow creeping of an ameba.

Beneath the pellicle, and embedded in the clear outer cytoplasm, are small oval bodies called **trichocysts.** These bodies reach the surface through pores and can be discharged to the exterior. During the process of discharge they become greatly elongated into fine threads. It is not clear what function the trichocysts serve, but they have been thought to be protective because they are discharged when the paramecium contacts injurious chemicals or is attacked by a predator. However, they have never been seen to deter a predator, even when massively discharged. Because they are found mainly in parameciums and other similar protozoans that feed on bacteria, it has been suggested that trichocysts anchor the paramecium while it is feeding.

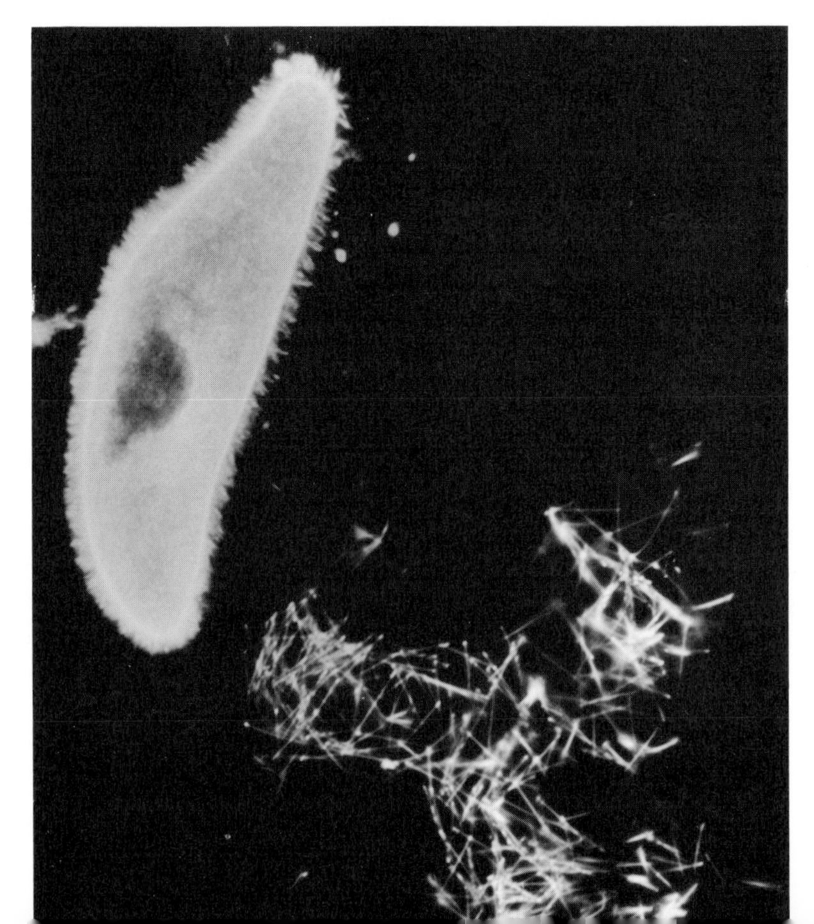

Trichocysts discharged from a paramecium irritated and later killed by a drop of ink added to water in which the protozoan was swimming. (P. S. Tice)

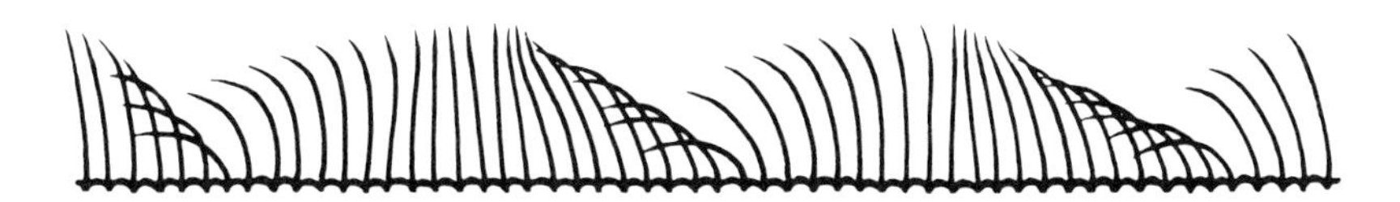

The cilia in a single row do not beat all at once but one after another, so that they appear to beat in waves. They beat so fast that all we ordinarily see under a microscope is a flickering at the edge of the paramecium. Their action has been analyzed by means of special lighting and photographic equipment. (After J. Gelei.)

Parameciums owe their speedy **locomotion** to accessory structures that are not unlike the oars of a racing shell. These short hairlike structures, numbering in the thousands, are cellular extensions called **cilia** (singular: *cilium*) that project through minute holes in the stiff pellicle. The cilia beat somewhat in the same manner as the arms are moved in the crawl stroke in swimming: each lashes back with a strong, stiff *effective stroke,* then more slowly reaches forward in a sweeping *recovery stroke.* The combined effect of the rhythmic beating of all the cilia is to drive the paramecium forward. Rows of cilia beat in slightly different phases, creating the appearance of waves passing over the surface of the paramecium. Moreover, the rows are arranged diagonally and the cilia beat obliquely rather than straight backward, causing the paramecium to revolve on its long axis and swim along a spiral path, so that it appears to screw its way through the water. By reversing the ciliary stroke, the paramecium can swim backward as well as forward; slight changes in the direction of the strokes can turn it in any direction.

In contrast to amebas, parameciums **feed** by means of a permanent, specialized food-catching apparatus. One side of the paramecium is strongly depressed, forming a concavity, the **oral groove,** which appears to be the opening into the slipper-shaped body. The oral groove leads to the so-called **cell mouth,** which is not really an opening but a portion of the surface that readily takes in food. From the cell mouth, the food passes through a region of cytoplasm that is reinforced with thin rods, the **cell pharynx,** at the end of which food vacuoles are formed. When a paramecium stations itself near a bit of decaying material, the beat of the oral cilia drives bacteria and other minute organisms deep into the funnel-like end of the oral groove, where special ciliary tracts whirl them around, concentrating them into a ball. The finished ball is then engulfed at the cell mouth to form a food vacuole. A paramecium that has found a suitable bit of debris and is feeding actively will soon become filled with food vacuoles. These vacuoles move through the interior in a more or less definite course by the slow circulation of the semifluid cytoplasm, and their contents are **digested** and **assimilated** essentially as described for amebas.

The few indigestible remnants in the food vacuoles are **eliminated** at a definite site on the surface of the body, called the **anal pore.**

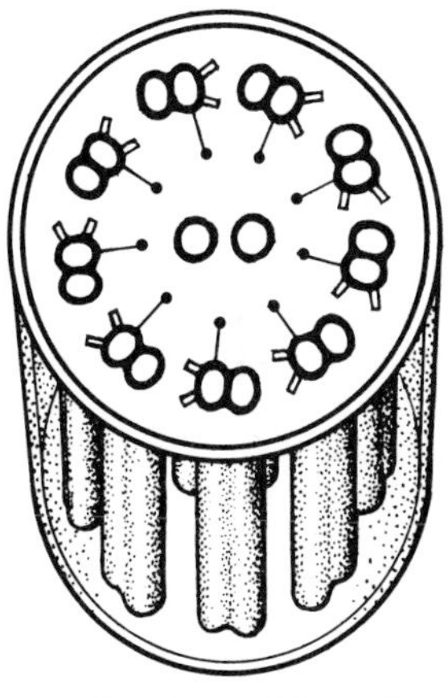

Diagram of a piece cut from the middle of a **cilium.** The cross-sectional face shows the "9 + 2" pattern of the microtubules that are responsible for the motion of the cilium: 9 pairs of tubules around the margin surround 2 single tubules in the center. This pattern is the same in virtually all motile cilia of protists, plants, and animals.

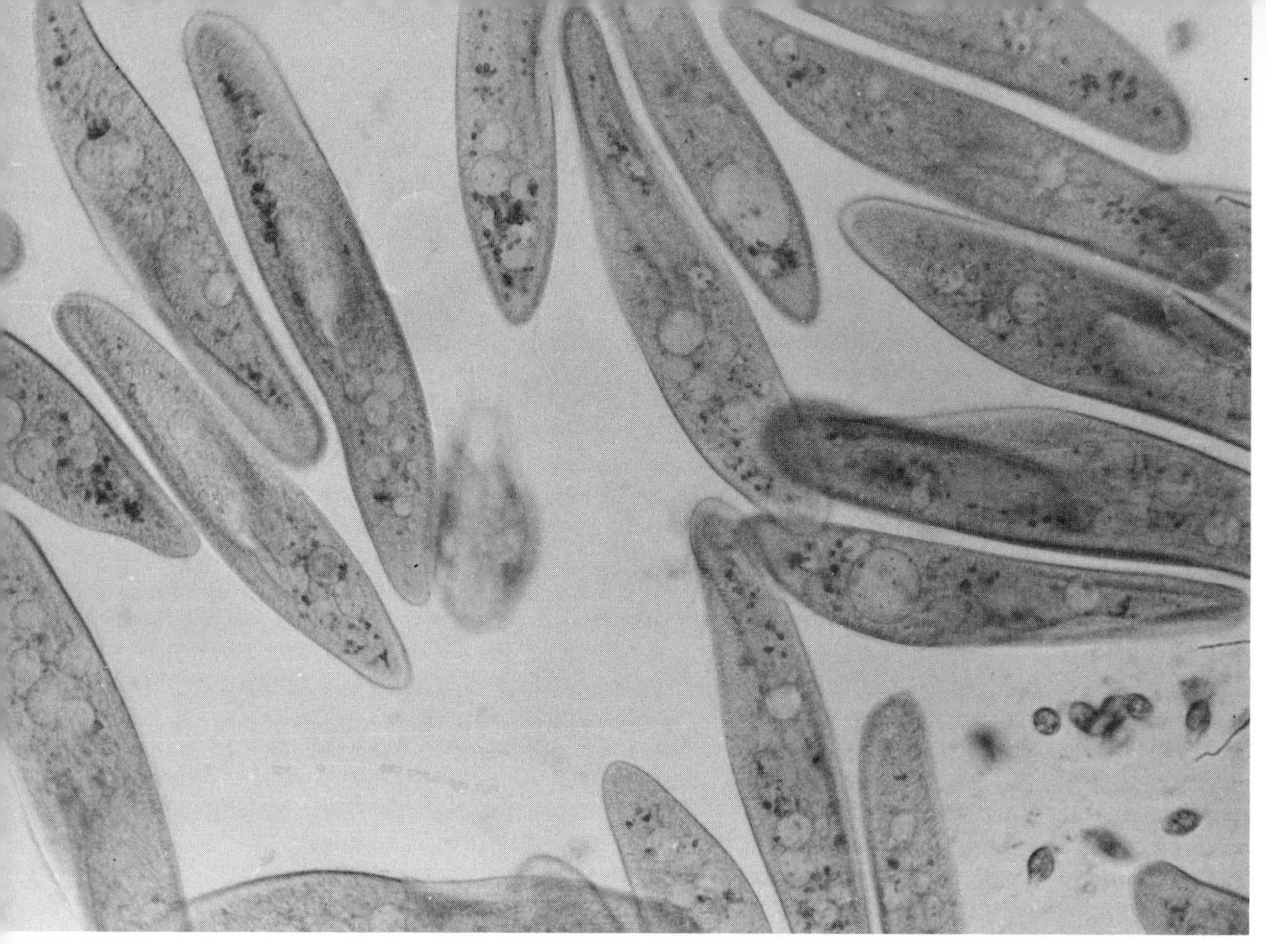

Swimming parameciums, "stopped" by a photomicrograph taken at 1/30,000 second, are seen in various positions. On the left are two individuals that present the full oral surface. The oral groove occupies the center of the upper portion of the body, and at its lower end can be seen the funnel-like part that leads to the cell mouth. All the individuals show food vacuoles in various stages of digestion. Several individuals show the canals of the contractile vacuole in the process of filling. (R. Buchsbaum)

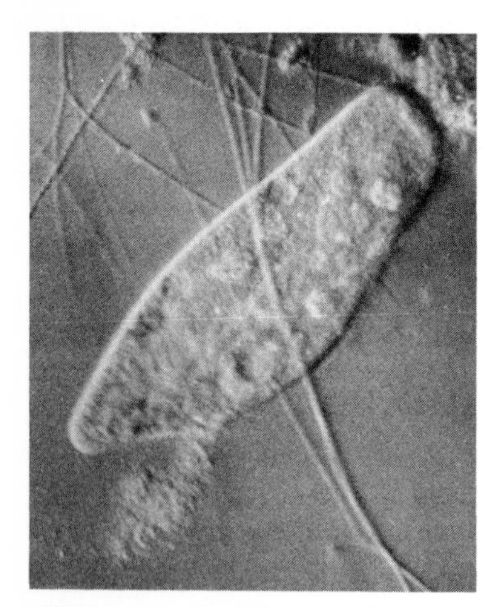

Elimination of solid wastes through the anal pore of a paramecium. The pore has a fixed position posterior to the oral groove. (Photomicrograph from a motion picture, P. S. Tice.)

Respiratory exchange and **excretion** take place, as in amebas, by diffusion through the surface. These processes are similar at the chemical level in all protozoans as well as in multicellular animals and fungi: oxygen is taken in and used to oxidize food and release energy, and carbon dioxide and nitrogenous wastes (mainly ammonia in protozoans) are given off. Water produced during oxidation adds to the water that is diffusing into the cell and must be actively expelled.

Water is expelled, as in amebas, through **contractile vacuoles.** There are two of these in a paramecium, one near the front end, the other near the rear, and they appear to be fixed and permanent structures. Each vacuole is surrounded by canals that radiate into the cytoplasm. At short intervals these canals fill with fluid, then discharge their contents into the vacuole, which swells in turn, then expels the fluid to the exterior through a discharge pore.

In a paramecium the two contractile vacuoles can eliminate a volume of water equivalent to its body volume in fifteen to twenty minutes, as compared to four to thirty hours in an ameba. A person eliminates a volume of urine equal to the body volume in about three weeks, but water is also lost through the lungs and sweat glands.

Paramecium dividing. The large nuclei appear as two large, dark, elongated bodies joined by a delicate strand. The small nucleus is already completely divided. The cytoplasm has begun to pinch in two. (R. Buchsbaum)

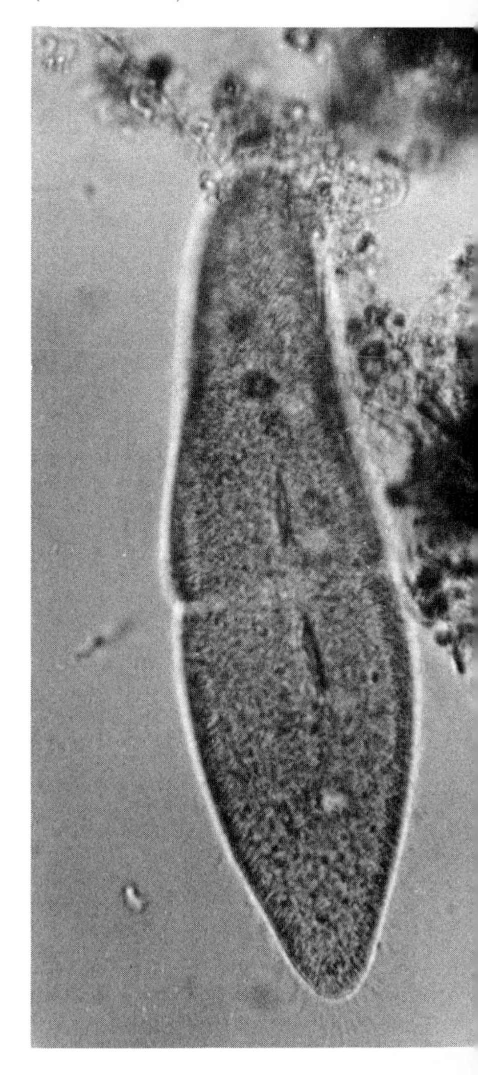

The high degree of cooperation displayed by the cilia when a paramecium is swimming and feeding suggests that some sort of **coordinating mechanism,** resembling nervous control in animals, is present. However, no such system has ever been definitely shown to exist. A system of fibrils that connect the bases of the cilia has long been thought to be a coordinating system but this has not been conclusively demonstrated; in other ciliated protozoans with cilia clumped into a few bundles, the fibrils can be experimentally cut, and afterward the cilia still move in a coordinated manner. It is thought that coordination is probably accomplished by electrical or mechanical events over the surface of the cell.

The innermost granular cytoplasm, as in an ameba, is more fluid than the surface layer, and contains food vacuoles, fat droplets, and other granules and bodies, as well as an elongate **large nucleus** (macronucleus). The large nucleus is concerned with directing the activities of the cell, in particular, the synthesis of proteins. After a paramecium has fed and grown for a time, it undergoes **asexual reproduction,** replicating by division in a manner similar to that of an ameba. However, whereas the nucleus of an ameba or most other eukaryotic cells contains only one complete paired set of genetic material (chromosomes), the large nucleus of a paramecium contains hundreds of sets. Therefore, instead of dividing by mitosis like the nucleus of an ameba, the large nucleus of a paramecium needs only to be pinched roughly in half when the cell divides and replicates. Under favorable conditions, parameciums divide several times each day, so that their numbers grow rapidly.

The large nucleus of parameciums and other ciliated protozoans is the only type of nucleus with the ability to regenerate. If a paramecium is cut into several pieces, each piece can regenerate into a whole paramecium provided it contains at least a small piece of the large nucleus.

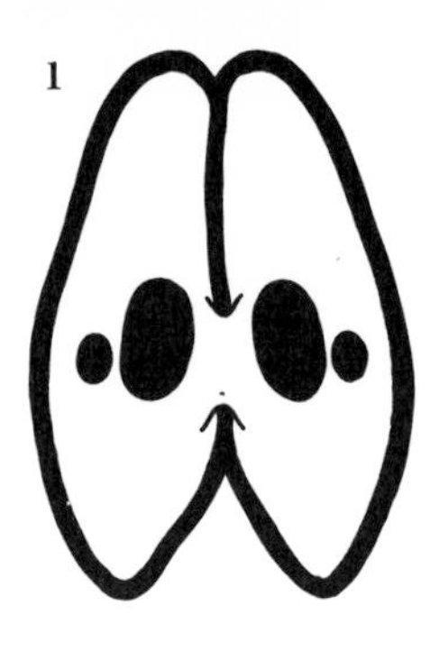

Two parameciums unite by their oral grooves.

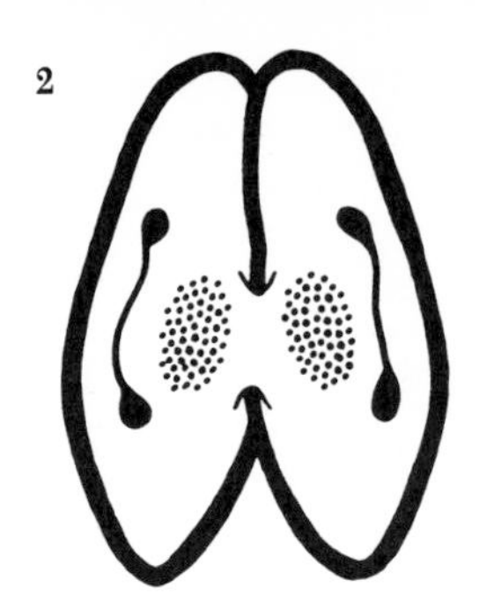

In each paramecium, the large nucleus begins to disintegrate. The small nucleus divides.

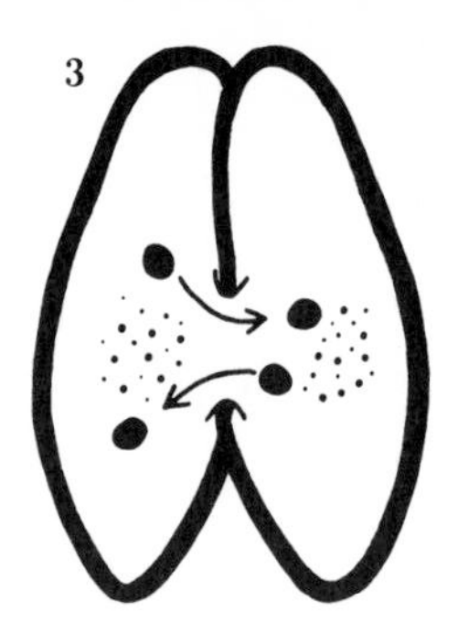

One of the two small nuclei in each paramecium migrates to the partner.

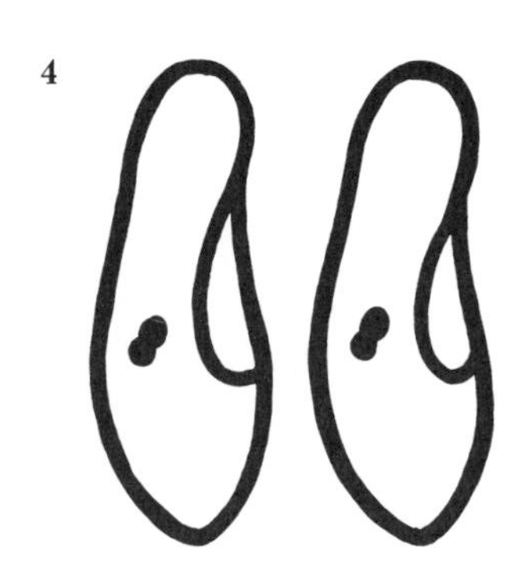

In each paramecium, the two small nuclei fuse into a new nucleus. The parameciums separate.

In each paramecium, the new small nucleus divides several times, eventually giving rise to both large and small nuclei.

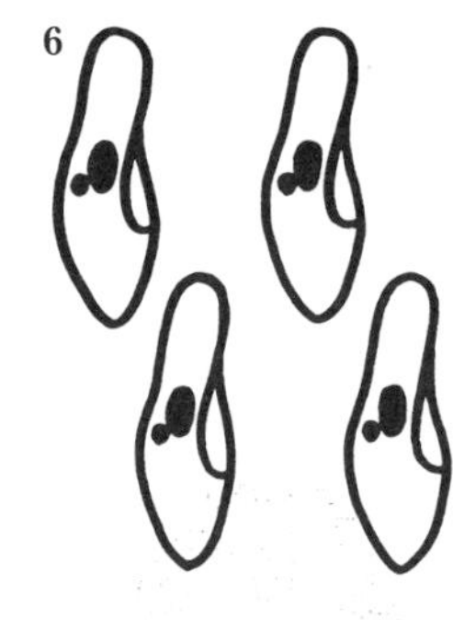

Each paramecium divides twice, resulting in four small parameciums.

Conjugation in one species of paramecium. For clarity, the large nuclei are omitted from **3;** actually they do not disintegrate completely until after the conjugants have separated. Not all stages are shown; in **2** and **5,** the small nuclei undergo extra divisions, some of the products of which disintegrate.

In addition to the large nucleus, a **small nucleus** (micronucleus) or sometimes several are present in most paramecium species. The small nucleus appears to have no function during the everyday activities of a paramecium. It divides by mitosis each time the paramecium divides, but it can be experimentally removed with little or no effect on the activities, growth, and division of a paramecium; some strains of paramecium have no small nuclei at all. The function of the small nucleus lies in **sexual recombination.** When two parameciums of certain different types are placed together the individuals recognize each other as "different" and temporarily fuse in pairs. The large nucleus disintegrates in both partners and each small nucleus undergoes meiosis, a series of divisions whereby the resulting small nuclei end up with only half the amount of genetic material originally present (see chapter 2). One of these reduced small nuclei in

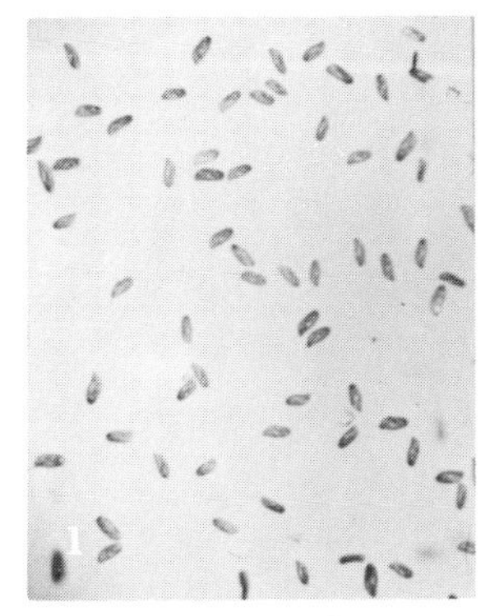
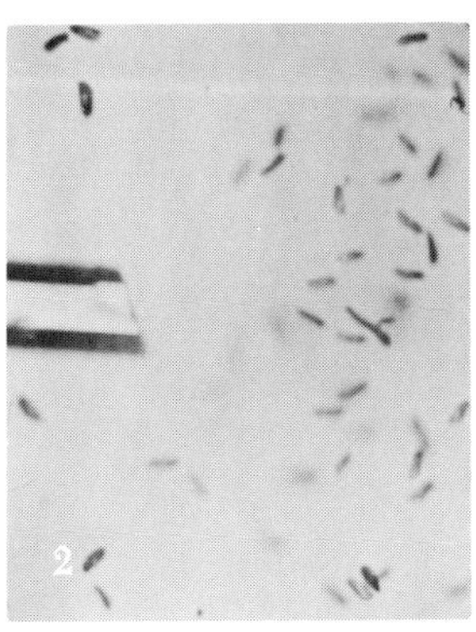
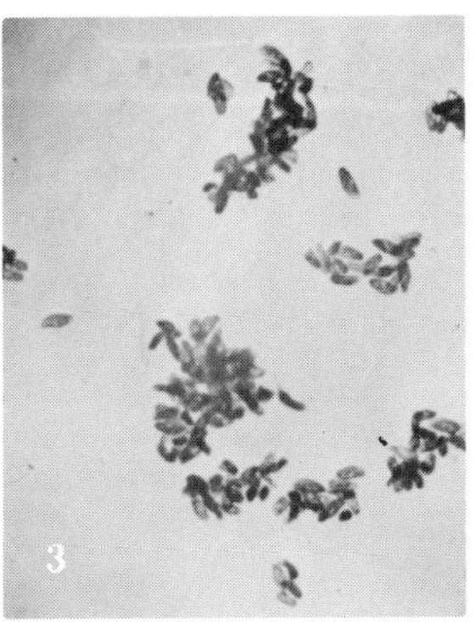
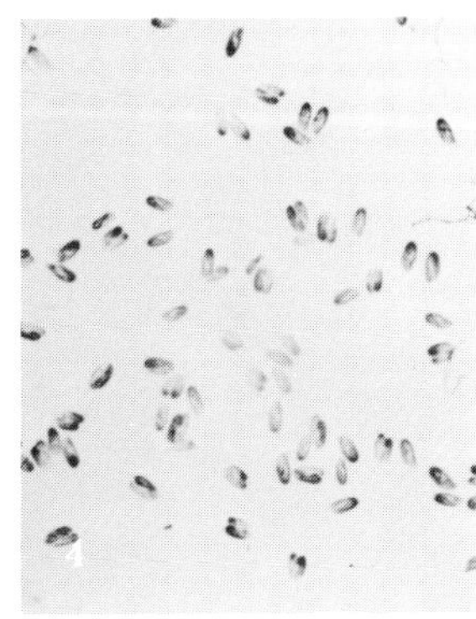

Mating reaction and conjugation. *From left to right:* (1) Individuals of one mating type swimming about. (2) A pipet introduces into the same drop of water individuals of another mating type. (3) Within 5 minutes after mixing, the parameciums are stuck together in large clumps. These clumps later become smaller, and after about 6 hours only small groups and conjugating pairs remain. (4) At 23 hours after mixing, there are only conjugating pairs and a few single individuals that did not obtain mates. Pairs usually remain joined for about 36 hours. Conjugation occurs only under certain conditions. The individuals must be mature, in a certain state of nutrition, and be mixed at a certain time of day. In the culture shown here the mating reaction occurred from about 10:00 A.M. to 3:00 P.M. In the species shown here *(Paramecium bursaria)*, several separate groups of mating types are known; there is no conjugation between types in the separate groups. (Photomicrographs from a motion picture, R. Wichterman.)

each partner migrates to the other partner; then the two small nuclei in each paramecium fuse to produce a single small nucleus that again contains the full normal amount of genetic material, but in a new combination to which both partners have contributed. The two parameciums then separate and swim away. The newly formed nucleus gives rise to both large and small nuclei.

The process of sexual recombination in parameciums and other ciliated protozoans is called **conjugation.** There are no visibly distinct males and females, but instead there is the equivalent of not just two but *several* different "sexes," or mating types. Parameciums will conjugate with members of other mating types but not with their own. Although more typical sexual reproduction, involving differentiated sperms and eggs, occurs in some other protozoans, conjugation in parameciums has the essential features of sexual recombination in most organisms, in that there is a *transfer of genetic material having new hereditary possibilities from one individual to another.*

The **behavior** of a paramecium is exactly what one would expect of an organism that has no specialized sensory structures to direct its movement. When not quietly feeding on bacteria, it roams about ceaselessly, bumping "head on" into obstacles in its path. After such a collision, a paramecium backs up by reversing the beat of the cilia, turns to one side, and goes off in a new

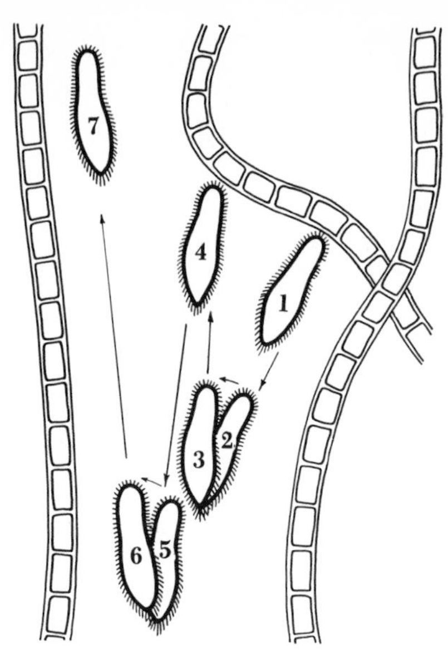

Avoiding reaction. 1: Paramecium encounters an obstacle. **2:** The paramecium backs up. **3:** It shifts its position. **4:** It again meets resistance. **5, 6:** It backs up and turns again. **7:** It finds a free path. (Based on H. S. Jennings.)

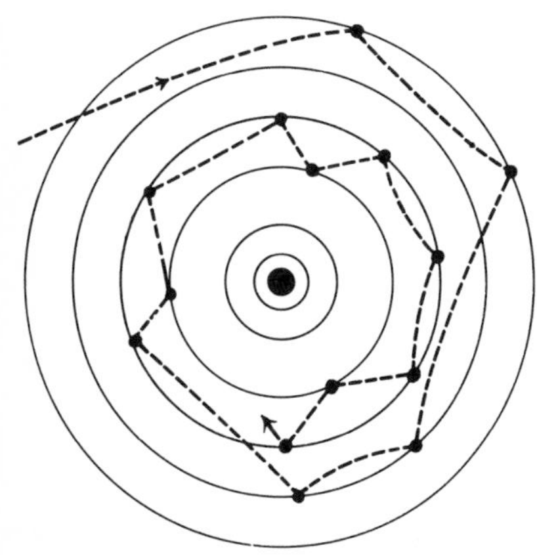

Trial-and-error behavior in relation to a chemical gradient. A small amount of a chemical is placed in the center of a drop of water, and it slowly diffuses outward. The concentric circles indicate zones of decreasing concentration of the chemical, from the center outward. The broken line shows the path of a paramecium placed in the drop. As the paramecium swims about, it gives an avoiding reaction whenever it enters a zone less favorable than the one it is in. It thus eventually enters and remains within the most favorable zone, in this case one with an intermediate concentration of the chemical. (After H. S. Jennings.)

direction. If this second path results in another collision with the obstacle, the set of movements is repeated. Finally the paramecium encounters a free path and continues on its course. The set of movements with which a paramecium backs up, turns, and swims off in a new direction is called the **avoiding reaction.** Physical obstacles, excessive heat or cold, irritating chemicals, unsuitable food, a potential predator—all elicit the avoiding reaction, which may be said to constitute most of the behavior of a paramecium.

In its constant explorations, a paramecium may swim by chance into a region rich in bacteria. Each time that it crosses the boundary of this region into a less favorable area, it gives the avoiding reaction; thus it remains in the more favorable region. This general method of finding the best conditions of existence is called **trial-and-error behavior** and is used by almost all protozoans, and to a greater or lesser extent by most animals, including people.

A paramecium need not actually enter an unfavorable region before reacting negatively. The beating of the cilia in the oral groove draws a constant stream of water, in the form of a cone, toward the oral groove. Thus a portion of the water ahead of the paramecium is sampled by the oral groove before the paramecium actually enters the new region. If there is an irritating chemical in the water ahead or if the water is hotter or colder, the paramecium has "advance information" and responds with the avoiding reaction.

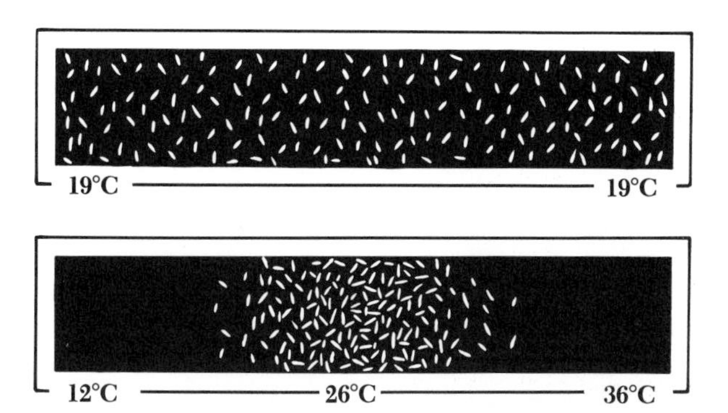

Behavior in relation to temperature. If temperature is uniform throughout a chamber, parameciums are uniformly distributed. If one end of the chamber is cooled to 12°C and the other end is heated to 36°C, the parameciums avoid these extremes of temperature and accumulate in the region of more favorable, intermediate temperatures. (After M. Mendelssohn.)

There have been many attempts to demonstrate "learning" in parameciums and other protozoans. Although these attempts have been largely unsuccessful, there is some evidence that the behavior of ciliated protozoans can be modified by *conditioning*, which is a form of learning. Some ciliated protozoans were exposed to repeated flashes of light, which normally elicit no response, at the same time that they were given electric shocks, which elicit the avoiding reaction; they became conditioned to respond with the avoiding reaction to flashes of light alone. By and large, however, parameciums and other protozoans depend simply on a continuing series of trials and errors to lead them to congenial conditions, and on repeated avoiding reactions to keep them out of trouble.

carbon particles

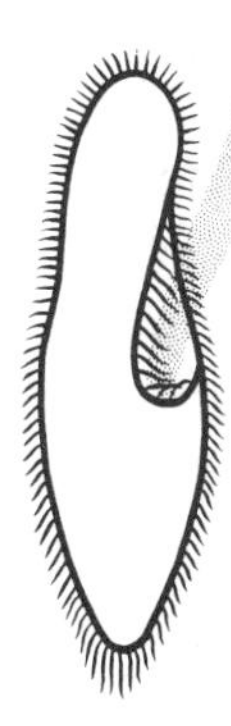

Paramecium samples the water ahead, as can be seen by placing in the path of the organism a drop of india ink containing visible particles. (After H. S. Jennings.)

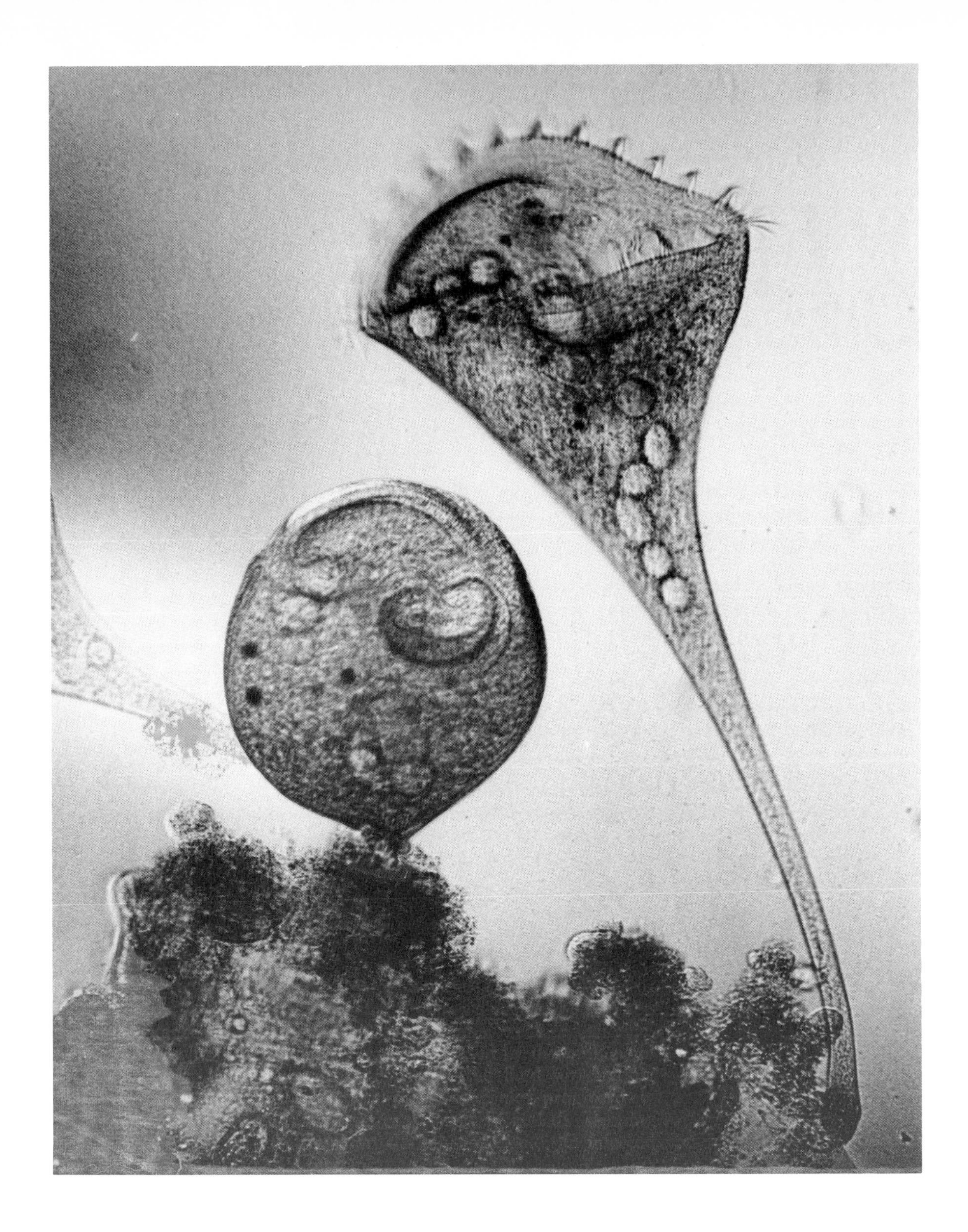

Chapter 4

Variations on a One-Celled Theme

Protozoans are cosmopolitan—the same species may be found on every continent and throughout the seas. In sharp contrast with the provincial habits of most animals, which are limited in their spread by water or land barriers, protozoans are readily swept along in ocean and river currents, or, when in an encysted state, are blown by wind or carried in the mud on birds' feet from pond to pond. Encystment also enables many protozoans to survive in habitats they otherwise could not invade. In the encysted state, they resist the heat and drought of summer in the desert, to break out after the first rains. Many kinds of protozoans are well suited to exploit the moist and nutritious interiors of plants and animals, including humans. Some cause no noticable harm; others are important and dangerous parasites.

About 30,000 living species of protozoans already have been described, of which about 30% are parasites. Almost all plants and animals harbor at least some parasitic species of protozoans in their bodies, and in many cases specific protozoans infect only specific hosts so we may expect many more species of protozoans to be discovered and described.

Protozoans are those organisms in the kingdom Protista that generally exhibit animal-like characteristics: mobility and ingestive feeding. All have the ability of absorptive feeding, some almost exclusively, and some also contain chlorophyll and are photosynthetic. There is no generally accepted higher classification scheme for protozoans. However, they may be conveniently divided into four groups, based mostly on their different methods of locomotion: flagellated protozoans, ameboid protozoans, spore-forming protozoans, and ciliated protozoans.

Opposite: **Protozoans** are protistans with the animal-like characters of locomotion and heterotrophic nutrition. The bodies of protozoans and other protistans are not divided into cells, and this limits their size. But these large ciliated protozoans, *Stentor coeruleus*, are about 2 mm long (when extended, as at right), easily visible to the naked eye and larger than some many-celled animals, which are included among its prey. (R. Buchsbaum)

Flagellated Protozoans

Flagellates are protozoans that have one or more long **flagella** (singular: *flagellum*), filamentous protoplasmic extensions by which the organisms swim. Flagella are like cilia in structure (covered by a cell membrane and containing a shaft of microtubules in the typical "9 + 2" arrangement) but are longer and exhibit different patterns of motion. Most flagellates have a more or less definite shape, commonly oval, and a permanent front end from which the flagella arise. Compared to parameciums and most other ciliated protozoans, flagellates are generally smaller and move more slowly and irregularly. There is usually only a single nucleus and never more than one kind. All of the photosynthetic protozoans are flagellates.

Members of the genus ***Euglena*** are among the commonest green flagellates of freshwater. They may be so numerous as to produce a green scum on the surface of ponds. The name applies to a number of species, all of which have an elastic pellicle and are able to contract and elongate the body in a characteristic squirming called **euglenoid movement.** At the front end there is a flask-shaped depression, called the flagellar pocket (or the reservoir), from which springs a long flagellum. By lashing this flagellum, the organism swims slowly forward, tracing a helical

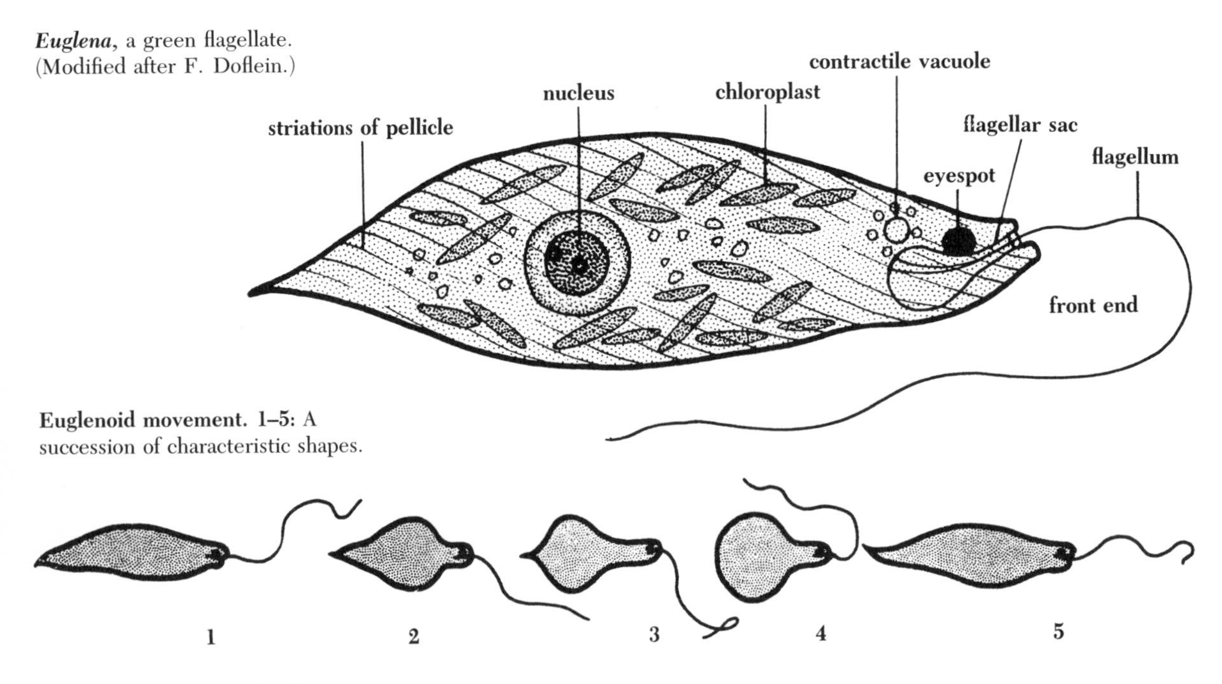

Euglena, a green flagellate. (Modified after F. Doflein.)

Euglenoid movement. 1–5: A succession of characteristic shapes.

path. A large contractile vacuole discharges its contents into the flagellar sac at frequent intervals.

Beside the flagellar pocket is an orange-red body, called the **eyespot** (or stigma), which acts as a pigmented shield for a light-sensitive receptor. The light-sensitive apparatus of a euglena permits it to detect not only the intensity of light but also the direction from which light comes. Placed in a dish, euglenas will move away from an area of darkness or intense light and aggregate in an area of moderate light. Light is important to euglenas as they depend mostly on photosynthesis for their nutrition. As long as they are exposed to light, they maintain themselves by photosynthesis and accumulate carbohydrates in storage bodies that are conspicuous in the cytoplasm. Euglenas can live in the dark if they are placed in a nutrient solution. Under these conditions, their chlorophyll degenerates, the organisms become colorless, and dissolved nutrients are absorbed through the surface membrane. However, the **chloroplasts,** the cytoplasmic organelles that bear the chlorophyll, are not destroyed by a brief dark period and will turn green again when exposed to light. If the chloroplasts are destroyed (by prolonged darkness or by treatment with antibiotics or ultraviolet radiation), the euglenas become permanently colorless and dependent on external sources of nutrients. Some closely related flagellates, which never have chloroplasts, were probably derived from photosynthetic ancestors.

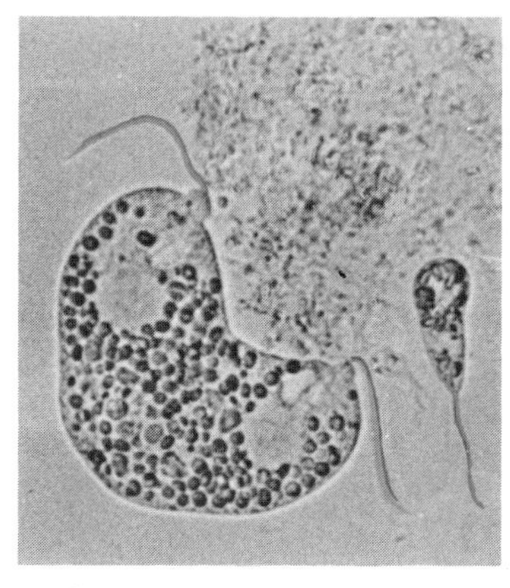

Euglena dividing. The flagellum and anterior part of the euglena have already split. The division is lengthwise, as in flagellates generally. Another small flagellate is seen at the right. (R. Buchsbaum)

Volvox is seen in freshwater ponds as a small green sphere that may be 1 or 2 millimeters in diameter. The sphere is composed of thousands of flagellated cells embedded in the surface of a gelatinous ball. Each cell has two flagella, a red eyespot, two contractile vacuoles, and a cup-shaped chloroplast. The sphere swims about, rolling over and over from the action of the flagella; but remarkably enough, the same end of the sphere is always

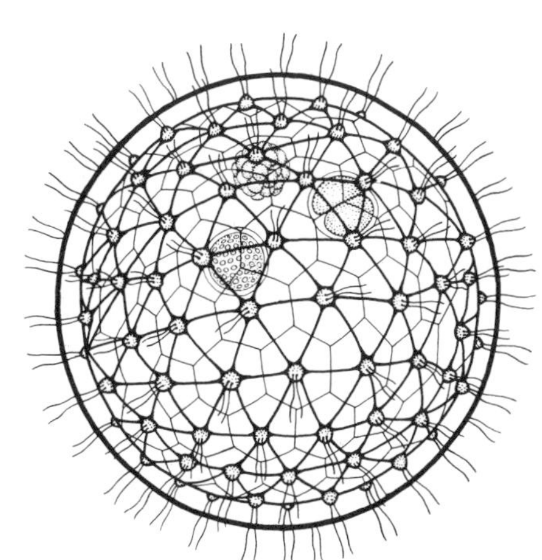

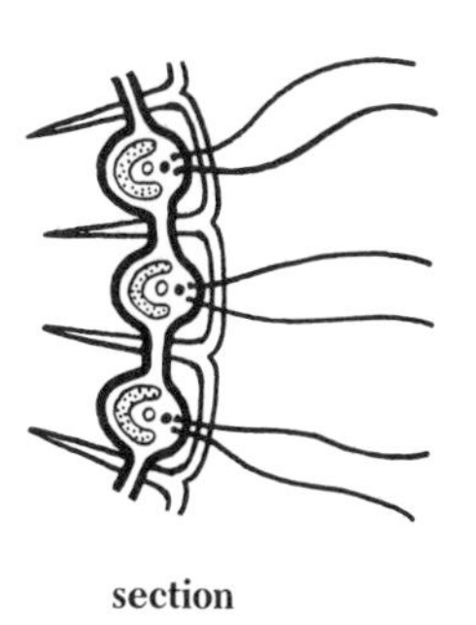

Volvox is photosynthetic and hardly qualifies as a protozoan, except by its active swimming. This colony contains three young developing colonies. The section (at the right) shows several individual cells, each with two flagella, an eyespot, and a cup-shaped chloroplast. A few elated flagellates lack chlorophyll nd feed by absorbing dissolved nutrients.

directed forward, and thus we can distinguish front and rear ends. Only the cells at the rear end are capable of reproduction, whereas those at the front end never reproduce but have larger eyespots and serve primarily in directing the course of the colony. Although the cells of some species are connected by cytoplasmic strands extending through the jelly, there is no evidence of coordination between the cells; each probably responds independently to stimuli such as light and mechanical contact, and the movement of the sphere is the result of their combined activities. Because of the apparent lack of coordination, and because there is relatively little differentiation between the cells, the sphere is usually considered a **colony** of cells rather than a multicellular organism. Nevertheless we see in *Volvox* the *beginnings of multicellularity* as it exists in animals, where each organism behaves as a single individual although it is composed of many cells, from hundreds to billions.

In **asexual reproduction,** one of the cells at the rear end of the colony enlarges, loses it flagella, and divides by mitosis a number of times until a small ball of cells is produced. A large sphere usually contains in its hollow interior several such new colonies in the process of formation. They are liberated by rupture of the parent colony.

Volvox displays a well-developed form of **sexual reproduction.** In the early stages of the evolution of sex, there is less sexual differentiation and the two fusing cells, or **gametes,** are alike; such a condition is seen in some related green flagellates. In *Volvox,* however, differentiated male gametes, or **sperms,** and female gametes, or **eggs,** are formed. In forming an egg, a cell of the colony increases greatly in size, takes on a rounded form, and becomes loaded with food, especially fatty substances. This food is contributed in part from adjacent cells and serves to give the young colony a good start in life. Another cell of the same or another colony, by repeated divisions, gives rise to numerous small flagellated sperms. These sperms swim about until they find an egg or die. Only one sperm fuses with the egg, and the combined cell, or **zygote,** secretes upon its outer surface a hard spiny shell that protects the organism during unfavorable conditions, such as drying or freezing. Within its shell, the zygote divides by meiosis. (As discussed in chapter 2, meiosis is a constant feature of sexual reproduction, but in *Volvox* it occurs at a time in the life history different from that in any animal life history.) One of the cells resulting from the meiotic division will form a new colony. With the return of favorable conditions of temperature and moisture, the shell breaks and the young colony, indistinguishable from an asexually produced colony, emerges.

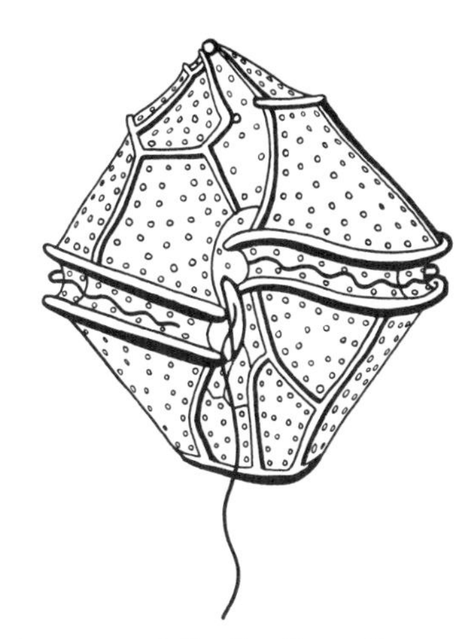

An armored dinoflagellate, *Gonyaulax*, sometimes becomes so abundant along coasts that it colors the ocean red for miles. These "red tides" cause the death of fishes and other animals, and the millions of decaying bodies cast up on beaches may produce a formidable stench. In some places (for example, Florida) red tides are caused by another toxic dinoflagellate, a species of *Gymnodinium*. Nontoxic dinoflagellates may cause red tides that kill fishes by depleting the oxygen in the water. (After C. A. Kofoid.)

Dinoflagellates occur in enormous numbers in the surface waters of the oceans, especially in warm seas; there are also freshwater species. They may occur as single cells or as colonies of cells. A typical dinoflagellate, *Gonyaulax*, is enclosed in an armor of cellulose plates and has two flagella, one lying in a groove encircling the organism and the other trailing backward. Many photosynthetic dinoflagellates are colored yellowish brown by pigments that they possess in addition to chlorophyll. Among these are the **zooxanthellas,** small dinoflagellates that live within the bodies of some other protozoans and within the cells of reef corals and many other invertebrates; the zooxanthellas release nutrients to their host organisms and use by-products of host metabolism (carbon dioxide, compounds of nitrogen and phosphorus) that would otherwise be excreted as wastes, so these associations are probably mutually beneficial.

A number of photosynthetic dinoflagellates (especially some species of *Gonyaulax* and *Gymnodinium*) produce a toxin that can kill fishes and other animals when these dinoflagellates become abundant in the water, causing "red tides." Certain animals that feed on them do not die but accumulate the toxin in their tissues. Bivalves such as clams, oysters, and mussels often become so poisonous that a person who eats even a few can die, and it is wise to be alert to the quarantines placed on these bivalves by health authorities, usually during summer months. "Ciguatera" poisoning results from eating tropical fishes that contain the toxin from bottom-living dinoflagellates. The toxic dinoflagellates live on the surface of seaweeds, and the toxin is accumulated both by herbivorous fishes that browse on the seaweeds and by carnivorous fishes that prey on the herbivores.

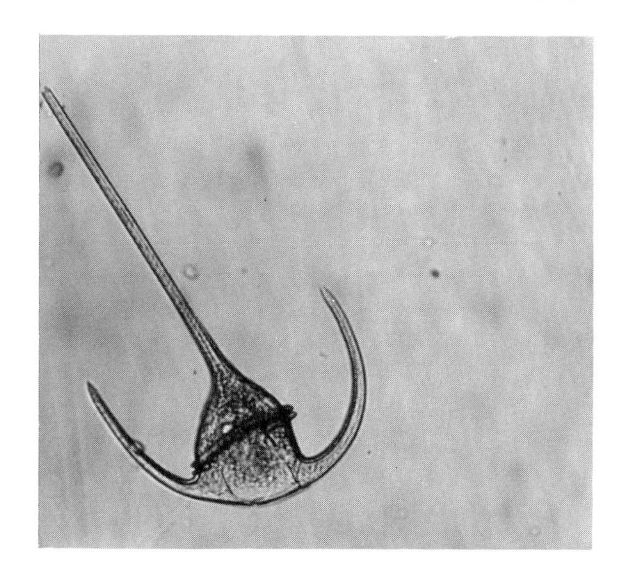

Photosynthetic dinoflagellate, *Ceratium*, has prominent projections, common among many organisms that swim or drift suspended in the upper, lighted layers of the open ocean. Such projections on a small organism slow its rate of sinking and may also deter some animals from eating it. (R. Buchsbaum)

Luminescent dinoflagellates, *Noctiluca.*

Other dinoflagellates are colorless and feed on minute organisms. *Noctiluca* is a large dinoflagellate, spherical and about a millimeter in diameter, with only one flagellum and a stout mobile tentacle used in catching prey. *Noctiluca* can produce light, as can many other dinoflagellates (as well as other marine and some land organisms; see *bioluminescence* in chapter 15). Luminescent dinoflagellates emit light when they are mechanically stimulated, and their tiny bright flashes are seen at night when a boat or swimmer cuts through the water or when the waves break against the shore. When abundant, they create spectacular displays of what looks like underwater fireworks. The light has been observed to inhibit feeding by small crustaceans (copepods; see chapter 17) that graze on dinoflagellates.

The mostly heterotrophic, animal-like flagellates are a diverse lot. Among them are the **collar flagellates,** which often live attached by a stalk to the substrate. In this group as well, the cells may occur singly or as small colonies. Each cell has a collar made up of many delicate cytoplasmic projections, from the center of which emerges a single flagellum. The beating of the flagellum draws a current of water toward the cell. Bacteria and other food particles in the current encounter the sides of the collar and main cell body, and are taken up into food vacuoles. The collar flagellates are of special interest because a strikingly similar type of cell occurs as the food-ingesting cell of sponges. Thus it is often suggested that sponges, which are many-celled animals, evolved from collar flagellates.

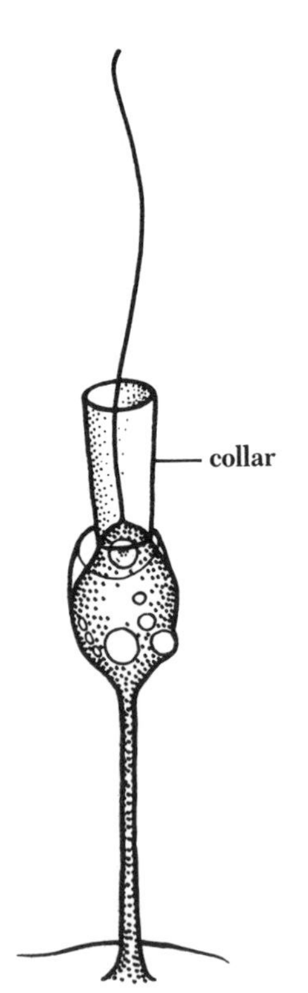

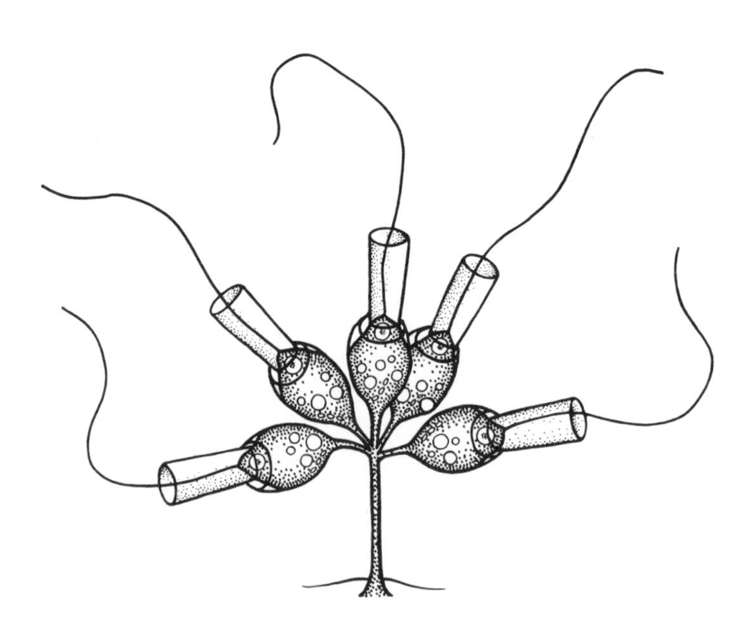

Collar flagellates may be found as solitary individuals, as pairs, and as small colonies. (After G. Lapage.)

Most notorious among the animal-like flagellates are the **trypanosomes** that cause Chagas' disease in South America and sleeping sickness in Africa (not the same as the sleeping sickness that is a viral encephalitis). The trypanosomes and their relatives live as parasites in insects, certain plants, and many vertebrates in Africa without causing much inconvenience to their hosts; over millions of years of exposure these hosts have presumably evolved the ability to resist or to tolerate the parasites. But when humans and their domestic animals become infected, incapacitating and sometimes fatal disease results.

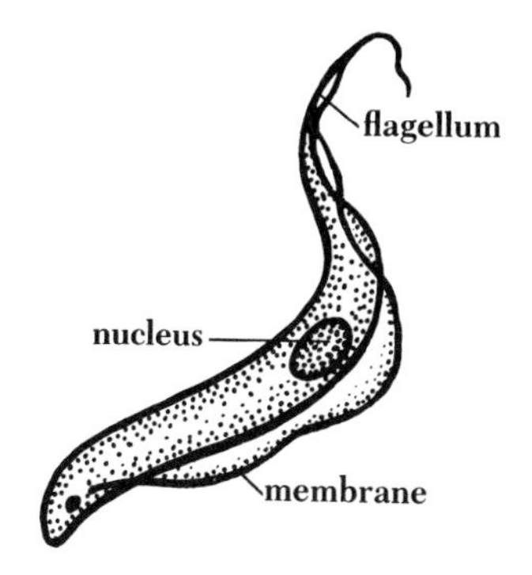

Trypanosome. The flagellum is attached by a membrane for most of its length but extends free at the anterior end of the organism.

The trypanosomes that cause African sleeping sickness are transmitted by blood-sucking tsetse flies. Practically all wild game in Africa harbor trypanosomes in their blood; and when a tsetse fly sucks the blood of an antelope, for example, or of an infected human, the blood taken into the intestine of the fly will contain these trypanosomes. From the intestine they invade the salivary glands of the insect, meanwhile multiplying and undergoing changes in form until they reach a stage in which they are infective to a vertebrate host. If a fly carrying such infective stages bites a human, the trypanosomes are injected into the blood of the victim with the saliva of the fly. In the blood they multiply rapidly and wriggle about among the blood cells, propelled by the undulations of the flagellum.

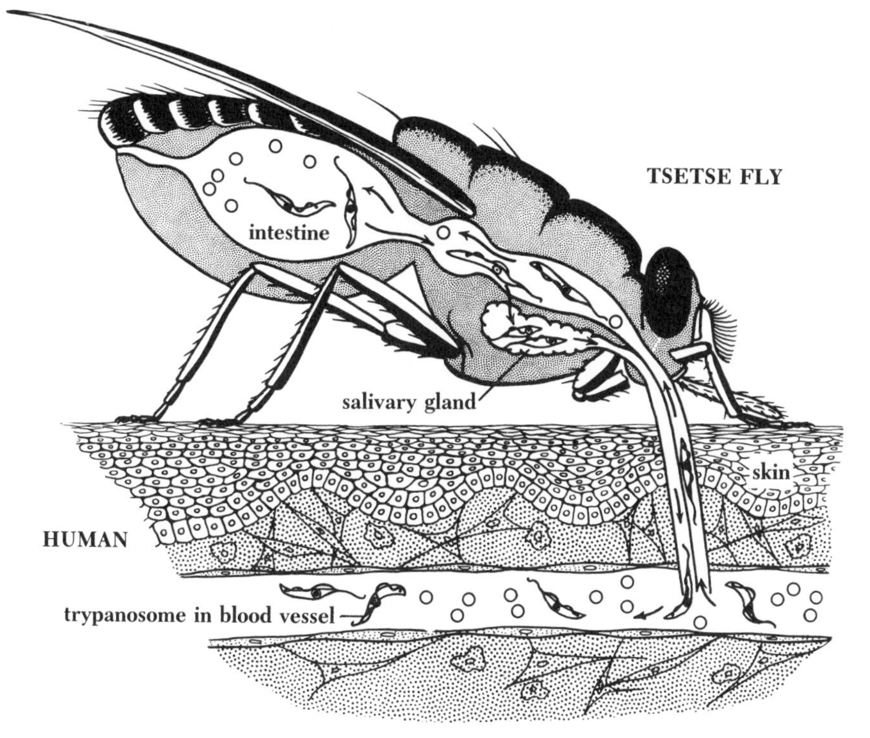

Two hosts are necessary to the life history of the trypanosomes that cause African sleeping sickness, a tsetse fly and a mammal. Here an infected fly is biting an infected human. Principal stages in the life history are shown: trypanosomes in the salivary glands of the fly being injected into the blood of the human, forms living in the blood and entering the sucking mouthparts of the fly, and those living in the intestine of the fly. The trypanosomes have no apparent ill effects on the fly. (Based on several sources.)

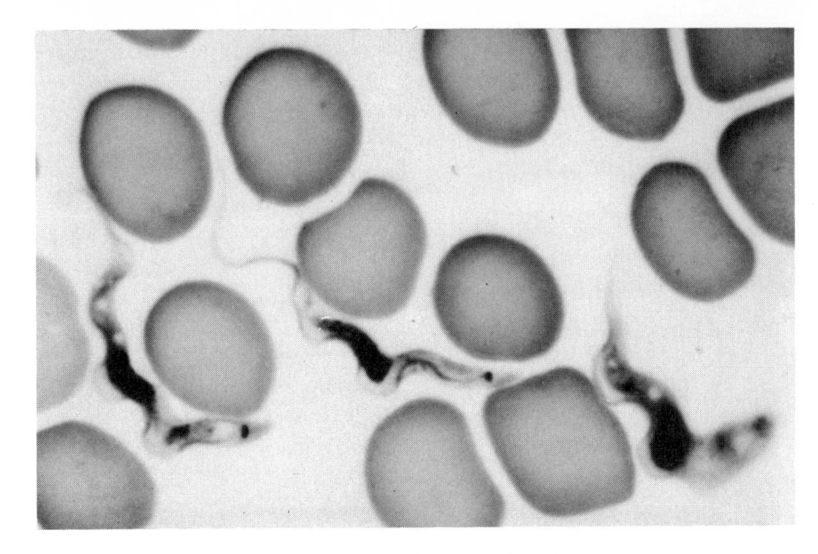

Trypanosomes that cause African sleeping sickness are seen here in a blood smear, among red blood cells. The parasites are about 25 μm (1/1,000 in) long. Trypanosomes probably lived originally in the intestine of insects, and when these hosts evolved blood-sucking habits, the flagellates were constantly exposed to vertebrate blood. Accidentally introduced into the blood stream of vertebrates when the insects were biting, the flagellates became adapated to this new environment and, finally, dependent for part of their life cycle on development in a vertebrate host. *Trypanosoma brucei gambiense.* (Stained preparation.)

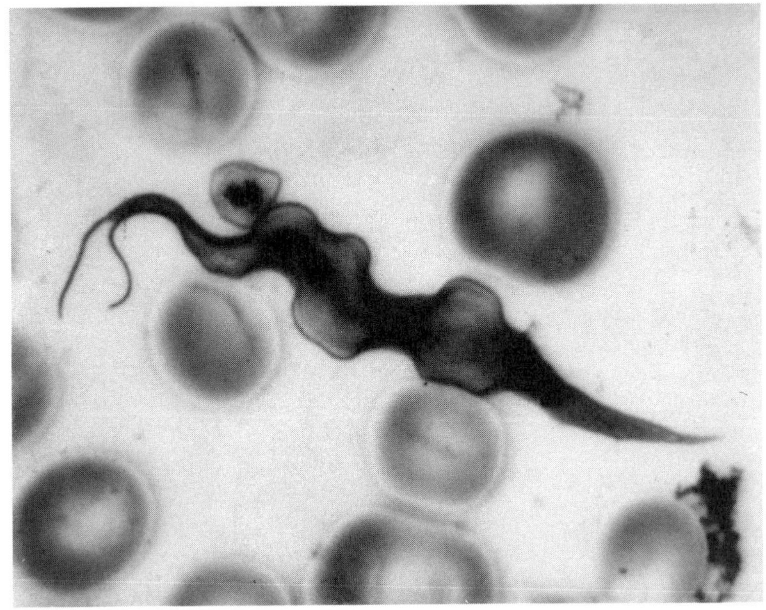

Dividing trypanosome in the blood of a rat. The flagellum is already divided but the nucleus is still single. This flagellate *(Trypanosoma lewisi)*, transmitted by fleas, does not harm the rat. The parasites multiply rapidly at first, but gradually the rat's immune system acquires the ability to inhibit their growth and finally destroys them. (Stained preparation.)

The bite of an infected fly does not cause fever for weeks or even months, while the flagellates increase in number. When the attacks of fever begin, the victim becomes weak and anemic, probably because of toxic substances released by the millions of flagellates. Finally, the parasites invade the fluid surrounding the brain and spinal cord, the person becomes progressively more apathetic and finally loses consciousness, and the sleeping sickness goes to a fatal end. At the other extreme, an infected person may show no symptoms.

Treatment with drugs and control programs directed against the tsetse flies have greatly reduced the impact of sleeping sickness since the middle of this century. Nevertheless, in areas where the disease occurs, one should observe measures to prevent infection, such as using mosquito netting and wearing clothing that covers the body to the wrists and ankles.

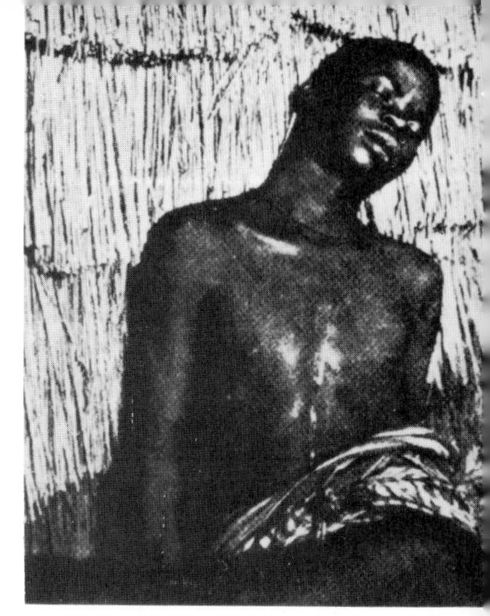

Victim of African sleeping sickness. Much of the "laziness" attributed to native Africans by early European explorers was no doubt due to this disease, and slave traders quickly learned not to accept as slaves any individuals having swollen glands in the neck, a symptom of trypanosome infection. (Zaire, Army Medical Museum.)

The **trichomonads** are flagellates common in the digestive tracts of vertebrates. They are pear-shaped protozoans with several flagella springing from the anterior end. One of these flagella extends backward and is connected to the body of the organism by a membrane. Through the body runs a rodlike bundle of microtubules, capable of active bending; it projects at the rear end and is used to anchor the organism to surfaces. In humans, *Trichomonas tenax* inhabits the mouth and several trichomonad species live in the intestine; *Trichomonas vaginalis,* which lives in the vagina of females and the urethra of males, is spread by sexual intercourse.

Some of the trichomonads and related flagellates may be extremely complex, having many flagella, complicated arrangements of associated structures, and sometimes multiple nuclei (from two to more than a thousand). One is *Giardia,* one of the diplomonads, so called because each individual has two sets of flagella and two nuclei; several species live in the intestines of vertebrates, including humans.

Giardia is widely distributed throughout the world and is probably a more common parasite than is realized, for many cases cause no symptoms. In others the victim suffers diarrhea, cramps, nausea, and other digestive symptoms. The flagellates interfere with both digestion and absorption of food. They mul-

Trichomonas vaginalis produces an itchy infection in women. Infected men may have no symptoms, but if the presence of this flagellate is suspected, both members of the couple should be checked and treated to prevent recurrence. Effective drugs are available; however, most drugs used to control parasitic protozoans, which are eukaryotes (as humans are) are more apt to cause side-effects than are the antibiotics effective against bacteria, which are prokaryotes. (After W. N. Powell.)

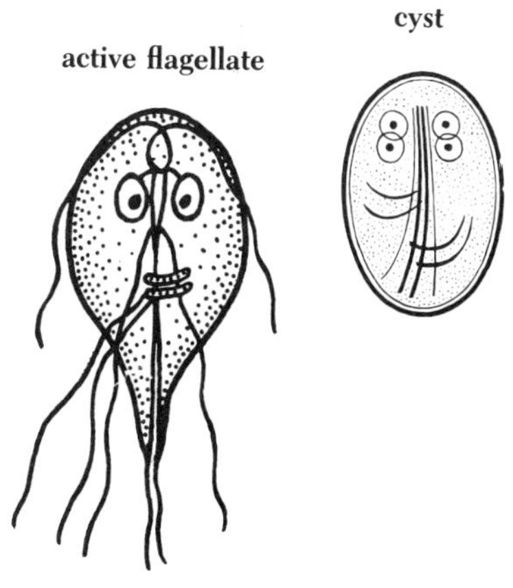

Giardia is a flagellate that lives as a parasite in the human intestine. (Modified after Kofoid and Swezy.)

Trichonympha extends a pseudopod (**1**) and engulfs a fragment of wood (**2**). In spite of this simple method of feeding, these flagellates are among the most structurally complex protozoans. (Modified after O. Swezy.)

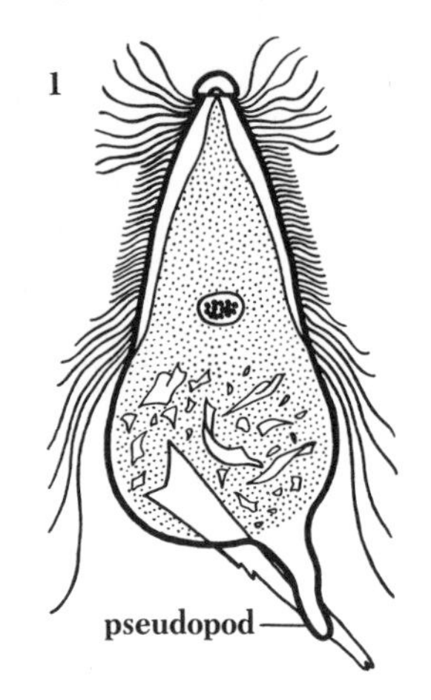

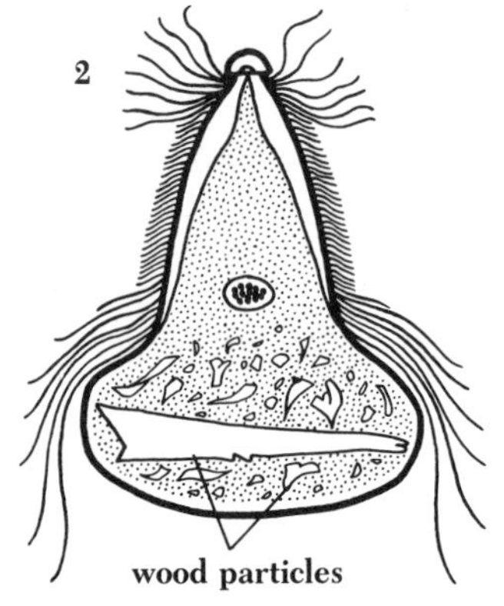

tiply in the intestine and pass out in the feces in the form of resistant cysts; a person becomes infected by ingesting the cysts (or occasionally the active flagellates), usually in food or water that has been contaminated with human wastes. Infections often spread through groups of small children who are not yet toilet trained and thence to their families. In the western United States, the flagellate is present even in some remote mountain streams, spread there either by people and their dogs or by wild mammals that are also hosts to this parasite. Even in apparently pristine areas, hikers and campers are wise to carry their drinking water or to boil any water taken from streams.

Trichonympha is a large flagellate that lives in the intestine of wood-eating termites. There is only one nucleus, but thousands of flagella cover the surface of the anterior end. The posterior end of the flagellate is ameboid and extends pseudopods to engulf bits of wood that the termite host has ingested. The termite, like most animals, lacks the necessary enzymes for digesting the chief constituent of wood, cellulose; if rid of its flagellates, it continues to eat but starves to death. The wood-eating insect depends entirely on its flagellates to digest the wood and transform it into soluble carbohydrate, a portion of which is released to the host. In return, the termite provides a moist environment and a steady supply of wood to the flagellates, and the relationship of the two organisms is mutually beneficial.

The **opalinids** are parasitic protozoans with from two to many nuclei and a dense covering of cilia. But unlike almost all the other ciliated protozoans, opalinids have nuclei of only one type (instead of large and small nuclei) and they reproduce sexually by producing two sizes of gametes (instead of by conjugating). Thus, although ciliated, the opalinids are usually classified with

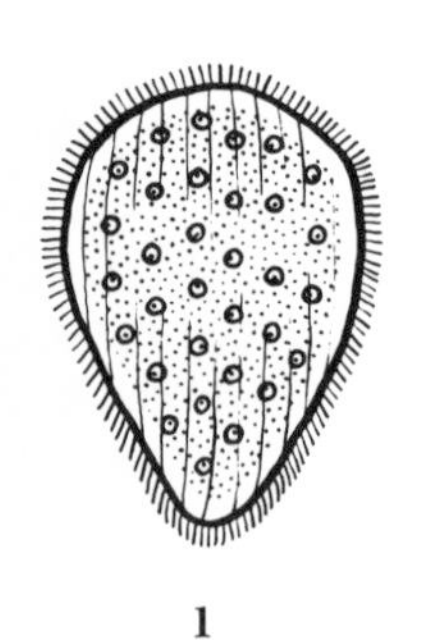

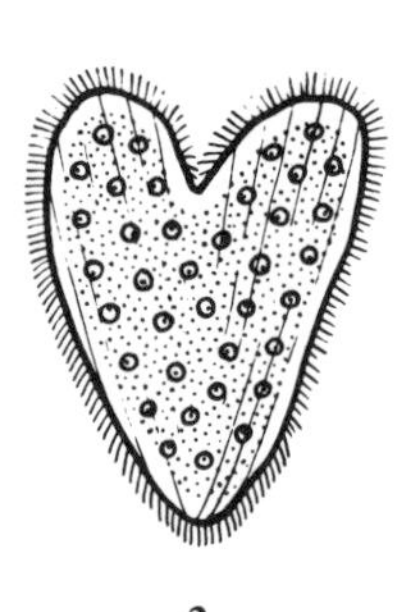

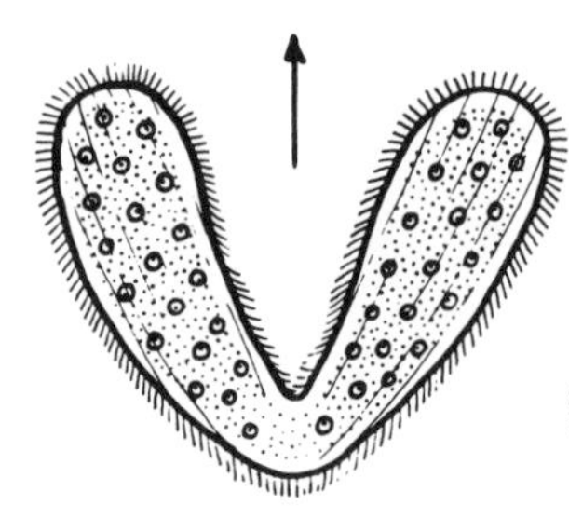

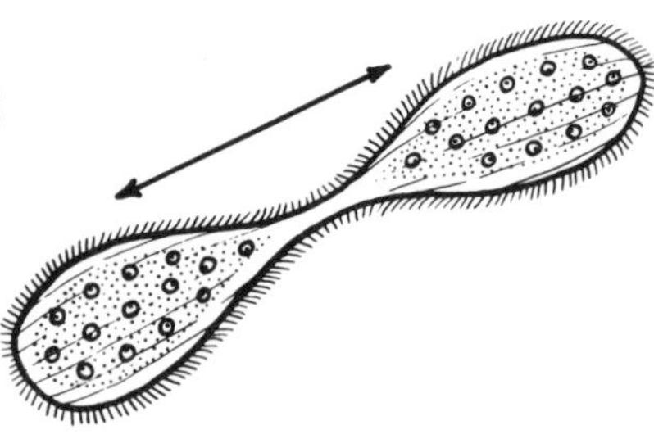

Opalina divides. **1:** Large individual with many nuclei. **2, 3, 4:** Stages in asexual division. Division may be longitudinal as in flagellates or transverse as in most ciliated protozoans. Arrows show direction of swimming.

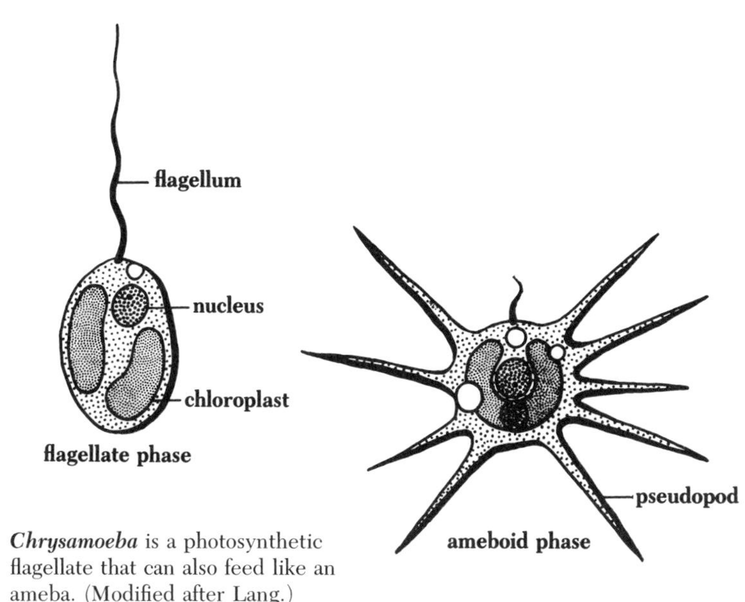

Chrysamoeba is a photosynthetic flagellate that can also feed like an ameba. (Modified after Lang.)

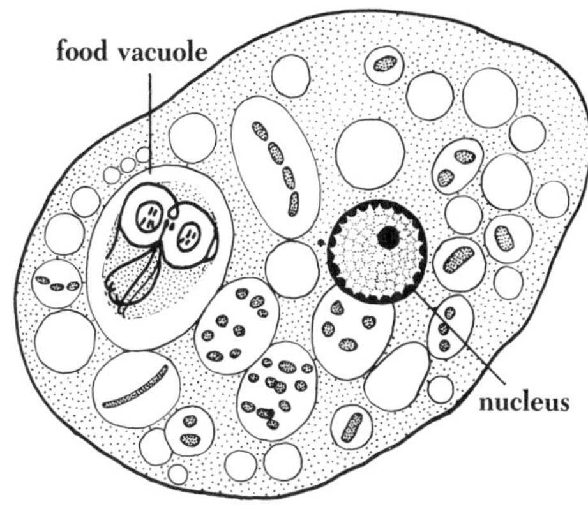

Intestinal ameba, *Entamoeba coli,* lives in humans and feeds harmlessly on particles in the intestinal contents. This particular specimen has just rendered a useful service to its host by engulfing a parasitic flagellate, *Giardia.* The smaller food vacuoles contain bacteria. (After F. Doflein.)

flagellates or in a group of their own. *Opalina* lives in the rectum of frogs, apparently without ill effects to the host. It takes up dissolved nutrients through its surface.

Ameboid Protozoans

The ameboid protozoans are those that move about and capture food by means of pseudopods. Some resemble the common amebas described in the last chapter and live free in fresh or marine waters or in damp soils. In addition to the free-living amebas, there are many species that live inside animals, particularly in the digestive tract. Most of them are harmless to the animal they inhabit, living in the intestine and feeding on bacteria and food fragments, at no expense to their host. A few species are definitely parasitic, that is, they feed on the host itself and cause it harm.

About half a dozen species of amebas that commonly live in humans are harmless or cause no serious ill effects. The **mouth ameba,** *Entamoeba gingivalis,* is found in a large part of the human population (perhaps 75% or more) in the mouth, where it feeds on bacteria and loose cells. This ameba does not form cysts and is spread from mouth to mouth in kissing or in sharing food.

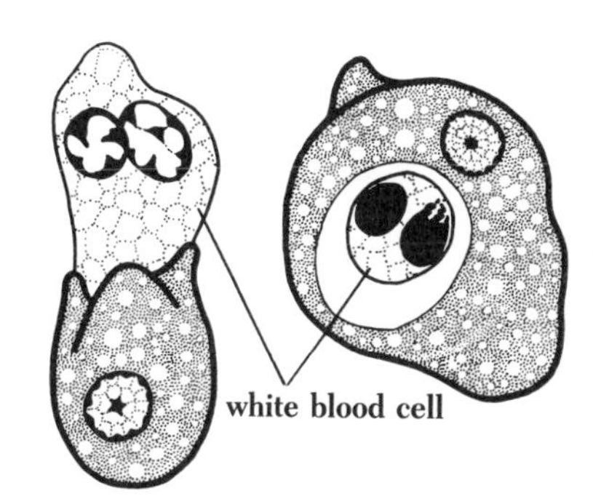

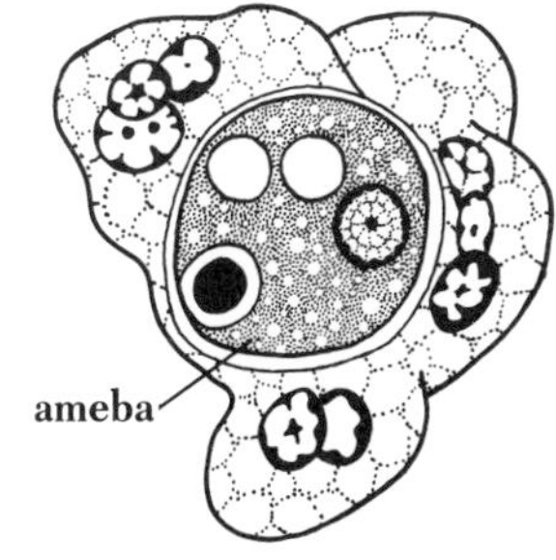

Above: **A mouth ameba,** *Entamoeba gingivalis,* engulfs a white blood cell. *Below:* The tables are turned; three white blood cells have joined forces and are engulfing an ameba. (Modified after H. Child.)

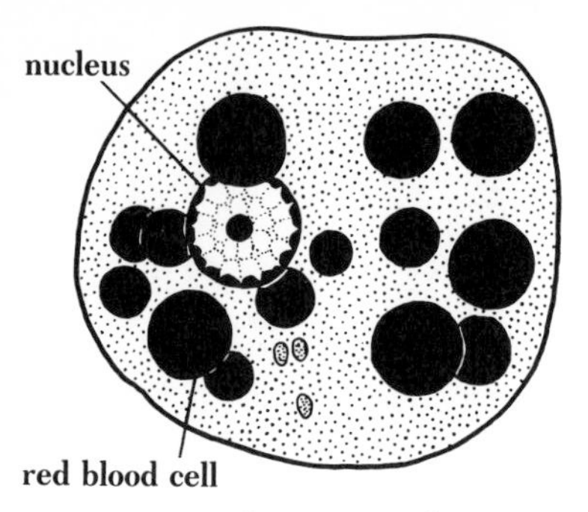

Dysentery ameba, *Entamoeba histolytica,* causes occasional epidemics in temperate regions, as when faulty plumbing allows sewage to get into the water supply. In tropical countries amebic dysentery is a constant menace and a serious drain on the energy of its victims. Travelers to parts of Asia and other regions where the disease is common should avoid eating uncooked vegetables or drinking unboiled water. (After F. Doflein.)

On the other hand, the **dysentery ameba,** *Entamoeba histolytica,* frequently causes severe disease. Dysentery amebas occur in about 10% of the people of the world, about 4% of the U.S. population. As many individuals harbor these amebas without experiencing obvious ill effects, carriers of amebic dysentery are more common in the population than is generally realized. Dysentery amebas live in the large intestine where they feed on the living cells and tissues, leading to the formation of abscesses and bleeding ulcers; they sometimes penetrate into the liver and other organs. The active amebas cannot live outside the body, and infection is spread from person to person by encysted forms that pass out with the feces and contaminate food or drinking water. The disease is most common in communities where people have poor sanitary habits or lack safe means of sewage disposal. Several drugs are effective in the treatment of amebic dystentery.

> A more dangerous but relatively rare ameba lives in hot springs and may infect bathers, with fatal results; this ameba was responsible for the closing of the famous baths in Bath, England. There have also been cases of certain normally free-living soil amebas *(Naegleria, Acanthamoeba)* invading the central nervous system and causing fatal encephalitis.

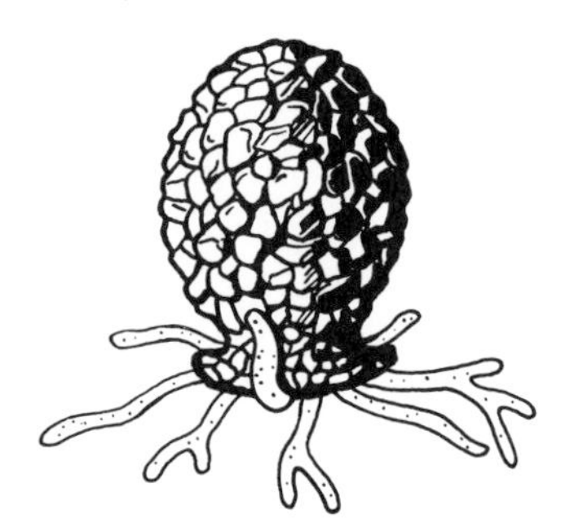

Difflugia constructs an opaque shell of organic material studded with sand grains or other particles. Only the pseudopods of the ameba, protruding from beneath the shell, are visible.

Some free-living freshwater amebas secrete a sort of shell, or test, which they carry about and into which they withdraw when disturbed. ***Difflugia*** constructs its test by gathering sand grains or other hard particles and cementing them together with a sticky secretion. The various species of this genus are recognized by the specific shapes of the tests that they construct. While we appear to be classifying these amebas on the basis of differences in structure, we are also classifying them by their differences in behavior. The common ***Arcella*** of freshwater ponds secretes a hard organic test about itself. When the ameba divides, it first constructs a second test, which is adopted by one of the two offspring; the other retains the original test.

The mainly marine **foraminiferans**—usually called by their nickname, **forams**—are ameboid protozoans with tests that range from simple organic coverings, sometimes reinforced with hard particles, to elaborate many-chambered calcareous (calcium carbonate) shells. A young chambered foram resembles an ameba and secretes a shell about itself; as growth continues, it adds a second chamber, and this process is repeated as the foram grows until sometimes there are more than a hundred communicating chambers. In many common species the chambers are added on in a spiral pattern, resulting in a shell that resembles a miniature snail shell.

Arcella secretes a hemispherical organic shell with an opening on the underside from which the pseudopods extend.

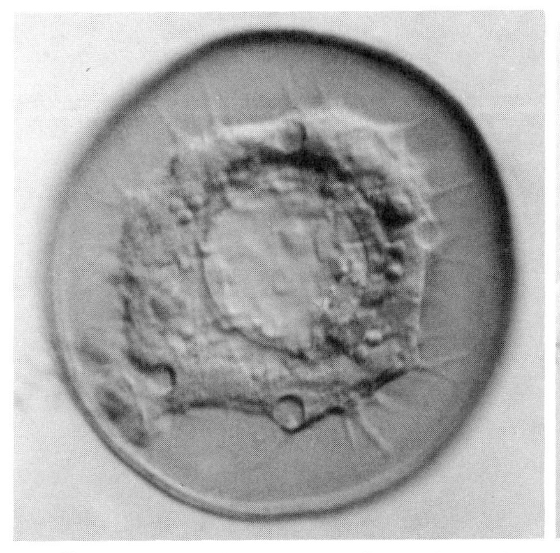

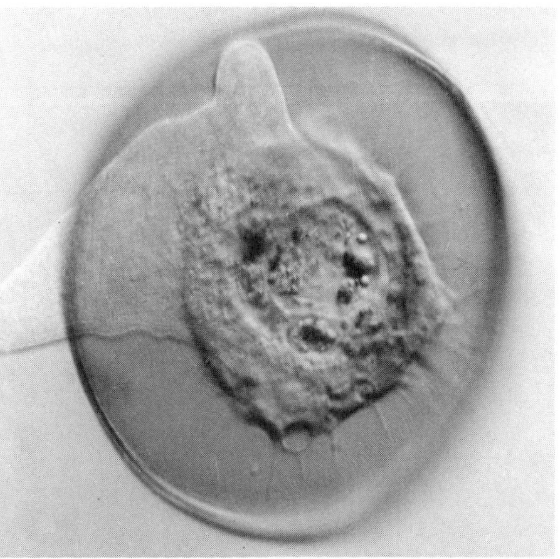

Arcella is a common freshwater ameba. Its brown hemispherical test has a round hole on the undersurface through which the pseudopods are extended. Seen here are two views of the undersurface of the test, focused at different levels. *Left:* Four contractile vacuoles and the two nuclei are visible; the pseudopods are withdrawn into the shell. *Right:* A large pseudopod is extended. (P. S. Tice)

The organism occupies all of the chambers and sends out long, delicate pseudopods in which cytoplasmic granules can be seen constantly streaming both toward and away from the tip simultaneously. The pseudopods unite into networks in which a variety of foods (algal cells, other protozoans, minute animals, and organic particles) are captured and digested. The pseudopods also serve in construction of new chambers of the test and in locomotion. Some forams float near the surface; most live on the mud of the ocean bottom, where they move actively about gathering food, or remain quietly attached to some surface waiting for food to come to them. Some branching plantlike forams live always anchored in mud by rootlike pseudopods that take up dissolved organic materials; upwardly extended pseudopods capture suspended food particles from the water. Other forams burrow or sift through sediments with their pseudopods, collecting small organisms to eat. Many shallow-water forams are nourished partly by zooxanthellas or other photosynthetic cells that they harbor in their cytoplasm.

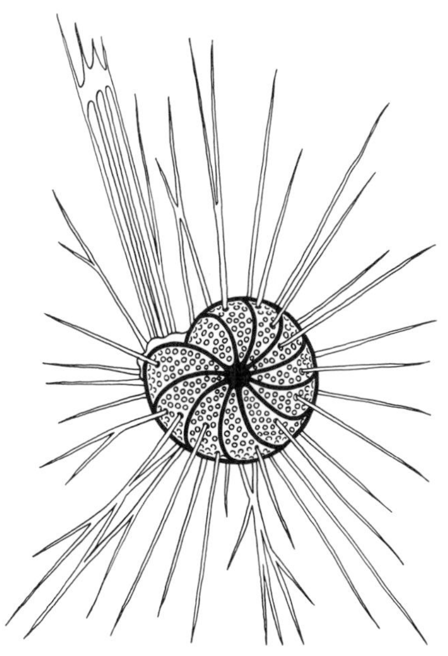

Foram with its slender pseudopods extending through pores in the walls of the shell as well as out of the main opening.

It is astounding to contemplate the number of individual foram shells composing the chalk cliffs of Dover, England, or the large chalk beds of Mississippi and Georgia, over 300 meters (1,000 feet) thick in places. The presence of chalk beds indicates that these regions were once covered by the sea, information

Foram shells, highly magnified under the microscope. The surface of many of the shells is perforated by pores, through which the living protozoan extended its slender pseudopods. The shell on the upper right has no pores; its former occupant protruded all the pseudopods through a single opening at one end. Aside from their architectural accomplishments, forams are remarkable among protozoans for the large size that some species attain; some living kinds may be 2 or 3 cm across, and on many warm beaches around the world, sifting the sand through one's hand will reveal large foram shells. Certain foram fossils measure nearly 15 cm (6 in). However, most forams are just visible to the naked eye. (R. Buchsbaum)

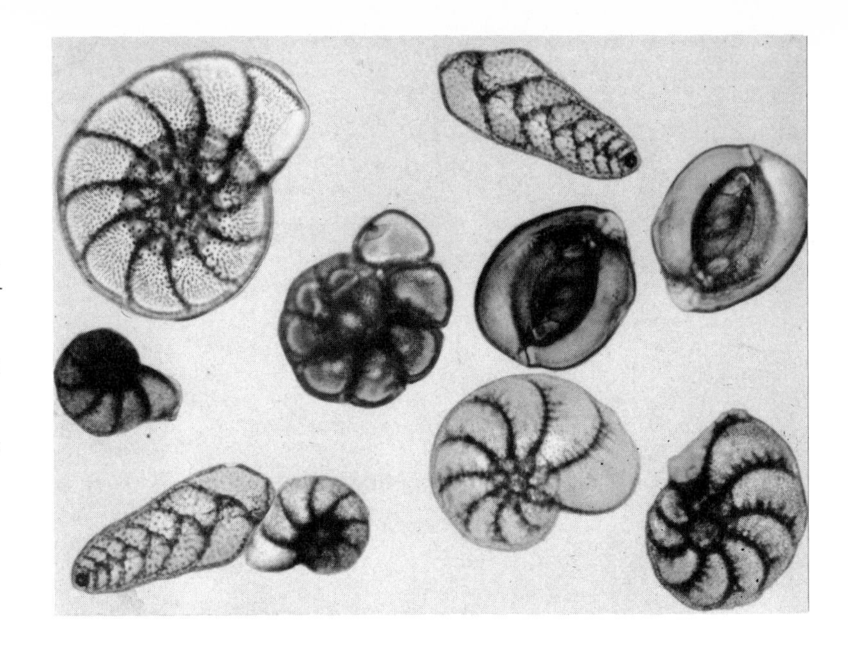

Globigerina is a common planktonic foram. The shorter radiating processes are spines on the shell that slow sinking. The pseudopods are mostly withdrawn. (R. Buchsbaum)

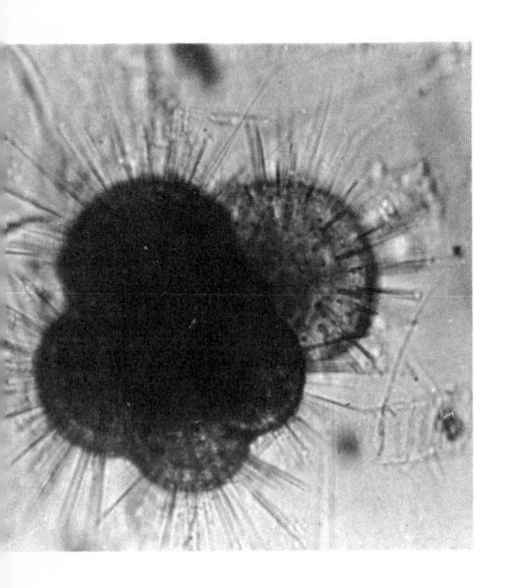

Limestone quarry yields great blocks of the fine stone that has gone into constructing the buildings of many institutions in the United States—evidence of a long history of flourishing foram populations in ancient seas. Indiana. (R. Buchsbaum)

useful to geologists. The rate of accumulation of shells, judging by present rates on the bottom of the Pacific, was about 60 centimeters in a hundred years. Most of the shells now being deposited are those of the foram genus *Globigerina* and related genera that float in the surface waters. As these forams die, their shells sink in a slow but steady rain to the ocean floor, where they form a gray sediment called "globigerina ooze." About 30% of the ocean floor (103,600,000 square kilometers, or 40,000,000 square miles) is covered with globigerina ooze. Some deposits form chalk; others form hard, fossiliferous limestone such as the famous Indiana building stone. Extinct forams preserved as fossils in rocks are of great value in developing oil fields. Borings at different depths are examined for their fossil forams, and the species present reveal a great deal about the history and structure of the underlying rock.

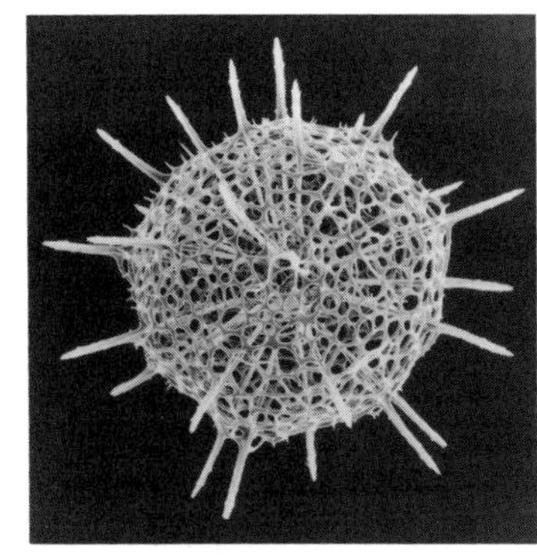

Skeleton of a radiolarian may be an almost unbelievably intricate latticework of silica, like this one, or it may be relatively simple, with only a few large spines. Scanning electron micrograph. (M. Gowing)

Radiolarians are ameboid protozoans that secrete elaborate skeletons composed mostly of silica (silicon dioxide). Stiff, slen-

Radiolarian skeletons sometimes settle out in great quantities, especially in certain deep regions of the oceans, where they may compose 20% or more of the sediment, called "radiolarian ooze." This ooze covers an area of 7,700,000 square kilometers (3,000,000 square miles) in the Pacific and Indian oceans. Radiolarian skeletons constitute a part of "Tripoli stone," used in abrasive powders for polishing metals. (R. Buchsbaum)

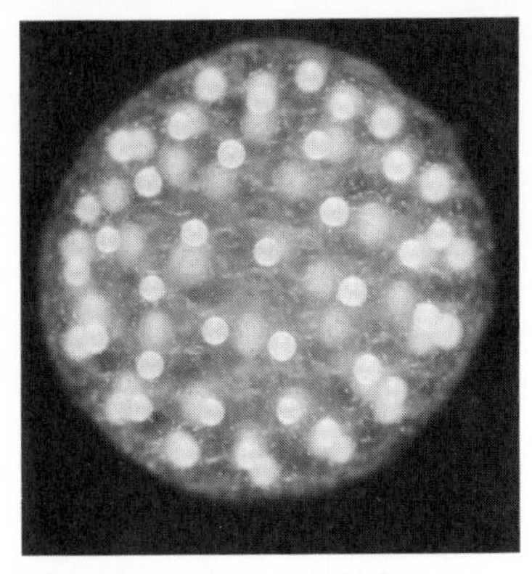

Colonial radiolarian with the individual members embedded in a gelatinous mass, was collected in a jar by a diver at about 50 m depth in Monterey Bay, California. This colony was about 2 mm in diameter, but in warm seas radiolarian colonies are measured in centimeters. (R. Buchsbaum)

der pseudopods, each supported by a central bundle of microtubules, radiate out through holes in the skeleton. The cytoplasm is divided by a membranous capsule into an inner region containing the nucleus and an outer region filled with food vacuoles that contain tiny organisms caught by the pseudopods. Often the outer region contains photosynthetic zooxanthellas that supplement the nutrition of the radiolarian as long it is exposed to light. The outer region also contains numerous vacuoles that give the cytoplasm a frothy appearance and enable the protozoan to adjust its buoyancy and so move up or down in the water. Radiolarians are especially abundant in warm seas, where colonial forms may produce gelatinous spheres or sausagelike masses 20 centimeters long.

Heliozoans are the only freshwater ameboid protozoans that are comparable to the exclusively marine radiolarians. Some heliozoans have perforated siliceous skeletons that resemble those of radiolarians, but others have only a gelatinous covering or a skeleton of loosely matted needles of silica. Freshwater species have contractile vacuoles. Most heliozoans float free in the water, but some live attached by a stalk. The radiating pseudopods are even stiffer than those of radiolarians and show little movement

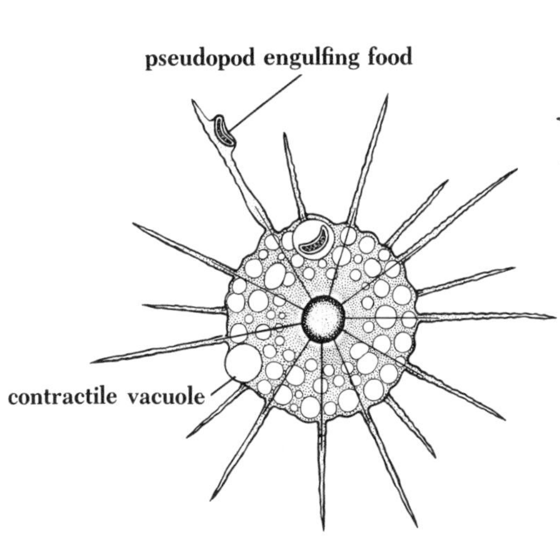

A common heliozoan, *Actinophrys sol.* Heliozoans were called "sun animalcules" because their stiff radiating pseudopods suggested the rays of light from the sun. The pseudopods are supported by axes of microtubules that run through the cytoplasm and converge on or near the centrally located nucleus.

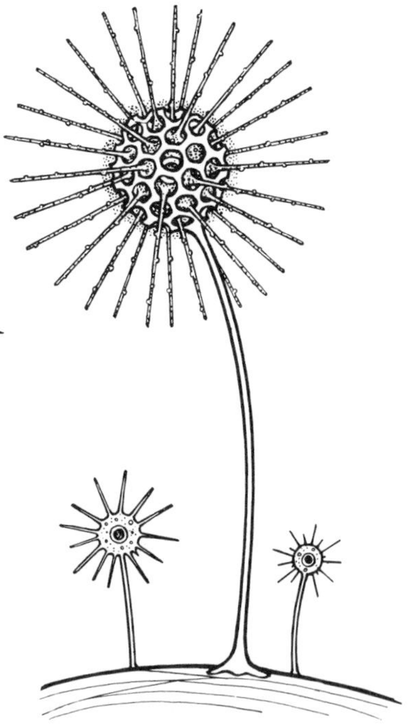

Stalked heliozoan, *Clathrulina elegans,* has a skeleton of siliceous latticework. (Modified after J. Leidy.)

besides the streaming of the granules in the cytoplasm. When a large organism is being engulfed, several pseudopods work together to draw it down into the main body of cytoplasm, where it is enclosed in a food vacuole.

Spore-Forming Protozoans

The spore-forming protozoans, or **sporozoans,** have no characteristic mode of locomotion, but are conveniently grouped together because all are parasites with complex life histories and all form "spores" or "sporozoites" by which they spread from host to host. In the typical life history, the nucleus of the protozoan divides many times without corresponding division of the growing cytoplasm; when cytoplasmic division finally occurs, many cells are produced, each containing one nucleus and a small amount of cytoplasm. Such multiple fission increases enormously the reproductive potential, a characteristic of parasites, in which those individuals lucky enough to reach a host must make up for all their unsuccessful relatives.

The several groups of sporozoans are probably not closely related but have arisen independently from separate protistan stocks. Most different from the other groups are the peculiar **cnidosporidians,** which infect various invertebrates and cold-blooded vertebrates, causing epidemic diseases of domestic honeybees and silkworms, and of salmon, halibut, and other fishes. Cnidosporidians have spores or other stages consisting of several differentiated cells, casting some doubt on their status as protozoans. **Gregarines** infest the digestive tract or body cavity of many invertebrates, especially annelid worms and insects. **Coccidians** parasitize the cells of the digestive tract in both invertebrates and vertebrates, sometimes the liver and blood of vertebrates. Only a few coccidians occur in humans, causing diarrhea

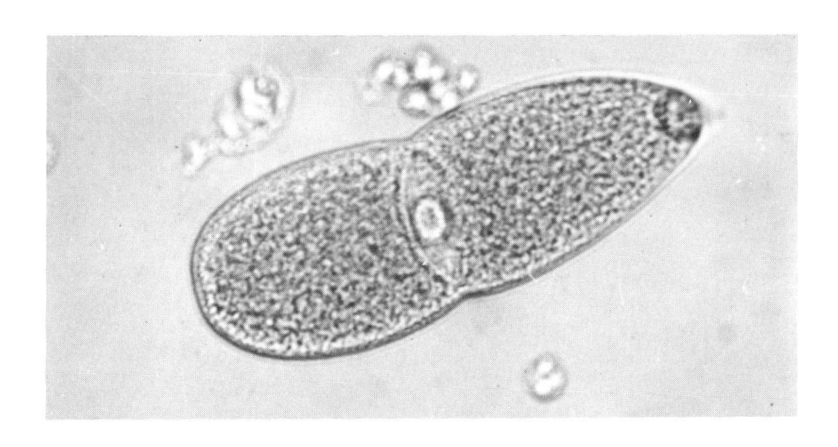

Gregarine. These sporozoan parasites inhabit the intestine, body cavity, and blood of many invertebrates. This one was found in the digestive gland of an acorn worm (see chapter 22) and was clinging to a host cell by the end seen here at the right. Bermuda. (R. Buchsbaum)

but seldom resulting in serious disease except in individuals with impaired immunity. The coccidian *Toxoplasma,* when acquired during pregnancy, may kill or blind the unborn child. Many domestic animals suffer from diseases caused by coccidians, and livestock or pets, especially cats, are sometimes sources of human infection.

Similar to coccidians but requiring two hosts (an arthropod and a vertebrate) are the **hemosporidians.** The most studied of the hemosporidians are species of *Plasmodium* that cause **malaria** in hundreds of millions of humans and in other warm-blooded vertebrates. The several species that affect humans are transmitted by anopheline mosquitoes. Stages both in the body of the mosquito and in the vertebrate host involve multiple fissions and these may be repeated before going on to the next stage in the life history.

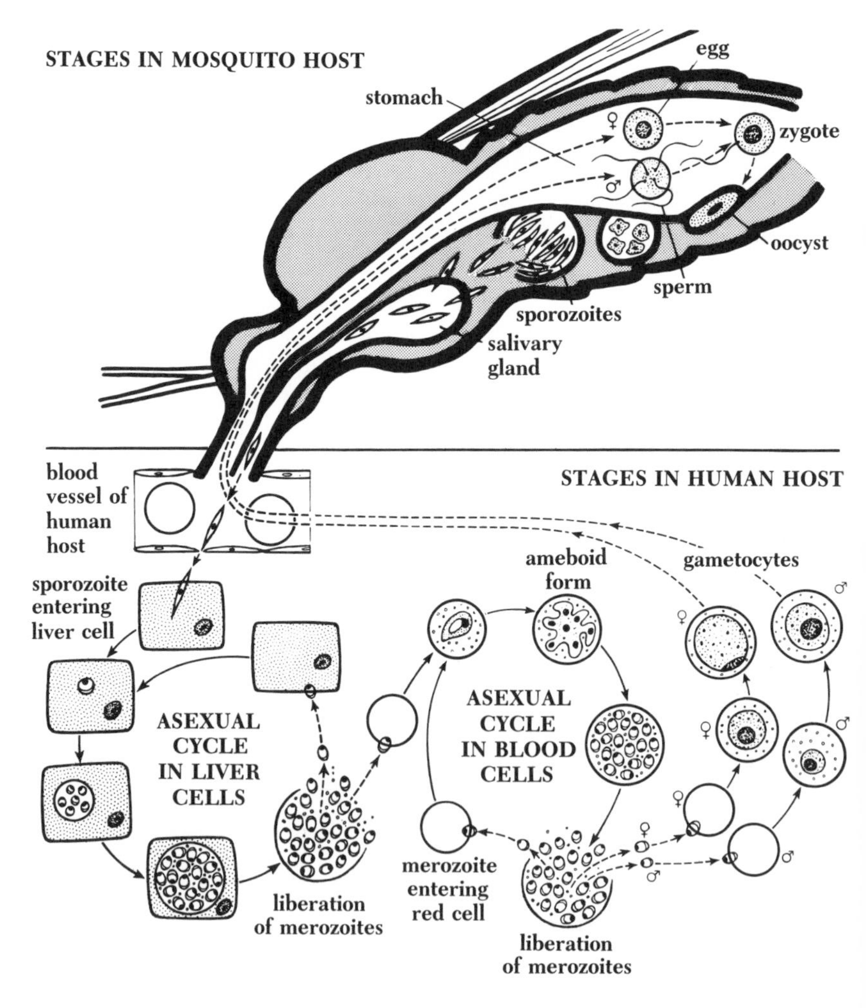

Life history of a malarial parasite, *Plasmodium vivax.* Egg and sperm unite to form a **zygote** in the stomach of the mosquito. The zygote encysts in the stomach wall as an **oocyst** that undergoes meiosis and multiple fission into large numbers of slender **sporozoites.** These invade tissues of the mosquito. Those that reach the salivary glands are carried, in the saliva of a biting mosquito, through the skin and into the bloodstream of a human host. The sporozoites leave the human bloodstream and enter tissue cells, such as the liver, where they grow and undergo multiple fission. Huge numbers of resulting **merozoites** are released. Some may enter new liver cells and repeat this cycle. Others now invade the bloodstream, enter red blood cells, and grow into ameboid forms that attack the hemoglobin of the red cell. The ameboid form undergoes multiple fission, the red cell is ruptured, and the released merozoites enter new red cells and repeat the cycle. Eventually some of the parasites develop into sexual forms, female and male **gametocytes,** which produce either eggs or sperms in the stomach of a mosquito that takes a blood meal from the infected human host at this stage.

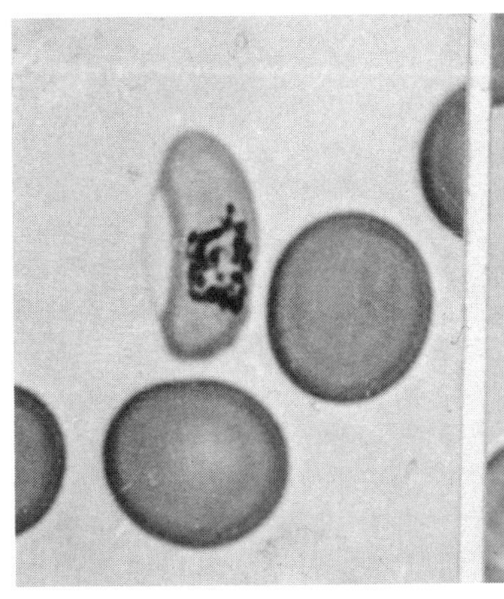
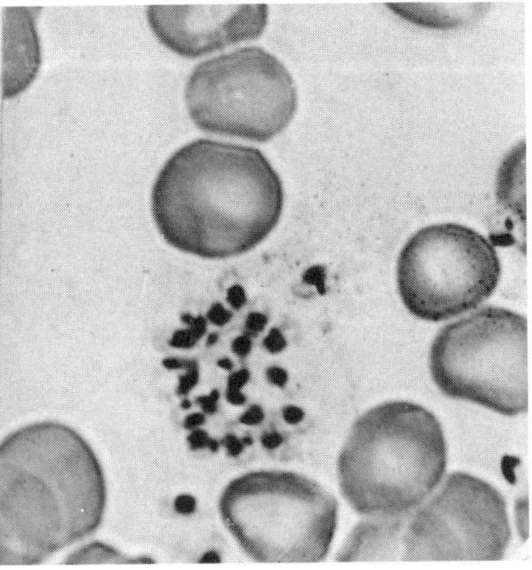

Malarial parasites. *Left:* Female sexual form of *Plasmodium falciparum* in a red blood cell. *Right:* Ruptured red blood cell that has liberated merozoites of *P. vivax*. Stained preparations. (Army Medical Museum)

Malaria in humans is initiated by the bite of a female anopheline mosquito which introduces the slender sporozoites of *Plasmodium* into the blood stream. The sporozoites enter liver cells so promptly that blood smears rarely reveal their presence at this stage and there are no symptoms. The shivering chills and burning fevers (up to 41.5° C or nearly 107° F) do not begin until large numbers of merozoites are liberated from the red blood cells, together with toxic materials. Individuals with sickle-cell trait of the red blood cells are resistant to malaria. In infections with *Plasmodium vivax*, a common form of human malaria, the parasites attack about 1% of the red blood cells. The cycles of multiple fission take about 48 hours, so that chills and fever recur at 48-hour intervals. The parasites may persist over many years, probably in liver cells; they emerge from time to time, causing repeated malarial episodes. The more severe disease caused by *Plasmodium falciparum* attacks 10% of the red blood cells, causing more violent paroxysms of chills and fever; clumping of red blood cells may result in coma or in death from obstruction of blood vessels in the brain. Most of the more than a million deaths each year are caused by *P. falciparum*.

After World War II, intensive programs of mosquito control were undertaken in many parts of the world; draining swamps and spraying insecticides (usually DDT) in breeding places and homes sharply reduced the incidence of malaria. Numbers of effective drugs for treatment of malaria were also developed. As time passed, however, strains of insecticide-resistant mosquitoes and strains of drug-resistant *Plasmodium* developed; and technical, social, and political barriers have combined with this biological difficulty to result in large-scale abandonment of control programs and a significant resurgence of malaria in many areas, especially of the dangerous form caused by *P. falciparum*. With the dream of eradicating malaria put aside, the fight currently continues to reduce mosquito populations by a mixture of measures and to develop new drugs that will successively overcome the resistance of the parasites. Travelers to malarial regions should take the preventive medicines available and avoid exposure to bites by staying in screened buildings after sunset. New technologies of genetic engineering are making progress toward a vaccine for inducing immunity to malaria.

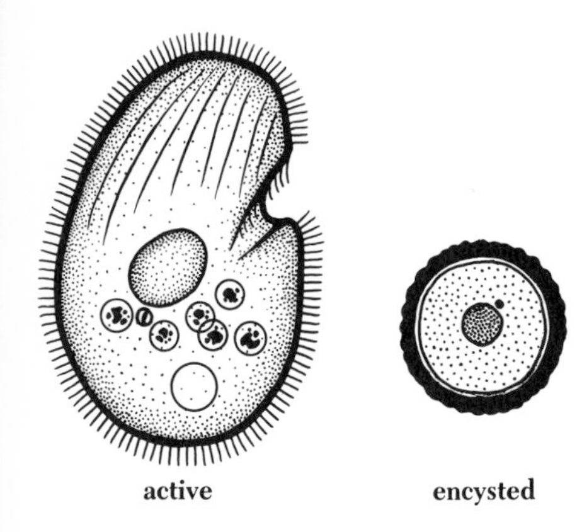

Colpoda. Next to parameciums, members of the genus *Colpoda* are perhaps the most common holotrichs of freshwater and soil. (Cyst, after Kidder and Claff.)

Ciliated Protozoans

The **ciliates** are among the most conspicuous of protozoans. A drop of freshwater or seawater that includes a bit of bottom sediment usually reveals a variety of ciliates crossing the microscope field so rapidly as to defy easy interpretation of their mode of locomotion. The cilia may cover the whole body in evenly spaced rows (as in parameciums) or may be limited to certain areas or may be clumped to form structures that act as little paddles or legs. There are even ciliates that have cilia only during part of their life history but are recognized as ciliates by the presence of structures usually associated with cilia and by the ciliate characters of having two types of nuclei and of undergoing conjugation.

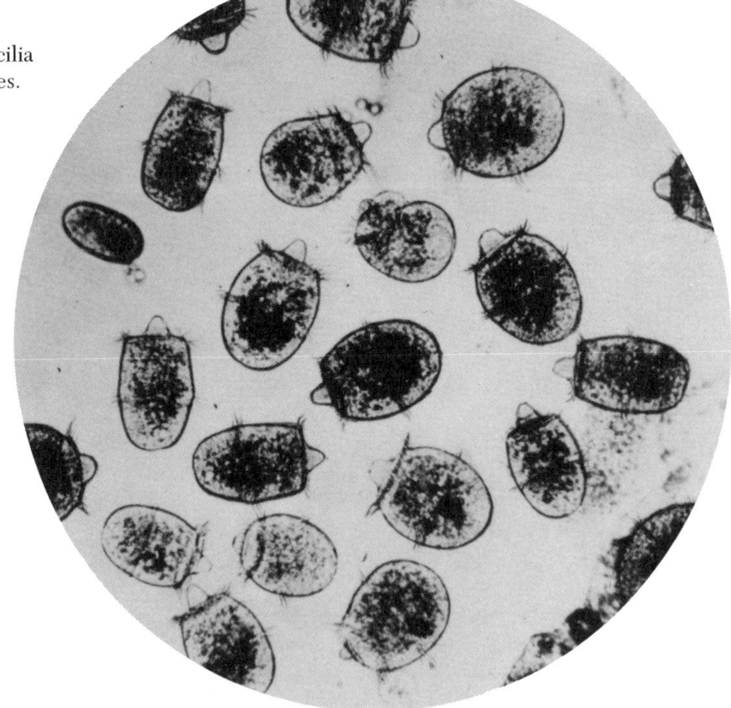

Didinium is a holotrich with cilia restricted to two beltlike zones. (C. Roemmert)

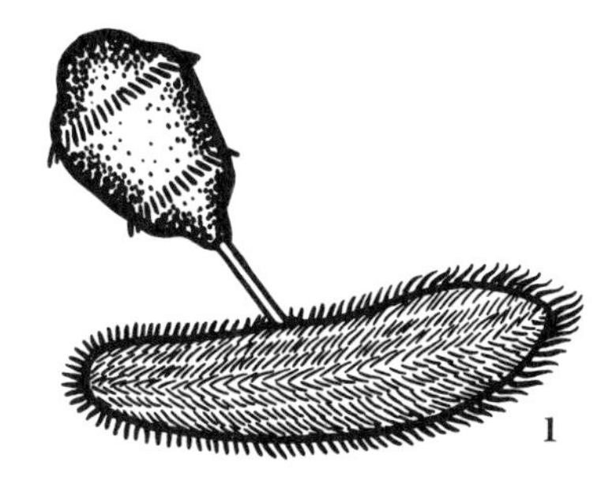

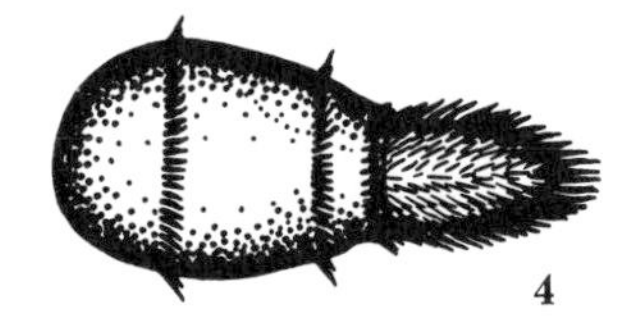

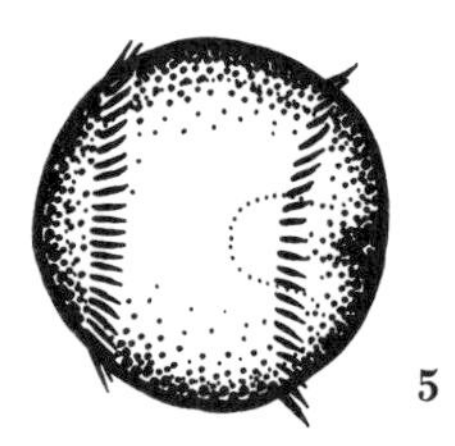

A didinium eats a paramecium. (Mostly after G. N. Calkins.)

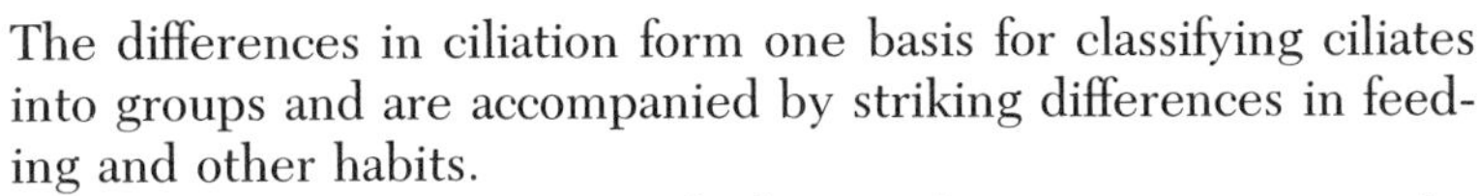

The differences in ciliation form one basis for classifying ciliates into groups and are accompanied by striking differences in feeding and other habits.

Of the two main types of ciliates, the HOLOTRICHS are the most diverse. *Paramecium* belongs to this group. The cilia are all short, fairly equal in length, and are evenly distributed over the surface in rows or restricted to certain regions. Many holotrichs have trichocysts. ***Didinium*** is a holotrich with two belts of cilia. Didiniums work hard for a living; they eat almost nothing but parameciums. From the center of the front end of a didinium projects a snout that is armed with rodlike structures and toxic trichocysts. The ciliate swims about at top speed, jabbing at everything with which it comes into contact—plants, other didiniums, even the glass walls of an aquarium. With these it has no success; but when it chances to strike a paramecium, the snout penetrates and the prey is swallowed whole. A small didinium can stretch to engulf a paramecium five times its own volume. Fortunately for the parameciums, the didinium is slowed down a bit by having to rebuild its snout armature after each meal. ***Balantidium coli*** is the only ciliate common as a human parasite.

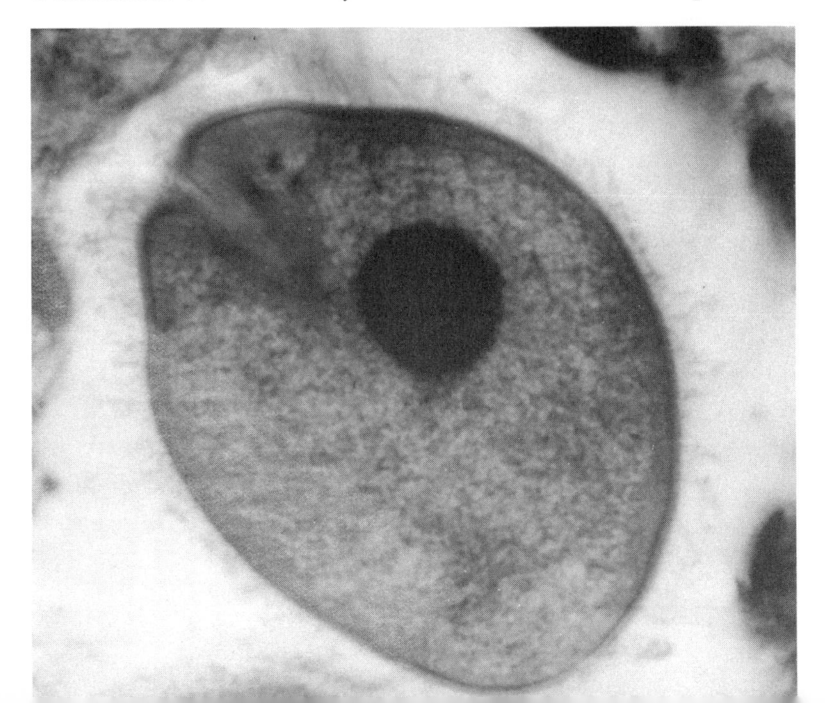

Parasitic ciliate, *Balantidium coli,* is the largest protozoan (60 μm) that lives in humans and the only ciliate that is a common human parasite. Infection results in abdominal pain and diarrhea, with discharge of mucus and blood. (Army Medical Museum)

Vorticella was so named because of the little vortex or whirlpool that it creates in the surrounding water by the action of the cilia around its mouth. Vorticellas have one sausage-shaped large nucleus and one small nucleus. (Based on Noland and Finley.)

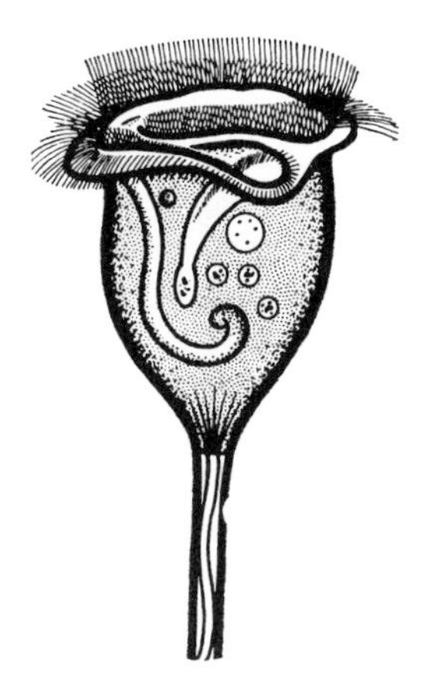

This large holotrich inhabits the large intestine, usually when the victim is already suffering from some other disease or infection; healthy people are not readily infected. Monkeys and pigs also harbor this ciliate, and the latter are the most common source of human infection.

Peritrichs are holotrichs with the cilia usually limited to a ring around the mouth. Some of these, such as *Vorticella*, live singly, while others form branching treelike colonies. The vorticellas have lost nearly all their cilia except the circlets around the broad end of the bell-shaped body. The body is firmly attached to the substrate by a slender stalk. When the peritrich is feeding undisturbed, the stalk is straight and extended, and at its upper end the cilia of the expanded bell beat actively. Vorticellas feed chiefly on bacteria brought to the mouth by ciliary currents. The least disturbance causes the stalk to shorten rapidly; spiral fibers that run its length contract like a coiled spring.

Vorticella is a genus of peritrichs, ciliates in which the cilia are limited to a ring around the mouth. They are included among the holotrichs. The bell-shaped bodies of these vorticellas are attached by their stalks to debris from the bottom of a pond. One is dividing. Several individuals with spiraled, partly contracted stalks can be seen among the extended ones. (R. Buchsbaum)

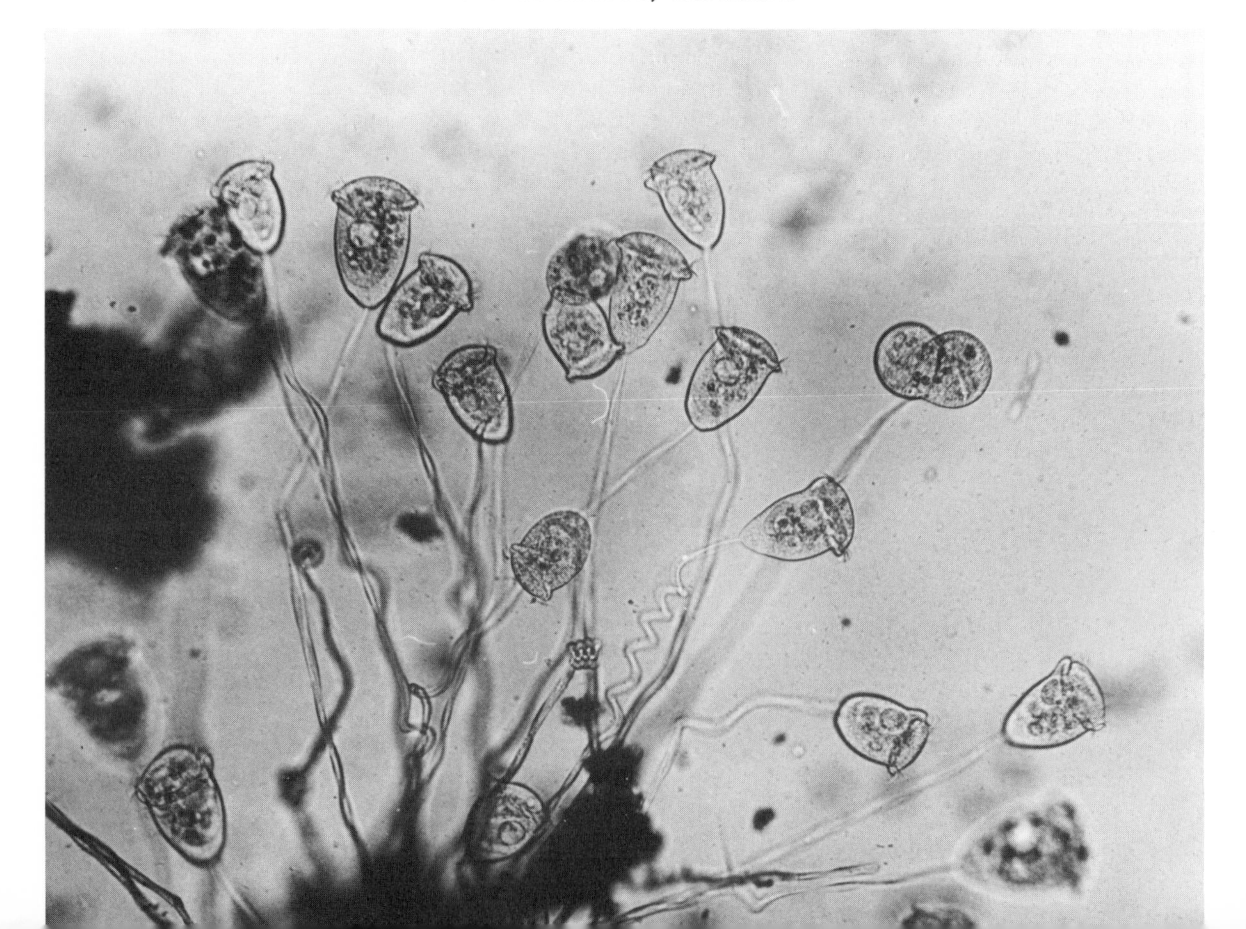

Suctorian with its distinctive clusters of tentacles. (R. Buchsbaum)

The surface layer of the bell also contains contractile fibers, which fold the bell edge over the circlets of cilia. Vorticellas reproduce asexually by dividing lengthwise; one of the offspring retains the stalk, while the other develops a girdle of cilia at its lower end and swims away, later developing a stalk and attaching itself. When conditions become unfavorable, many bells may develop similar girdles of cilia, break loose from their stalks, and swim away, later to reattach elsewhere. Sexual reproduction involves conjugation, as in other ciliates, but is a one-sided affair. The two conjugating vorticellas are produced by unequal division and are of very different size; a smaller, free-swimming one attaches to a larger stalked one and is resorbed by it as conjugation ends.

In **suctorians,** another group of sessile holotrichs, two conjugating individuals bend toward each other, and one may finally

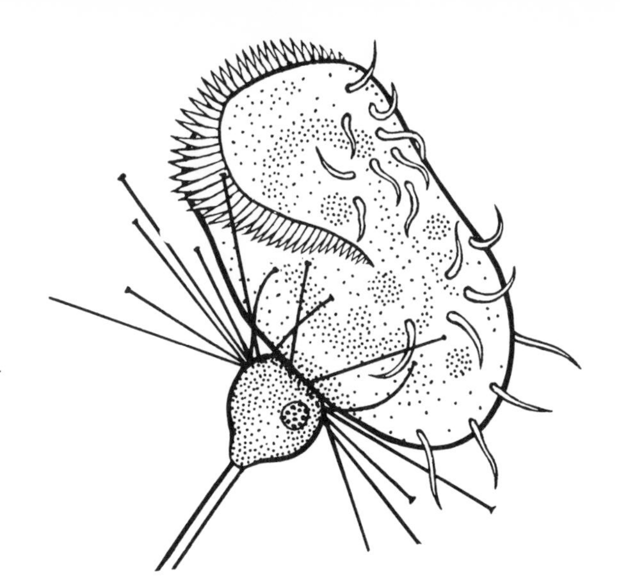

Suctorian, *Tokophrya,* feeds on another ciliate *(Euplotes)* many times its own size. It will take the suctorian about 15 minutes to finish its meal and by that time its body will be stretched to several times the normal size. (After A. E. Noble.)

pull the other from its stalk and resorb it, but in most suctorians the pair part peacefully. Because they conjugate and have two types of nuclei, suctorians are recognizable as ciliates, although they have no cilia during most of their life history. They sit in one place, attached to the substrate directly or by a stalk, and capture their prey (mostly other ciliates) by means of long cytoplasmic projections called tentacles. When a prey organism happens to come into contact with the tentacles, it is immobilized and held. The enlarged tentacle tips contain structures that penetrate the surface of the prey, and the suctorian takes in the prey cytoplasm through the tentacles. Suctorians reproduce asexually by dividing or by releasing short-lived ciliated stages that swim off and soon attach, lose their cilia, and grow tentacles.

The SPIROTRICHS include a variety of ciliates that have a spiral zone of special long cilia near the mouth. These long cilia form **membranelles,** triangular plates formed from many cilia bound together. The zone of membranelles serves primarily to create the water currents that bring food to the mouth but may also play a part in locomotion. Two of the groups most commonly seen in freshwater may be mentioned.

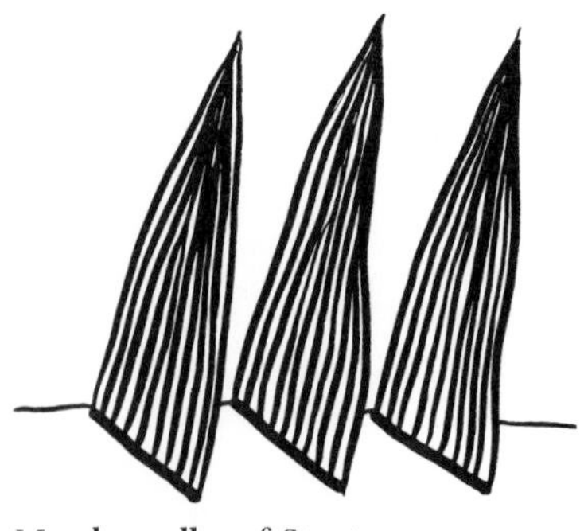

Membranelles of *Stentor* are triangular plates formed by many cilia. (Modified after F. Doflein.)

Heterotrichs have, in addition to a spiral of membranelles, a covering of short cilia like that of many holotrichs. One of the most familiar heterotrichs is ***Stentor,*** a trumpet-shaped ciliate that can swim about freely but, when feeding, attaches by its lower end to a water plant or similar object, stretches out to full length, and vibrates its circlet of membranelles so rapidly that they look like a swiftly rotating wheel (see the figure at the beginning of this chapter). The water current thus produced sweeps small food organisms toward the mouth. *Stentor* has a remarkable type of large nucleus, resembling a string of beads, and may have up to eighty small nuclei. Beneath the surface there are lengthwise contractile fibers that shorten the body when they contract. The body of heterotrichs is usually round in cross-section.

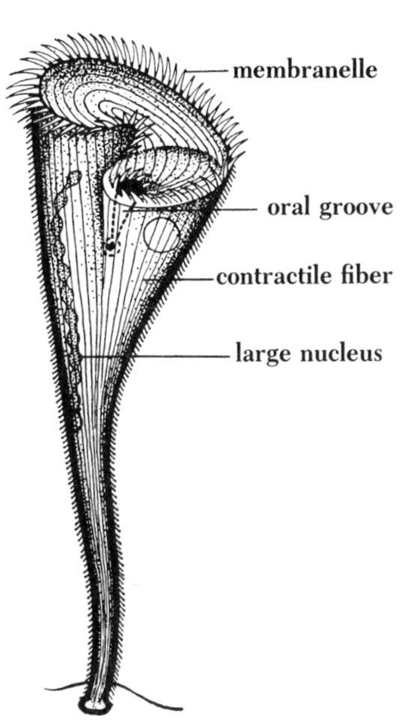

Stentor, a heterotrich, is a giant among protozoans; some specimens may be 2 mm long. An investigator who placed a hungry stentor in a thick suspension of euglenas estimated that the flagellates were taken in at a rate of about 100 per minute.

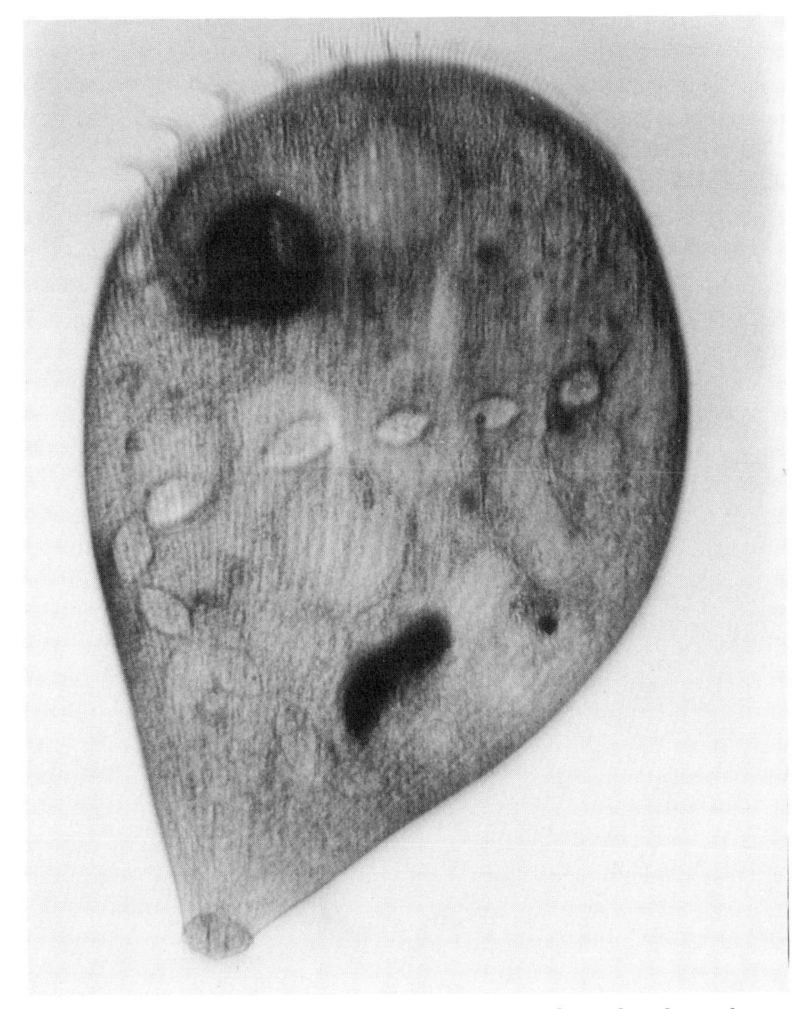

Stentor can swim about freely, as the one seen here is doing, but these ciliates mostly remain attached to some solid object. Members of this species, *Stentor coeruleus*, are blue in color. The membranelles of many cilia bound together can be seen at the edge of the area surrounding the mouth. Also visible are the contractile fibers, which enable the protozoan to change shape, and the large nucleus, which looks like a string of beads. Photomicrograph taken at 1/30,000 second. (R. Buchsbaum)

A hypotrich swims in rapid jerks and is more easily observed when it is crawling about on a surface with its cirri, as shown here.

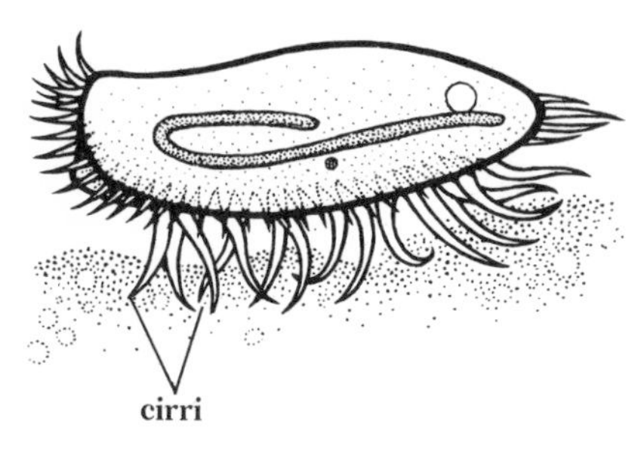

The differentiation of cilia seen in heterotrichs is carried to curious extremes in **hypotrichs,** which are flattened and have the cilia confined almost entirely to the lower surface, though the upper surface may bear a few short bristly cilia. On the lower surface, some of the cilia are bound into stout bundles called **cirri** (singular: *cirrus*). The cirri do not beat rhythmically like ordinary cilia, but are used like legs as the hypotrich crawls about on vegetation. Hypotrichs are common both in freshwater and in seawater.

Hypotrichs scurrying about on plant material collected from the seashore in Bimini, Bahamas. (R. Buchsbaum)

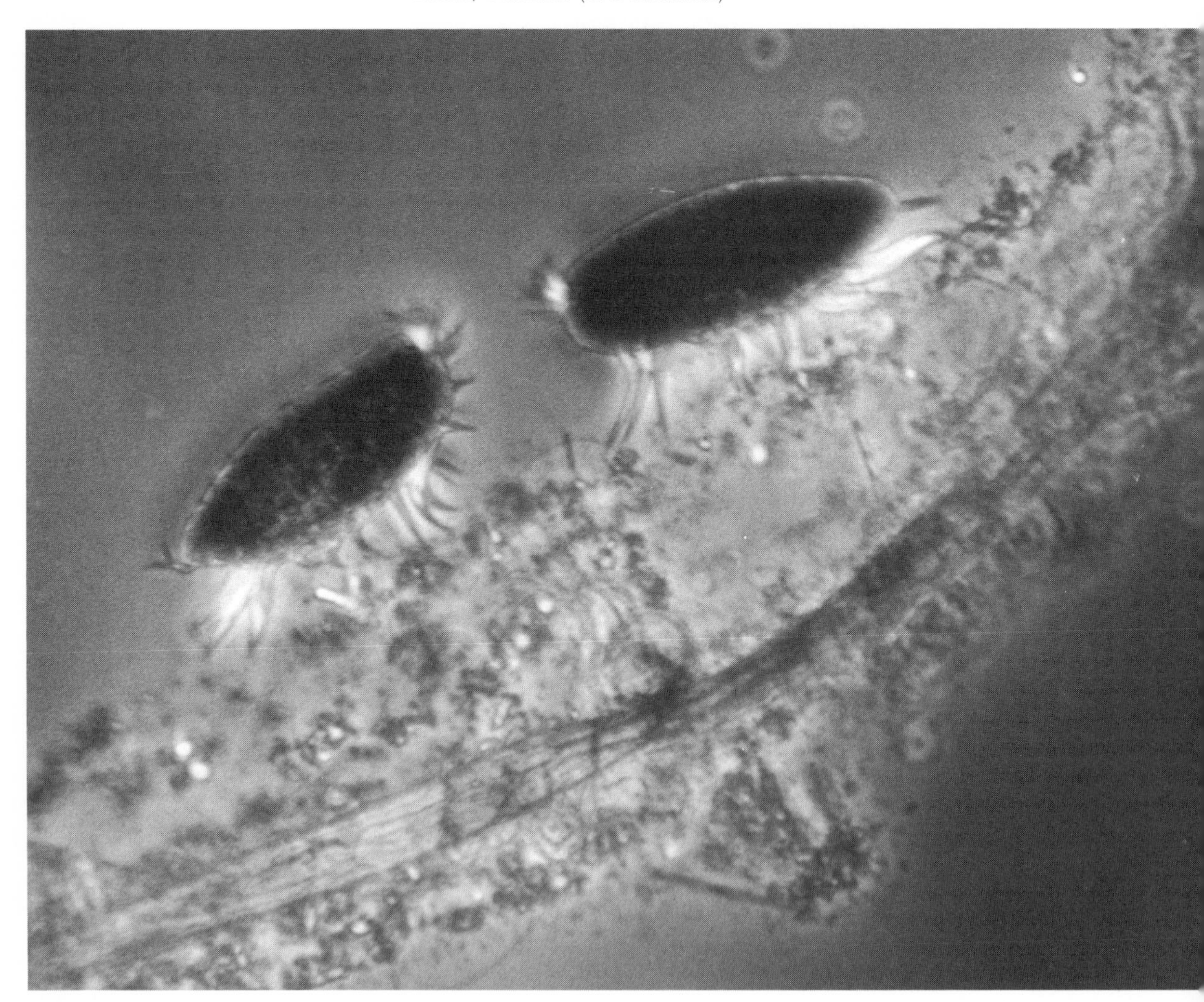

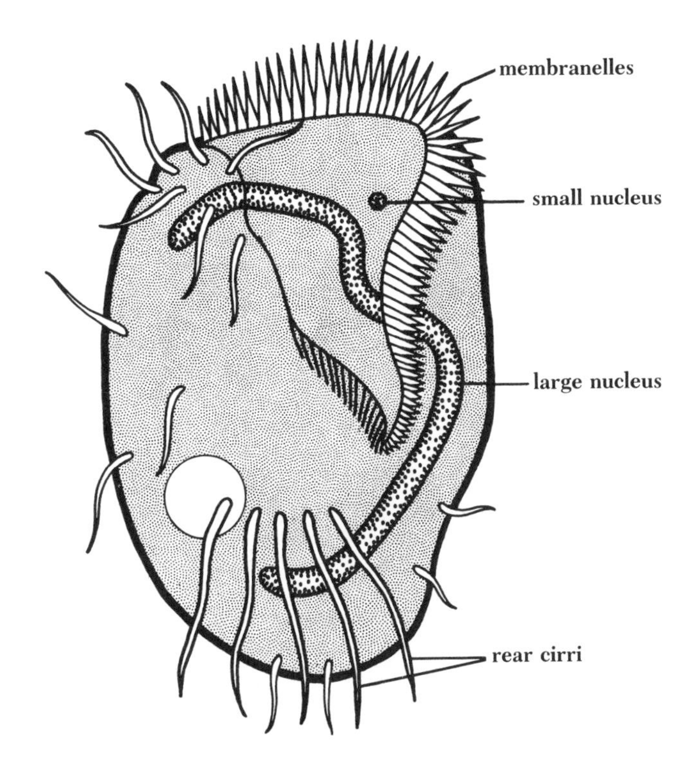

Euplotes is a hypotrich with almost all of its cilia concentrated into a row of membranelles near the mouth and a few large cirri on the lower surface. (Based on a photomicrograph by C. V. Taylor.)

Groupings within the ciliates continue to undergo revision. A recent authoritative reclassification has proposed three major groups of ciliates to replace the two used here: the **kinetofragminophorans,** corresponding to the more primitive holotrichs and including the suctorians; the **oligohymenophorans,** corresponding to the more specialized holotrichs and including the peritrichs; and the **polyhymenophorans,** corresponding to the spirotrichs. We have retained the groupings of holotrichs and spirotrichs here as easy and useful informal designations that will probably continue to provide continuity in ciliate classification for some time.

In this chapter we have seen that the limitations of differentiation within a single, small cytoplasmic mass have not prevented the protozoans from becoming enormously varied and abundant organisms. But among them we have also seen forms having many cells or many nuclei, with some differentiation among these elements and with an accompanying increase in size. These suggest the beginnings of a different level of organization—multicellularity—which distinguishes the animals that are the subject of the rest of this book.

Sponge vendor on a street in Athens, Greece. (M. Buchsbaum)

Sponges vary in shape with species and with exposure to surf or bottom current. Seen here are the dried skeletons of sponges from shallow, quiet waters of the Caribbean and Mediterranean. The rectangular "sponge" is synthetic. (R. Buchsbaum)

Chapter 5

A Side Issue: Sponges

Sponges, or rather the skeletons of sponges, were commonly used by the ancient Greeks for bathing, for scrubbing tables and floors, for padding helmets and leg armor, and—when the armor failed—for stopping blood flow from wounds. The Romans fashioned them into paintbrushes, tied them to the ends of wooden poles for use as mops, and made them serve, on occasion, as substitutes for drinking cups. In the first part of the twentieth century, sponges had an even wider variety of uses, and sponge fishing was an industry that every year harvested a million kilograms (over a thousand tons) of sponges. Today household "sponges" are usually synthetic, but the superior qualities of certain natural sponges are still prized by professional car or wall washers, and by leather workers, potters, silversmiths, and lithographers.

Commercial sponges grow only in warm shallow seas; but many other kinds live in both warm and cold seas, from intertidal shores to abyssal depths around the world, and a few flourish in freshwater. The skeletons of most of these sponges are too hard, scratchy, or brittle, or too small to be useful as "sponges." However, some sponges are important to humans in other ways. Many produce substances that are toxic to predators or to microbes, and these species are being studied as possible sources of new drugs for treatment of cancer and other diseases. For sessile animals such as sponges, which cannot bite or run away or withdraw into a shell, prickly skeletons and toxic or bad-tasting substances are important defenses against predators; but even these do not deter the various marine snails, sea stars, fishes, and other animals that feed on living sponges.

A living bath sponge looks more like a slimy piece of raw liver than like the familiar sponge of the bathroom. It grows attached to the substrate like a plant and, to the casual observer, shows the same kind of unresponsive behavior. For a long time sponges

Sponge fishing in the United States has its center at Tarpon Springs, Florida, where the sponge boats bring in their "catch." The sponges are brought up by helmet divers and after preliminary cleaning are hung on the rigging, where they die and decay. The skeletons undergo further cleaning and preparation when brought ashore. The sponge industry is now much reduced, as growth of sponges in Tampa Bay has been markedly decreased by pollution and by mechanical damage from the nets of fishermen trawling for bottom fishes. (R. Buchsbaum)

Hooking sponges was a method used in shallow water, common in the days before overfishing decimated shallow-water populations of commercial sponges. While one person rows, another scans the bottom through a glass-bottomed bucket (which cuts off surface reflections) and pulls up the sponges with a two-pronged hook on the end of a long pole.

A natural sponge holds more water and stands hard wear better than do synthetic "sponges." Millions of sponges were once harvested and sold every year for use in households and in industry; but overfishing, an epidemic disease of sponges in Florida and the Caribbean, and the availability of inexpensive synthetic sponges have together made natural commercial sponges relatively rare. A sponge of the type and size shown here would today sell for $30 or more. (R. Buchsbaum)

Living sponge, hooked and pulled from its attachment on the sea bottom at Batabanó, Cuba. (R. Buchsbaum)

Preparing sponges for the market. The sponges are washed, sorted, and strung into convenient bundles. Before they can be used, they must be further cleaned to remove the cellular debris, pounded with a mallet to break up the shells or skeletons of various invertebrates that lived in the cavities of the living sponge, and then trimmed to a regular shape with a shears. Tarpon Springs, Florida. (R. Buchsbaum)

The compressibility of sponges is used to advantage in shipping them. Here a large bundle of sponges is loaded into a baling machine, which compresses them into a compact package. Highly elastic, the sponges will spring back to their original dry volume when unpacked. Batabanó, Cuba. (R. Buchsbaum)

Cut open, the sponge looks like a piece of raw liver. In this gross aspect the supporting framework of spongin fibers is not distinguishable from the mass of living cells. The cut has exposed several excurrent channels, each leading to a vent. (R. Buchsbaum)

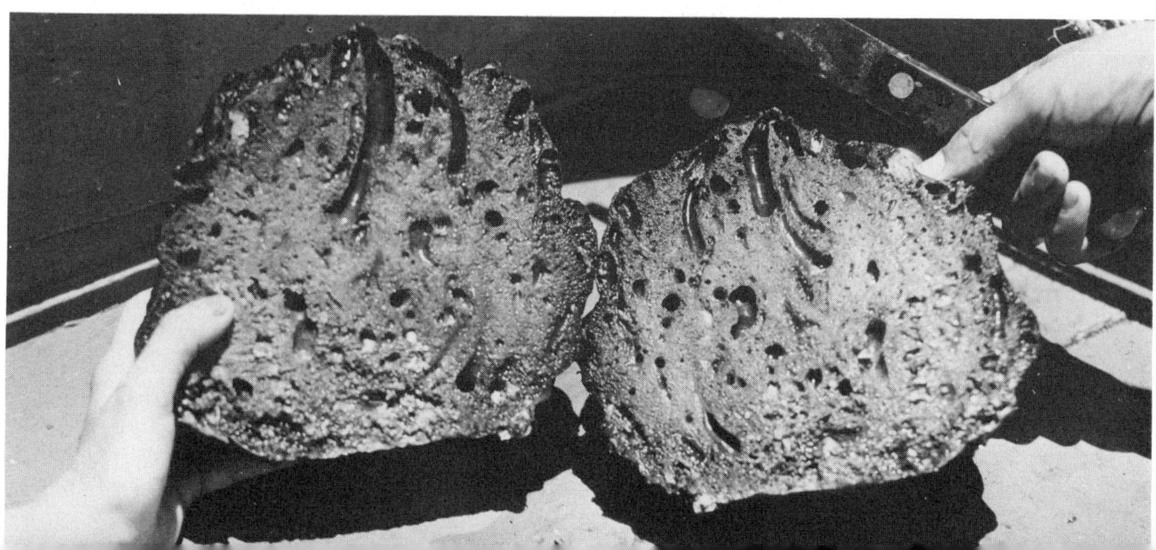

were variously described as animals, plants, both animal and plant, and even as nonliving substances secreted by the many animals that take shelter in the cavities of a sponge. It was not until about the middle of the nineteenth century that the last skeptics were finally convinced that sponges were really animals. In fact, sponges are organized so differently from other animals that move about and respond rapidly to conditions around them that it was difficult to understand how they could capture food.

A clue to this mystery is readily obtained by adding a suspension of colored particles to the water near a sponge, thus disclosing a great deal of unsuspected activity. A steady jet of water is seen to issue from one or more large holes at the top of the animal. Closer inspection reveals that water is entering through microscopic pores that riddle the entire surface. A sponge lives like an animated filter, straining out the minute organisms contained in the stream of water that passes constantly through its body. From the possession of the millions of pores, this phylum of animals has been called the **PORIFERA,** or "pore bearers."

Up to now, most of the organisms that we have considered were microscopic masses of living matter. Larger animals, such as the sponges, are not merely larger masses, but their bodies are subdivided into microscopic units, or cells. This many-celled structure almost always accompanies increase in the size of animals chiefly because the diffusion of oxygen and of other metabolic substances is such a slow process that the interior of a large solid animal would not receive oxygen or dispose of wastes fast enough to support life. Cellular construction divides a large mass into a great number of small masses, or cells, making it possible to have spaces between them and thereby exposing an aggregate surface area many times that of the undivided bulk. This increases the surface through which substances can diffuse in and out. Also, the distance that they must travel by diffusion is reduced because substances can be brought to nearly every cell in water currents. A glance at the diagram of a simple sponge shows that much more surface is exposed than if the same amount of living matter were in a simple, solid mass.

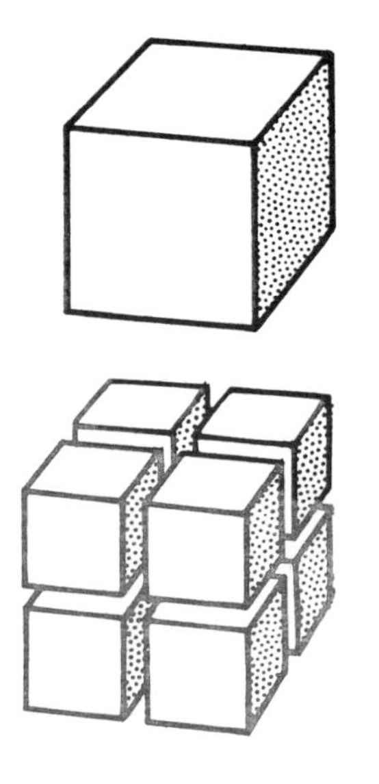

Dividing a mass into smaller units greatly increases the surface exposed to the environment.

In an organism that is not divided up into many cells, specialization is restricted to that of cytoplasm, nucleus, and other organelles. Consequently, a protozoan is limited not only in size but also in degree of specialization. In contrast with this acellular level of organization, the sponges may be said to be constructed on a **cellular level of organization.** No one cell must carry on all of the life activities, but different cells may become specialized for different functions. The various kinds of cells are not rugged individualists like the protozoans but show definite socialistic tendencies. Only certain types feed; and these pass on some of the food to cells that specialize in protection, mechanical support, or reproduction. This division of labor among cells increases the possibility of modes of life not available to simpler organisms.

Just how animals came to be composed of many cells is a question to which no definite answer can be given. One view is that as organisms grew larger, the number of nuclei increased and cell membranes were formed between them. Another view is that the many-celled condition resulted from the failure of single protozoan-like cells to separate completely from each other, or to disperse, after division. An example of this is seen in organisms such as colonial collar flagellates or *Volvox*.

A simple sponge such as *Leucosolenia* is vase-shaped with a large **excurrent vent** at the top and microscopic **incurrent pores** perforating the sides. The vase is covered and protected on the outside by flattened **covering cells,** which fit together like the tiles in a mosaic, forming a sheet of cells called an **epithelium.**

Three kinds of sponges growing together in a cluster, which was attached to a pile of a harbor wharf and has just been removed from the water. The main mass is a much-branched growth of the simple calcareous sponge *Leucosolenia* ("white tubes"), described in the text. At the left protrude three large and two small vase-shaped individuals of *Sycon*, which also has calcareous spicules but has a more complex internal structure. At the top, surrounding the stalk by which the cluster was attached, is a flat encrusting sponge, *Halichondria*, which has siliceous spicules. Plymouth, England. (D. P. Wilson)

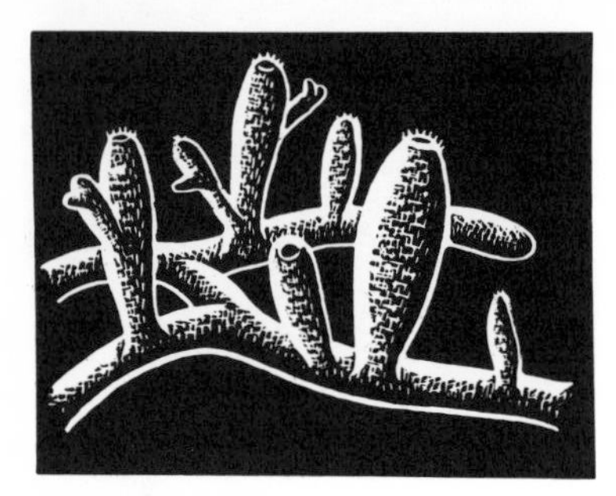

A simple sponge, *Leucosolenia,* natural size.

The large internal cavity of a sponge is lined by a special kind of cell, called a **collar cell** because its free end is encircled by a delicate collar of slender fingerlike projections; this free end also bears a long flagellum whose base passes through the collar. The beat of the flagella of the collar cells creates the water current that passes through the sponge. Water enters by the minute pores, flows upward through the main cavity, and leaves by way

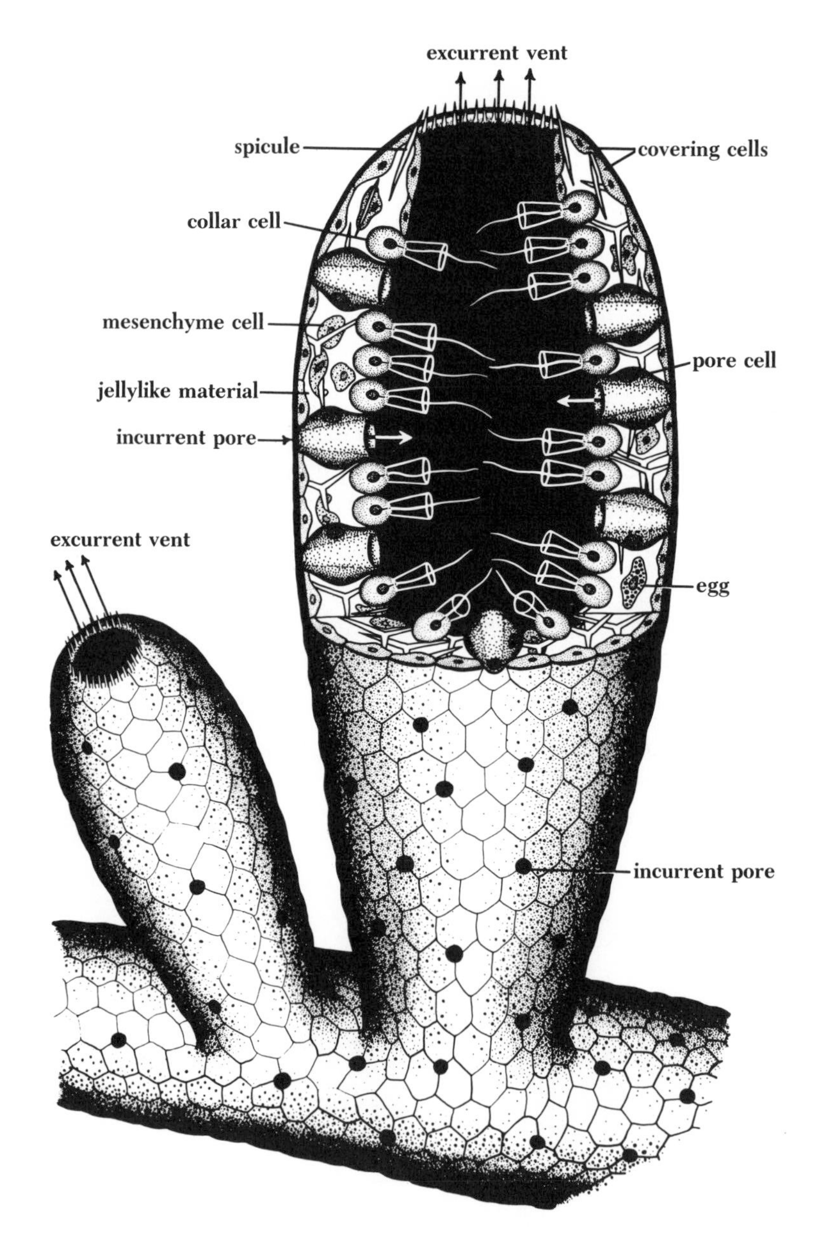

Diagram of a simple sponge. The upper part is cut away to show the structure.

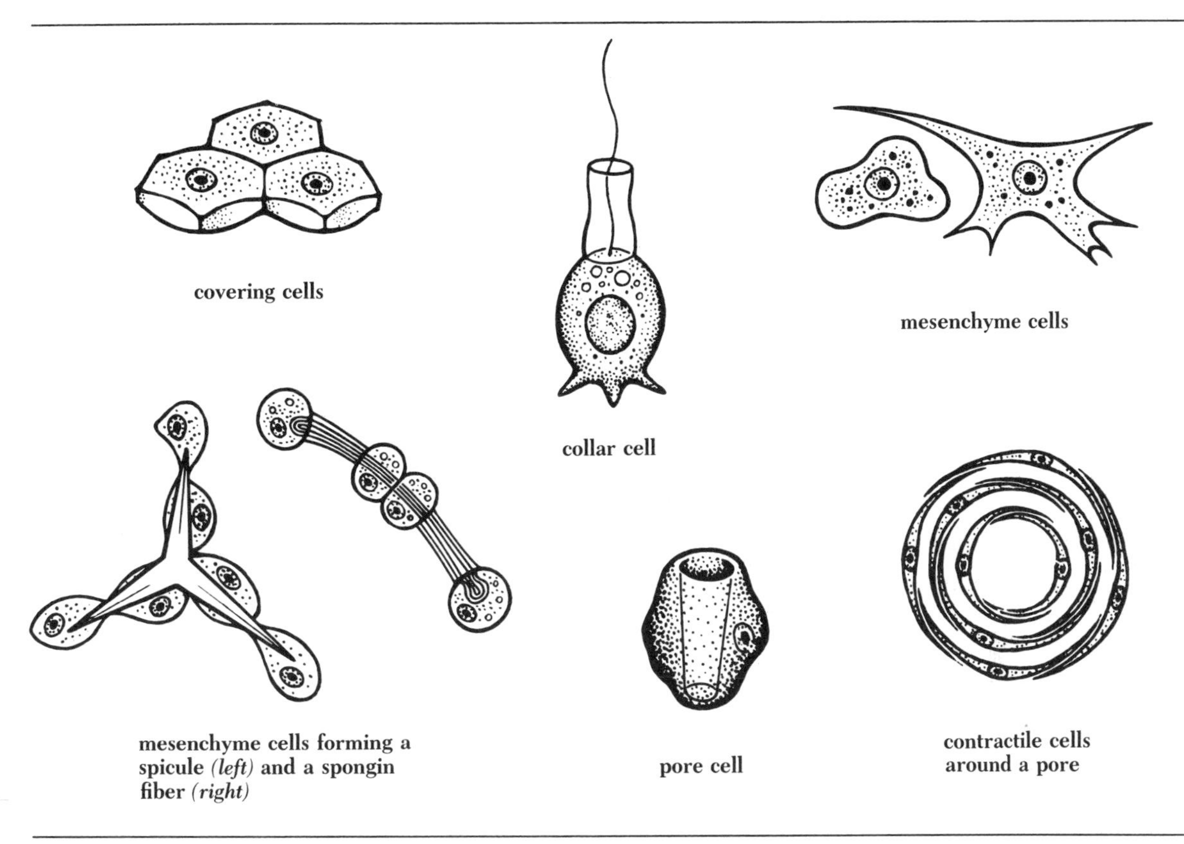

Types of sponge cells.

of the large hole at the top. As the water current passes through, the collar cells capture and ingest food organisms in the same way as do certain flagellate protozoans. The spiral beat of the flagellum, from base to tip, causes a stream of water to flow away from its tip and brings particles toward the base of the cell. The water passes between the slender projections that form the collar, which thus acts as a sieve. Particles stick to the outside of the collar and pass down its outer surface into the cytoplasm at the base, where they are engulfed. The collar cells digest food in food vacuoles or pass it on to certain other cells that carry on digestion.

The collar cells look and behave almost exactly like the collar flagellates described earlier. For this reason it is believed that sponges evolved from the same group of ancestral protozoans that also gave rise to the modern collar flagellates.

Between the outer epithelium of covering cells and the inner epithelium of collar cells is a nonliving jellylike material containing living **mesenchyme cells** that move about in ameboid fashion. They receive partly digested food particles from the collar cells, complete the digestion, carry the digested food from one place to another, and store food reserves. They also transport waste material to surfaces from which it can be carried away by the outgoing current of water. Certain mesenchyme cells are unspecialized and can develop into any of the more specialized cell types in the sponge.

One of the chief functions of some of the mesenchyme cells is to secrete the **skeleton,** which, together with the jellylike material, supports and protects the soft cellular mass and enables the sponge to grow to considerable size. The skeleton consists of hard needlelike **spicules,** or of tough fibers of **spongin,** or of both.

The spicules of sponges come in a variety of intricate forms, and one sponge may have spicules of several shapes. As the shapes are fairly constant for any particular species, they serve as an important criterion in identification.

The simple sponge *Leucosolenia* belongs to the class Calcarea, composed of small chalky or **calcareous sponges** with spicules of calcium carbonate. The **glass sponges** are mostly deep-water sponges with exquisite glasslike skeletons. They comprise the class Hexactinellida, in which all the members have six-rayed (hexactine) siliceous spicules, and most have, in addition, a latticework of fused siliceous spicules. Unlike other sponges, glass sponges do not have discrete cells (each surrounded by a complete membrane) but consist of a continuous network of cytoplasm containing many nuclei. The class Demospongiae, by far the largest class, includes all the so-called **siliceous sponges** in which there are siliceous spicules (not six-rayed), sometimes combined with fibers of spongin. Most of the sponges in temperate marine waters, and all of the freshwater sponges, are siliceous sponges. This class also includes the **horny sponges,** which have no spicules at all but have a skeleton composed entirely of fibers of spongin, which is a collagenous protein related to those in horn and in the connective tissues of most animals. Commercial sponges are all horny sponges. Finally, the **coralline sponges,** class Sclerospongiae, have a skeleton of siliceous spicules and organic fibers within the body of the sponge, overlying a basal skeleton of dense calcium carbonate. Coralline sponges are found in deep water or in submarine caves and other dark places.

Another type of mesenchyme cell is the **pore cell,** shaped like a short thick-walled tube. In a simple sponge the pore cells lie with their outer ends opening among the covering epithelial cells and their inner ends opening among the collar cells, so that they form the pores through which water is drawn into the sponge.

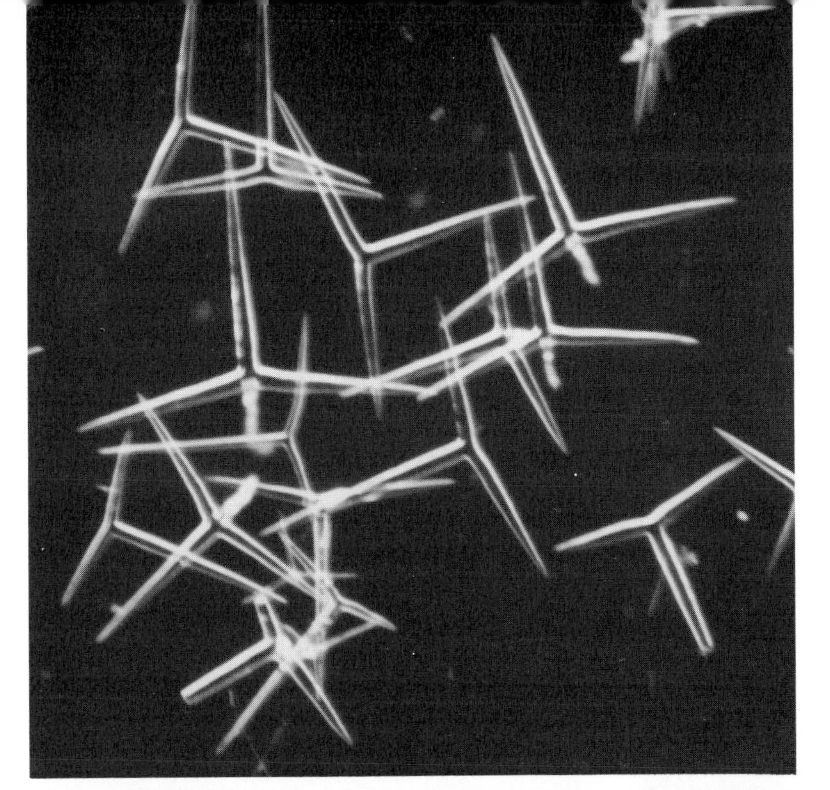

Triradiate spicules of calcium carbonate from a calcareous sponge. (R. Buchsbaum)

Spongin fibers form the elastic, interlacing skeleton of a horny sponge. *Below left:* Photomicrograph of a bit of dry sponge skeleton. *Below right:* Water has been added and is taken up by the fibers as well as into the spaces between fibers. (R. Buchsbaum)

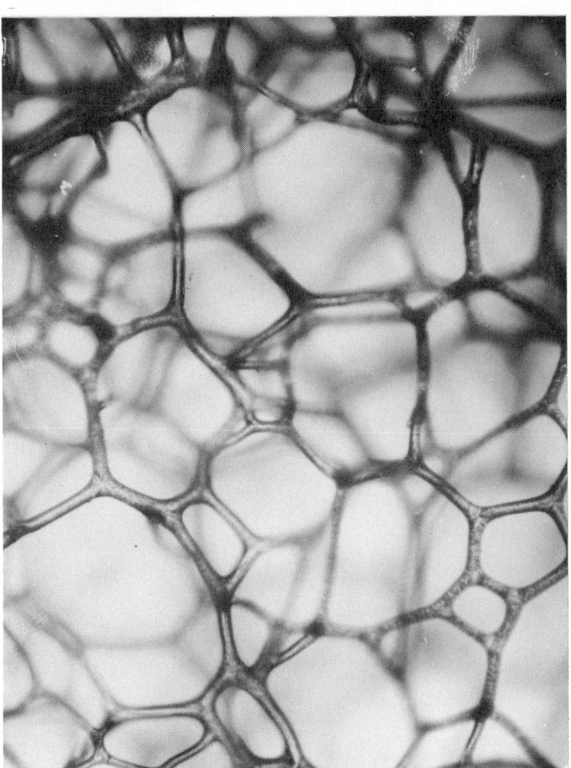

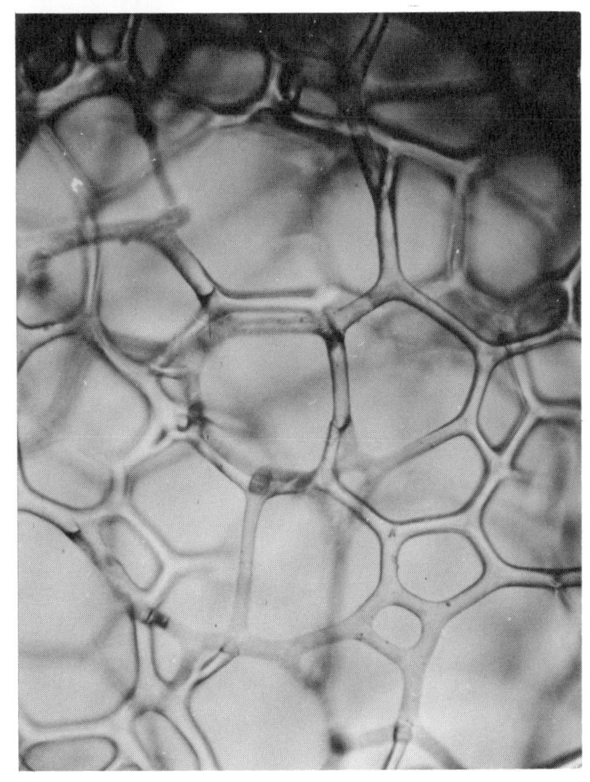

The pore cells are contractile and can vary the size of the pores or even close them completely.

In some sponges there are special elongated **contractile** or **muscle cells,** which produce movement by becoming shorter (and thicker), thus drawing closer together adjacent structures. They are arranged around openings, and, when irritating substances are present in the water, the contraction of the muscle cells narrows the openings. There are no identifiable sense cells that receive stimuli from the environment and no nerve cells to transmit them to other parts of the sponge. A mechanical stimulus, such as a touch or cut, usually produces only a local response, which may be a closing of pores or contraction around the immediate area of the stimulus. Closure of pores and stopping of water flow throughout the sponge can usually be accounted for by direct stimulation of pore or muscle cells and of collar cells by the circulating water. Thus most sponge **behavior** apparently results when individual cells are directly affected; the responses observed are localized, uncoordinated, and slow. In a few cases, a local mechanical or electrical stimulus has been shown to produce a gradual spread of coordinated behavior throughout a sponge, but the rate of conduction is slow compared to the speeds with which stimuli travel in nerves.

In addition to bringing a constant supply of food, the current of water passing through a sponge furnishes an ample supply of oxygen for all the cells and carries away carbon dioxide and nitrogenous waste substances. Consequently, even very large sponges do not have any special mechanisms to aid in respiration or excretion, but simply have a larger and more elaborate system of canals.

The evolution from small, simple sponges to large, complex sponges revolves chiefly about the problem of increasing the surface in proportion to the volume. If a simple sponge such as *Leucosolenia* were to enlarge indefinitely without modification in form, it would soon reach a point at which the inner surface available for the location of collar cells would not be large enough to bear the number of collar cells necessary, either to take care of the food demands of all the other cells composing the large bulk, or to move the great volume of water contained in the central cavity. In some sponges this problem has been solved by a simple folding of the body wall, which increases the surface available for the location of collar cells and strengthens the wall. In most sponges, like the bath sponges, there has been a further increase in the folding of the walls, resulting in intricate systems of canals with innumerable chambers lined with flagellated collar cells.

Complexity increases with size for another reason. A sessile

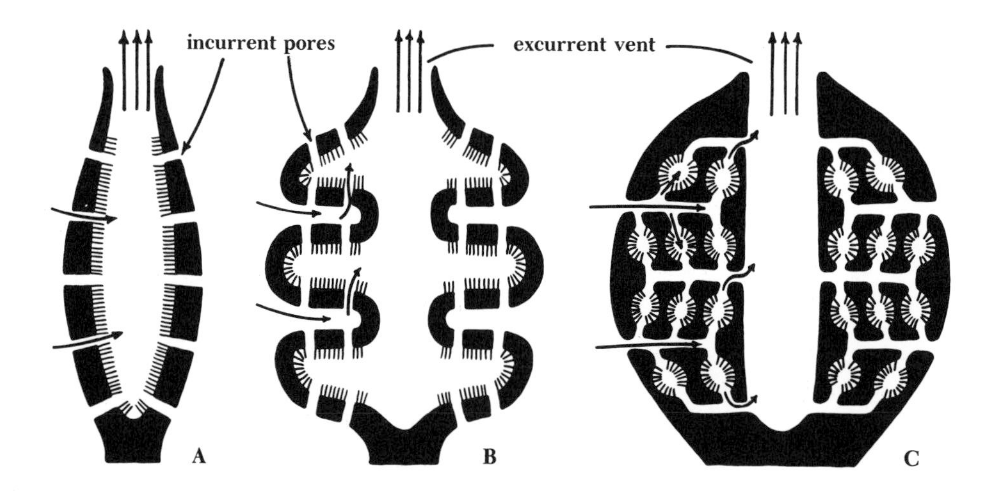

Types of sponge structure. A: Simple sponge. B: More complex sponge with folded wall. C: Complex sponge, such as a bath sponge, with elaborate system of canals and flagellated chambers. (Modified after various sources.)

animal cannot get up and leave when the food or oxygen supply of its environment runs low. The water that leaves a sponge has been filtered of much of its food and oxygen and is loaded with toxic wastes resulting from metabolism. If this water is not thrown sufficiently far away it will enter the animal over and over again. For a large complex sponge, with greater numbers of metabolizing cells for its volume, the amount of waste water discharged at the excurrent vent becomes a more significant problem. The evolution of the structure of sponges, then, especially large ones, has been in increasing the rate of water flow through the sponge and in separating as completely as possible the outgoing from the incoming water. In the same way, small isolated communities of humans have little problem disposing of their wastes, but large, concentrated urban communities require increasingly complex systems for disposing of sewage and toxic wastes.

Sponges engage in **sexual reproduction,** usually at well-defined seasons. Certain cells of the mesenchyme, or of other types, first divide and then enlarge greatly with reserve food to become female gametes, or **eggs;** other cells divide to form male gametes, or **sperms.** In some species both kinds of gametes may arise in one individual, which is then known as a **hermaphrodite.** In others they occur in different individuals. Both eggs and sperms may be shed into the sea, and then **fertilization,** the fusion of sperm and egg, is *external.* But in most sponges only the sperms are shed; they are brought into another sponge in its water current, and *internal* fertilization occurs.

In sponges and in nearly all other many-celled animals, the fertilized egg, or **zygote,** begins its development by dividing into

Glass sponge from deep waters of the Pacific, *Euplectella,* often called "Venus's flower basket." The skeleton (here shown cleaned and dried) consists of separate siliceous spicules plus an interlacing lattice framework. Long spicules at the base anchor it in the soft bottom. (R. Buchsbaum)

Siliceous sponge *(Microciona prolifera),* a bright red sponge much used in regeneration experiments. It lives in shallow water and grows up to 20 cm high. Common on the East Coast of the United States, this species was probably introduced to the West Coast by ships. (American Museum of National History)

Horny sponge, the elephant's-ear sponge, from the Mediterranean Sea, has numerous excurrent vents on the inner wall of the cup. Easily cut up into flat pieces and with a fine texture, it is valued by potters. Dried skeleton. N. Africa. (R. Buchsbaum)

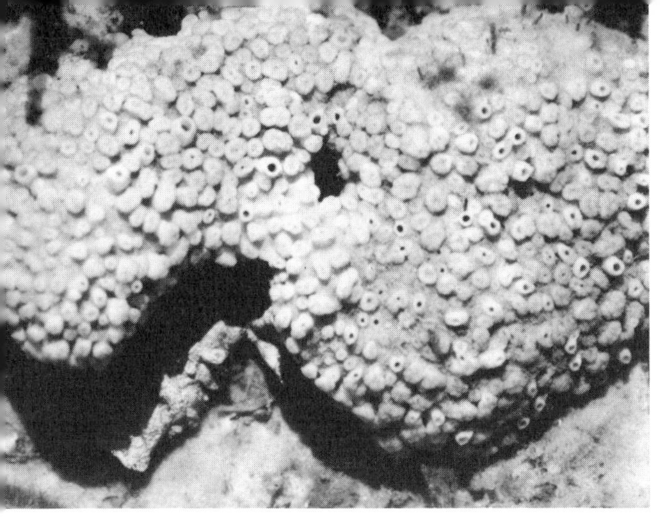

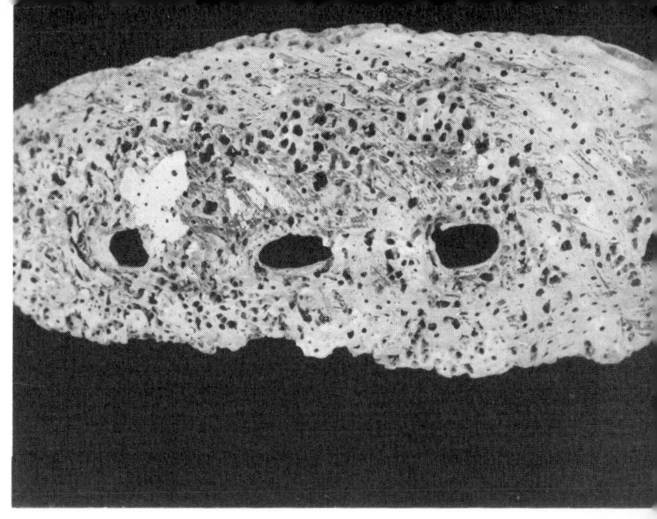

Boring sponge, *Cliona,* is seen *(left)* as bright yellow encrusting masses with many excurrent vents, each topping a little mound. Not seen is the part of the sponge growing in the tunnels that it has bored within a calcareous substrate such as a shell *(right),* coral, or limestone rock. Boring sponges play a significant role in the recycling of calcareous structures in the sea. (R. Buchsbaum)

Freshwater sponges are found in unpolluted streams, lakes, and ponds. They sometimes grow in large mats covering many square meters, but more often occur as small encrusting masses or fingerlike branches on sticks and stones. In the shade they are yellowish, gray, or brown in color; in strong sunlight they are usually colored green by algal cells that live and photosynthesize among and within the sponge cells. *Spongilla,* shown here, has a sparse horny skeleton combined with many needlelike siliceous spicules. (American Museum of Natural History)

Sponge spawning. Like smoke from a chimney, sperms were observed being released from the excurrent vent of this large *Verongia* in a column 3 m high for at least 10 minutes. Jamaica, West Indies. (H. M. Reiswig)

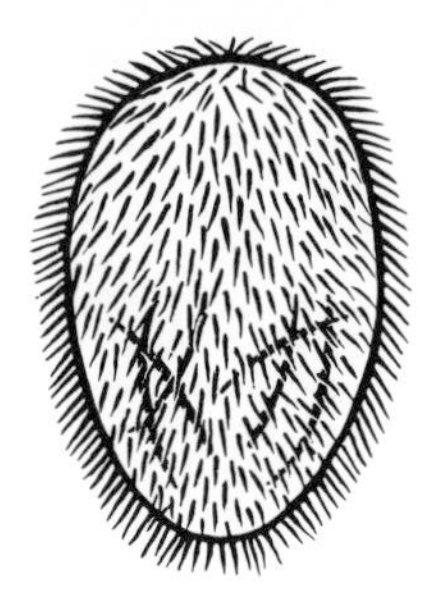

Sponge larva.

two cells. These promptly divide, forming four cells. Continued division results in a ball of cells, either solid or hollow, known as a **blastula.** Few animals are hatched into the cold cruel world at this tender stage. But in some simple sponges the blastula, a simple hollow ball one cell layer thick, becomes a flagellated **larva,** which escapes from the parent body and swims about. A larva is any young free-living stage that is not just a miniature adult but will undergo dramatic changes in structure, or **metamorphosis,** before attaining the adult form. A familiar example is the sequence of caterpillar (larva) and butterfly (adult). The sponge larva, after swimming about for a short time, settles down, becomes firmly attached, and metamorphoses into a young sponge. Through their larvas the sessile sponges are able to spread geographically and to send some of their offspring far enough away from home so that they do not set up business in direct competition with their parents.

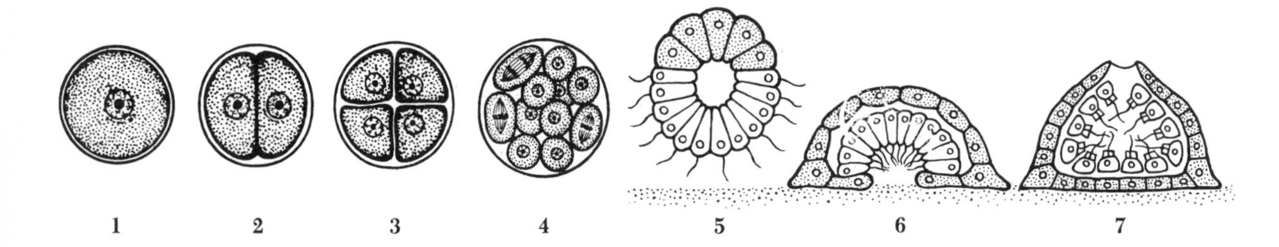

Early development and metamorphosis of a simple sponge. **1:** Fertilized egg. **2:** Two-celled stage. **3:** Four-celled stage. **4:** Continued cell division produces a ball of cells. **5:** Larva settles on flagellated end. **6:** Flagellated cells are inverted into interior. **7:** Young sponge with lining of collar cells; an excurrent vent has opened at the free end. (Adapted from P. E. Fell.)

Sponges grow by budding and branching, somewhat like plants. Species of *Leucosolenia* sprout horizontal branches that extend over the rocks and give rise to an extensive mass of vase-shaped uprights that may branch again. Some sponges grow as irregular encrustations and increase their mass indefinitely over the surface of rocks, vegetation, wharf pilings, or even on the backs of crabs. Each of the many excurrent vents in these sponges, or each upright in *Leucosolenia,* is sometimes viewed as defining a sponge "individual." Or, a new "individual" may be considered to arise only when pieces break off and continue to grow; the "individual" is then being defined as any physically separate sponge. That there is little difference between **growth** and **asexual reproduction** of such new "individuals," by either definition, reflects the *low level of individuality* in sponges.

Many sponges, especially freshwater ones, produce asexual reproductive units known as **gemmules,** which consist of a mass of food-filled mesenchyme cells, often surrounded by a heavy protective coat strengthened with spicules. The gemmule survives drying and freezing and carries the sponge over the winter

Encrusting sponge, *Halichondria panacea*, sometimes called the "breadcrumb sponge" because of the way it crumbles when broken. It grows as a flat adhering crust on the underside of overhanging rocky ledges or attached to any hard surface where it is protected against direct sunlight when the tide is out. The sponge has numerous excurrent vents, each at the tip of a conical projection. The soft parts are supported by interlacing needlelike siliceous spicules. Brittany, France. (R. Buchsbaum)

Sponge growing on a hermit crab is *Suberites*, also called the sea orange from its color and smooth round shape. It grows on the hermit crab's adopted shell, and when the shell later dissolves, the hermit is sheltered by the sponge alone. This sponge has both siliceous spicules and spongin fibers, and it is distasteful to most predators. Banyuls, Mediterranean coast of France. (R. Buchsbaum)

season or a period of drought. Under favorable conditions, the sponge cells emerge through a thin spot in the gemmule coat, aggregate in a small mass, and grow into a new sponge. Or, a gemmule may develop into a swimming larva.

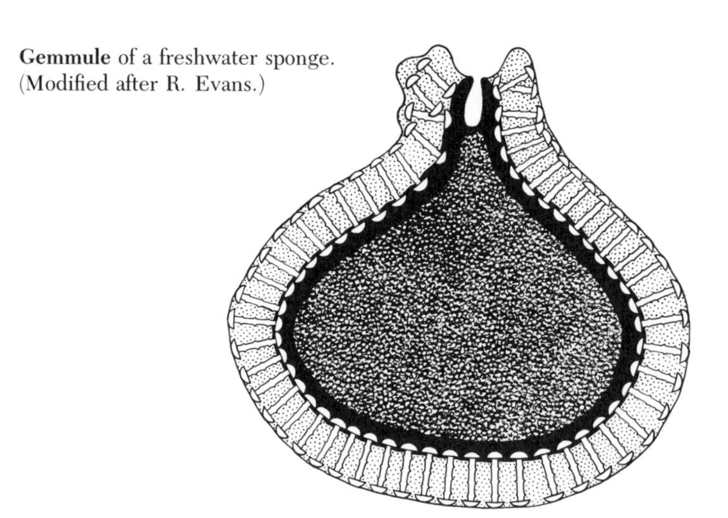

Gemmule of a freshwater sponge. (Modified after R. Evans.)

Sponge cells emerging from top of gemmule. (Modified after A. Wierzejski.)

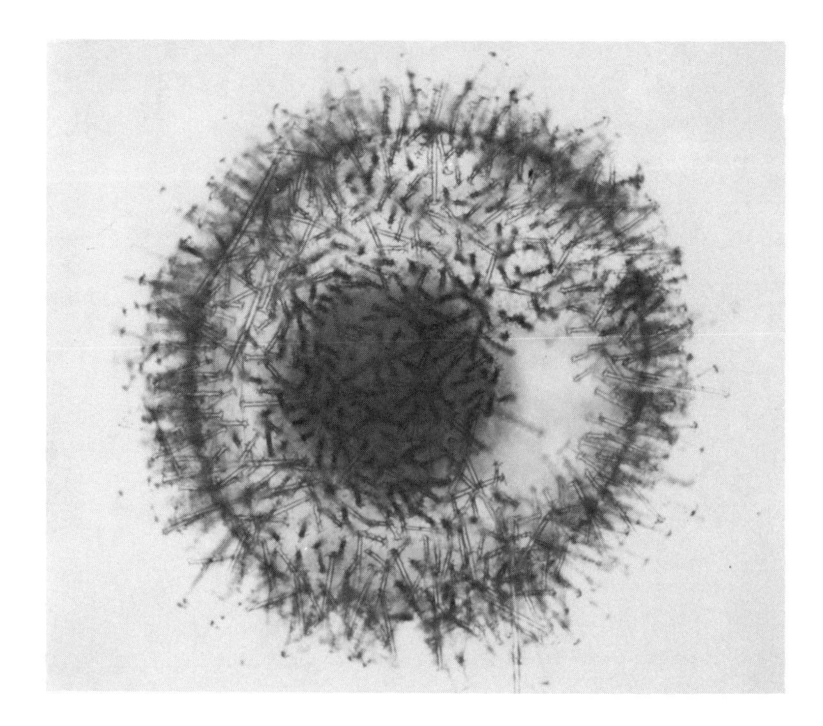

Gemmule has anchor-shaped siliceous spicules, which help protect the cells within. The gemmule withstands the winter; in the spring the cells emerge and grow into a new sponge. *Anheteromeyenia* from a Wisconsin lake. (Courtesy, J. R. Neidhoefer and F. A. Bautsch.)

All animals—but particularly the less complex ones—have some capacity to replace lost or injured parts, a process called **regeneration.** Some sponges are noteworthy in this respect, and under a certain kind of treatment their cells display a behavior that reminds us strongly of the absurd cartoons showing a dog being ground to bits in a meat grinder, only to emerge as small, compact sausages that still retain some of the behavior of the original dog. When certain sponges are pressed through fine silk bolting cloth, the cells are separated from one another and come through singly or in small groups. In a dish of seawater these separated cells creep about on the bottom in ameboid fashion. When they happen to come in contact with each other, they stick together; and after some time most of the cells are found to have united into small masses. Finally, these masses of aggregated cells grow up into new sponges.

The sponge body plan is unique. No other many-celled animals have the principal opening as a vent instead of a mouth; or feed by means of collar cells; or show so low a degree of coordination among the various cells. Hence it is thought that sponges evolved from a group of protozoans different from the ones that gave rise to any of the other many-celled animals. And the phylum Porifera is sometimes set aside in a separate subkingdom of animals. Yet the sponges are an abundant and widespread phylum, and the sponge body plan is of interest to us because it illustrates the *cellular level of organization:* cell differentiation without much cell coordination. But in the general trend of animal evolution, the sponges are little more than a side issue.

Cells separated by pressing a living sponge through fine silk bolting cloth.

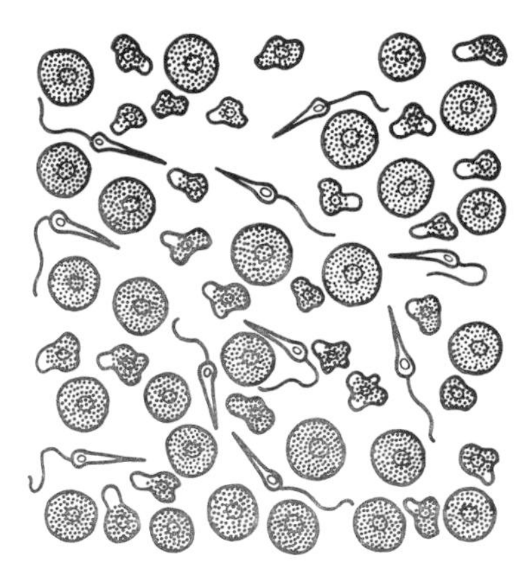

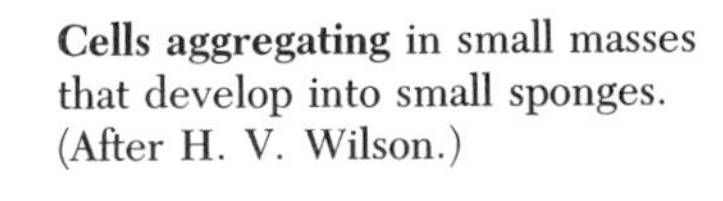

Cells aggregating in small masses that develop into small sponges. (After H. V. Wilson.)

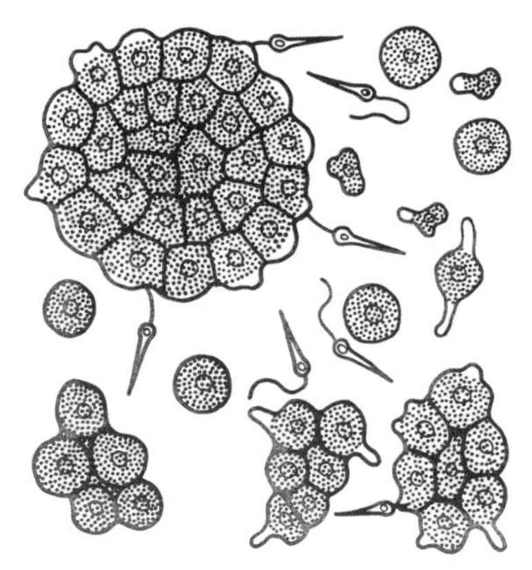

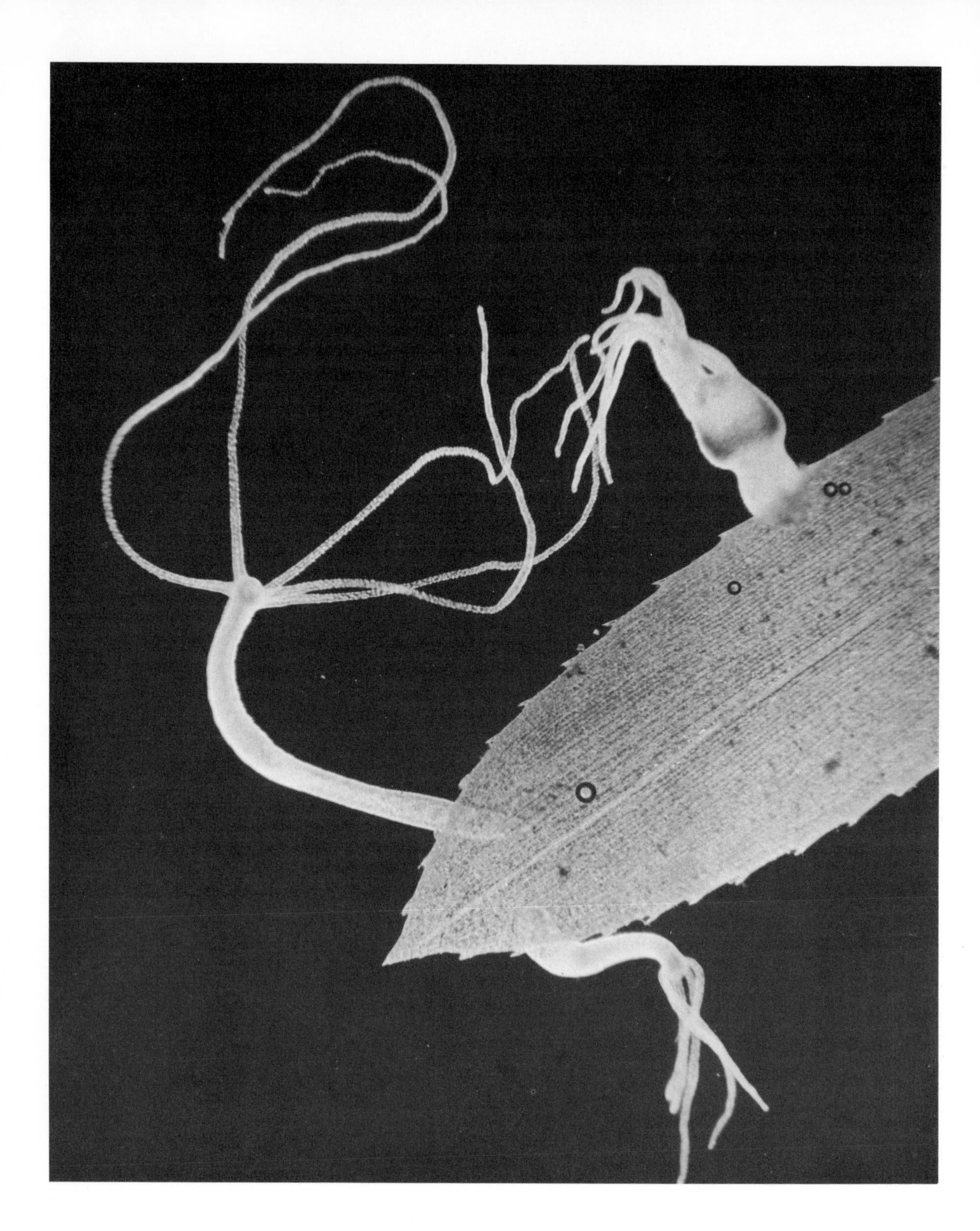

Chapter 6

Two Layers of Cells

A hydra, with its circle of slender tentacles, looks like a centimeter of string with one end frayed out into several strands and the other attached to a rock or water plant by its disklike base. Because of their small size, translucency, and habit of contracting down into a little knob when disturbed, hydras are readily overlooked. Yet these animals are abundant and widely distributed in ponds, lakes, and streams around the world, being among the few members of their phylum that have invaded freshwater. The marine relatives of the hydras—the jellyfishes, sea anemones, and corals—are the more conspicuous members of the large and varied phylum CNIDARIA (ny-DAIR-ee-uh).

The body of a hydra has a large central cavity that leads to the outside through an opening, surrounded by tentacles, at the upper end of the animal. But any resemblance to the central cavity and excurrent vent of a sponge is misleading, for a hydra is constructed in quite a different way: the spacious central cavity is a **digestive cavity** and the opening is a **mouth.** Beginning with the cnidarians, almost all the invertebrate animals to be described in this book, as well as vertebrates, have a digestive cavity that connects with the outside through a mouth.

Although, in the course of evolution, a great many different types of animal structures have arisen, their cells may be classified into a few main categories—about five: **epithelial, mesenchyme** or **connective, muscular, nervous,** and **reproductive.** All but nerve cells have already been encountered in the sponges. Nerve cells are added by the cnidarians.

An association of cells of the same kind that work together to perform a common function is called a **tissue.** Thus a mass of mesenchyme cells or some other type of connective cells is known as a "connective tissue," a bundle of muscle cells is spoken of as "muscular tissue," and a group of nerve cells as "ner-

Opposite: **Hydras,** extended and contracted, on a water plant. *Hydra littoralis.* (P. S. Tice)

A hydra consists of **two layers of cells.** *Left:* A hydra in longitudinal section. *Right:* A hydra in cross-section. Between the two cell layers is the jellylike mesoglea.

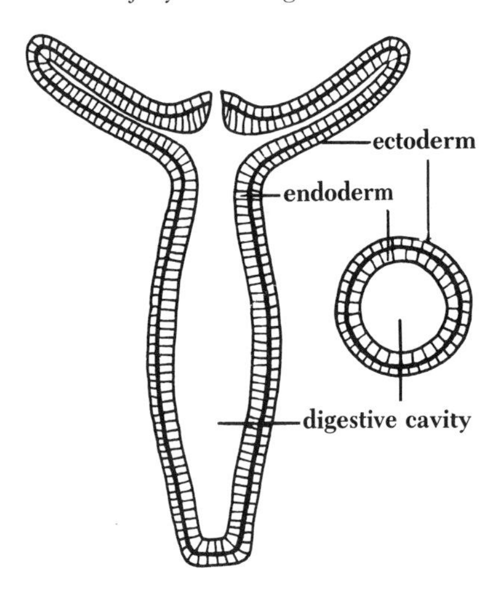

vous tissue." Sponges were presented as animals organized on a cellular basis, but they have some beginnings of tissue formation. For instance, the flattened cells covering the exterior of a sponge fit closely together to form an epithelium.

The more complex animals—humans, for example—have many more different kinds of cells than a hydra, but they are all modifications of these same basic cell types. Epithelial tissue covers the exterior surface of a human, lines the mouth and digestive tract, lines the heart and blood vessels, and is folded in various places to form glands. Liver and thyroid cells are epithelial; blood and bone cells are mesenchyme or connective-tissue types.

The coordination of cells into tissues is a distinct advance, since various functions can be performed better by a group of cells of the same kind acting together than by separate cells. Scattered muscle cells responding individually to external conditions could not produce much movement, while a sheet or bundle of muscle cells contracting together can exert great force, quickly changing the shape of an animal or propelling it through the water. Because their cells act together in a more coordinated fashion than do the cells of sponges, the cnidarians may be said to have reached the **tissue level of organization.**

Cross-section of a hydra shows the two cell layers (ectoderm and endoderm) with the thin layer of jellylike mesoglea sandwiched between and a spacious central digestive cavity. (Photomicrograph of a fixed and stained section, A.C. Lonert.)

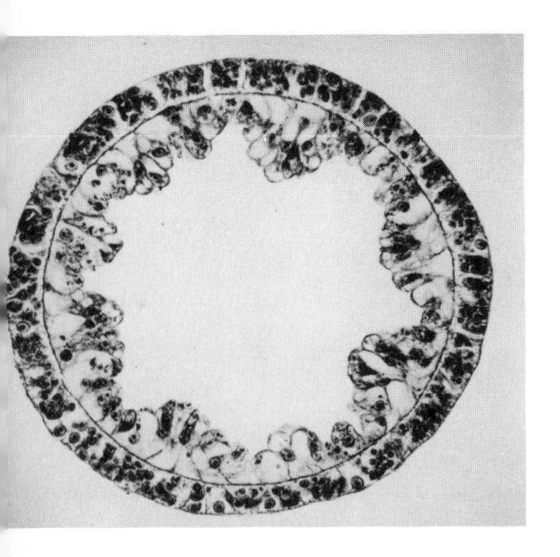

A hydra consists of **two layers of cells.** The outer layer, or **ectoderm** (literally "outside skin"), is a protective epithelium, as in sponges; but it contains several other kinds of cells. The "inner skin," or **endoderm,** lines the internal cavity and is primarily a digestive epithelium. The cells of both layers differ from the epithelial cells of most animals in that their drawn-out bases contain long contractile muscle fibers. The muscle fibers of the ectoderm run lengthwise. When they contract equally on all sides, the body is shortened. When they contract more on one side than on the other, the body is bent in the direction of greatest contraction. The muscle fibers in the endoderm run circularly; and when they contract, the body becomes narrower and longer. There is no separate muscular tissue in a hydra; the muscle fibers occur only in the bases of epithelial cells, which also perform other functions. Since they cannot be strictly classed as either epithelial or muscular, these cells of dual function are named **epithelio-muscular cells.**

Between the two layers of cells is a thin layer of jellylike material, the **mesoglea,** produced by both ectoderm and endoderm. Among the bases of the epithelial cells of both layers, but particularly in the ectoderm, are small mesenchymal cells called **interstitial cells.** The interstitial cells of a hydra, like the mesenchyme

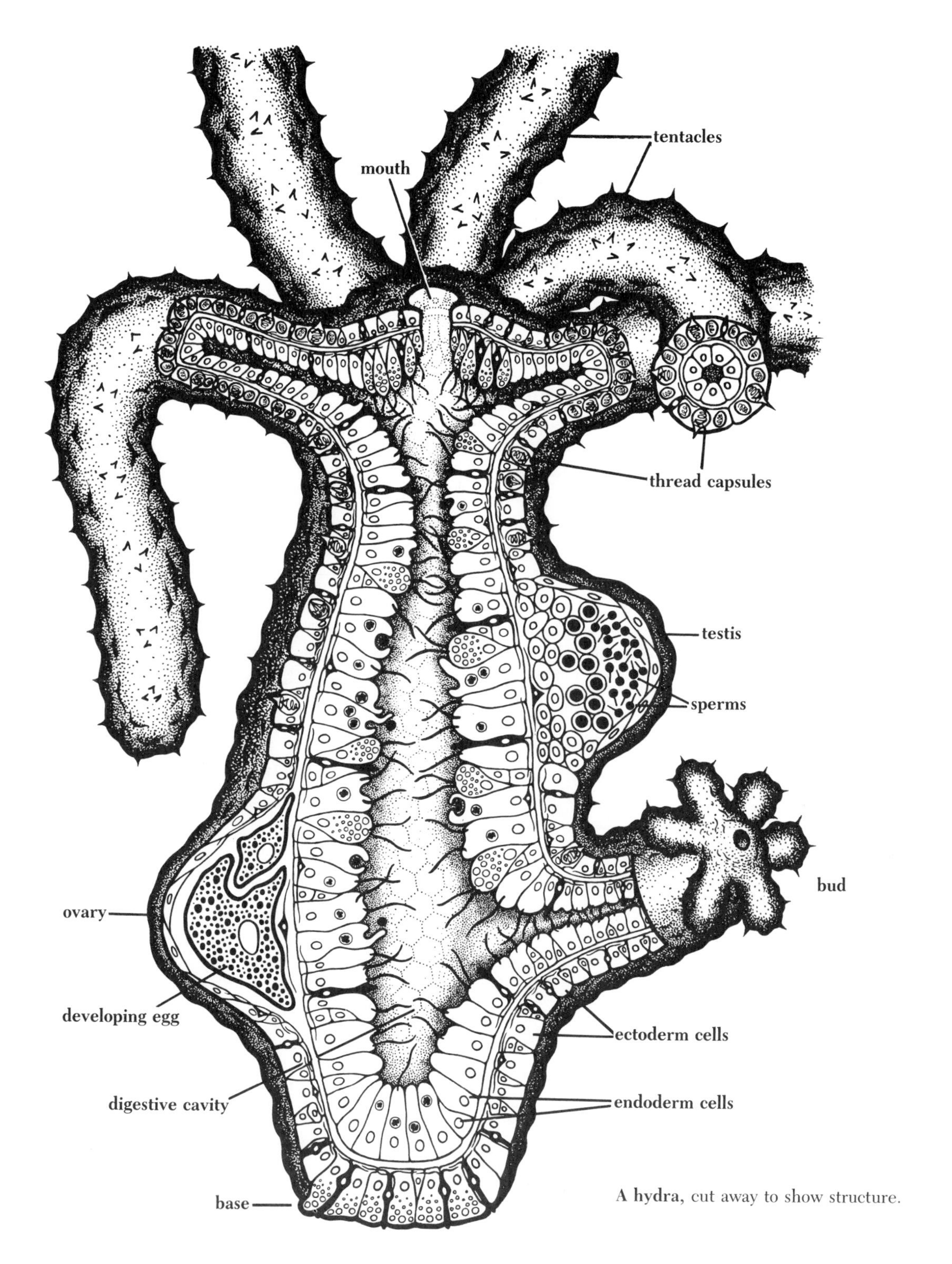

A hydra, cut away to show structure.

cells of a sponge, are the least specialized cells and are the only ones normally capable of developing into some of the other kinds of cells, which will be described in connection with the hydra's activities.

In feeding, a hydra does not chase its prey, but remains attached to the substrate, with the almost motionless tentacles trailing in the water. The **radial symmetry** of hydras and other cnidarians—with tentacles radiating from around the mouth and with no differentiated sides—permits them to welcome prey or repel attackers that approach from any direction. When a small crustacean or worm brushes one of the tentacles in passing, the unlucky victim is suddenly riddled with a shower of toxic, numbing threads shot out from certain of the **thread capsules,** effective weapons with which the body, and particularly the tentacles, are heavily armed. (A thread capsule is also called a *nematocyst* or a *cnida*, and from this last name comes the phylum name Cnidaria.) There are four kinds of thread capsules in hydras, all produced within specialized thread cells that are derived from interstitial cells and come to lie in the ectoderm. Each fluid-filled thread capsule contains a long spirally coiled hollow thread and lies within a thread cell that has a fine hairlike trigger projecting from the cell surface. When the trigger is stimulated, a physiological change occurs, such that the pressure within the capsule is suddenly increased and the coiled hollow thread is turned inside out. This has been compared to the way in which one everts the finger of a rubber glove by blowing into the glove.

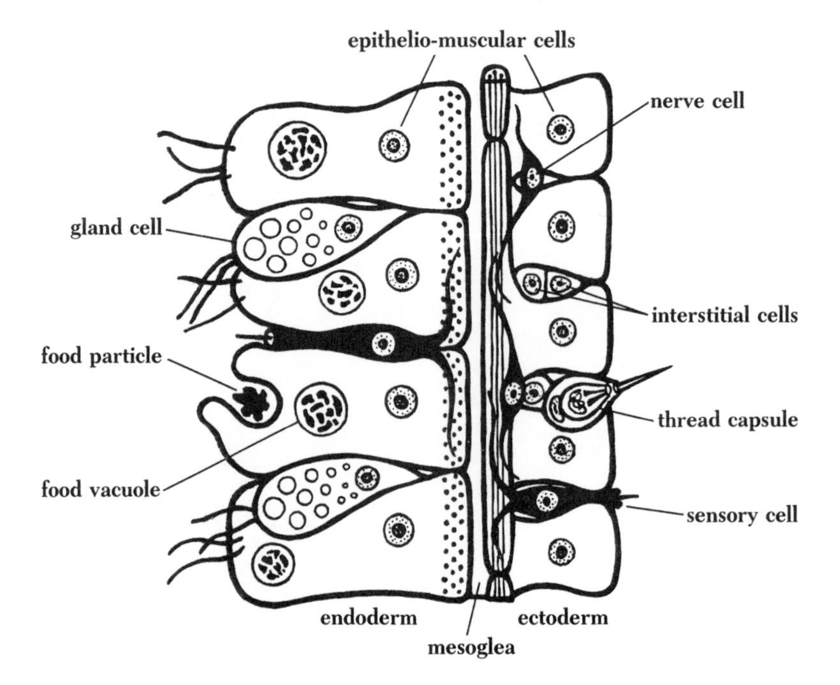

Portion of body wall of a hydra enlarged to show **cell types.** The muscle fibers of the endoderm, which run circularly, appear in this *longitudinal section* as black dots. The muscle fibers of the ectoderm, which run lengthwise, meet end to end.

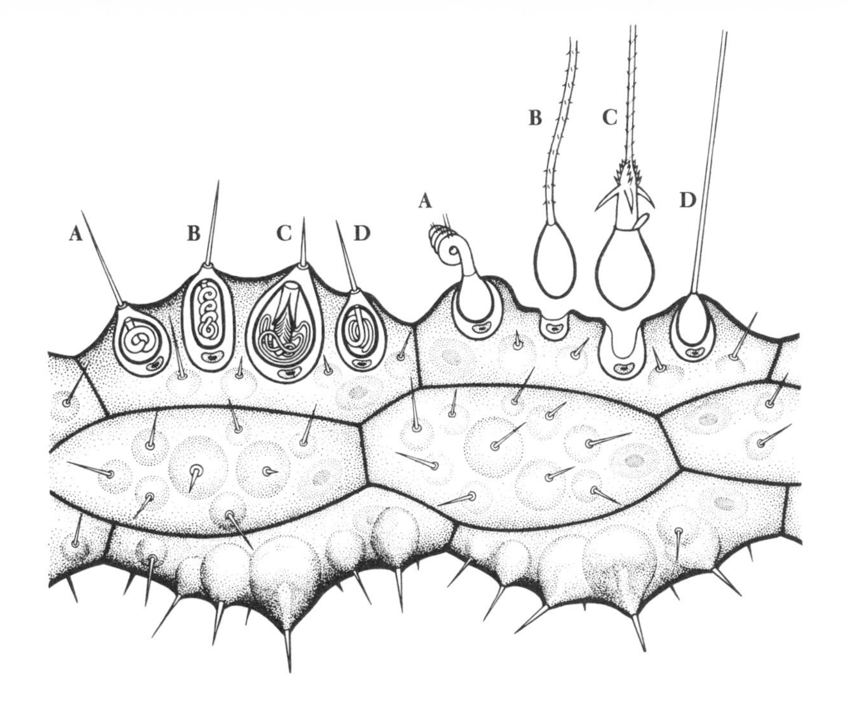

Portion of a tentacle showing the **batteries of thread capsules.** Each capsule lies within a thread cell, and the clusters of thread cells are enveloped by large epithelial cells. *Left:* Undischarged capsules. *Right:* Discharged capsules. **A:** Aids in prey capture by winding about bristles or other projecting parts. **B:** Discharged in defense against nonprey animals. **C:** Pierces and paralyzes prey. **D:** Fastens tentacles to solid objects during locomotion. Note that **B** and **C** are released from their thread cells immediately after discharge, whereas **A** and **D** must be held in place at least briefly. (Based on various sources.)

Thread capsules of a hydra. (Photomicrograph by Nomarski optics, R. D. Campbell.)

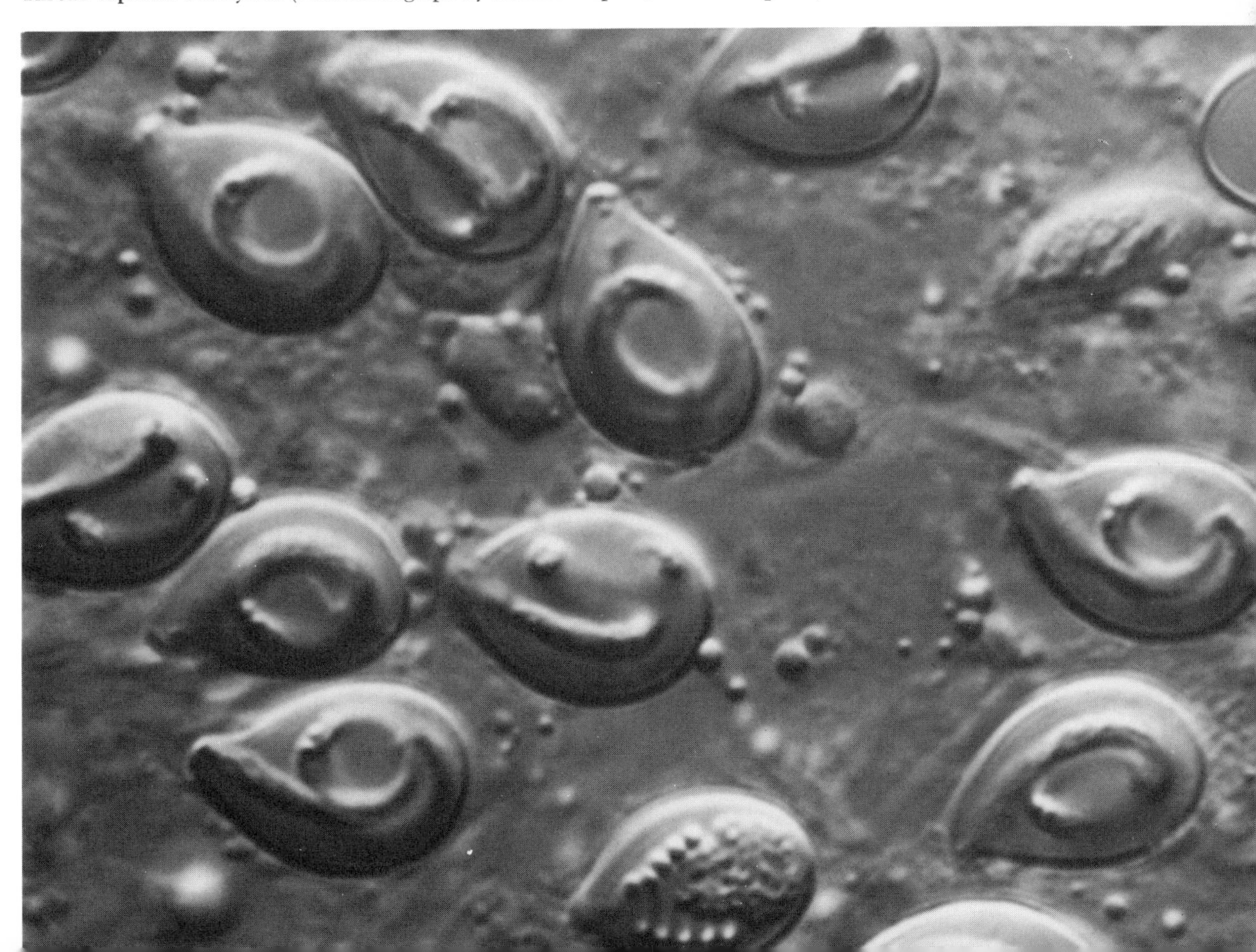

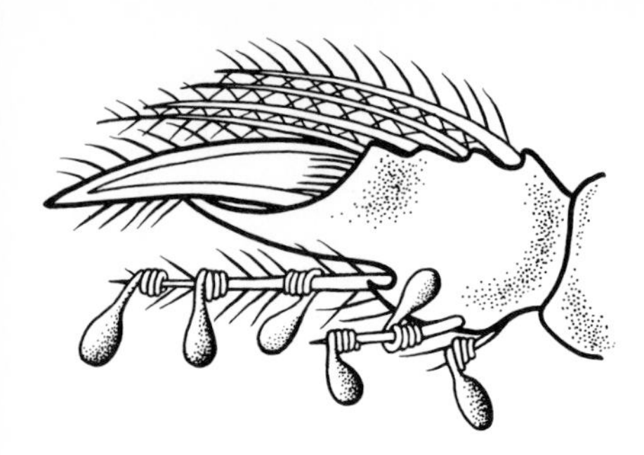

Two kinds of thread capsules equip hydras to prey on the small bristly crustaceans (see chapter 17) that make up much of their diet. *Above:* Thread capsules observed clinging to the bristles of a crustacean. *Below:* Large spines puncture the hard covering of a crustacean, making a hole through which the thread enters and injects a paralyzing toxin. (Both after O. Toppe.)

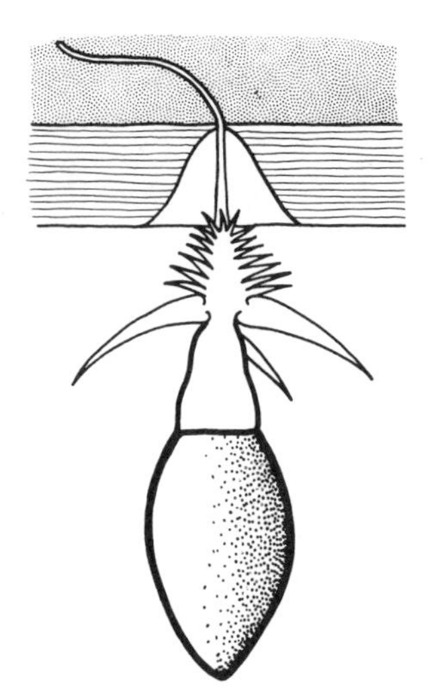

The largest and most conspicuous type of thread capsule, distinguished by large spines at the base of the thread, is discharged with such explosive force that the spines pierce the body of the prey and the thread injects a toxic substance contained in the capsule. After the toxin from a number of these stinging capsules has paralyzed a small animal, the tentacles are wrapped around the prey and contract, drawing the prey toward the mouth, which opens widely to receive it. The victim is swallowed by means of muscular contractions of the body wall, aided by a mucus secretion from gland cells lining the inside of the mouth region. A thread capsule can be discharged only once. Used ones are discarded and are replaced by new ones in new thread cells, derived from interstitial cells.

Digestion begins in the interior cavity. Gland cells in the endoderm secrete enzymes, chiefly of the protein- and fat-digesting types, which reduce the digestible parts of the prey to a thick suspension containing many small fragments. This material is then engulfed by pseudopods of the epithelio-muscular cells of the endoderm. The process of digestion is completed within food vacuoles in these cells, for the hydra has retained, in part, the protozoan method of food ingestion and digestion. Since preliminary digestion takes place in the large digestive cavity, where enzymes poured out by many cells act together to disintegrate a food organism, a hydra can eat animals that are very large as compared with those that can be taken by a sponge. In a sponge,

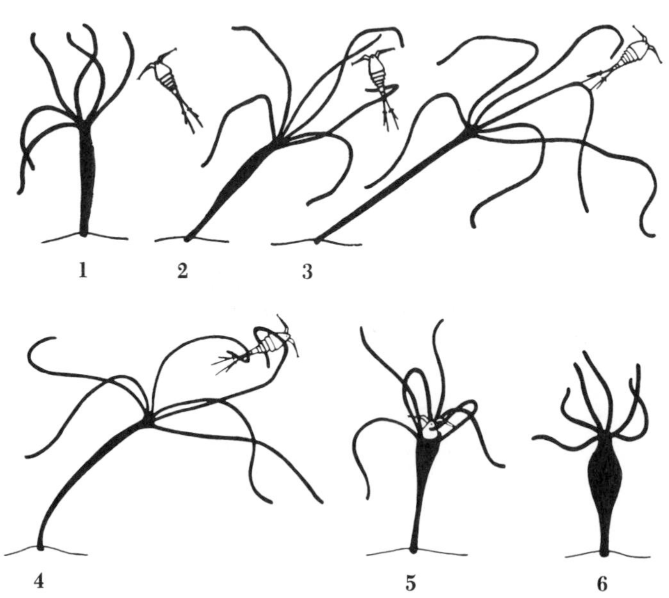

A hydra catches (1–4) and eats (5–6) a copepod (minute crustacean).

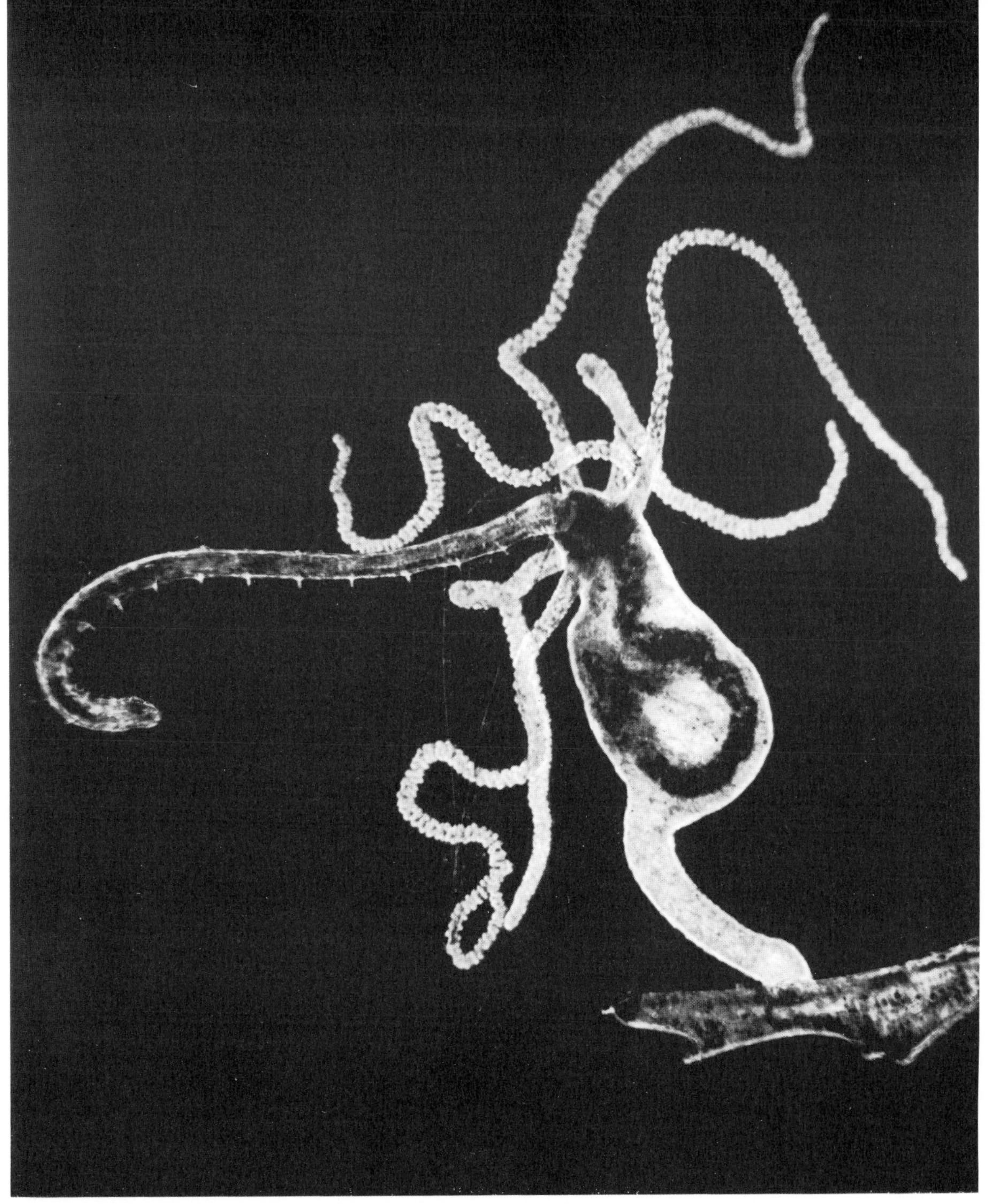

Hydra eating a worm, part of which is already in the digestive cavity but still visible through the translucent body wall. Besides these freshwater earthworm relatives (oligochetes), hydras eat small crustaceans, very young fishes, and any other small animals that can be snared by the long tentacles. The bumps on the tentacles are batteries of thread capsules. (P. S. Tice)

as in a protozoan, the prey must be of a size that can immediately be engulfed by a single cell.

Elimination of indigestible remnants left in the central cavity is through the mouth, which serves as both entrance and exit. The digested food is passed by diffusion from cell to cell. Currents, set up by muscular movements of the body and by the beating of long flagella on the endoderm cells, circulate the food throughout the cavity of the body and of the hollow tentacles. The cavity thus has the double function of digestion and circulation, and it may be termed a **gastrovascular cavity** ("stomach-circulatory" cavity).

Respiratory exchange and **excretion** of wastes take place by diffusion, as in protozoans. The ectoderm is freely exposed to the surrounding water and, because of the circulation in the gastrovascular cavity, the endoderm is also constantly exposed to circulating fluids. However, through this extensive surface, which facilitates exchange and excretion, excess water constantly enters the tissues by osmosis. The endoderm of a hydra secretes salts into the central cavity, thus drawing excess water from the tissues into the cavity. The fluid within the gastrovascular cavity is periodically expelled through the mouth by contractions of the body. Salts must be replaced by taking them up from the water and from food. The problems of salt and water regulation may account for the small number of cnidarians that have managed to invade freshwaters.

When well fed and healthy, hydras **reproduce asexually by budding.** The buds occur about one-third up the length of the body from the base. Here both ectoderm and endoderm hump up, forming a projection that elongates. The gastrovascular cavity of the bud, at its base, is continuous with that of the parent, and in this way the bud receives a supply of nourishment. It soon sprouts tentacles and a mouth at its far end, and in two or three days the bud looks like a little hydra. It begins to feed on its own, and shortly after this it constricts off from the parent and takes up the serious responsibilities of an independent life.

Regeneration would seem to be an easy matter for these simple animals, and hydras do show a marked capacity for replacing tentacles or speedily repairing the more serious injuries likely to be incurred by so delicate an animal. Even if a hydra is cut into a number of pieces, all of the pieces that are not too small will grow the missing parts and will become complete and independent hydras (see chapter 10).

The ability of these animals to replace lost or injured parts won for them the name "hydra." An early naturalist saw in this a resemblance to the regenerative powers of the mythical water monster, Hydra (from the Greek word for water). The Hydra,

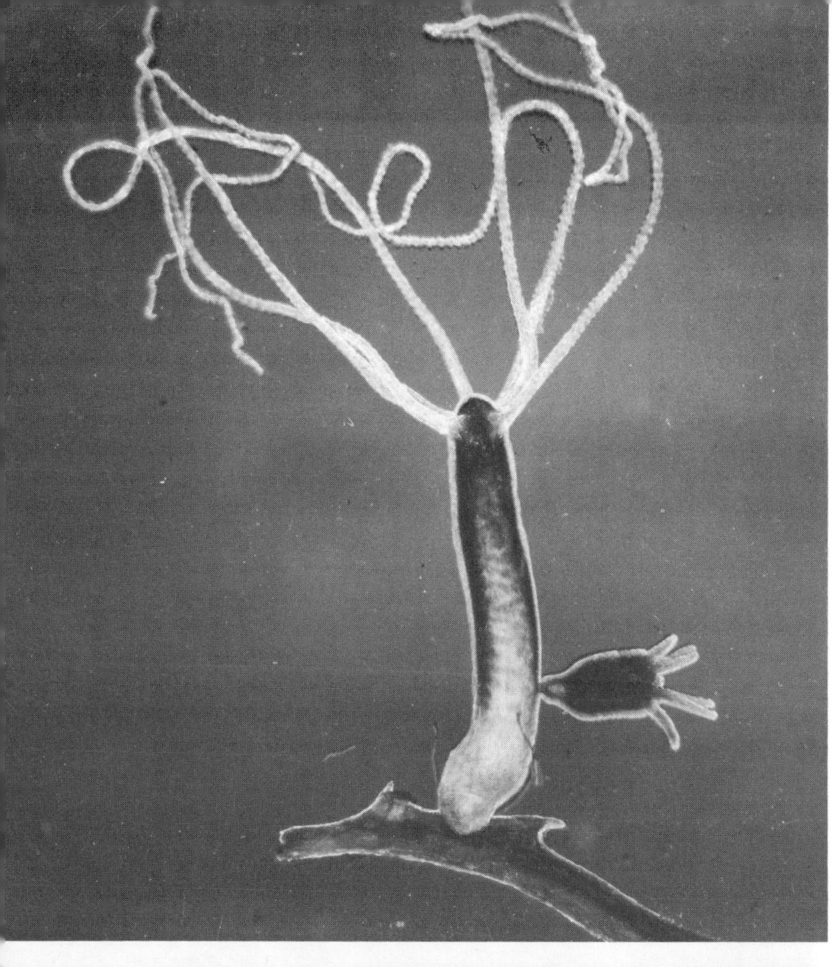

Hydra with a bud. This bud is relatively advanced, probably feeding on its own with the well-developed tentacles and already constricted at the point of attachment. It will soon detach and become fully independent. (P. S. Tice)

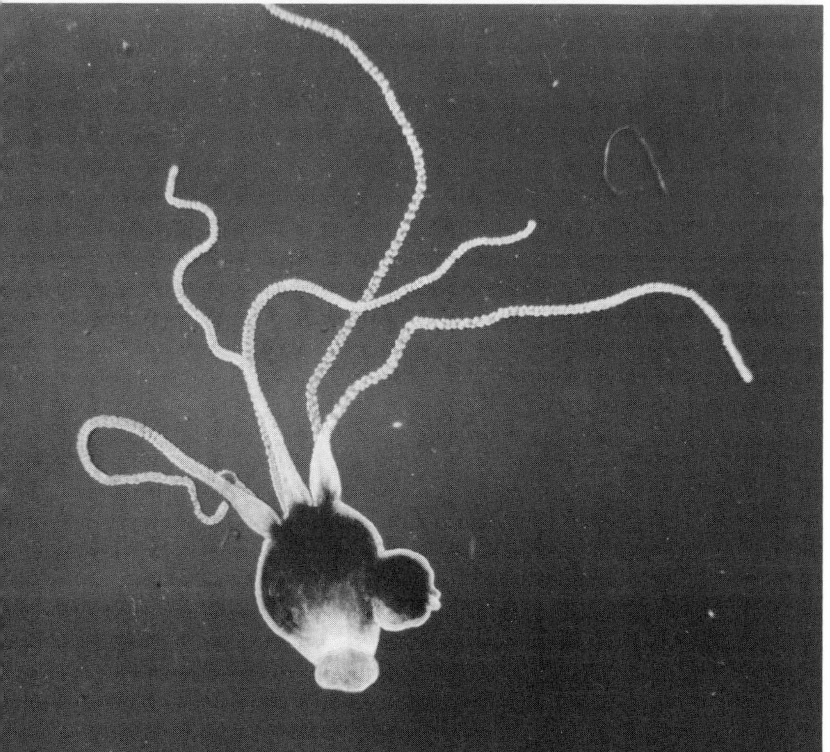

Parent and bud are here both contracted, but at times they may show quite independent behavior. (P. S. Tice)

which Hercules undertook to slay, had nine heads; and as Hercules cut each one off, two grew in its place.

Hydras **reproduce sexually** at certain times of the year, generally in the fall or winter. Some species are hermaphroditic, with both eggs and sperms produced in each individual. In other species, there are separate male and female individuals. The gametes come from interstitial cells in the ectoderm. In certain regions these cells suddenly start to grow rapidly, causing the body wall to bulge locally. Such bulges are known as **testes** when filled with sperm-forming cells and as **ovaries** when filled with egg-forming cells. Either kind of gamete-producing structure may be called a **gonad.** In each testis the interstitial cells first enlarge and then divide a number of times to form many sperms.

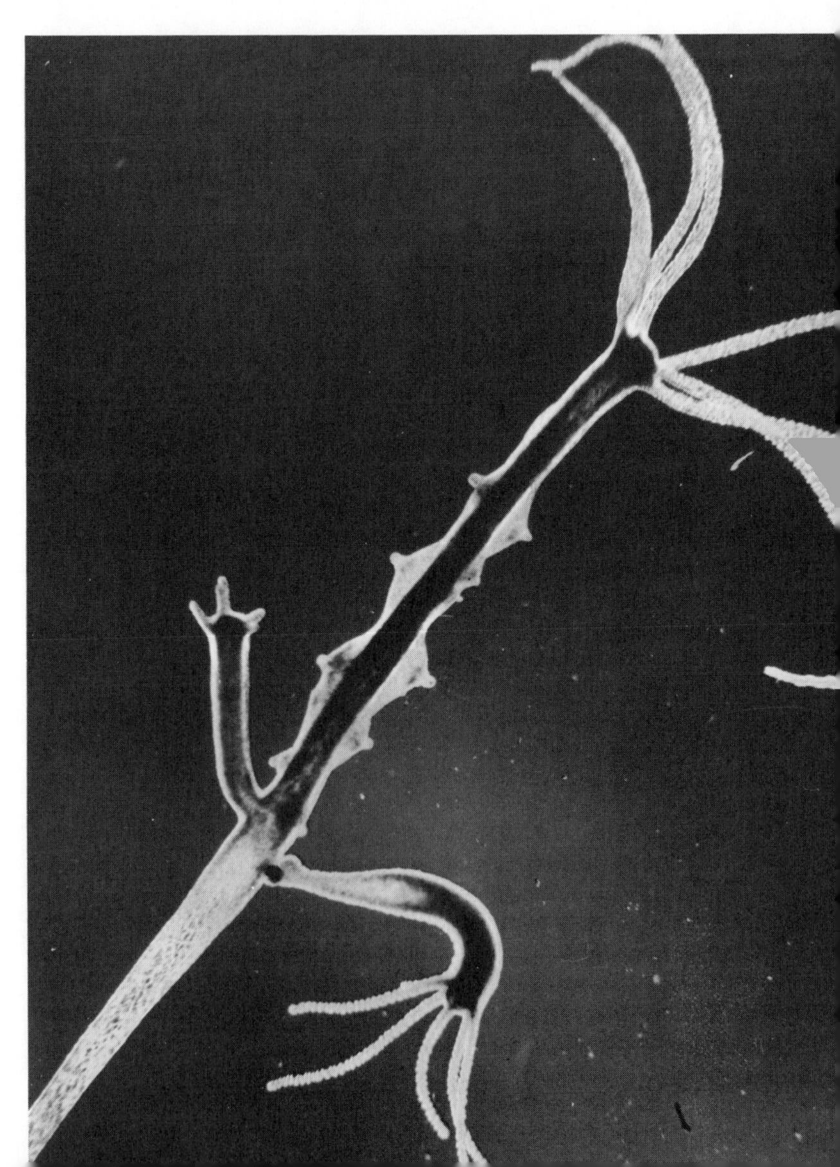

Male hydra with rows of testes along the body between the mouth end and the budding zone. This hydra has two buds, a younger one on the left and an older one on the right. (P. S. Tice)

An ovary also contains many interstitial cells at the start; but several of these fuse, all the nuclei but one degenerate, and the result is a single large ameboid cell. This cell finally incorporates the remaining yolk-filled interstitial cells and becomes the spherical **ripe egg,** packed with food reserves that will later nourish the developing embryo.

Just how the formation of sex cells is initiated is not definitely known. That one of the factors may be low temperature is suggested by the fact that some species of hydras will produce testes and ovaries if kept in a refrigerator for two or three weeks. Sometimes abundant food appears to stimulate the development of sexual maturity. Or, hydras become sexually active when the concentration of carbon dioxide increases, such as may occur naturally under conditions of stagnation and crowding. A hydra

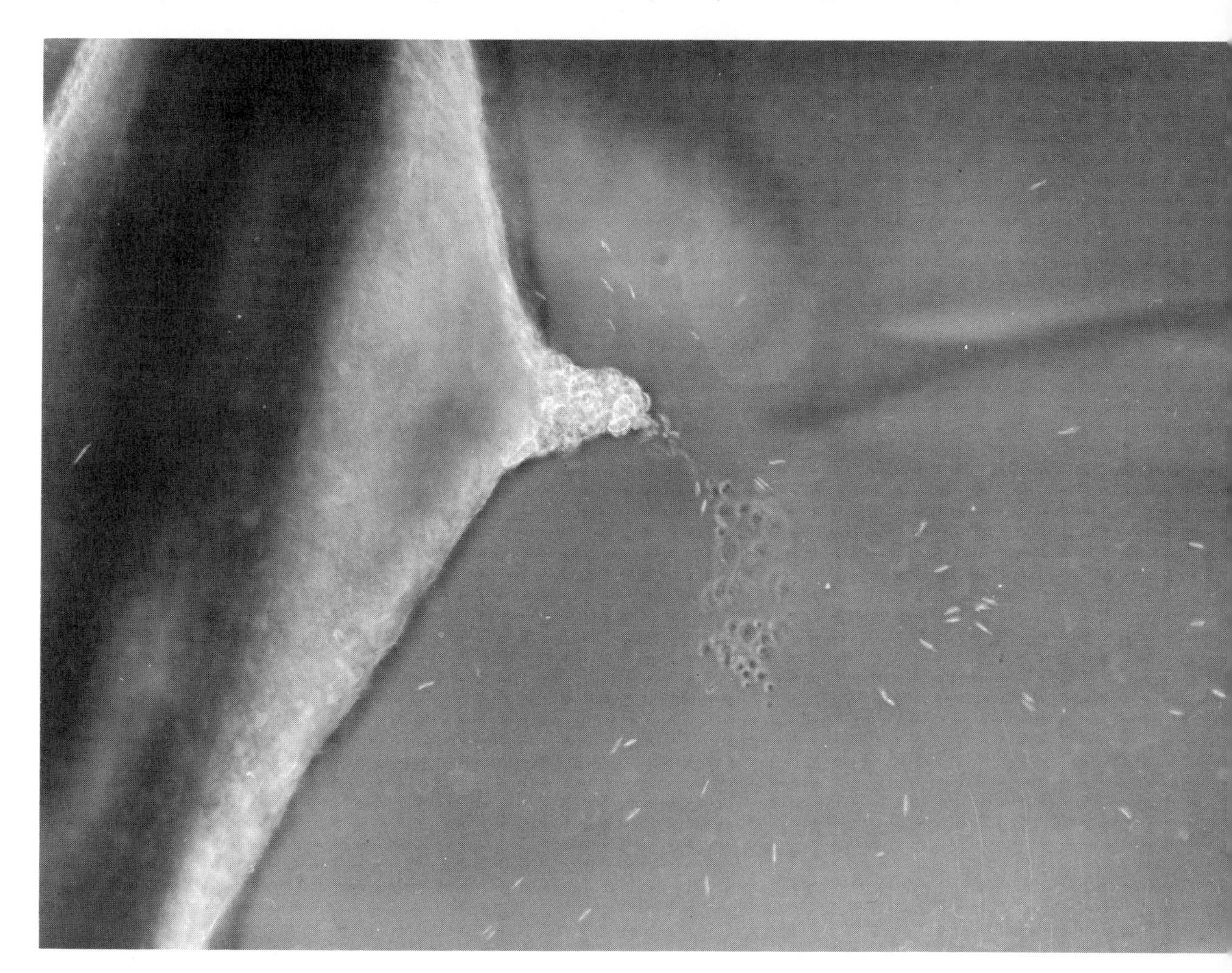

Sperms being discharged from a testis. The ectoderm is ruptured, liberating sperms that swim through the water, a few of them finally reaching female hydras bearing ripe eggs. (P. S. Tice)

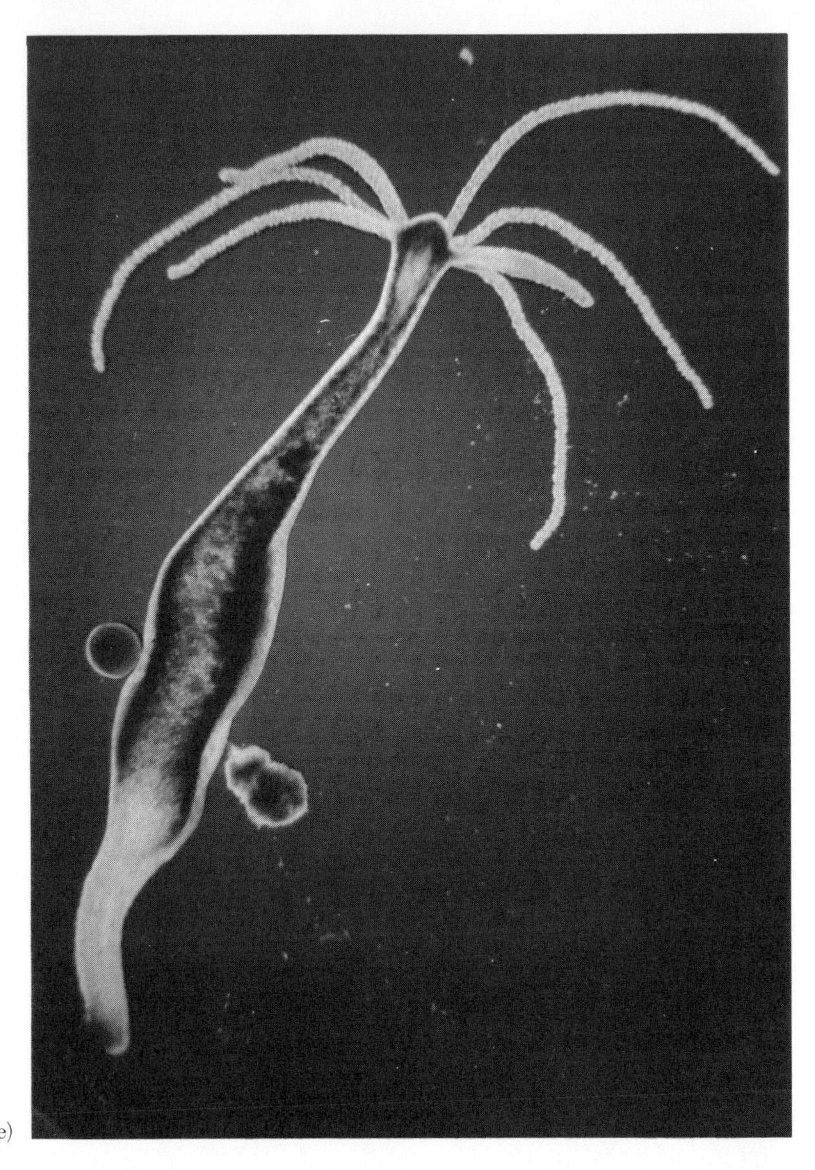

Female hydra with two eggs. One is extruded and ready for fertilization; the other has failed to become fertilized and is disintegrating. (P. S. Tice)

may also be induced to produce gametes by a small graft from another, already sexually active hydra, suggesting that hormones may be involved.

The ripe egg breaks through the covering ectoderm and projects with its outer surface freely exposed to the water. Sperms, discharged from a testis, swim through the water and surround the egg; one enters and **fertilization** occurs. If not fertilized within a short time after it is first exposed, the egg dies and distintegrates.

Development proceeds, as in sponges and in nearly all other many-celled animals, with repeated divisions of the fertilized egg, and there results a hollow blastula. In the blastula of a hydra, the cells composing the single layer now divide, and some

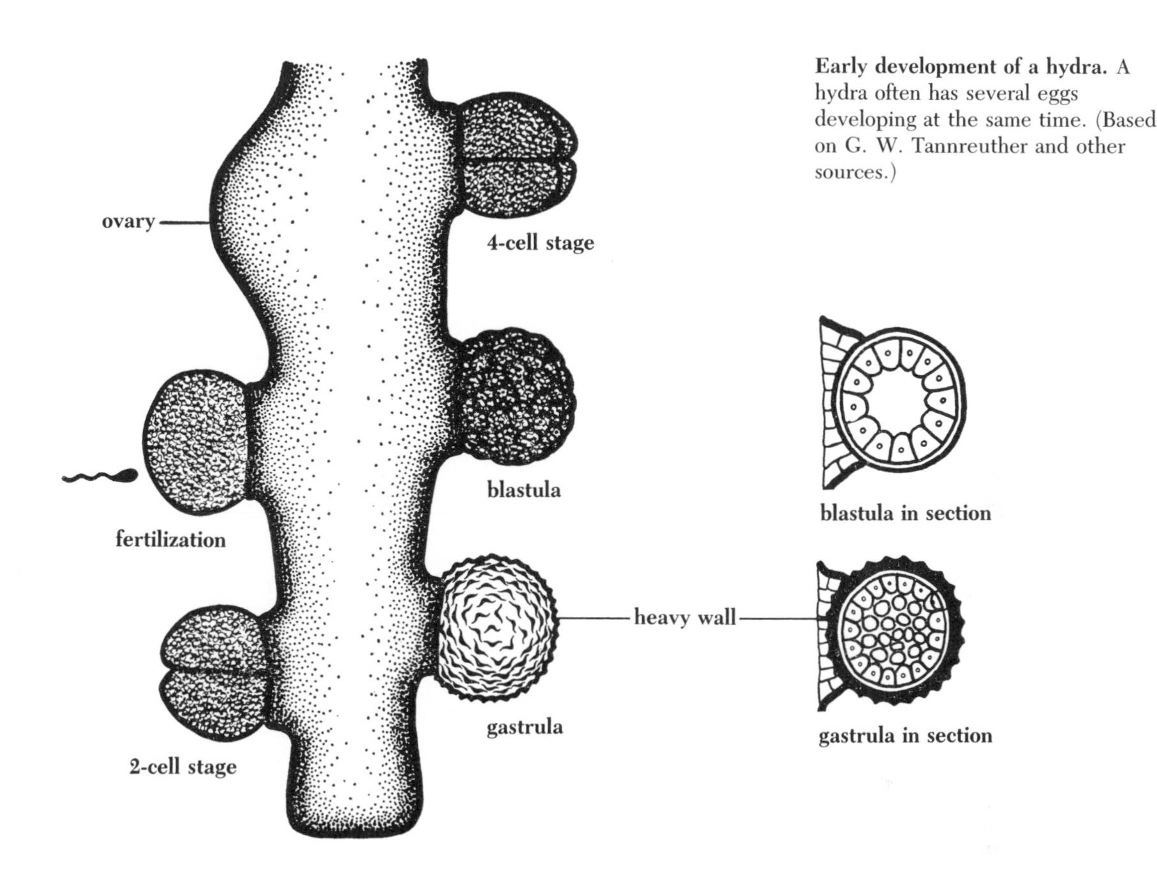

Early development of a hydra. A hydra often has several eggs developing at the same time. (Based on G. W. Tannreuther and other sources.)

of the surface cells migrate inward so that cells accumulate in the interior cavity, forming a two-layered stage, known as a **gastrula.** The outer layer of cells produces a heavy covering membrane, or shell. The gastrula usually drops from the parent and becomes fastened, by a sticky secretion, to the substrate. Under favorable circumstances a young hydra may hatch from the shell after a week or more; in winter the egg may lie dormant until the following spring. When development is resumed, the outer layer, or ectoderm, becomes differentiated into a protective epithelium. The inner mass of cells, or endoderm, becomes hollowed

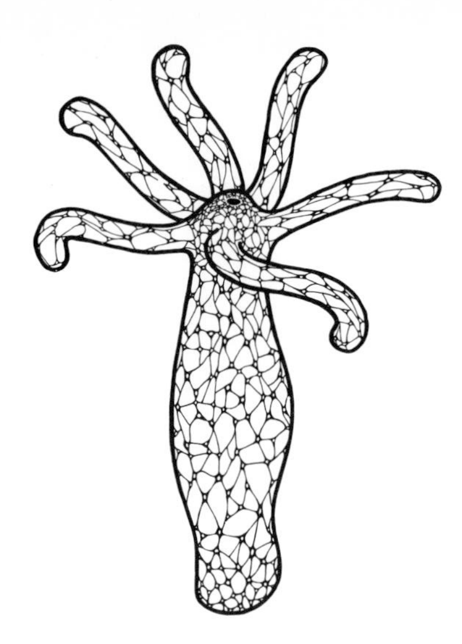

A nerve net of separate nerve cells extends throughout the ectoderm of a hydra.

out and differentiates as a digestive epithelium. Tentacles develop, a mouth breaks through, and the young hydra hatches out by rupture of the shell.

The **behavior** of a hydra is far more varied and complex than that of a sponge but perhaps not much more so than that of a complex protozoan. A hydra has all of the essentials of the kinds of **nervous mechanisms** present in any of the more complex many-celled animals, but often in a simpler, less developed form. There is a **nerve net** of separate nerve cells extending throughout the ectoderm of a hydra, as well as a smaller number of endodermal nerve cells. The ectodermal nerve net is more concentrated around the mouth and base than elsewhere, but there is little evidence of any specialized controlling group of nerve cells or brain such as characterizes the nervous systems of more complex animals. As the nerve net is composed of separate cells, a traveling nerve impulse, in passing from one nerve cell to another, must pass across definite gaps at the junctions between nerve endings. Such junctions, or *synapses*, are characteristic of the nervous systems of more complex animals also, and in these the junctions are usually so constituted that impulses can pass across them in one direction only. In cnidarians some synapses are of this type, but others may transmit impulses both ways, permitting diffuse conduction in all directions, with *few definite pathways*. Also, the impulses travel much more slowly than in the more specialized nerves of most other animals; but the nervous mechanisms of a hydra serve well enough for the limited activities of a small and sedentary animal. Even a single, simple nerve net allows for some flexibility of response. A weak stimulus applied to the tip of one of the tentacles initiates few nerve impulses and results only in the contraction of the single tentacle stimulated. A very strong stimulus to a tentacle tip initiates many impulses and causes the whole animal to contract.

Stimuli are received by slim, pointed **sensory cells,** which are sensitive to touch or to chemical substances in the water. Those of the ectoderm are more numerous and lie with their pointed ends projecting to the outside; those of the endoderm are sparser and have their pointed ends directed toward the digestive cavity. Impulses initiated by the sensory cells are passed on to the nerve cells, which lie at the bases of the epithelio-muscular cells, and are then transmitted to the muscle processes of these cells, to thread cells, and to gland cells.

Together, the sensory cells and nerve net **coordinate the activities** of a hydra, enabling an animal composed of many thousands of cells to react as one integrated individual. Moreover, they work together in such a way that the hydra's response is not

Hydras on a water plant, as a close and careful observer, with some luck, might find them in a suitable pond or lake. Because of their habit of budding, hydras may quickly become abundant under favorable circumstances. However, these hydras were reared at Loomis Laboratory. (R. Buchsbaum)

Feeding behavior (modified after H. M. Lenhoff). **1:** Hydra with mouth closed and tentacles quietly extended. **2:** Glutathione has been added to the water; the tentacles writhe above the mouth. **3:** The mouth begins to open and the tentacles bend toward it. **4:** The mouth opens widely, ready to receive prey.

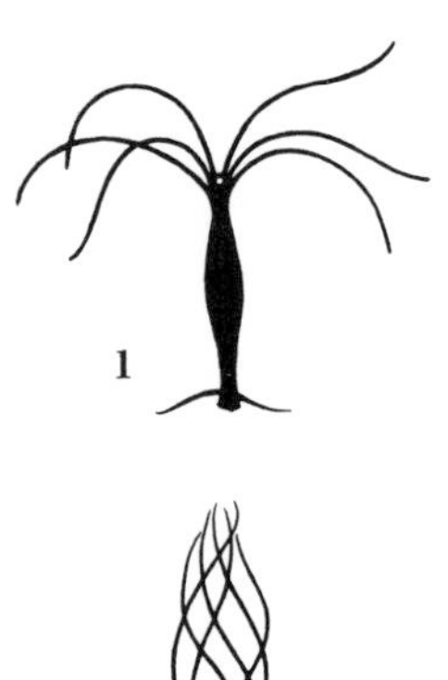

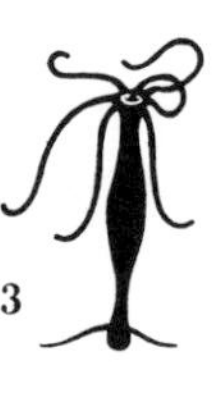

random but is usually *adaptive*, that is, the response results in a favorable adjustment of the animal to the changing circumstances of its environment. A strong mechanical stimulus, such as would cause the whole body to contract, might be delivered by a potential predator or a heavy object; shortening the hydra and getting it out of the way would be an adaptive response. A weak stimulus is more likely to represent the passing of some harmless animal or a bit of drifting vegetation, and the simple withdrawal of a tentacle is more appropriate. If a weak mechanical stimulus is combined with a chemical stimulus indicating the presence of food, as when a small prey animal touches a tentacle and is punctured by thread capsules, still more coordination is called for. The other tentacles join in grasping the prey and work together to cram it into the mouth, which has already opened in response to chemicals leaking from the prey. The nerve cells likewise coordinate the muscular contractions involved in swallowing food and, later, in ejecting indigestible remains.

The importance of coordinated activity is emphasized by what happens when it fails. Sometimes a hydra swallows its food so rapidly that it takes in one or more of its own tentacles, and a hydra has even been observed to swallow its own base along with the prey. Fortunately, it did not digest its own cells; and after a time the swallowed parts emerged, apparently uninjured.

Behavior varies with the physiological condition of the animal. A satiated hydra usually remains attached, with the tentacles quietly extended, and does not react to food when it is presented. It will exhibit a feeding response—opening the mouth and bending the tentacles toward it—only when an artificially high concentration of the chemical feeding stimulus (glutathione) is added to the water. When the hydra is ready to feed again, its behavior changes. At intervals the body and tentacles suddenly contract and then slowly extend in a new direction. Presumably, this increases the amount of territory controlled by the animal. If a food organism does not appear after some time, the tentacles begin to wave more actively, and the body contracts and expands in a new direction more frequently. A "hungry" hydra will exhibit a feeding reaction in the presence of a very small amount of glutathione, such as would be given off by animal prey. Eventually, if food is still not forthcoming, the hydra may move off to new hunting grounds.

The simplest method of **locomotion** in a hydra is a gliding on the base by means of a creeping ameboid movement of the basal cells. The most rapid method is a kind of somersaulting. The animal bends over, attaches its tentacles to the bottom by means of the adhesive thread capsules, loosens its base, swings the base

A hydra somersaulting. This is its most rapid method of locomotion.

over the mouth, and attaches it to the bottom; then it loosens the tentacles and repeats. In species of hydras in which the tentacles are two to five times longer than the body, the animals can move by throwing out the extended tentacles and catching hold of some object, then loosening the base and contracting the tentacles until the body is pulled up to the object—as if the animal were "chinning" itself. Hydras have also been observed floating free in the water, transported by water currents.

Hydras react to a variety of stimuli. A hydra will move away from a region in which the temperature rises above 25° C, or will move from the bottom of a dish to the top if carbon dioxide or products of decay accumulate. Many species tend to move toward lighted areas, where there are usually more food organisms. Such movements are executed by a kind of trial-and-error procedure, but almost always in a direction likely to lead to the continued existence of the animal. Eventually the hydra settles on a new site, and its patient life of trapping and digesting is resumed.

Obelia

The nearest marine relatives to the hydras are cnidarians known as **hydroids,** which are usually seen as delicate plantlike growths on kelps, rocks, and wharf pilings along seacoasts. Some of the commonest of these belong to the genus ***Obelia,*** colonies of which often reach several centimeters in height. The colony arises by budding from a single hydralike individual. The buds fail to separate; and after repeated budding, there results a branching growth, permanently fastened to some object and consisting of numerous members united by stems. As with sponges, one may ask whether an *individual* is represented by the whole colony or by each bud. Because the activities of the member buds are subordinate to the colony as a whole, they are sometimes thought of as *subindividuals.* Each is referred to as a **polyp,** a name applied to any tubular cnidarian that bears a whorl of tentacles around the free end of the body and is attached at the other end.

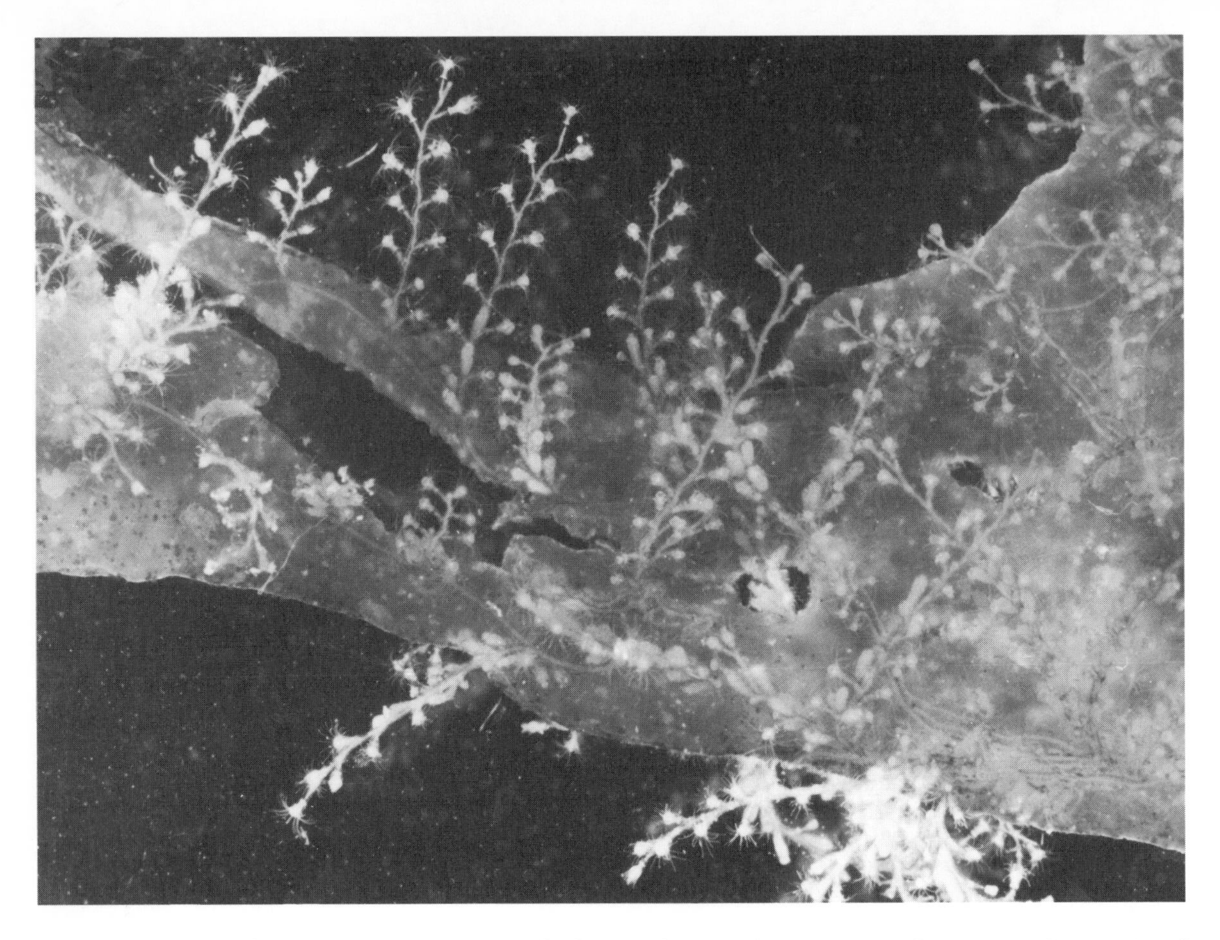

Hydroid colonies, *Obelia,* growing on seaweed. The upright stems and the horizontal stolons are both visible. Central California. (R. Buchsbaum)

A hydra is also a polyp, a name that means "many feet." Polyps use their "many feet" chiefly for feeding and only occasionally for moving about. Hence, the name is not particularly appropriate but comes to cnidarians by an indirect route. It was derived from *poulpe,* the French word for octopus, because an early French naturalist thought that cnidarian tentacles resembled the "feet" of an octopus.

The polyps and stems of *Obelia* are protected and held erect by a chitinous **horny covering,** secreted by the ectoderm. This covering encloses all the stems and extends around each polyp as a transparent cup. When irritated, a polyp can withdraw into its cup; and the rapid contraction and slow expansion of the polyps are about the only movements that can be seen in a colony of *Obelia.* The stems are unable to move because of the rigidity of the covering; but at certain points the covering is arranged in rings that allow for flexibility as the stems are swayed by water currents.

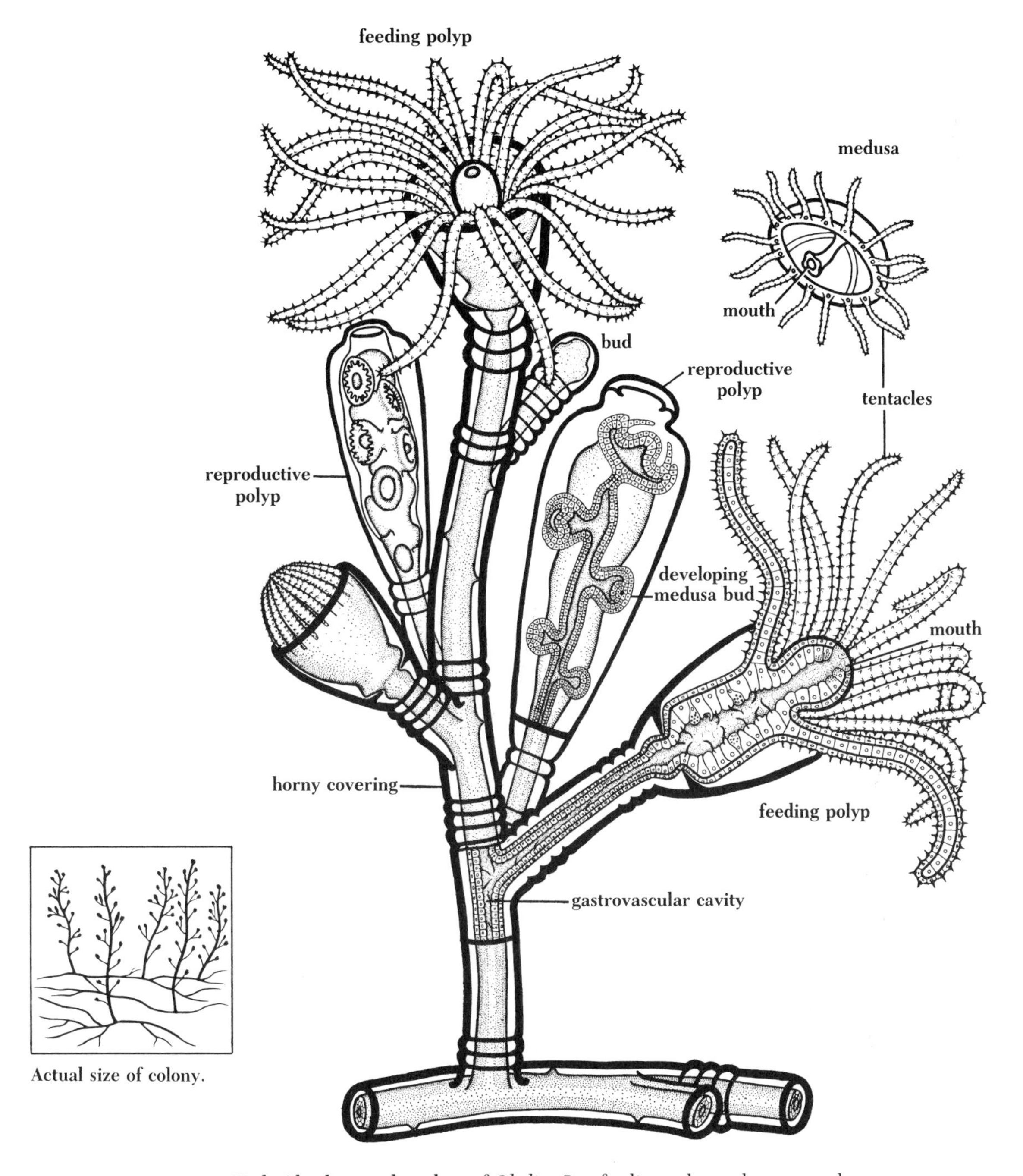

Hydroid colony and medusa of *Obelia.* One feeding polyp and one reproductive polyp have been drawn in section, showing the two layers of cells characteristic of cnidarians.

A hydroid polyp is built on the same plan as a hydra and consists of the same two cell layers, ectoderm and endoderm. These are composed of cell types similar to those of a hydra.

Feeding by polyps of *Obelia* is like that of hydras. They capture small prey by means of tentacles armed with thread capsules. The tentacles are not hollow, as are those of a hydra, but are solid, having a central core of large endodermal cells. The polyps and stems are hollow, and the **gastrovascular cavity** of every polyp is continuous with that of every other polyp in the colony. The food is partly digested in the cavity of the polyp, and the resulting fluid is circulated about through the stems by the beating of the flagella of the digestive epithelium. Thus, food is distributed throughout the colony in thoroughly cooperative fashion, and digestion is completed in food vacuoles within the cells lining the gastrovascular cavity.

Asexual reproduction by budding is the usual method of increasing the number of polyps. In addition to the continuous budding on any vertical axis, horizontal stolons sprout from the base and grow over the substrate, like the runners of a strawberry plant. The stolons give rise to a series of upright stems, so that the entire colony may, after a time, consist of hundreds of polyps.

If we examine older colonies carefully, we see that the polyps are not all alike. Those we have already described have tentacles with which they catch prey, and these may be called **feeding polyps.** In some of the angles where feeding polyps branch from a stem, there occur polyps that have no tentacles and cannot feed but are nourished through the activities of the feeding polyps. These are called **reproductive polyps,** but we could search them in vain throughout the year for any signs of testes or ovaries; the polyps of *Obelia*, unlike hydras, never produce gametes. The reproductive polyps are specialized instead for **asexual reproduction** of a particular type. Each is enclosed by a transparent, vase-shaped covering and consists of a stalk on which are borne little saucerlike buds, the largest and most completely developed near the top, the smallest and least developed near the base. If live colonies are kept in a dish of seawater for a time and observed, it is possible to watch the topmost "saucer" escape through an opening at the upper end of the vase-shaped covering and swim about as a tiny form called a **medusa,** a name applied to any free-swimming "jellyfish" type of cnidarian. (The name is derived from a fancied resemblance of the waving tentacles of jellyfishes to the snaky tresses of the mythical Gorgon Medusa.)

The **medusa of *Obelia*** looks like a tiny bell-shaped piece of clear jelly. From the middle of the under surface, where one expects to find the clapper of a bell, hangs a tube that bears, at

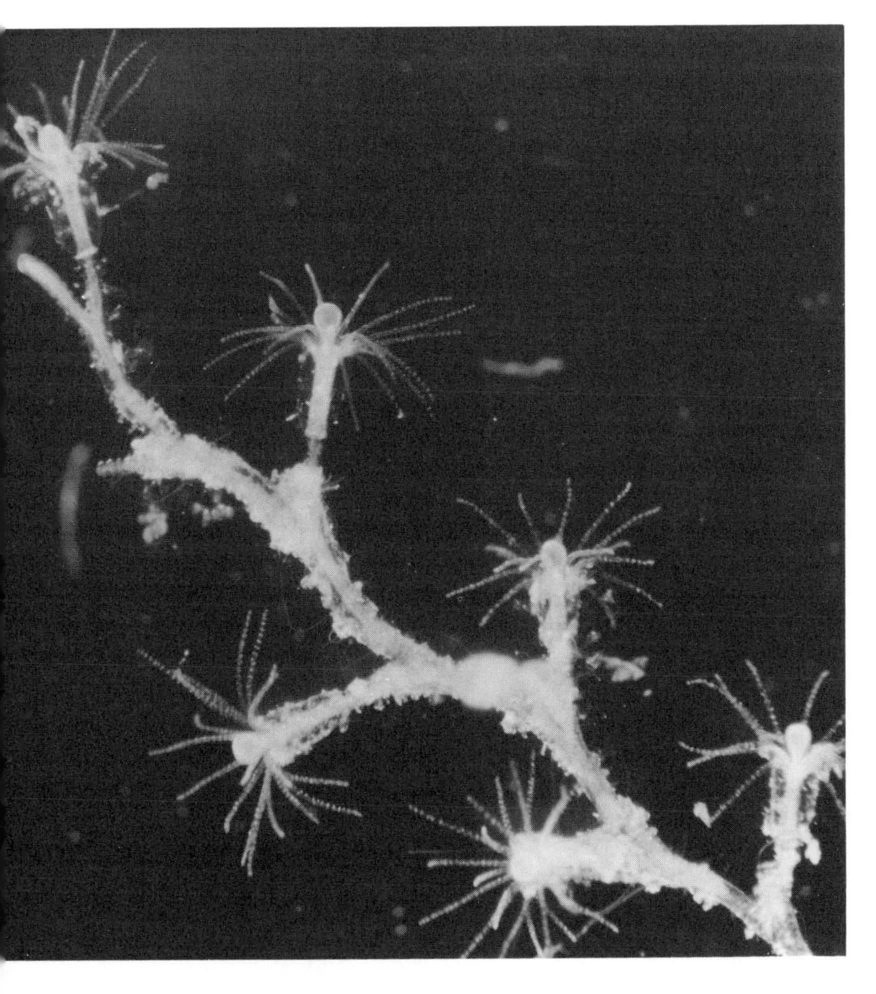

Feeding polyps of *Obelia,* each with the extended tentacles radiating about the rounded mouth end. (R. Buchsbaum)

Medusa of *Obelia*, photographed with the bell expanded *(above)* and contracted *(below)*, is only about 1 mm in diameter. This medusa was set free when a piece from a polyp colony was placed in a drop of seawater on a glass slide. The number of tentacles increases as the medusa grows. (R. Buchsbaum)

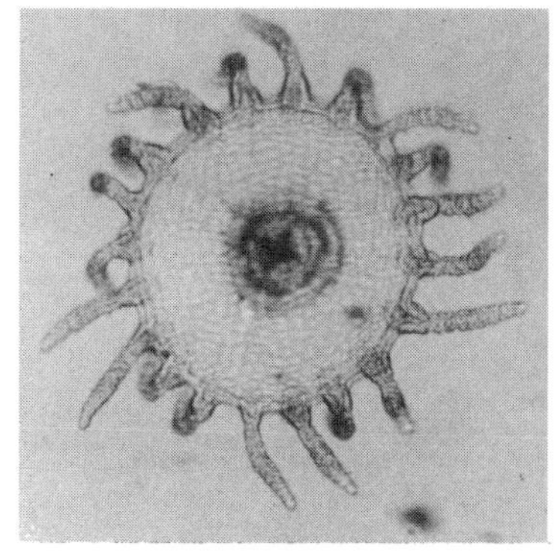

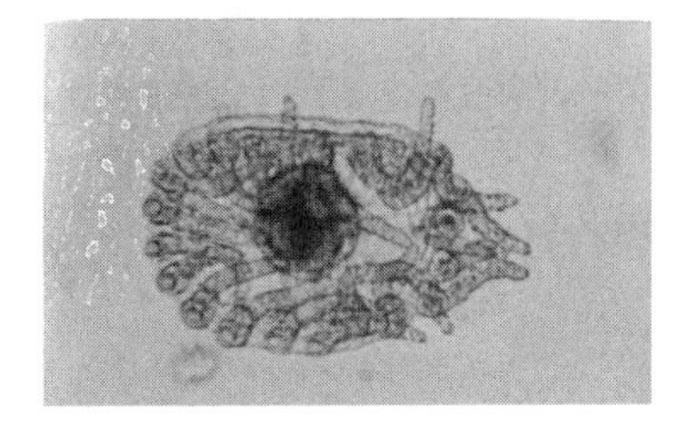

its free end, the **mouth.** The mouth leads up this hollow tube into the **gastrovascular cavity,** which branches out into canals that carry food to all parts of the medusa. From the margin of the bell hangs a circlet of **tentacles,** well armed with thread capsules. The tiny animal swims by alternately contracting and relaxing the muscle cells of the umbrella-like **bell.** As it swims or drifts with the current, the trailing tentacles catch small organisms.

The primary function of the medusa is **sexual reproduction.** From the under side of the bell hang four **gonads.** In female medusas they are ovaries and produce eggs; in male medusas they are testes and produce sperms. Eggs and sperms are shed into the seawater, where fertilization takes place. The zygote develops into a free-swimming larva called a **planula,** with ectoderm and endoderm. The beating cilia of the ectoderm propel

Obelia, life history. Sperm and egg come from separate male and female medusas.

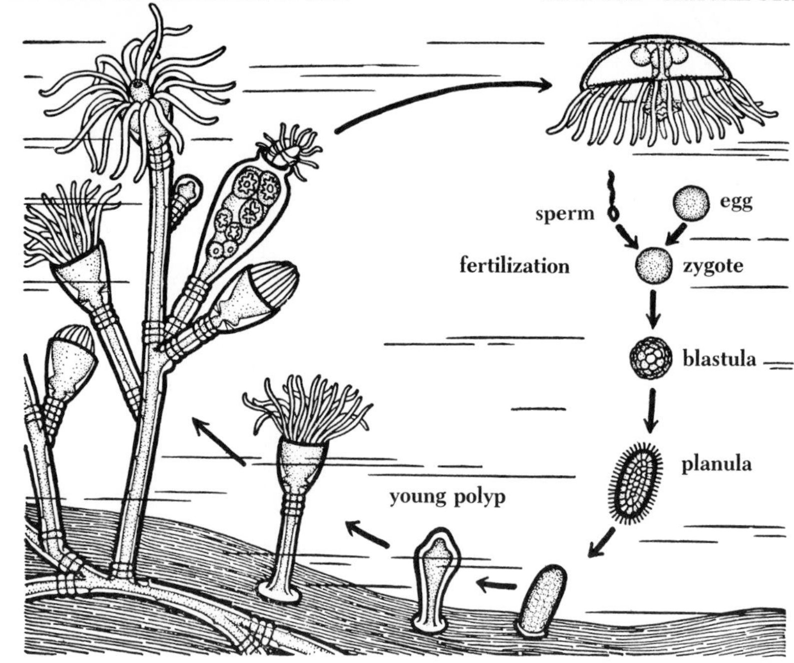

Planula is the larval type typical of cnidarians. The outer ectodermal and inner endodermal layers are conspicuous in this one. Ciliated planulas may swim or may creep along the bottom. Central California. (Photomicrograph by R. Buchsbaum.)

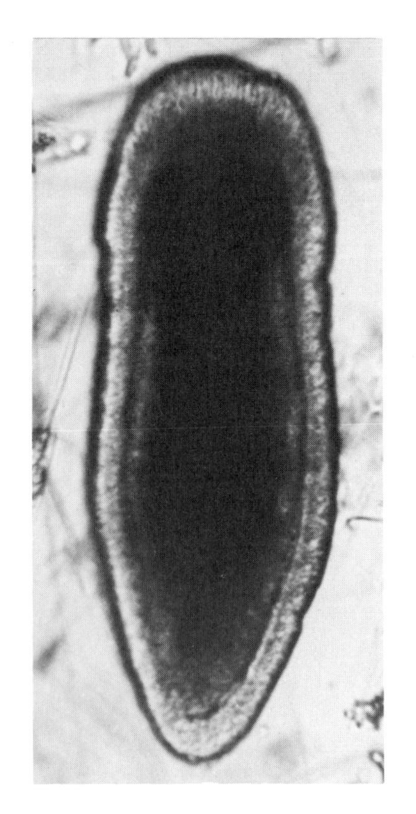

the planula through the water. It swims about for a time, then settles on a rock or on a piece of kelp, becomes fastened at one end, produces tentacles and a mouth at the other end, and develops into a polyp that, by asexual budding, finally gives rise to a new colony of sessile polyps. The free-swimming larva and medusa both serve as a means of spreading to new localities.

Although a medusa seems very different from a polyp in general appearance, the two forms are really similar in construction. Both are composed of ectoderm and endoderm, and both consist of similar types of cells. In the medusa, ectoderm covers the entire surface of the bell and the tentacles, while endoderm lines the various parts of the gastrovascular cavity. One difference from the polyp is the greater thickness of the jellylike mesoglea between the two layers of cells. In the absence of supporting structures like the spicules of sponges or the connective tissue of other animals, the mesoglea gives a firm consistency to the otherwise fragile body. Correlated with its greater locomotor activity, the medusa has a more highly developed **nerve net** and a marginal **nerve ring** that acts as a controlling center. There are also more specialized sensory cells. The polyp and medusa may be regarded as having adapted the same general pattern to two different ways of life—attached and free-swimming.

The occurrence of cnidarians in two forms, the medusa and the polyp, is a phenomenon termed **polymorphism** ("many

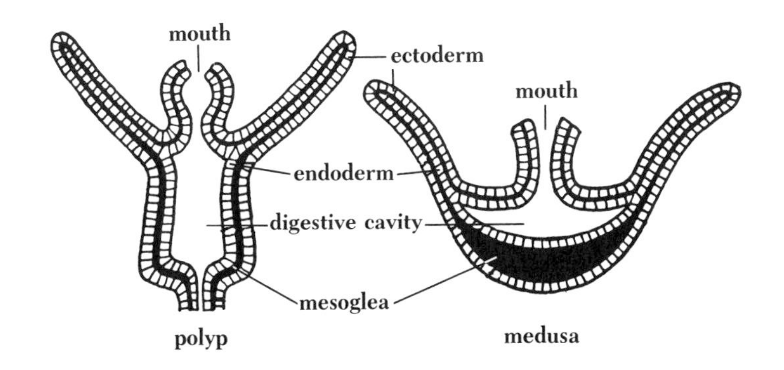

Polyp and medusa are constructed on the same general plan.

forms"). Polymorphism is not unique to cnidarians, for many other groups show structural differentiation of individuals—of the same or different parts of the life history—that fits them for different roles in the life of the species. *Obelia* may be said to consist of three kinds of individuals, each with its own functions: a sexually reproductive medusa; a feeding polyp that snares prey and buds off new polyps; and a nonfeeding reproductive polyp that buds off medusas. The work that in most other animals is done by every individual of the species is performed in these cnidarians by different kinds of individuals. In some colonial cnidarians, still other kinds of individuals are found, such as stinging polyps that do not feed or reproduce but are heavily armed with thread capsules.

Cnidarians have only **two layers of cells**—ectoderm and endoderm. Their most characteristic structures are the **thread capsules** with which they catch their prey; these are produced by no other phylum of animals. Because they have a nervous network, and because their cells act together in more coordinated fashion than do the cells of sponges, the cnidarians are said to have reached the **tissue level of organization.** Because very few animals living today illustrate this simple level of construction, the cnidarians are described early in our story; but from the predatory way of life of all cnidarians, we would have to guess that swimming prey, probably belonging to more complex phyla, were already present at the time that cnidarians evolved. The variations among cnidarians result chiefly from a shifting emphasis in the evolution of the several groups—first on the medusa stage, and then on the polyp stage of the life history. Some, such as *Obelia*, have polyp and medusa almost equally developed. In others the medusa stage is less prominent or wholly lacking, as in hydras. And in still others, large and well-developed medusas predominate while the attached polyp stage is very brief or small or is absent. Some of these variations are described in the next chapter.

Chapter 7

Stinging-Celled Animals

Now that television routinely brings colorful images of tropical seas into our living rooms, millions of viewers who live far from seacoasts are more familiar with pulsating jellyfishes, flower-shaped sea anemones, latticed sea fans, and massive coral domes than with many of the drab little invertebrates that live in their own gardens. On the television screen, divers brush past tall sea plumes swaying with the currents. They poke at fleshy masses of lobed soft corals or steady themselves by grabbing a convenient branch of hard coral. But all of these are treated by both camera and narrator as little more than a colorful seascape, and the zoom lens and script quickly close in on a variety of small brilliant fishes or an octopus eating a crab. Even those viewers who are stimulated to seek a first-hand view, and who go equipped with snorkel or scuba gear into warm seas, often fail to realize that most of the seascape is composed of living cnidarians. Only those who have paused first to learn something about the many kinds of stinging-celled animals are prepared to understand what they see underwater. Though many cnidarians look like flowers or fragile bells of jelly, these animals are carnivores, well equipped to sting, capture, and swallow their prey.

Opposite: **Sea anemones** may seem to be only lovely parts of the seascape until you stop to notice their activities. The ones with a fluffy crown of numerous delicate tentacles *(Metridium)* catch mostly minute organisms instead of the larger prey such as fishes on which other anemones, like the one on the lower right, feed. If disturbed, these expanded anemones will quickly contract down into little mounds. Or they may seek another spot; one was observed to move half a meter in 24 hours. Helgoland Aquarium. (F. Schensky)

Cnidarians occur in two forms: polyps and medusas. In the last chapter we introduced the cnidarians with the hydras, which are polyps, and with the hydroid *Obelia,* which has a small short-lived medusa and a more conspicuous polyp phase. Yet, after comparing the great diversity of cnidarians, biologists have concluded that the ancestral cnidarian was probably a medusalike form. This medusa would produce other medusas directly from eggs and sperms, by way of a swimming larval stage resembling a polyp. Eventually this larval polyplike stage became more and more important in the life history and took on an independent existence. At first the polyps were only a young transitory stage in the development of the medusa; they were incapable of sexual reproduction and would grow up into medusas, or bud off medusas, which produced eggs and sperms. But some cnidarian polyps eventually developed the capacity for forming gametes and dropped out the medusa stage altogether, as we see not only in hydras but also in sea anemones and corals and their many marine relatives.

A cnidarian that follows a pattern close to the presumed ancestral ones is ***Gonionemus***, which has a well-developed medusa and a simple inconspicuous polyp. The gelatinous bowl-shaped bell of the medusa of *Gonionemus* has a convex outer surface and a concave under surface. From the center of the under surface hangs a tube, the **manubrium**, with the **mouth** at its tip. The other end of the manubrium leads into four **radial canals** that traverse the mesoglea to the margin of the bell. There they join a **ring canal**

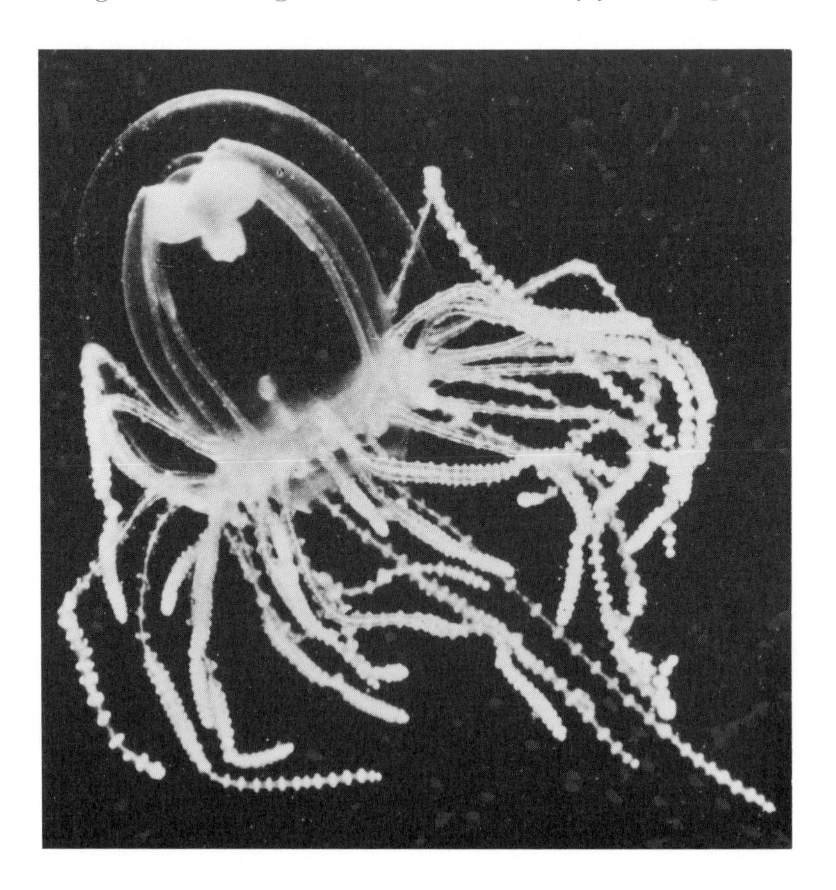

A hydrozoan jellyfish, *Gonionemus*, common in shallow marine waters. England. (D. P. Wilson)

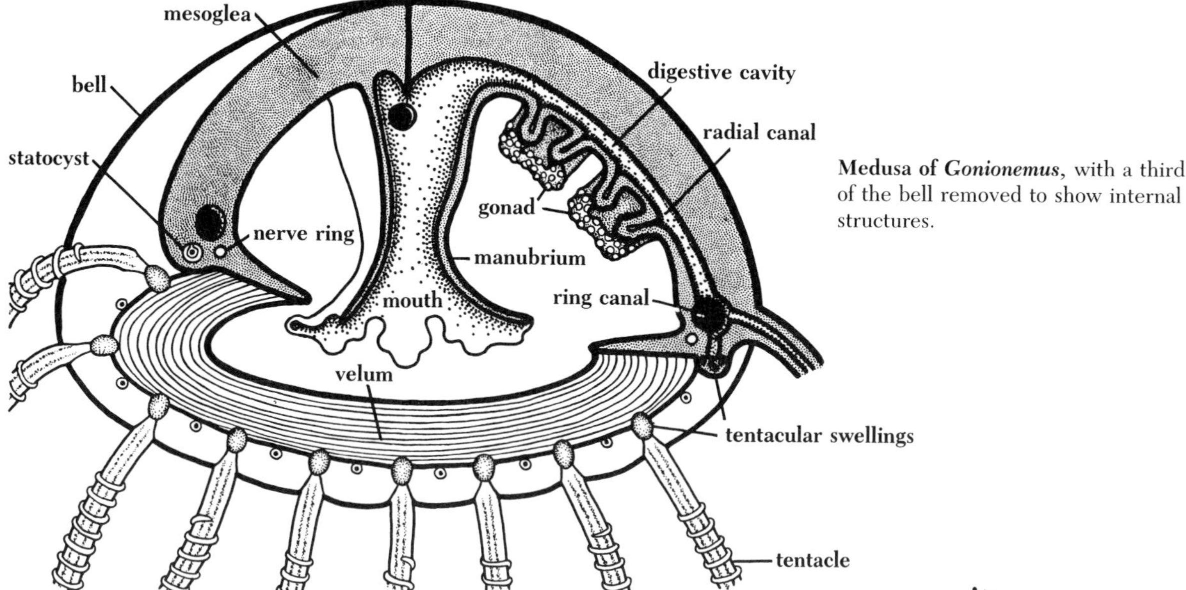

Medusa of *Gonionemus*, with a third of the bell removed to show internal structures.

that runs around that margin and connects with the cavities of the hollow tentacles. This continuous cavity through manubrium, radial canals, ring canal, and tentacles is the **gastrovascular cavity;** it distributes partly digested food to all regions of the body.

The medusa **swims** slowly by rhythmic pulsations of the bell. From the margin of the bell, a muscular shelf, the **velum,** projects inward. The contraction of muscle fibers in the velum and bell forces water out from the concavity of the bell and jet-propels the animal in the direction opposite to that in which the water is expelled. Between contractions the elasticity of the mesoglea restores the original shape of the bell.

The medusa of *Gonionemus* **feeds** while actively swimming about or by a kind of "fishing" technique. The medusa swims upward, turns over on reaching the surface of the water, and then floats slowly downward with the bell inverted and the tentacles spread out in a wide snare, from which passing worms, shrimps, or small fishes seldom escape. When at rest, the medusa attaches to the bottom or to vegetation by adhesive pads near the tips of the tentacles.

The free-swimming habit of *Gonionemus* requires greater activity and a more elaborate nervous mechanism than that of sedentary polyps like hydras and hydroids. The **nerve network** that runs beneath the ectoderm of the under surface of the bell is concentrated around the margin of the bell into a **nerve ring,** which controls and coordinates the animal's behavior. Specialized **sensory structures** occur around the bell margin. Embed-

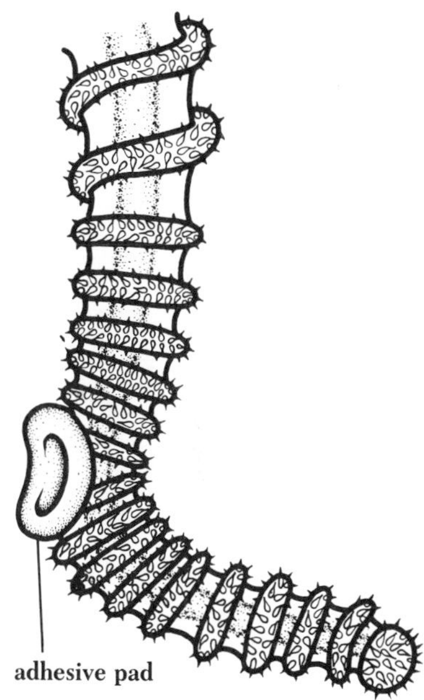

End of **tentacle of *Gonionemus*** showing adhesive pad and batteries of stinging thread capsules.

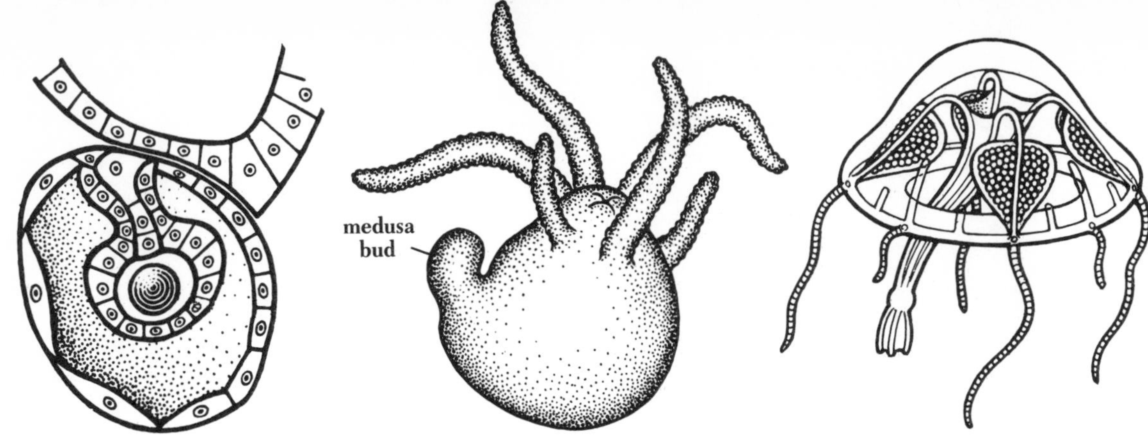

Statocyst of *Gonionemus.* Motion of the hard mass within its sac gives information about the movements of the medusa and about its position with respect to gravity. (Modified after L. J. Thomas and after H. Joseph.)

Polyp of *Gonionemus* with medusa bud. (Based on H. Joseph.)

***Liriope*,** a medusa that develops directly from a free-swimming larva without a fixed polyp stage. (After A. G. Mayer.)

ded in the mesoglea between the bases of the tentacles are **statocysts,** each a small sac containing a hard mass. As the medusa swims about, motions of the mass within the sac act as stimuli that modify the swimming movements. In addition, there are prominent swellings at the bases of the tentacles, abundantly supplied with sensory cells and lined with pigment. These tentacular swellings may be special light receptors, although the whole epithelium of the lower surface is generally sensitive to light. The swellings are a chief site for the formation of thread capsules, which from there migrate out along the tentacles and take their places in the stinging batteries.

The **ovaries** and **testes** occur on separate female or male individuals and appear as folded ribbons that hang from beneath the four radial canals. The eggs or sperms break through the surface ectoderm and are shed directly into the water, where fertilization takes place. The zygote divides many times to form a ciliated **planula.** The planula swims about for a time, then finally settles down, loses its cilia, and develops an internal cavity. A mouth breaks through at the unattached end, tentacles push out around the mouth, and the young polyp of *Gonionemus* soon resembles a minute, squat hydra. The polyp feeds and buds off little creeping larvas, which also become feeding polyps. Finally,the polyps begin to produce buds of a different kind, which detach and develop into adult medusas.

The life history of *Gonionemus* resembles that of *Obelia*, with the emphasis shifted to the medusa. Many cnidarians closely related to *Gonionemus* have an adult medusa and no attached polyp at all. One of these, *Liriope*, has a larva that sprouts tentacles before the bell develops and that looks like a free-swimming polyp. If this larva were to settle down and become attached, it would resemble the minute polyp stage of *Gonionemus*. In some hydroids this juvenile fixed polyp stage has been elaborated into a relatively large and flourishing colony, while the medusa is very small, as in *Obelia*. In other hydroids the medusa stage has been reduced still further. The medusa buds of *Hydractinia*, for example, begin to develop in the usual way but never have any medusalike features and fail to detach from the colony. They are reduced saclike structures that shed eggs and sperms into the water. The final step in this direction is the complete elimination of the medusa—a condition illustrated by hydras.

In *Obelia* we saw division of labor not only in the life history but also in the composition of the polyp colony. *Hydractinia* car-

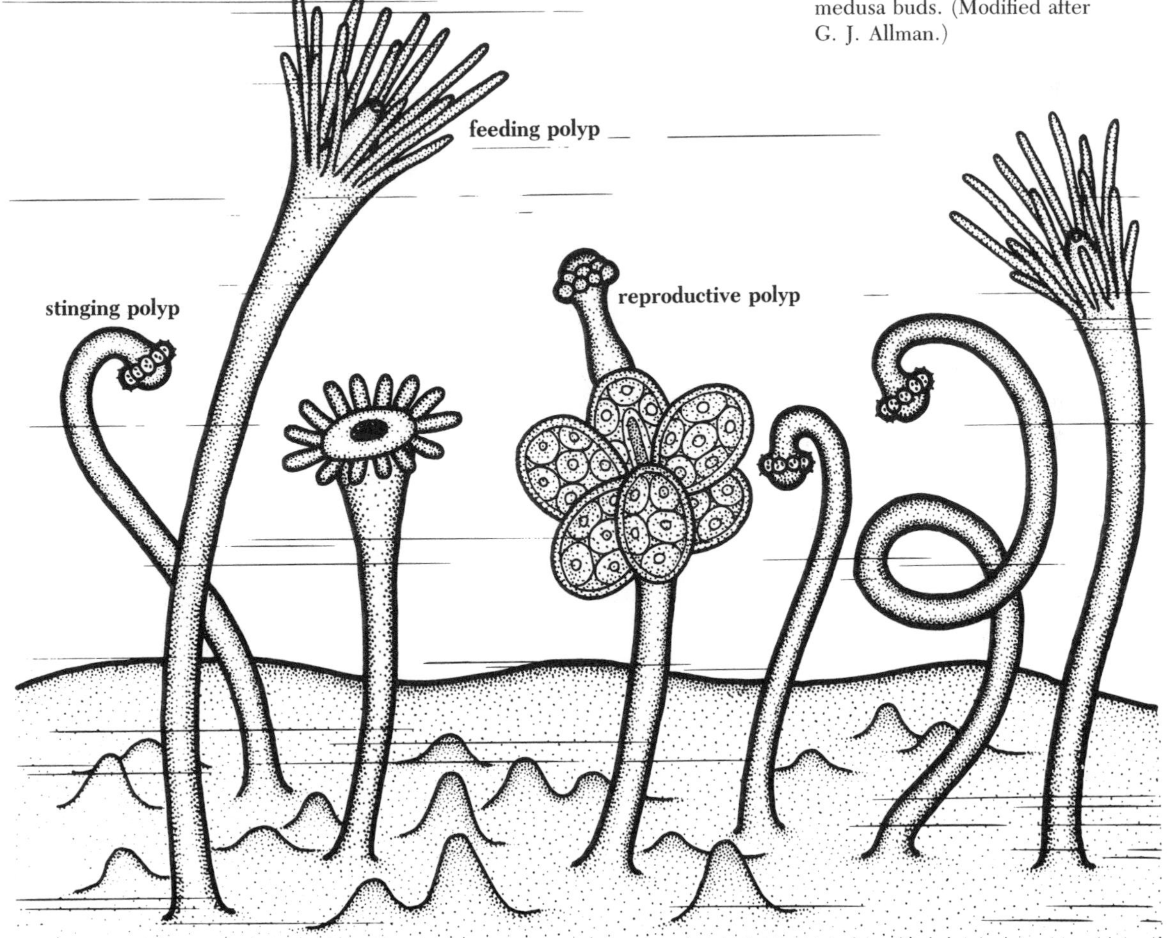

Hydractinia, a polyp colony that shows polymorphism and reduced medusa buds. (Modified after G. J. Allman.)

ries this **polymorphism** a step further. The colony has *feeding polyps*, with mouth and long tentacles; *reproductive polyps*, which have only knoblike tentacles and cannot feed but bear the medusa buds; and *stinging polyps*, which have knoblike tentacles and cannot feed but are richly supplied with thread capsules. The stinging polyps capture food and transfer it to the feeding polyps and also help to protect the colony.

It is among the **siphonophores,** however, that we find the extremes to which colonial organization can be carried. These complex floating colonies have not only more than one kind of polyp but also more than one kind of medusa—and all of these simultaneously. In addition to the sexual medusas (either free or attached), there may be numerous modified medusas called swimming bells, which cannot feed or reproduce but serve to propel the colony. *Physalia*, often called the Portuguese man-of-war, has no swimming bells and is driven about by the action of the wind on its crested float. From the underside of the float there hang down into the water several kinds of specialized polyps, clusters of attached medusas, and tangles of long tentacles that may reach a length of 20 meters or more and are armed with

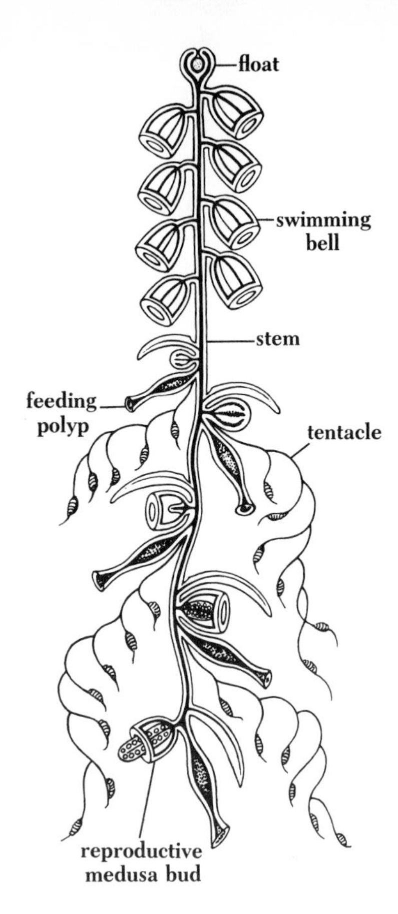

Siphonophore is a floating, polymorphic colony. A continuous gastrovascular cavity runs the length of the stem, which forms the axis of the colony, and connects all the members. (Mostly after C. Chun.)

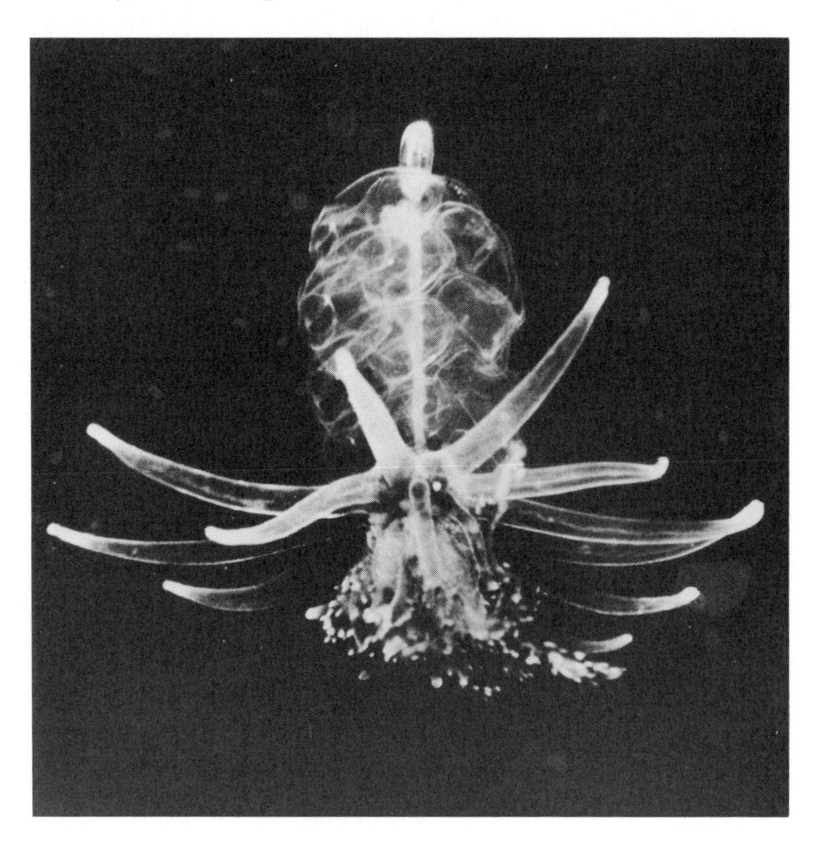

Siphonophore, *Physophora*. Beneath the float that tops this colony is a cluster of swimming bells (modified medusas) and beneath these are feeding, reproductive, and stinging polyps. Monterey Bay, California. (D. Wobber)

Portuguese man-of-war, *Physalia,* is the most notorious of the siphonophores. The colony shown here has just caught a fish. About 1/2 natural size. (Courtesy of the New York Zoological Society.)

Portion of a tentacle of *Physalia* seen through a microscope and enlarged 180 times. The bottom half of the microscope field is occupied by the edge of the tentacle, whose surface appears to be almost solid with undischarged, and a few discharged, thread capsules of several kinds. The largest ones clearly show the hollow coiled thread within. Above are a number of discharged capsules with long hollow threads extending out of the field. The unusual size of these capsules, and the great length of their threads, suggests why this siphonophore is among the most feared cnidarians. Plymouth, England. (D. P. Wilson)

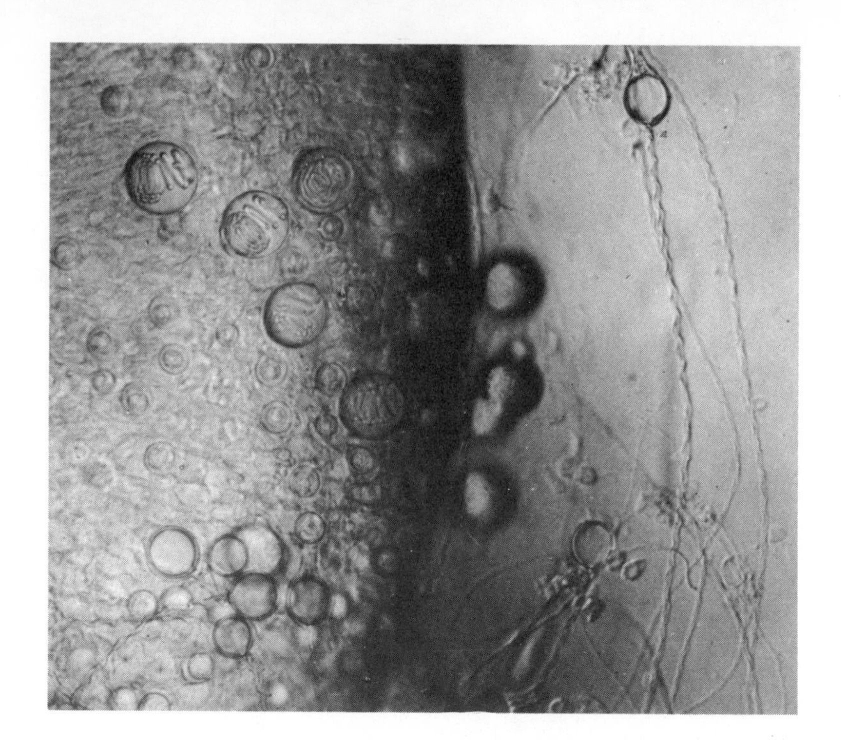

especially large and powerful thread capsules that can readily paralyze a large fish. The vivid blue float is a familiar and a beautiful sight on the surface of warm seas all over the world—but it is not a welcome one to swimmers who know that the trailing tentacles can inflict serious and sometimes fatal injury to humans.

In the siphonophores it is frequently difficult to tell where one polyp ends and the next begins. In some types the subindividuals of the colony seem not to be single members but small groups of polypoid and medusoid members that are regularly set free to live independent of the parent colony. On the other hand, many siphonophores possess a well-developed colonial nervous system, and the colony members act together in a highly coordinated fashion, perhaps less like members of a colony than like the organs of a single animal.

The phylum Cnidaria is divided into three classes. All of the cnidarians mentioned thus far belong to the class **Hydrozoa** ("hydralike animals"). Most hydrozoans are polyp colonies that give rise to free medusas *(Obelia)* or to reduced, attached medusas *(Hydractinia)*. But there are exceptions, such as *Hydra* and *Gonionemus*. Both polyps and medusas are small, delicate, and much simpler in structure than the members of the other two classes.

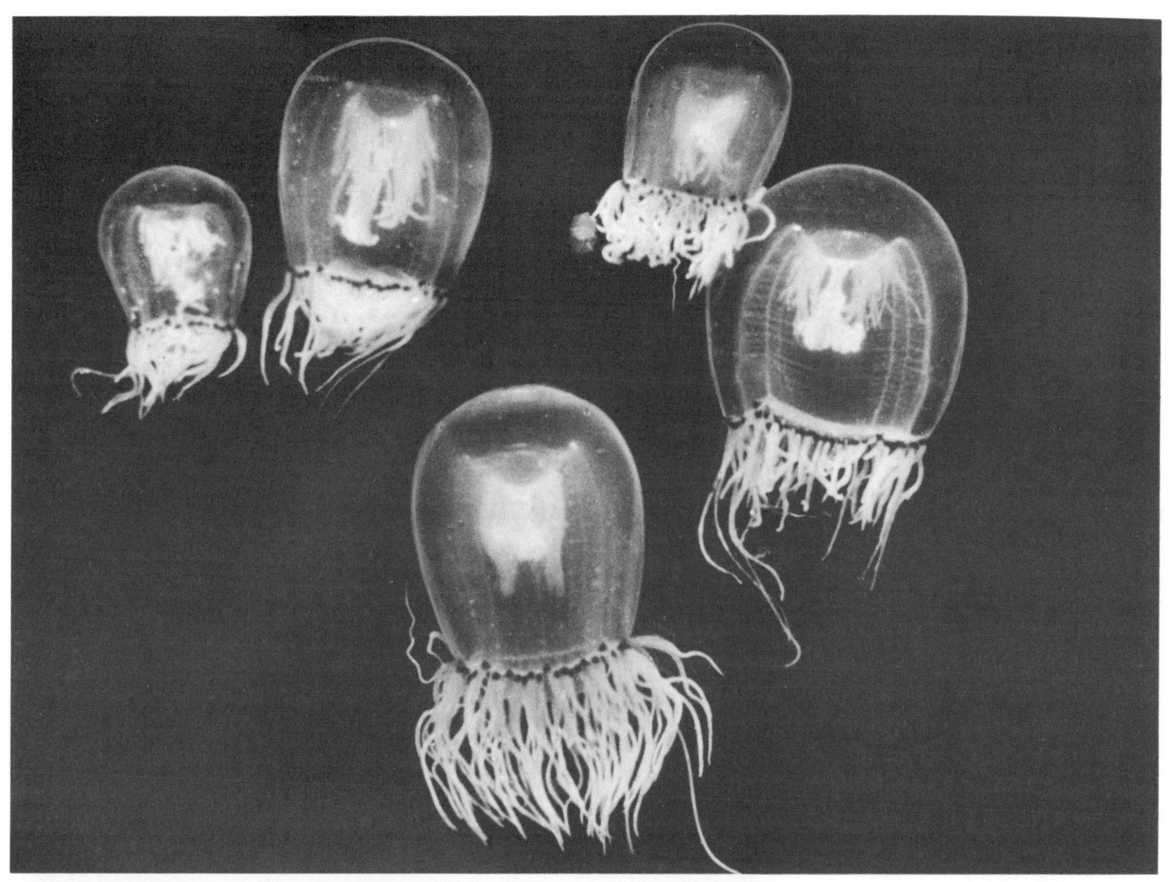

Polyorchis, a hydrozoan jellyfish common in bays along the West Coast of North America. The mouth is at the end of the long trumpet-shaped manubrium, surrounded by the gonads, which hang like bunches of string. At the base of each tentacle is a dark eyespot ringed with red at the base. The tiny, obscure polyps were only recently discovered. About natural size. Monterey Bay, California. (R. Buchsbaum)

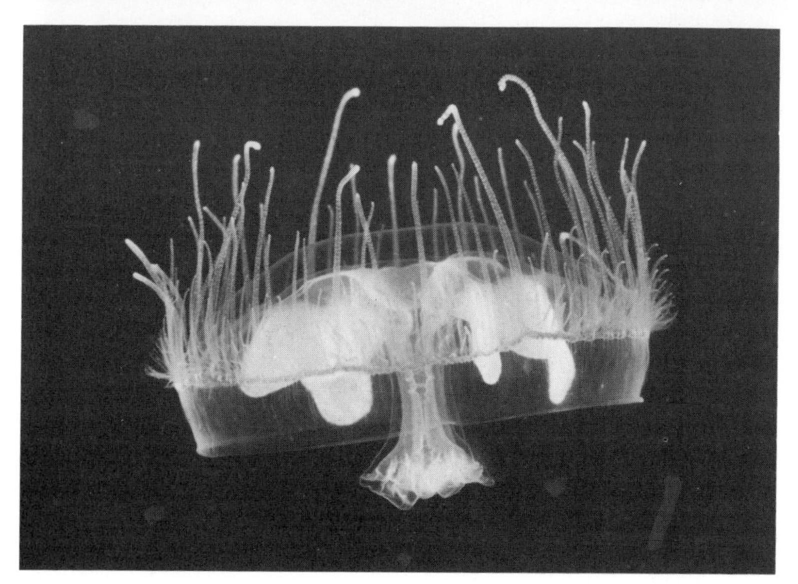

Freshwater jellyfish, *Craspedacusta,* from a lake in Pennsylvania. When fishing, it swims to the surface of the water and then sinks down with tentacles outspread. Diameter about 3 cm. (R. Buchsbaum)

Above left: **Ostrich-plume hydroid,** *Aglaophenia,* is commonly found cast up on beaches. About natural size. Monterey Bay, California. (R. Buchsbaum)

Above right: **Portion of hydroid colony,** *Plumularia.* Tentacles covered with thread capsules give small animals little chance of slipping through. Bermuda. (R. Buchsbaum)

"Air ferns," sold in stores as carefree miracle plants that need no soil or water, are dried hydroid colonies that have been dyed green. From the Thames Estuary, England. (R. Buchsbaum)

Naked hydroid, *Tubularia,* has no protective cup, though a stiff covering surrounds the vertical stems and horizontal connecting stolons. (P. S. Tice)

Velella ("little sail"), a common **chondrophore.** Here three of these large polyps (up to 6 cm in widest diameter), float at the surface mouth downwards, like medusas. An erect sail projects from the upper surface. Around the margin is a single row of stinging tentacles. From the under surface hangs a large central mouth; it is surrounded by numerous reproductive stalks (each with a small mouth) that bud off tiny, simple free-swimming medusas. (R. Buchsbaum)

Opposite: **Fire coral** gets its common name from its fiery sting. Swimmers in tropical waters around the world soon learn to recognize and avoid this **hydrozoan coral.** It grows in a variety of branching or platelike or encrusting forms, but has a characteristic tan color and fine-textured surface, densely covered with tiny holes from which protrude the long slender polyps (hence its generic name *Millepora*, "thousand pores"). Fiji. (J. S. Pearse)

Beached velellas, driven ashore by a spring storm. Their bodies littered hundreds of meters along a central California beach. Such large-scale strandings are fortunately not common, but individuals are often found washed up on shores. (R. Buchsbaum)

Below: **Medusa of *Aurelia*,** a scyphozoan jellyfish, occurs in the coastal waters of all seas. In colder waters, it is sometimes seen in shoals so dense that the water appears to be almost solid with them. It is easily recognized by the shallow, saucer-shaped bell, the fringe of numerous small tentacles, and the four large horseshoe-shaped gonads. Here viewed from the oral surface, the jellyfish displays the four delicate, transparent mouth lobes whose cilia maintain a steady food-bearing current of minute organisms into the centrally located mouth. Medusas of *Aurelia* usually range from 10 cm to 25 cm in diameter, occasionally reaching a meter. Plymouth, England. (D. P. Wilson)

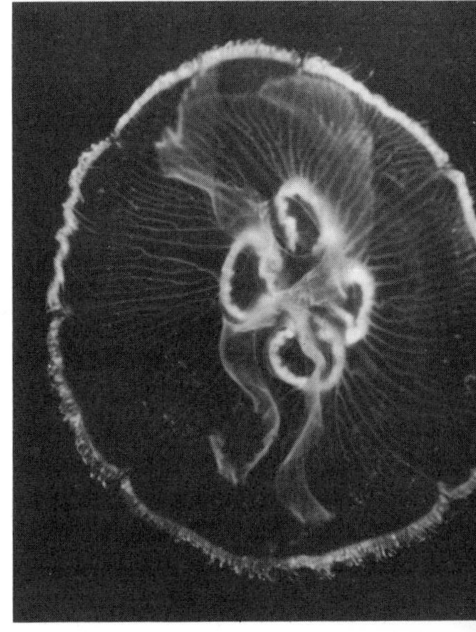

A second class of cnidarians, the **Scyphozoa** (sy-fo-ZO-a), includes the larger jellyfishes, all marine. They can be roughly distinguished from hydrozoan jellyfishes by their large size and by the absence of a velum. Moreover, the polyp stage, which is usually lacking or is very small, differs in structure from a hydrozoan polyp.

Aurelia is one the commonest of the scyphozoan jellyfishes and occurs all over the world. From ships one often sees large shoals of them drifting along together or swimming slowly by rhythmic contractions of the shallow bell, about the same size and shape as a saucer.

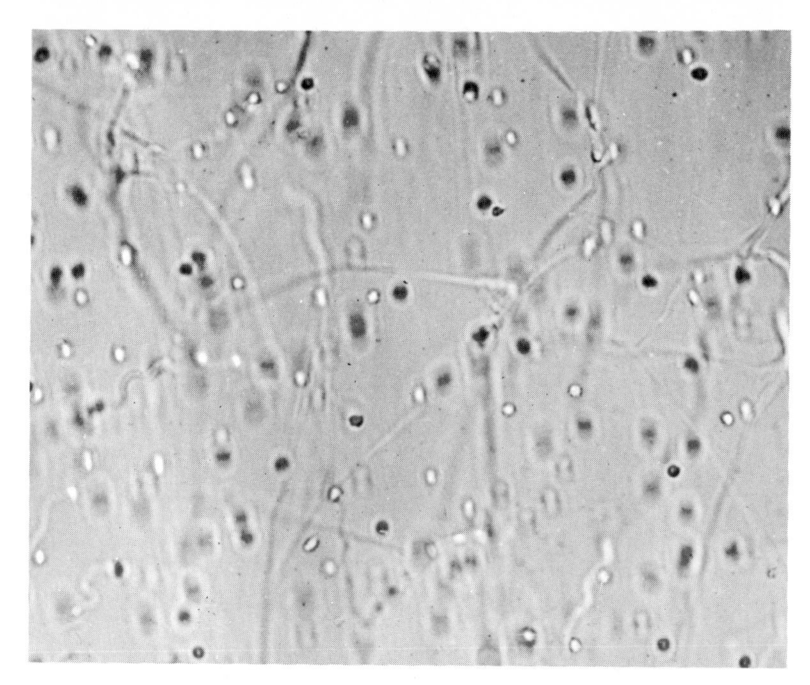

A bit of mesoglea, highly magnified, from the bell of an *Aurelia*. The streaks are fibers that strengthen the jelly. The dots are ameboid cells. Mesoglea is about 96% water, 3% salts, and 1% organic matter. (R. Buchsbaum)

At the end of a short manubrium is a square mouth, the corners of which are drawn out into four trailing **mouth lobes,** each with a ciliated groove. Thread capsules in the lobes paralyze and entangle small animals, which are then swept up the ciliated grooves and through the mouth into a spacious stomach cavity in the center of the bell. Partially digested food is distributed through numerous radial canals to the margin of the bell. Cilia lining the entire gastrovascular cavity maintain a steady current of water, which brings food and oxygen to, and removes wastes from, the internal parts of this large animal.

The margin of the bell bears a fringe of short and very numerous tentacles, set closely together except where they are interrupted by eight equally spaced notches. In each notch lie **sensory structures,** including two pigmented eyespots, sensitive to light; a hollow sac, containing hard particles whose motions set up stimuli that direct the swimming movements; and two pits, lined with cells that are thought to be sensitive to food or to other chemicals in the water.

The **gonads** of *Aurelia* are four horseshoe-shaped bodies on the floor of the large central part of the gastrovascular cavity. Testes or ovaries occur in separate male or female individuals. In a male medusa the sperms are discharged into the cavity and are shed to the outside through the mouth. In a female the eggs are shed into the cavity and are fertilized there by sperms that enter with the food current. Then the normal feeding current is re-

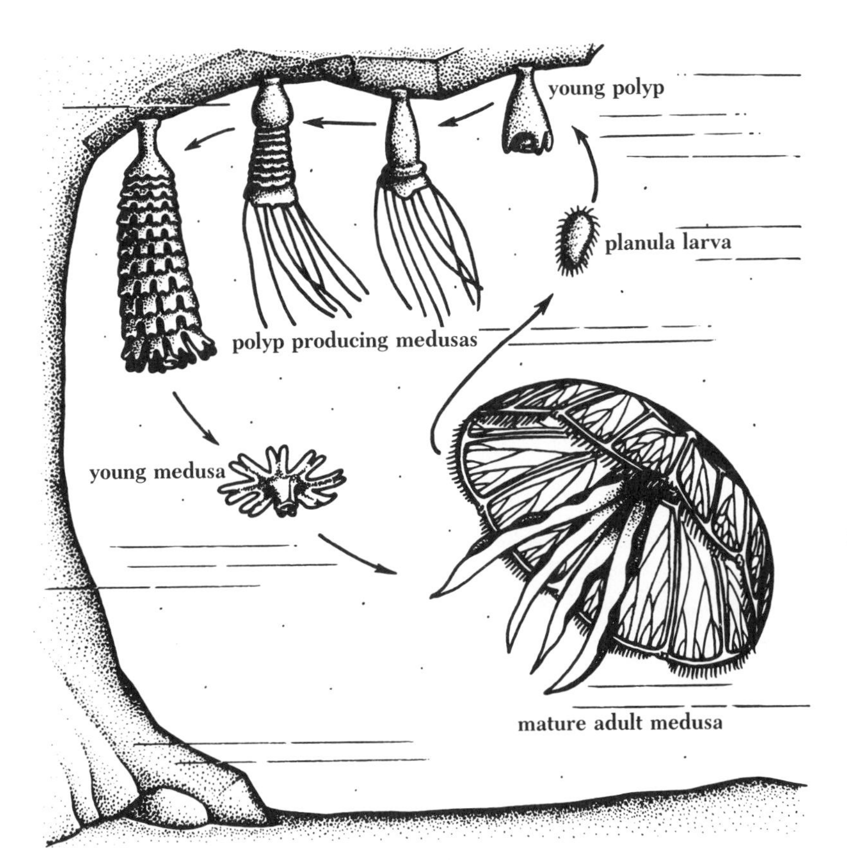

Life history of *Aurelia.*

versed briefly, and the **fertilized eggs** are swept out through the mouth to lodge in the folds of the mouth lobes, where they continue to develop. The ciliated **planula larva** swims away and eventually settles on an overhanging rock ledge or other firm surface. There it develops into a small **polyp** with long tentacles. The polyp feeds and stores food, and may survive in this stage for many months, meanwhile budding off other small polyps like itself. During cold seasons, fall to spring, it develops a series of horizontal constrictions that gradually deepen until the polyp resembles a pile of saucers. One by one the saucers pinch off and swim away as little eight-lobed medusas, which gradually develop into adults. This type of development is characteristic of those scyphozoans that have polyps in the life history and does not occur in the other two classes of cnidarians.

The stinging thread capsules of *Aurelia* usually produce little sensation in human skin. But even a small *Cyanea* can raise huge weals on the arms and legs, and the pink and blue monsters of the north Atlantic (sometimes over 2.5 meters across the bell with tentacles 40 meters long) are a real danger to any swimmers who brave those cold waters. Such huge masses of jelly are among the largest of the animals without backbones.

Polyps of *Aurelia* growing on an overhanging rock surface. Seen here are several small polyps and many others in various stages of constriction. The longest measure about 15 mm. A young jellyfish released only a few seconds before is swimming away. Plymouth, England. (D. P. Wilson)

Beached jellyfishes are often found in great numbers after a storm. A large jellyfish is firm enough to support the weight of a person. Monterey Bay, California. (W. K. Fisher)

Cassiopeia is a scyphozoan jellyfish in which there are no tentacles, and the mouth lobes are divided, their grooves narrowed or fused. The original mouth opening is small or absent, and food organisms are taken in through numerous small openings on the mouth lobes. *Cassiopeia* has the lazy habit of lying in shallow sunlit waters with the mouth side up, as seen here. Symbiotic photosynthetic dinoflagellates (zooxanthellas) in the tissues of the jellyfish are thus exposed to the sun, and a portion of the food they make goes to supplement the nutrition of their host. Bimini, Caribbean. (R. Buchsbaum)

A jellyfish swims by alternately contracting and relaxing the bell. *Left:* The bell is contracted, forcibly expelling the water from its concavity and so jet-propelling the animal in the direction opposite to that in which the water is expelled. *Right:* The bell is relaxed to admit water again. *Chrysaora hysoscella,* called the "compass jellyfish" because of its markings, is one of the commonest of scyphozoan medusas. It occurs in great numbers toward the end of summer along the Atlantic coasts of Europe; related species are found on North American coasts. This jellyfish was in the Helgoland Aquarium, West Germany. (F. Schensky)

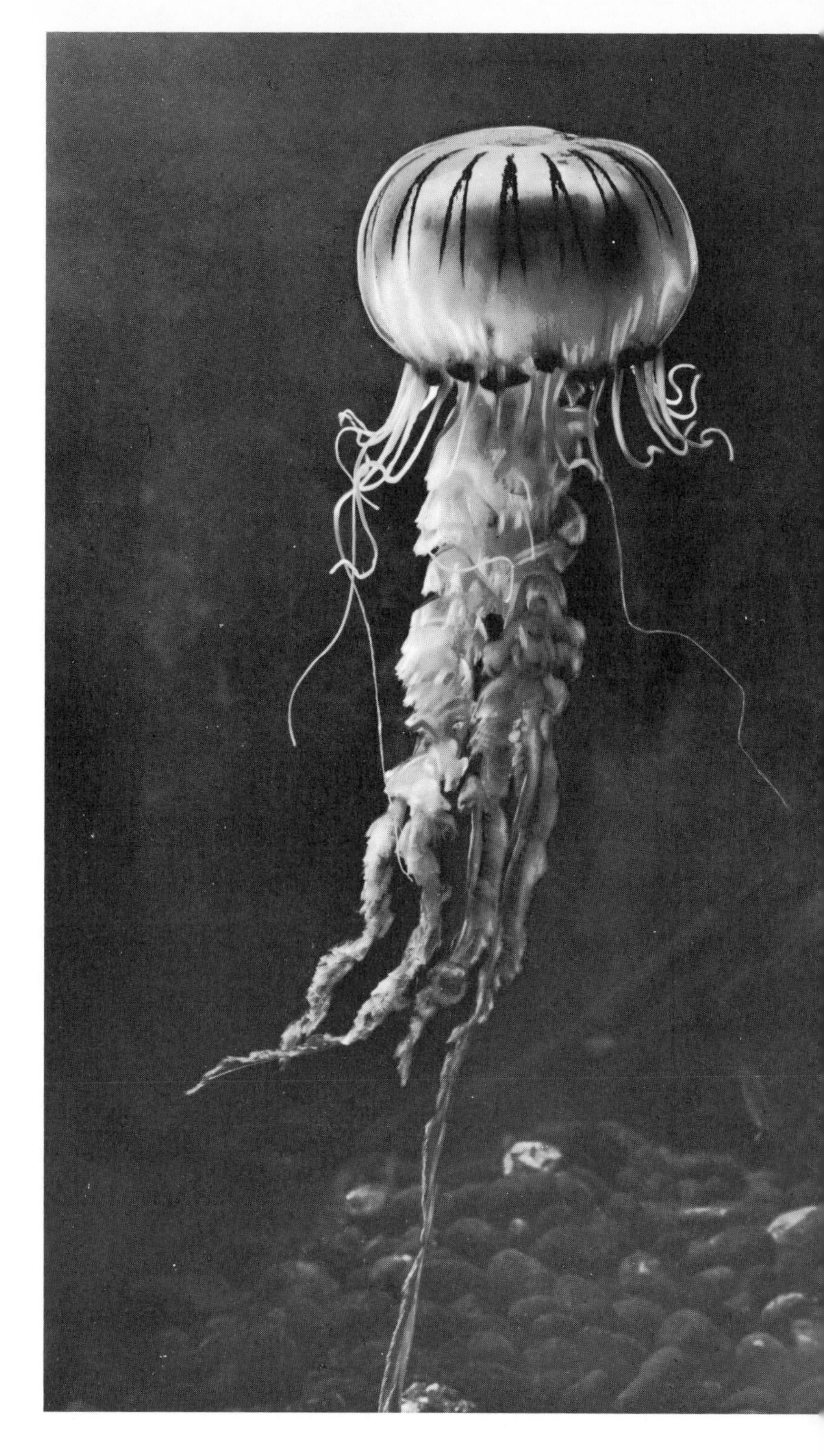

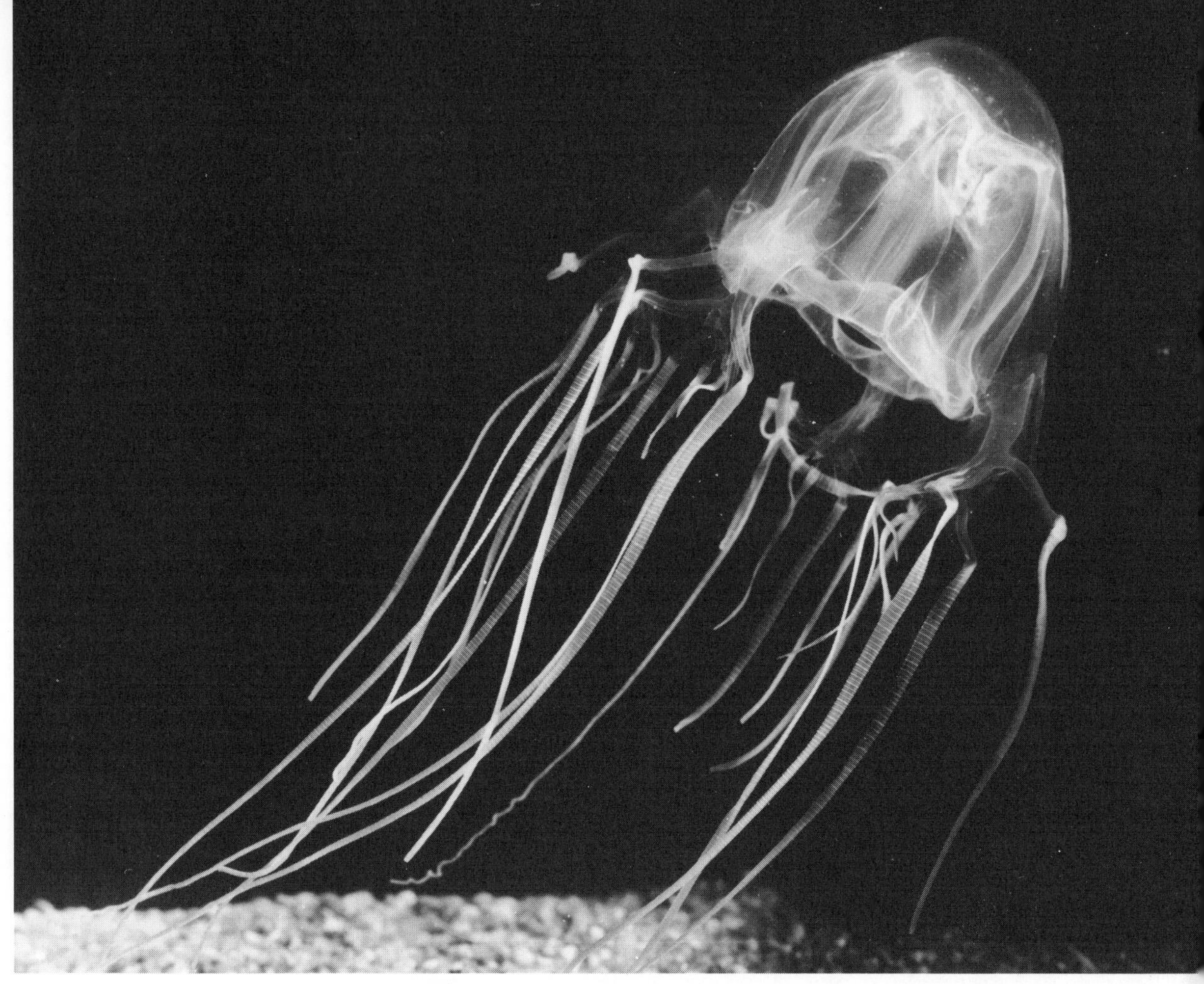

Box jelly, or cubomedusa, is the name given to certain jellyfishes with four-sided symmetry. Cubomedusas are strong, active swimmers with large, complex eyes and are notorious for their sting. Seen here is the sea wasp, *Chironex fleckeri,* which is among the largest of cubomedusas (bell up to 25 cm high) and is common in northern coastal waters of Australia during the rainy summer months. Only a small minority of stingings are fatal, but extensive contact with the tentacles of a large sea wasp can cause death, sometimes within 3 minutes or less, apparently from heart failure. (K. Gillett)

The third and largest class of cnidarians, the class **Anthozoa** ("flower animals"), consists of marine polyps that have no medusa stage. Anthozoan polyps are technically distinguished from hydrozoan polyps by the fact that the gastrovascular cavity is divided up by a series of vertical partitions, and the body wall turns in at the mouth to form a tube lined with ectoderm. But superficially there is no difficulty in telling the large fleshy sea anemones or the stony corals from most of the small fragile hydrozoan polyps.

A **sea anemone** (a-NEM-o-nee) is a solitary polyp. The body consists of a stout muscular cylindrical **column,** expanded at one end into a flat **oral disk** having a central **mouth** surrounded by several to many circlets of hollow **tentacles.** The other end forms a smooth, muscular **basal disk** on which the anemone can slide about very slowly and by which the anemone can hold to rock so tenaciously that one is likely to tear the animal in trying to pry it loose.

The tube that hangs down from the mouth into the gastrovascular cavity is called the **pharynx.** From the body wall to the pharynx, and attached to both, stretches a series of vertical **partitions.** Between these primary partitions is a secondary set of narrower ones, which extend only part way from the body wall to the pharynx; and between these are still narrower tertiary and sometimes quarternary sets. The chambers between the primary partitions are in open communication with each other below the pharynx; but above the level at which they attach to the pharynx, they communicate only through one or more holes in the wall of each partition.

The partitions are double sheets of endoderm held together by a central layer of mesoglea, and they serve to increase the digestive surface of the cavity, making it possible for a sea anemone to digest a relatively large animal, such as a fish or a crab. The free edges of the partitions are expanded into convoluted

Sea anemones, so called from their resemblance to flowers, are among the most familiar cnidarians because they are easily seen in tide pools on rocky shores. The one on the right has just captured a small fish, which is held by thread capsules on the tentacles. Helgoland. (F. Schensky)

Senior citizens of the tidepool are the long-lived anemones. Animals like these *(Anthopleura xanthogrammica)* have been observed to occupy the same spot for over thirty years. Anemones have lived in captivity for close to a century, and some large individuals in nature could be even older. Pacific Grove, California. (W. K. Fisher)

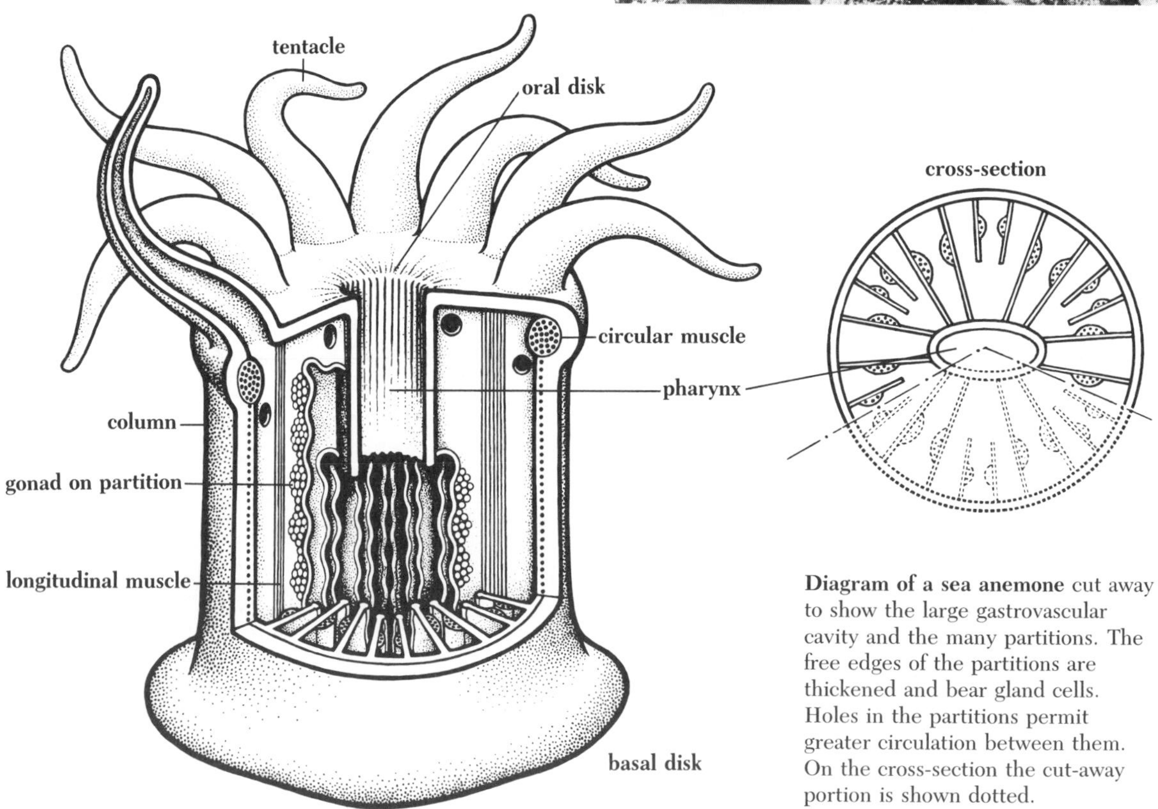

Diagram of a sea anemone cut away to show the large gastrovascular cavity and the many partitions. The free edges of the partitions are thickened and bear gland cells. Holes in the partitions permit greater circulation between them. On the cross-section the cut-away portion is shown dotted.

Oral disk of *Metridium* shows the elongated mouth with ciliated groove (light-colored) at one end. Both are modifications of the basic radial symmetry of cnidarians (see chapter 9). Maine. (R. Buchsbaum)

Basal disk of *Metridium* (seen through the glass wall of an aquarium) is marked by radiating lines where the partitions are attached and reflects the basic radial symmetry of cnidarians. Maine. (R. Buchsbaum)

thickenings, or **digestive filaments,** which bear the gland cells that secrete digestive juices. Digestion is completed, as in other cnidarians, by endoderm cells lining the gastrovascular cavity.

The pharynx is not cylindrical but is somewhat flattened, and at one or both ends of its long axis is a longitudinal groove lined with cilia that are much longer than the ones lining the rest of the pharynx. The cilia in these grooves beat downward, drawing a current of water into the gastrovascular cavity and providing the internal parts of the sea anemone with a steady supply of oxygen. At the same time, the shorter cilia lining the rest of the pharynx beat upward, creating an outgoing current of water that takes with it carbon dioxide and other wastes. During feeding, these shorter cilia reverse their beat, and the food is swept down the pharynx and into the digestive cavity.

Sea anemones are among the most structurally complex of the polyp types of cnidarians. They have a well-developed **nerve net,** mesenchyme cells in the mesoglea, and several sets of special-

Sea anemones expanded *(Metridium)* in a darkened aquarium. Most anemones are negative to light and in nature expand fully and feed only in partial or complete darkness. Maine. (R. Buchsbaum)

Same anemones contracted after being illuminated. Anemones contract from time to time, even in the dark and especially after a meal, effectively cleaning out the gastrovascular cavity. Maine. (R. Buchsbaum)

ized **muscles.** A layer of circular muscles serves to narrow, and therefore to extend, the body. The longitudinal muscles are concentrated into prominent bands that run, one on each partition, from oral disk to basal disk. Their contraction pulls down the oral disk, with its tentacles. A strong circular muscle at the outer edge of the oral disk then closes over the mouth, much as a pouch is drawn closed by a string. In the contracted condition anemones resist drying or mechanical injury during low tide.

Some sea anemones **reproduce asexually** by pulling apart into two halves. In other species fragments of the basal disk are pulled off and left behind as the animals slide about, and these fragments regenerate into tiny anemones. In **sexual reproduction** the eggs or sperms form on the partitions of the gastrovascular cavity and are ejected through the mouth, or the eggs may be fertilized internally. The fertilized egg develops into a planula, which finally settles down in some rocky crevice and grows into a single anemone.

Anemone dividing, *Anthopleura elegantissima*. The halves pull apart until the thin strand of tissue connecting them gives way, and each half heals. Tens or hundreds of animals, asexually produced in this way from a single original anemone, remain together in dense aggregations on rocky shores of the west coast of North America. Puget Sound, Washington. (L. Francis)

Tube anemone (cerianthid) has 2 whorls of slender, graceful tentacles. It has no basal disk but a rounded basal end with which it burrows. Strong longitudinal muscles in the column allow it pull down and disappear into its tube in the blink of an eye. It belongs to an order of anthozoans separate from that of the sea anemones (actinians). Banyuls, Mediterranean coast of France. (R. Buchsbaum)

Black coral colony has tiny six-tentacled polyps that are supported by a whiplike or treelike skeleton of a hard horny material. Black corals are also called thorny corals or antipatharians. New Zealand. (K. Gillett)

Zoanthids are anemone-like but belong to another anthozoan order. They live permanently attached, usually as encrusting colonies in which the individual polyps are united by a mass of tissue containing hollow connecting tubes of endoderm. Santa Catalina Island, southern California. (R. Buchsbaum)

Black coral jewelry is made of the cut, polished skeleton. Heavy collecting in some areas has greatly reduced populations of black corals. Fortunately, some colonies persist in waters deeper than most coral divers venture. (R. Buchsbaum)

The **stony corals** are like small sea anemones but are mostly **colonial** and secrete a hard, protective **calcium carbonate skeleton** with cups or grooves into which the delicate polyps can retract. From the wall of the cup a series of radially arranged vertical plates project inward between the digestive partitions. The stony cup and its plates are outside the polyp and in contact only with the ectoderm that secretes them.

Many kinds of small cup corals grow in temperate waters along marine shores. Colonies of five to thirty individuals of *Astrangia* encrust shells and rocks as far north as Cape Cod on the Atlantic coast of North America. And even the cold, deep waters of Norwegian fjords support great banks of a colonial branching coral, *Lophelia*. But the great majority of corals—those of the **coral reefs**—flourish and build reefs only in sunlit tropical or subtropical waters, where the average annual sea temperature is above 23°C (73°F). In the endoderm of reef-building corals there live single-celled photosynthetic dinoflagellates (zooxanthellas), which contribute to their hosts' nutrition and metabolism and without which the corals cannot grow fast enough to build reefs. The need of these dinoflagellates for light limits reef building to shallow waters, less than about 30 meters (100 feet).

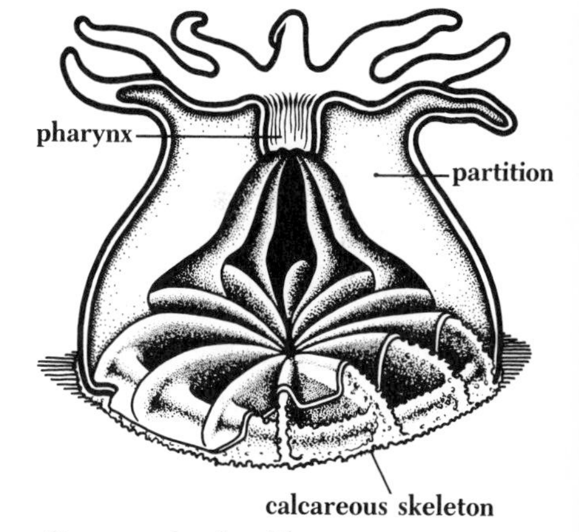

Stony coral polyp. This young animal is shown as if partly cut away to reveal the gastrovascular cavity and, beneath the polyp, the beginning of the calcareous skeleton. (After Pfurtscheller.)

Solitary corals. *Left:* Expanded polyp (*Balanophyllia*) is bright orange to yellow; spots on the tentacles are batteries of thread capsules. *Right:* Solitary coral (*Paracyathus*), seen from above, illustrates the radial symmetry of cnidarians. Monterey Bay, California. (R. Buchsbaum)

Mushroom corals are so named because the skeletal plates of these solitary polyps look like the gills of mushrooms. A common genus is *Fungia*. Unlike other corals, mushroom corals do not live permanently attached to a hard surface, but can slowly move around over sand or other substrates. A mushroom coral may even succeed in righting itself if overturned. Shown here are (left) a polyp with its disk and tentacles expanded and (right) with tentacles contracted. Oahu, Hawaii. (R. Buchsbaum)

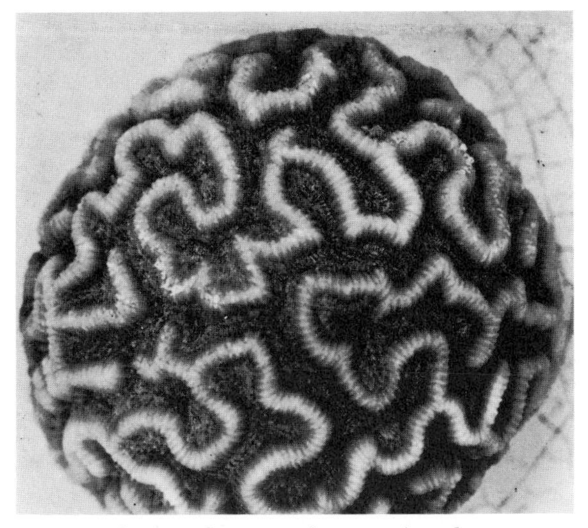

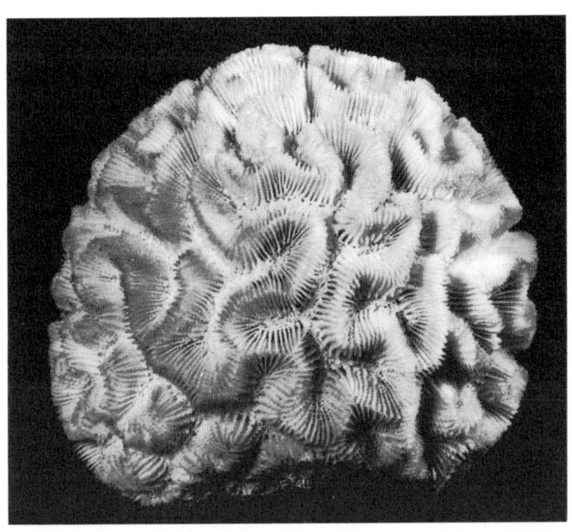

Brain coral. The soft green or brown polyps do not sit in separate cups but in long, winding grooves. Separate mouths occurring at intervals mark the positions of the polyps, which appear to be continuous with each other. In the photo above left, taken by day, the polyps and tentacles lining the grooves were contracted. Most corals remain contracted during the day and expand fully to feed only at night. **Cleaned skeleton of brain coral** above right shows the grooves the polyps once occupied and the delicate plates of hard, white calcium carbonate. (Slightly less than natural size.) Bermuda. (R. Buchsbaum)

Coral blocks for building houses in Bermuda are cut from the hilltops with a saw. The material consists of the calcareous skeletons of corals and other organisms, ground fine by wave action, deposited on the beaches, and then cemented into the soft coral rock that forms the surface layers of these volcanic islands. (R. Buchsbaum)

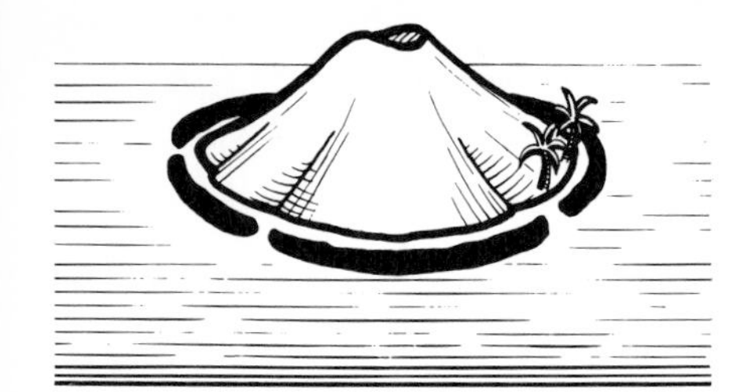

Coral reefs. *Top:* A fringing reef growing around an oceanic island. *Middle:* A small barrier reef widely separated from the island. *Bottom:* An atoll. (Based on various sources.)

Three main types of coral reefs are recognized. **Fringing reefs** border coasts closely or are separated from them at most by a narrow stretch of water. **Barrier reefs** also parallel coasts but are separated from them by a channel deep enough to accommodate large ships and are many kilometers wide. Captain Cook sailed within the Great Barrier Reef of Australia for over 1,000 kilometers (600 miles) without even suspecting its presence until the channel narrowed and he was wrecked on the reefs. **Atolls** are ring-shaped coral islands enclosing central lagoons. They are quite isolated from any land, and thousands of them dot the tropical Pacific.

Charles Darwin reasoned that if an island, surrounded by a fringing reef, were to subside very slowly, so slowly that the reef could grow upward at about the same rate, the island would grow smaller and smaller, and the fringing reef would become separated from it by a wide, deep channel, eventually becoming converted into a barrier reef. If this process were to continue, the island would finally disappear entirely beneath the surface of the water, and the rising barrier reef would become a ring-shaped island, or atoll. This theory is still the most widely accepted one, though changes in sea level during and after glacial periods, which Darwin did not know about, may also have played a role in shaping reefs. In some cases, an atoll may have been formed directly, without going through a fringing-reef and a barrier-reef stage, upon a submarine platform built up close to the surface by volcanic activity.

Great Barrier Reef of Australia at low tide. The reefs, which form a barrier 2,000 km (1,260 mi) long, and many kilometers wide, are a serious hazard to ships. This view is unusual in that it consists almost entirely of stony corals of a single genus, *Acropora,* in the growth form called staghorn coral. (W. Saville-Kent)

Coral skeletons of many kinds form the bulk of the material of the Great Barrier Reef. The living corals differ even more because the polyps show striking variety in form and in their brilliant colors. These dried and bleached skeletons are beautiful and are used as ornaments, but they give about as good an impression of the exquisite beauty of living corals as one would get of the beauty of a bird from its whitened bones. Heavy collecting has seriously damaged many reefs around the world. (W. Saville-Kent)

The sea anemones, corals, and their relatives presented so far are **hexacorallians,** anthozoans in which the tentacles and internal partitions are numerous and often arranged in multiples of six. In another large group of anthozoans, the **octocorallians,** the polyps always have eight branched tentacles and eight internal partitions. Almost all members of this latter group are colonial, and the body cavities of the polyps are in communication with each other through endodermal canals, which penetrate the whole colony. The polyps are remarkably similar, but the skeletons of the various groups are strikingly diverse. In the fleshy **soft corals** (alcyonaceans), the skeleton consists of spicules of calcium carbonate, which lie scattered in the soft tissues. In the **horny corals** (gorgonians), such as sea fans and sea whips, there is a branching skeleton of a proteinaceous hornlike substance; and there may be scattered spicules also, or a solid calcareous core as in the **precious coral** with its hard pink, red, or "coral"-colored skeleton, used in jewelry. Other octocorallians with colorful calcareous skeletons are the **organ-pipe coral,** with a brick-red skeleton of closely packed tubes, and the **blue coral.**

In the warm clear waters off many tropical shores it is easy to don mask and snorkel or scuba gear and descend into a world in

Soft corals, or alcyonaceans, are abundant and prominent members of Pacific reefs. The many small eight-tentacled polyps, when expanded, give the fleshy colonies a fluffy appearance. When the polyps are contracted (as in parts of this photo), soft corals look like masses of leathery seaweed. Madang, Papua New Guinea. (J. S. Pearse)

Sea pens *(Ptilosarcus)* live with the stalk embedded in the mud. The featherlike upper portion of the colony bears the feeding polyps that capture tiny animals from seabottom currents. Besides the feeding polyps there are ones that serve to circulate water through the colony. Sea pens come in many different forms, and many are luminescent. Oregon. (R. Buchsbaum)

Horny corals, or gorgonians, are often called sea whips, sea fans, or sea plumes, according to their shape. Bahama Islands.

Sea whip brought up from 6 m (20 ft) of water. The dark spots represent the positions of retracted polyps. The skeleton of gorgonians is not rigid as in stony corals but is quite flexible. Colony is about 30 cm (1 ft) high. Bermuda. (R. Buchsbaum)

Sea fan, *Eunicella verrucosa*, has narrow branches all in one plane. Colony is about 30 cm (1 ft) high. Plymouth, England. (D. P. Wilson)

Portion of a sea fan *(Eunicella verrucosa)* with fully expanded polyps. As in other octocorallians, the polyps have eight feathery tentacles and share the food they catch with the rest of the colony. Plymouth, England. (D. P. Wilson)

Sea plumes, like sea whips and sea fans, are gorgonians with a flexible hornlike skeleton. Caribbean. (J. S. Pearse)

Precious coral, *Corallium rubrum,* is a gorgonian with a calcareous skeleton. The bright, permanent colors of the skeleton, so valued in jewelry, show through the pale tissues of living precious coral. (In stony corals it is the living tissue that is colorful, while the skeleton is white.) Precious coral is best known from Mediterranean waters. (P. S. Tice)

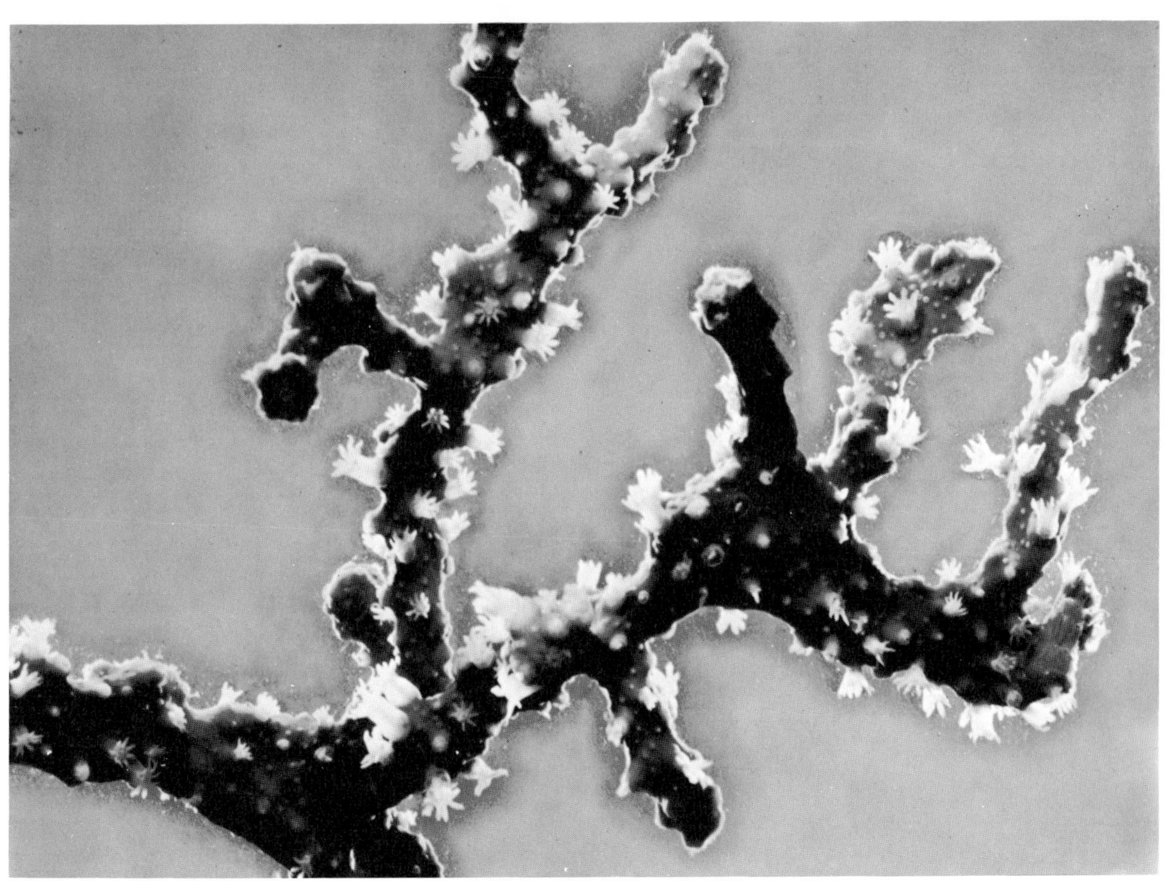

Polyp of precious coral has eight feathery tentacles surrounding an elongate mouth, as is typical of octocorallian polyps. (P. S. Tice)

Skeleton of organ-pipe coral, *Tubipora.* In life, sage-green polyps inhabit the uppermost portion of the brick-red calcareous tubes of the skeleton. (R. Buchsbaum)

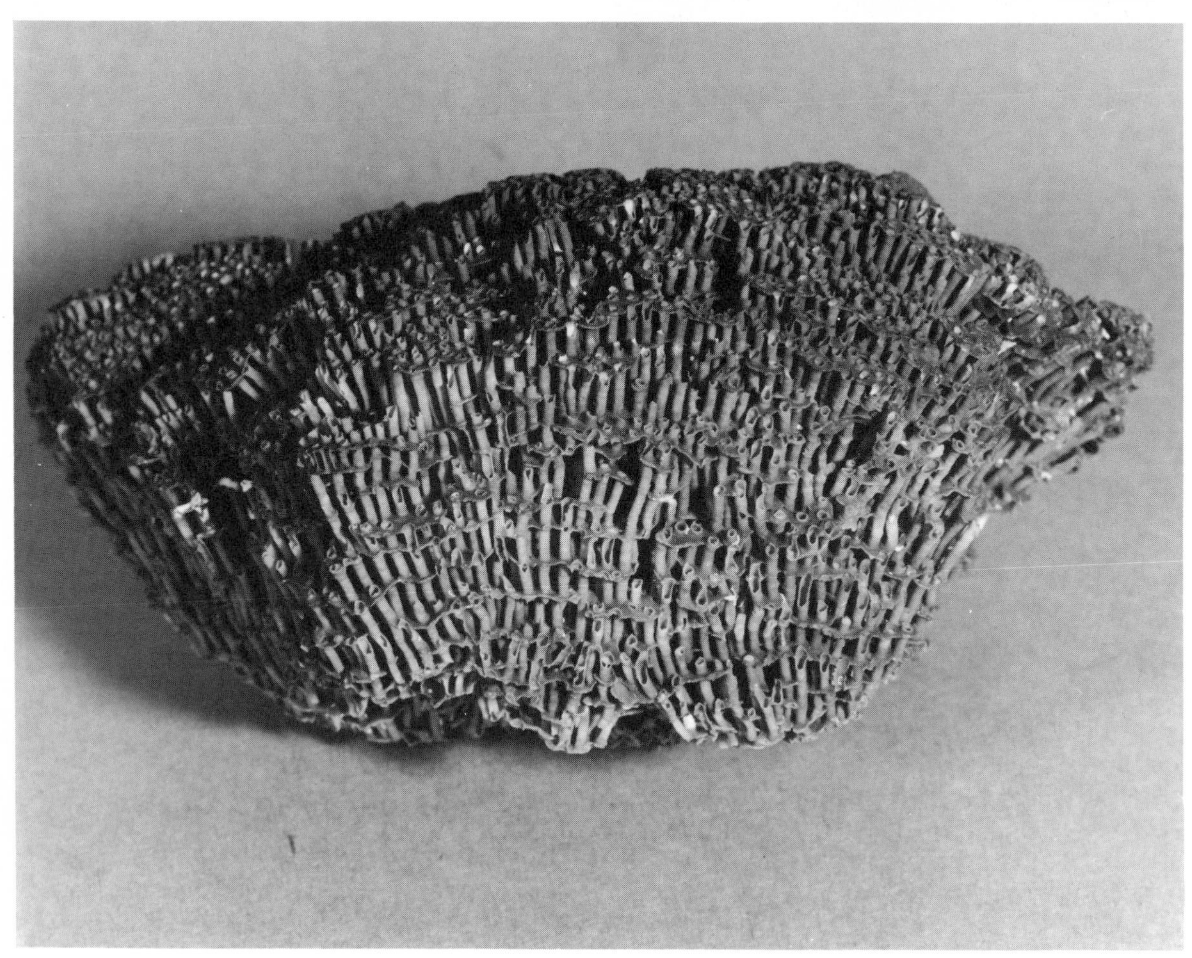

which the entire landscape consists of anthozoans. Tall purple gorgonians tower overhead and low bushy ones spring from all sides like blossoming shrubbery. Among them are massive dome-shaped colonies of stony corals, hundreds of years old. A carpet of huge sea anemones with their tentacles waving gently in the currents looks like a meadow stirred by the wind. And even the brilliant white sand is found, if you look closely, to consist of calcareous grains from the eroded skeletons of corals that were once a living part of this cnidarian jungle.

Massive brain coral. Heron Island. Great Barrier Reef, Australia. (R. Buchsbaum).

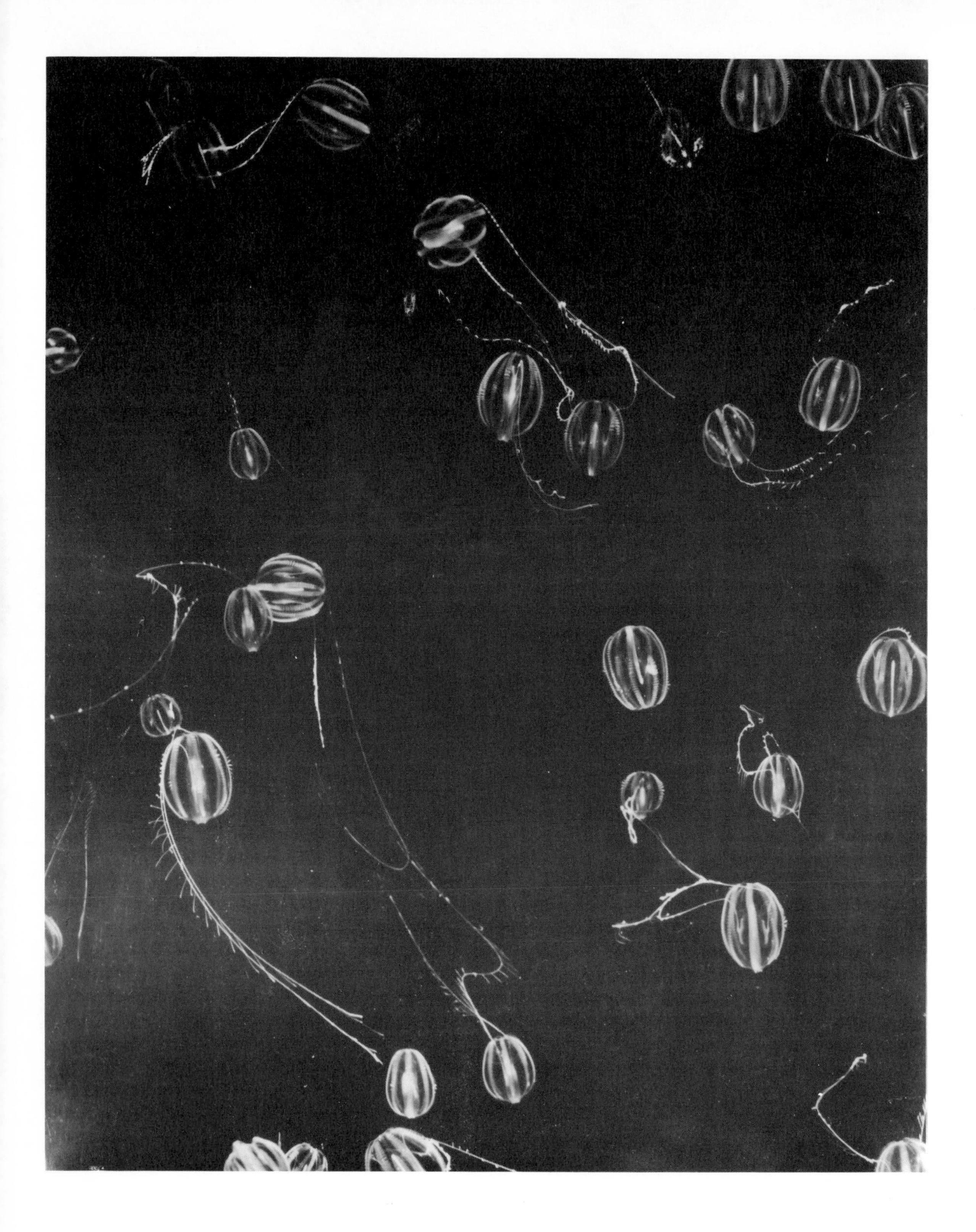

Chapter 8

Comb Jellies

Comb jellies are transparent gelatinous animals that float in the surface waters of the seas, mostly near shores. Graceful but not powerful swimmers, they are unable to resist the movements of currents and tides, so that they often accumulate in vast numbers in some bay where wind-driven waves have carried them. During a storm their fragile bodies are swept ashore by the high waves and are strewn about on the beaches. They are, therefore, not an unfamiliar sight to people who live on the seacoasts; and they have been given many common names, such as "sea gooseberries" and "sea walnuts." These names describe the shape and size of two of the most common types; but they give no suggestion of the unique character from which has been derived the common name "comb jelly" and the name of the phylum, **CTENOPHORA** ("comb bearers"). A ctenophore (TEN-o-for) swims through the water by means of eight rows of **ciliary combs.** Each row consists of a series of little plates formed of large cilia fused together at their attached ends like the teeth of a comb. The rows radiate over the surface of the animals from pole to pole, like the lines of longitude on a globe.

The combs are lifted rapidly in the direction of the upper pole, then lowered more slowly to their relaxed position. Those in each row beat one after the other, in sequence from the upper toward the lower pole. All eight rows do not always beat at the same rate; those on one side may beat faster, allowing the animal to turn and maneuver as it is slowly, smoothly propelled through

Beached comb jellies are common, especially after storms. *Pleurobrachia.* San Francisco, California. (R. Buchsbaum)

Diagram of a typical comb jelly, *Pleurobrachia.*

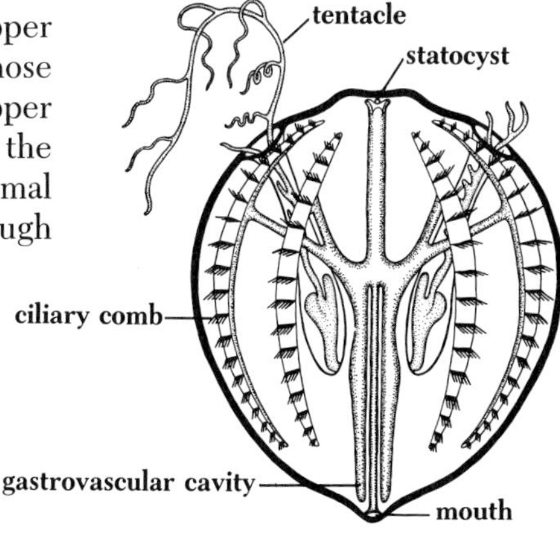

Opposite: **Sea gooseberries,** *Pleurobrachia,* swim by means of eight rows of ciliary combs. Two long tentacles sweep the water for small food organisms. About natural size. Helgoland. (F. Schensky)

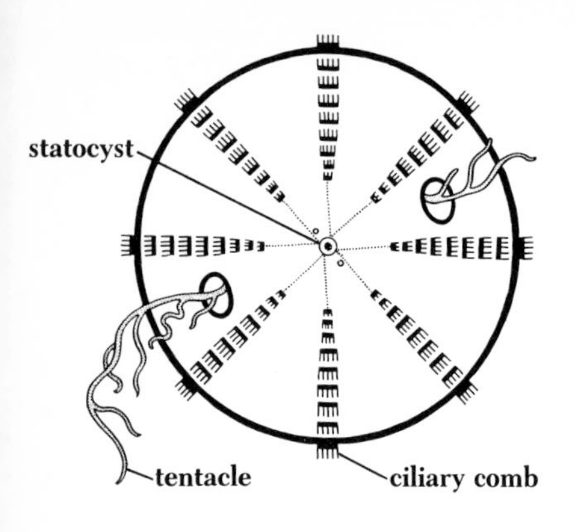

Upper pole of *Pleurobrachia.* The 8-parted symmetry of comb jellies is modified by the two tentacles, by various structures at the upper pole, and by the flattening of mouth and pharynx in the vertical plane perpendicular to that passing through the tentacles. Such modified radial symmetry is called **biradial symmetry** (see end of chapter 9).

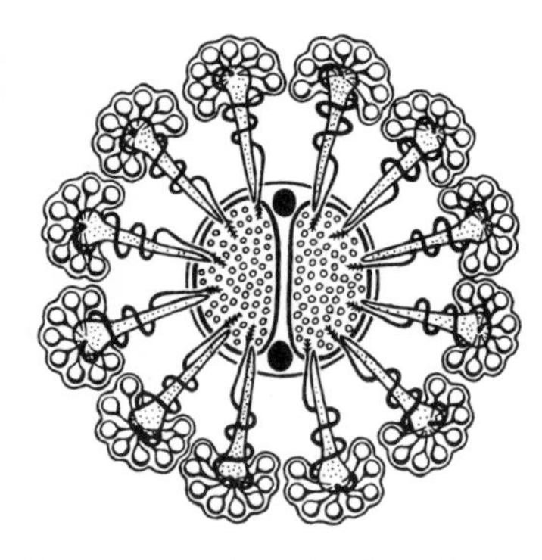

Cross-section through a branch of a tentacle. The outer surface is covered by the sticky heads of the **adhesive cells.** Each cell has a coiled filament (anchored in the central muscular axis of the tentacle branch) that prevents the cell from being wrenched off by the struggling prey. Two black dots in the axis represent nerve strands.
(Based on P. V. Fankboner.)

the water with the lower pole, the mouth end, in front. The rapidly beating combs refract light and produce a constant play of changing colors. Comb jellies are noted for the beauty of their daytime iridescence, but this is certainly matched at night by those species that are luminescent. When the animals are disturbed as they move slowly through the dark water, they flash along the eight rows of combs.

At the upper pole of a comb jelly is a **sensory area** composed of nerve cells and sensory cells. In the center of this area is a covered pit containing a sensory structure (statocyst) that consists of a little mass of calcareous particles supported on four tufts of cilia connected with sensory cells. It acts as a sort of balancing device or "steering gear."Any tilting or turning of the body causes the calcareous mass to bear more heavily upon the ciliary tuft of one side or another. This stimulates the sensory cells, and the stimulus is carried by nerve cells to the swimming combs of the appropriate side, causing them to beat slower and thus righting the animal or otherwise adjusting its orientation in the water. From the polar sensory area a **nerve net** extends all over the body and is concentrated into **eight nerve strands,** one under each row of combs. This system regulates and coordinates the activity of the eight comb rows, for, if the polar area is removed, the combs still beat but the animal's movements become uncoordinated.

The general **body plan** of a ctenophore resembles that of a cnidarian jellyfish. There is an epithelium of ectoderm covering the outer surface, an epithelium of endoderm lining the gastrovascular cavity, and a thick jellylike mesoglea between. The mesoglea contains ameboid mesenchyme cells and long delicate muscle cells that run from one part of the body to another. The mouth is situated in the center of the lower pole and leads through a pharynx into the branched gastrovascular cavity. Undigested food is expelled through the mouth.

Those ctenophores thought to be more primitive have globular or pear-shaped bodies with a branched muscular **tentacle** on each side that can be withdrawn into a pouch. The tentacles have no thread capsules (these are unique to cnidarians), but the branches are covered with special **adhesive cells** that stick to and entangle the prey. Such comb jellies catch small shrimps or fishes by extending their tentacles full length and then curving and looping through the water, with the sticky tentacles sweeping a wide area. Some other kinds of comb jellies have the tentacles reduced to short filaments; they feed mostly on minute planktonic organisms that are caught by the ciliated grooves and swept toward the mouth, or on larger prey captured between large lobes that surround the mouth. Still other comb jellies lack

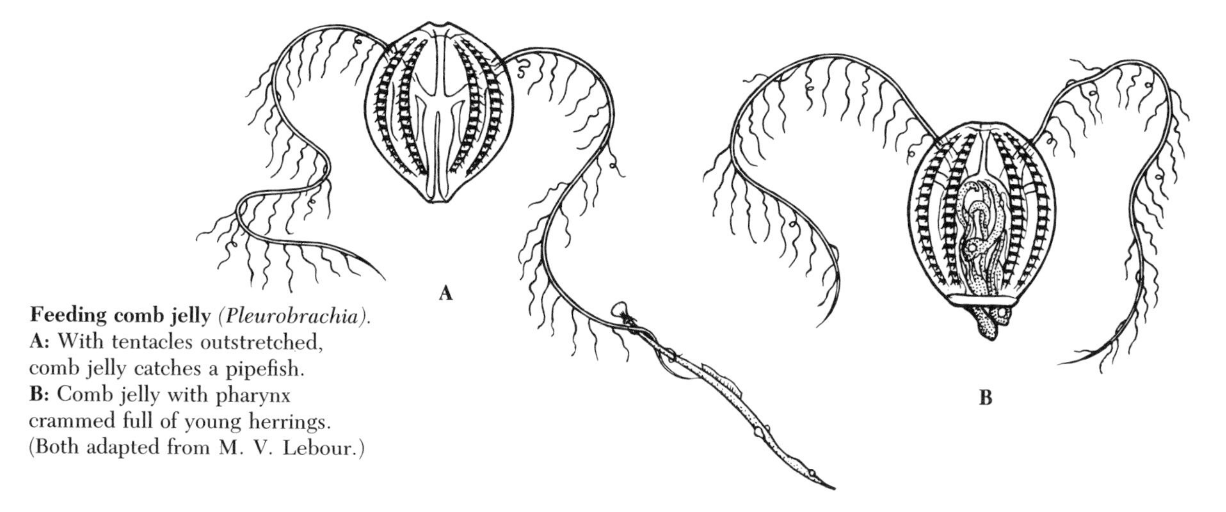

Feeding comb jelly *(Pleurobrachia).* **A:** With tentacles outstretched, comb jelly catches a pipefish. **B:** Comb jelly with pharynx crammed full of young herrings. (Both adapted from M. V. Lebour.)

Lobed comb jelly, *Mnemiopsis,* called the "sea walnut" on the U.S. East Coast, where it is often seen in large numbers near the surface. The comb rows are conspicuous in this photo. Rows of cilia around the mouth end (toward the lower right in this photo) direct minute organisms or food particles to the mouth, and lobes around the mouth enclose larger prey. These comb jellies are luminescent, and when disturbed at night, as by the passing of a boat, they light up along the comb rows. About twice natural size. Woods Hole, Massachusetts. (R. Buchsbaum)

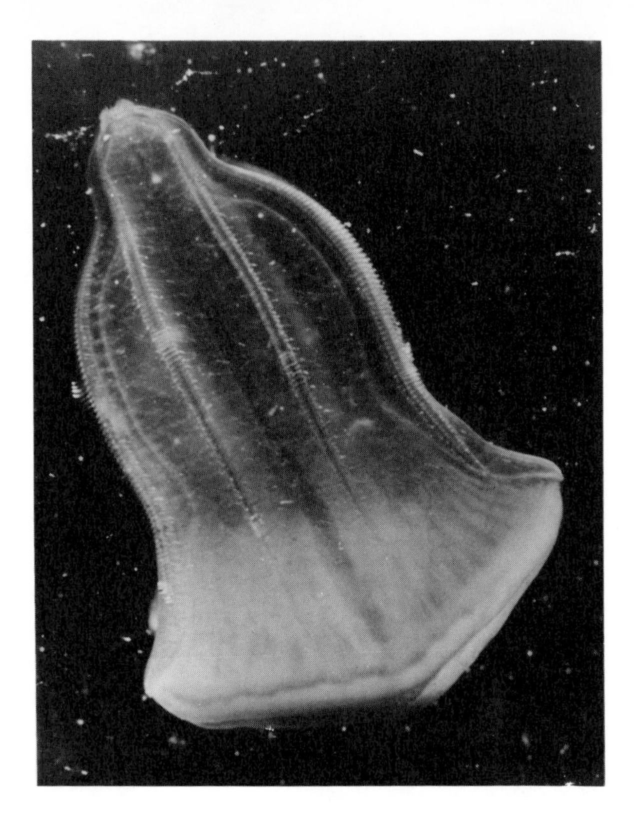

Eight-celled stage of a developing comb jelly already shows the biradial symmetry characteristic of adults (see end of chapter 9). Monterey Bay, California. (R. Buchsbaum)

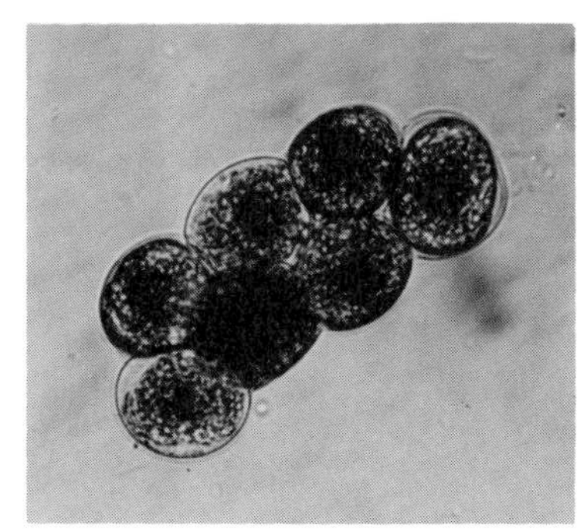

Beroe belongs to a group of ctenophores that lack tentacles. They feed on a variety of soft-bodied prey, especially ctenophores, which they capture in the wide mouth. Gulf of Mexico. (R. Buchsbaum)

tentacles, but have a huge extensible mouth with which they seize and swallow soft-bodied prey, especially other comb jellies.

All ctenophores are **hermaphrodites.** Both ovaries and testes occur on the walls of each of the gastrovascular branches that run beneath the comb rows. Eggs and sperms are shed to the outside through pores in the ectodermal epithelium, and the parent usually dies soon after. The young embryos or larvas develop free in the sea.

A number of bizarre forms occur among the comb jellies. *Cestum* is flattened and elongated from side to side and reaches a length of over a meter. It swims by muscular undulations of the ribbonlike body. *Coeloplana* is flattened in the other direction, so that the two poles, bearing mouth and sensory area, are brought close together. It has two typical ctenophoran tentacles, but has lost its combs. Extensive development of muscle fibers enables it to creep about over surfaces like a worm. Such an animal illustrates how round, free-swimming organisms can become flattened, bottom-creeping forms. The next level of increasing complexity among animals is represented by such flattened creepers, the flatworms. And it is difficult to resist the

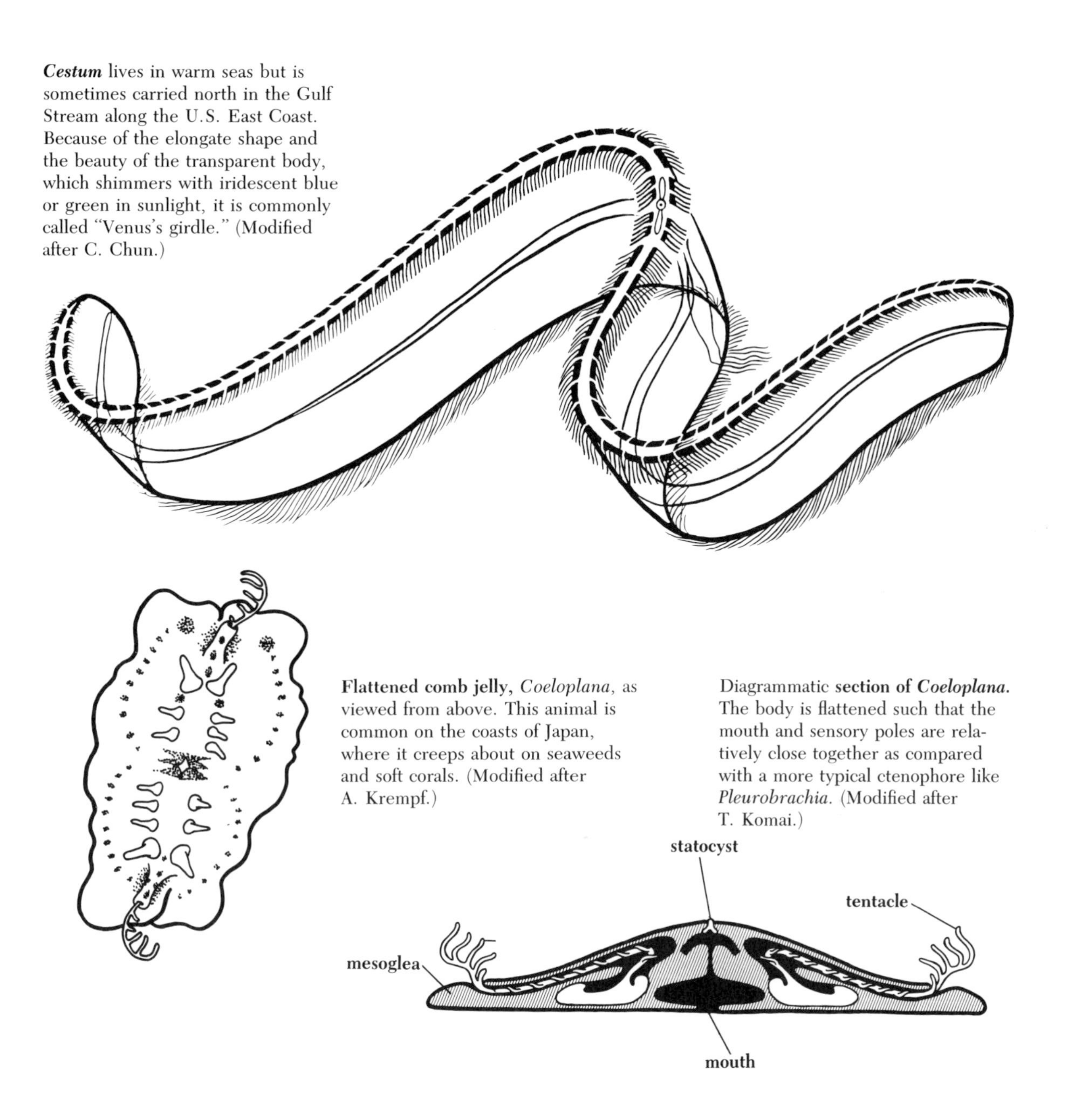

Cestum lives in warm seas but is sometimes carried north in the Gulf Stream along the U.S. East Coast. Because of the elongate shape and the beauty of the transparent body, which shimmers with iridescent blue or green in sunlight, it is commonly called "Venus's girdle." (Modified after C. Chun.)

Flattened comb jelly, *Coeloplana,* as viewed from above. This animal is common on the coasts of Japan, where it creeps about on seaweeds and soft corals. (Modified after A. Krempf.)

Diagrammatic **section of *Coeloplana.*** The body is flattened such that the mouth and sensory poles are relatively close together as compared with a more typical ctenophore like *Pleurobrachia.* (Modified after T. Komai.)

temptation to derive flatworm ancestors from flattened comb jellies such as *Coeloplana.* However, the preponderance of facts does not support the attractive idea that flattened comb jellies are a connecting link between animals at the tissue level of organization (cnidarians and ctenophores) and the more complex animals such as the flatworms described in the next chapter.

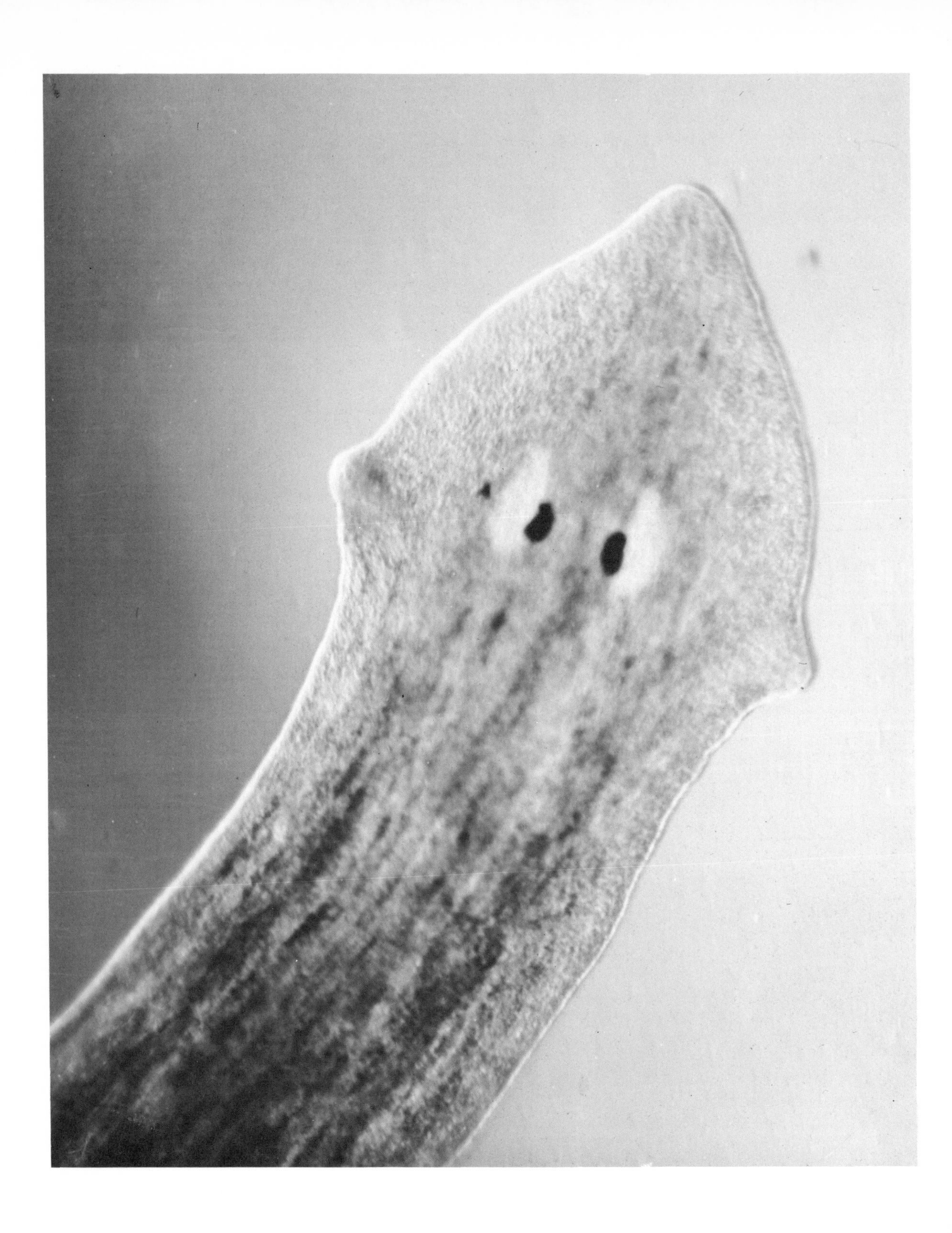

Chapter 9

Three Layers of Cells

Put a piece of raw meat into a small stream or spring and after a few hours you may find it covered with hundreds of black worms that are feeding upon it. These worms, each about 1 or 2 centimeters long, are called **planarians.** When not attracted into the open by food, they live inconspicuously under stones and on vegetation.

Planarians belong to the phylum PLATYHELMINTHES ("platy," flat; "helminthes," worms), which also includes many free-living marine species and two important groups of parasites, the flukes and tapeworms. There are many species of planarians, just as there are many species of amebas and hydras.

A planarian differs from a hydra in that one end of the body has a definite **head,** with eyes and other sensory structures. The head is always directed forward in locomotion; and the body is differentiated into front, or **anterior,** and rear, or **posterior,** ends. A planarian has an elongated flattened body; and when it moves along a stream bed, one surface of the body always remains upward while the other is kept against the bottom. The upper surface is termed **dorsal** (meaning back), and the lower surface, **ventral** (meaning belly). The eyes and other external structures are symmetrically arranged on the two sides, and this is also true of the internal structures.

A planarian **moves** about in a characteristic slow, gliding fashion, with the head bending from side to side as though testing the environment. This gliding is both *ciliary* and *muscular*. If we prod the animal, it hurries away by marked muscular waves. The surface of a planarian consists of an ectodermal epithelium, or **epidermis.** The epidermis is ciliated, particularly on the ventral side, and numerous gland cells, which secrete a mucous material, open on the surface of the body and pour out mucus upon which the worm moves. The cilia obtain traction on this bed of

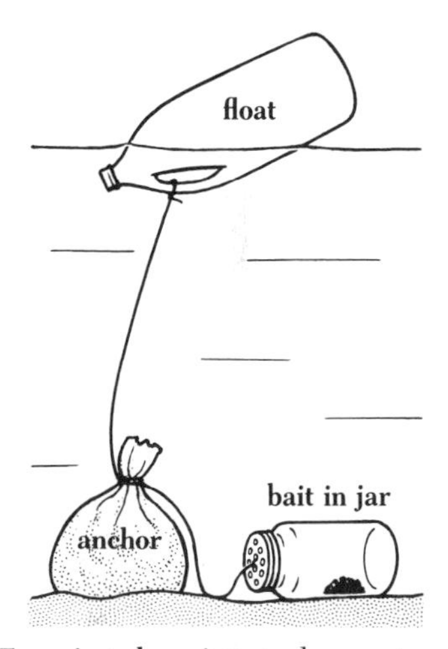

Trapping planarians in deep water is easily done by placing bait (pieces of liver or fish) into a jar with a perforated screw top. The jar protects the bait from large scavengers such as crayfishes, while the small holes in the cover permit bait juices to leak slowly out and small planarians to enter. The float is a plastic bottle; the anchor is a canvas bag filled with stones. The cord from float to anchor can be adjusted for various depths. (After R. Kenk.)

Opposite: **Head of a planarian** *(Dugesia tigrina),* a freshwater flatworm. (P. S. Tice)

Planarian, *Dugesia tigrina,* is bilaterally symmetrical. (R. Buchsbaum)

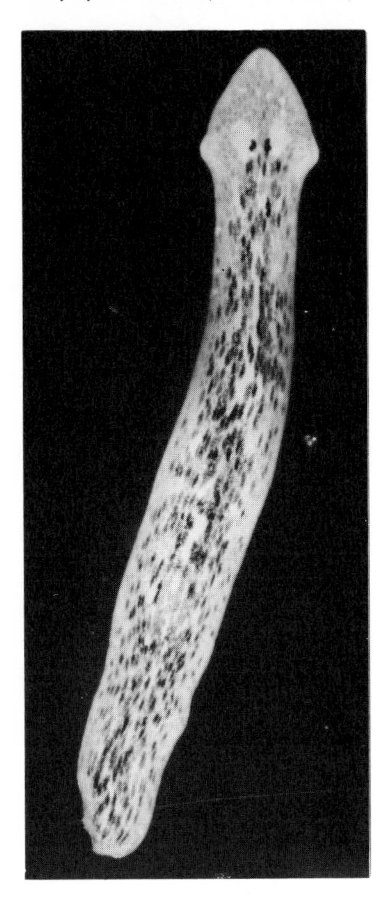

mucus, and as they beat backward, they help to move the animal forward. Planarians do not swim freely through the water, but move only in contact with a solid object or on the underside of the surface film. When a worm leaves the surface, it glides down attached to a thread of mucus. Just underneath the epidermis are layers of muscle cells. The outer layer runs in a circular direction, and the inner layer in a longitudinal direction. Muscles also run between dorsal and ventral surfaces and help to make possible all sorts of agile bending and twisting movements. The muscles are not part of the epithelial cells, as in a hydra, but are independent muscle cells specialized for contraction. Also, they are not developed from ectoderm or endoderm but arise in a different way.

Beginning with flatworms, all more complex animals have a mass of cells between the ectoderm and the endoderm, appropriately called the **mesoderm** ("middle skin"). This layer gives rise to muscles and to other structures that make possible an increasing complexity in animal activities. Like almost all characters of animals, the mesoderm does not appear suddenly in fully developed form. Its early beginnings are perhaps comparable with the ameboid mesenchyme cells in the mesoglea of sponges and of many cnidarians. We recognize it as mesoderm when, as in the flatworms and in all more complex animals, it is more massive than either ectoderm or endoderm and gives rise to definite structures, such as the reproductive structures.

The largest two-layered animals are certain sea anemones and jellyfishes that have attained great size and some degree of body firmness by the secretion of great quantities of jellylike mesoglea in which may be found a sparse population of mesenchyme cells. In three-layered animals the mesenchyme has been increased from a scattered group of wandering ameboid cells to a many-layered structure of tissues that gives firmness and bulk to the body.

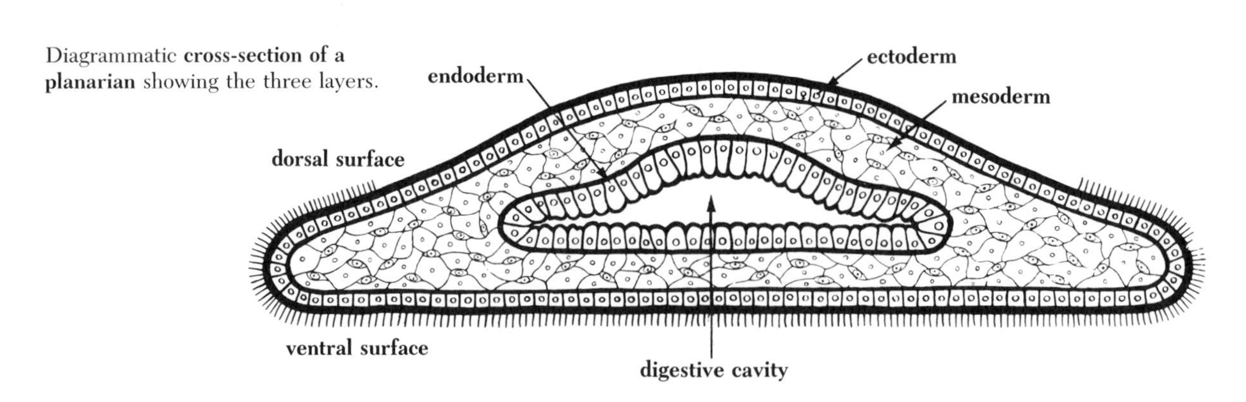

Diagrammatic **cross-section of a planarian** showing the three layers.

The cnidarians are animals organized mostly on the tissue level. From planarians to humans, animals are constructed on a more complex level of organization. Not only do cells work together to form tissues, but tissues of various kinds are closely associated to form one structure, called an **organ,** adapted for the performance of some one function. Thus the human stomach is an organ composed of epithelium, connective tissue, muscle layers, and nervous tissue. The epithelium lines the cavity and contains the gland cells that secrete gastric juices; the muscle layers produce the stomach contractions; the nervous tissue coordinates the muscle contractions and relates stomach activity to the whole body; and the connective tissue binds the various layers together. An organ usually cooperates with other organs or parts in the performance of some life activity, and such a group of structures devoted to one activity is termed an **organ system.** Thus, the stomach is part of the digestive system; and all the other parts of this system, such as the esophagus, liver, and intestine, are necessary to the proper performance of digestion. The more complex animals are made up of a number of such organ systems, as the digestive system, excretory system, circulatory system, nervous system, reproductive system, and so on. Flatworms have simpler versions of most of these, and they are the simplest phylum of animals built on the **organ-system level of construction.**

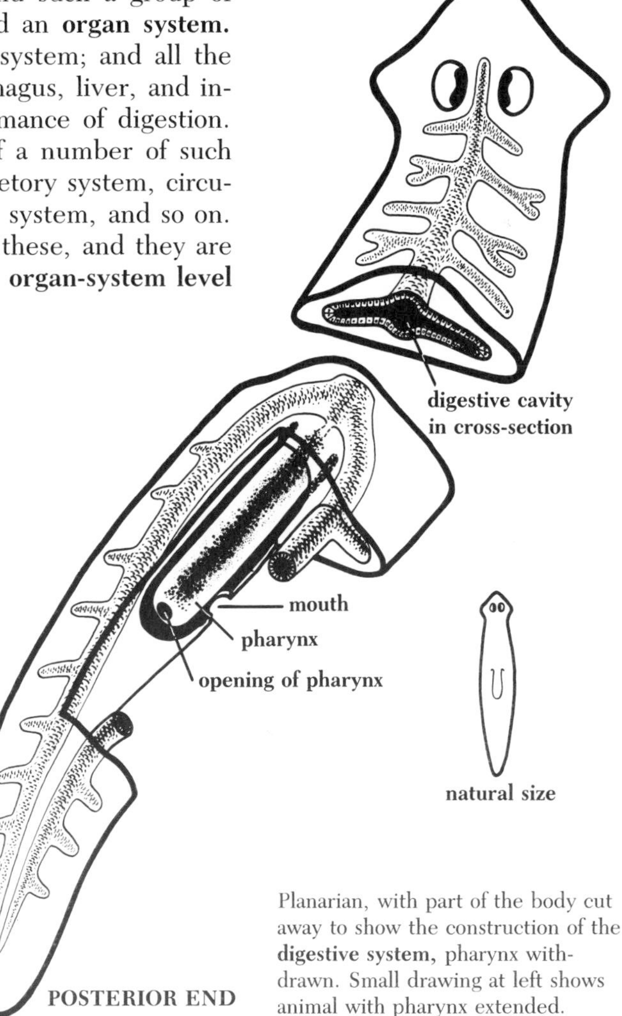

Planarian, with part of the body cut away to show the construction of the **digestive system,** pharynx withdrawn. Small drawing at left shows animal with pharynx extended.

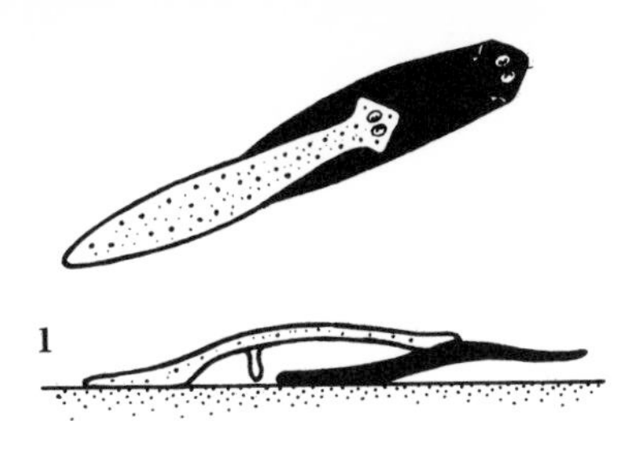

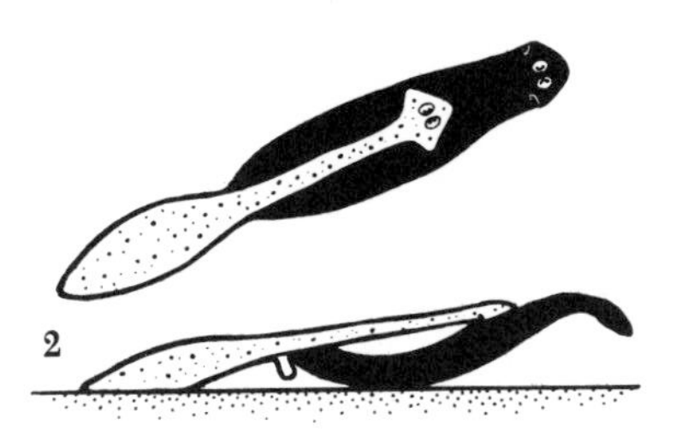

A planarian preys on another. (Stages 1 and 2 are shown in both top and side views.) **1:** A spotted planarian lunges and attaches its sticky head to the back of a blackish planarian. **2:** The tail of the spotted predator sticks to the bottom, as the worm shortens and hauls in its prey. **3:** The predator loops its body around the prey and begins to feed on it. (Modified after J. B. Best.)

In the **digestive system** of a planarian the mouth, curiously enough, is not on the head but near the middle of the ventral surface. It opens into a cavity which contains a tubular muscular organ, the **pharynx,** attached only at its anterior end. The pharynx contains complex muscle layers and many gland cells. By means of the muscles, the pharynx can be greatly lengthened and is protruded from the mouth for some distance during feeding. Planarians feed on small live animals or on the dead bodies of larger animals. They can detect the presence of food from a considerable distance by means of sensory cells on the head. They move toward their food, mount upon it, and press it against the bottom with their muscular bodies. Struggling prey can be held in this way more easily after they have become entangled in slimy secretions from the worm. The pharynx is protruded posteriorly through the mouth and inserted into the prey. Enzymes secreted by the pharynx soften the prey tissue while sucking movements of the muscles of the pharynx tear the tissue into microscopic bits which are then swallowed, along with the juices of the prey.

From the anterior attached end of the pharynx the rest of the digestive system extends as a branching **digestive cavity** throughout the interior of the animal. It consists of three main branches, one that runs forward and two that pass backward, one on either side of the pharynx, to the posterior end. Numerous and fairly regularly spaced side branches provide for the distribution of the food to all parts of the body; thus, as in cnidarians, it is a gastrovascular cavity. The digestive epithelium consists simply of the endoderm and corresponds to the endoderm of a cnidarian.

There is little **digestion** of food in the gastrovascular cavity, for the food is mostly broken down into small particles before it enters the cavity and is thus ready to be taken up by the epithelial cells in ameboid fashion and incorporated into food vacuoles. The digested food is absorbed and passes by diffusion throughout the tissues of the body. There is only one opening to the gastrovascular cavity; indigestible particles are eliminated through the mouth.

Experiments on a common species of planarian that was fed on liver showed that, after a meal, all the ingested liver was taken into the epithelial cells in about eight hours, and that three to five days were required for the complete digestion of the food vacuoles so formed. Much of the food was found to be converted into fat, which was stored in the digestive epithelium.

Practically all animals can store food reserves upon which they draw in time of need. A small animal such as an ameba stores very little and, unless it goes into the inactive encysted state, will die after about two weeks without food. Starved hydras survive much longer periods. But planarians are peculiarly adapted to go for many months unfed while remaining active. During starvation they first use the food stored in the digestive epithelium, whole cells breaking down. Later they begin to digest other tissues, the reproductive organs usually going first. Externally one can observe only that the worms grow steadily smaller though retaining the same general appearance. A worm starved for six months may shrink from a length of 20 millimeters to 3 millimeters. Because of their ability to go for months without food, planarians make ideal household pets for busy or absent-minded people.

The region between the outer protective epidermis (ectoderm) and the inner digestive epithelium (endoderm) is filled with various organs surrounded by **mesenchyme,** mostly in the form of a solid tissue although some mesenchyme cells are free and move about. The muscle layers already mentioned are embedded in this mesenchyme, and it also contains many gland cells that open to the surface and secrete mucus or sticky substances. The gland cells are largely derived from the ectoderm; but the organs, muscles, and mesenchyme are mesodermal.

Flatworms have *no special respiratory or circulatory systems.* Respiratory exchange occurs through the epidermal epithelium of the whole body surface, and probably also through the digestive epithelium of the gastrovascular cavity. For enough oxygen to penetrate the mesenchyme (or for carbon dioxide to pass out) by the slow process of diffusion through the tissues, the distance to be traveled cannot be very great. This may explain the flatness of flatworms.

A new system that is not found in any of the forms already studied is the **excretory system.** This lies in the mesenchyme and consists of a network of fine tubules that run the length of the animal on each side and open to the surface by minute pores. Side branches from these tubes originate in the mesenchyme in tiny enlargements known as **flame bulbs.** Each flame bulb has a fluid-filled center in which beats a tuft of flagella that suggests a flickering flame. The center of the flame bulb is continuous with the cavity of the tubules of the system, and the flagella drive a current of fluid along the tubules to the pores. As with the contractile vacuoles of freshwater protozoans, the primary function of the flame bulb system in these freshwater flatworms is apparently to remove excess water from the tissues.

A planarian has a highly complicated **reproductive system** for sexual reproduction, in contrast to the simple aggregations of eggs or sperms found among cells of other tissues in the sponges and cnidarians. Discrete ovaries and testes arise in the planarian mesenchyme; and there is a system of tubules and chambers in

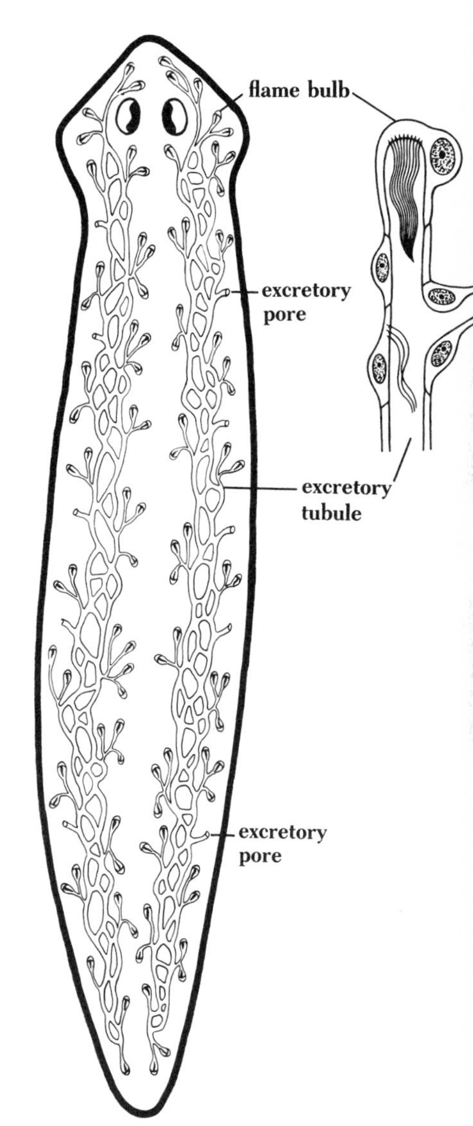

Excretory system of a planarian. At right is shown a single flame bulb and a portion of an excretory tubule.

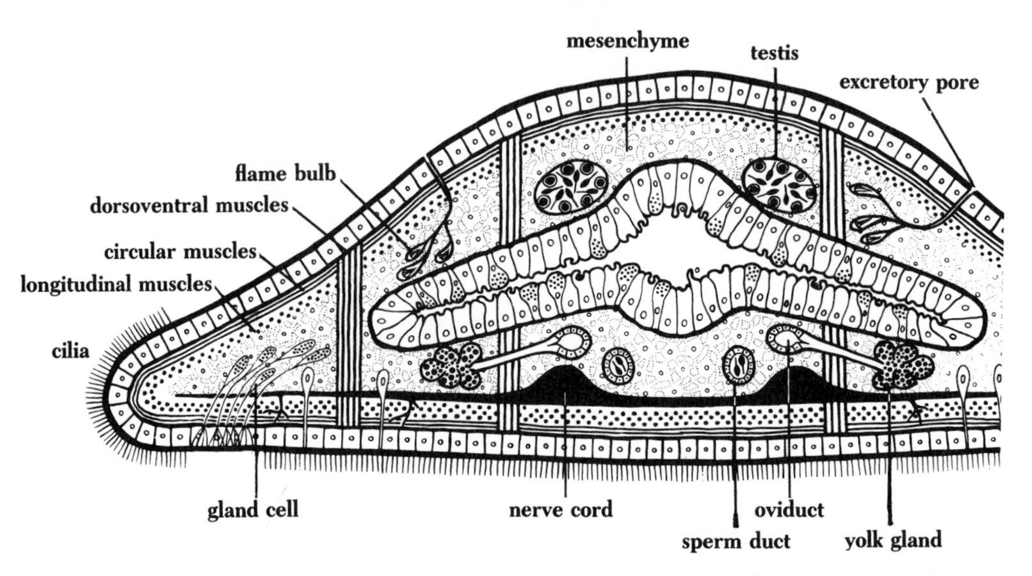

Cross-section through a **sexually mature planarian,** showing internal organs.

which fertilization occurs, as well as complex sex organs for the transfer of sperms. Planarians are hermaphrodites, forming both male and female reproductive organs in every individual, but exchange of sperms takes place and cross-fertilization is effected. After the breeding season, the reproductive system degenerates and is regenerated anew at the beginning of the next sexual period.

When sexually mature, each worm has a pair of **ovaries** close behind the eyes. From each ovary a tube, the **oviduct,** runs backward near the ventral surface. Multiple **yolk glands,** consisting of clusters of yolk cells, lie along the oviduct, into which they open. From each of the numerous **testes** along the sides of the body leads a delicate tube, and all these tubes unite on each side to form a prominent **sperm duct,** which runs backward near the oviduct. The sperm ducts packed with sperms during the time of sexual activity, connect with a muscular protrusible organ, the **penis,** which is used for transfer of the sperm to another planarian. The penis projects into a chamber, the **genital chamber,** into which there also open the oviducts and a long-stalked sac called the **copulatory sac.** The genital chamber opens to the exterior by a **genital pore** on the ventral surface behind the mouth.

Although each planarian contains a complete male and female sexual apparatus, self-fertilization does not occur; instead, two worms come together and oppose their ventral surfaces. The penis of each is protruded through the genital pore and deposits sperms in the copulatory sac of its partner. After copulation, the worms separate. The sperms soon leave the copulatory sac and travel up the oviducts until they reach the ovaries, where they fertilize the ripe eggs as they are discharged. The fertilized eggs

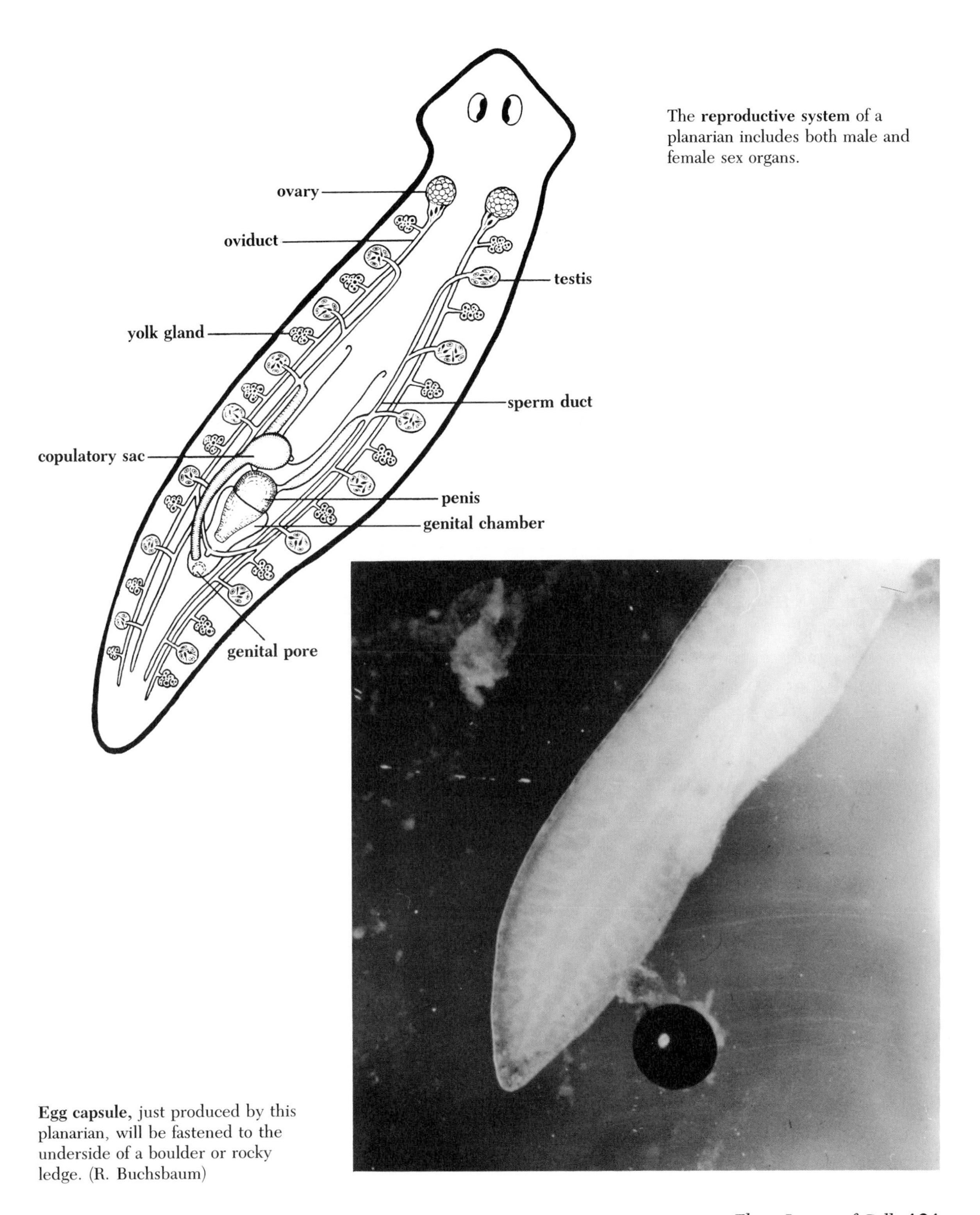

The **reproductive system** of a planarian includes both male and female sex organs.

Egg capsule, just produced by this planarian, will be fastened to the underside of a boulder or rocky ledge. (R. Buchsbaum)

pass down the oviducts, and at the same time yolk cells are discharged from the yolk glands into the oviduct. When eggs and yolk cells reach the genital chamber, they become surrounded by a shell to form an **egg capsule.** The eggs of most flatworms are peculiar in that food reserves do not occur in the eggs themselves but are in yolk cells that accompany the eggs. The capsules (each containing fewer than ten eggs and thousands of yolk cells) are passed out through the genital pore and are fastened to objects in the water. They hatch in two or three weeks into minute worms, like their parents except that they lack a reproductive system.

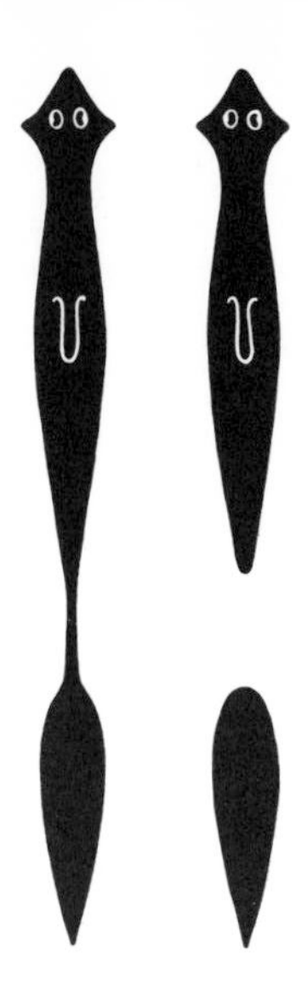

Asexual division. *Left:* Just before division. *Right:* Just after. The rear piece will soon develop a head, pharynx, and other structures.

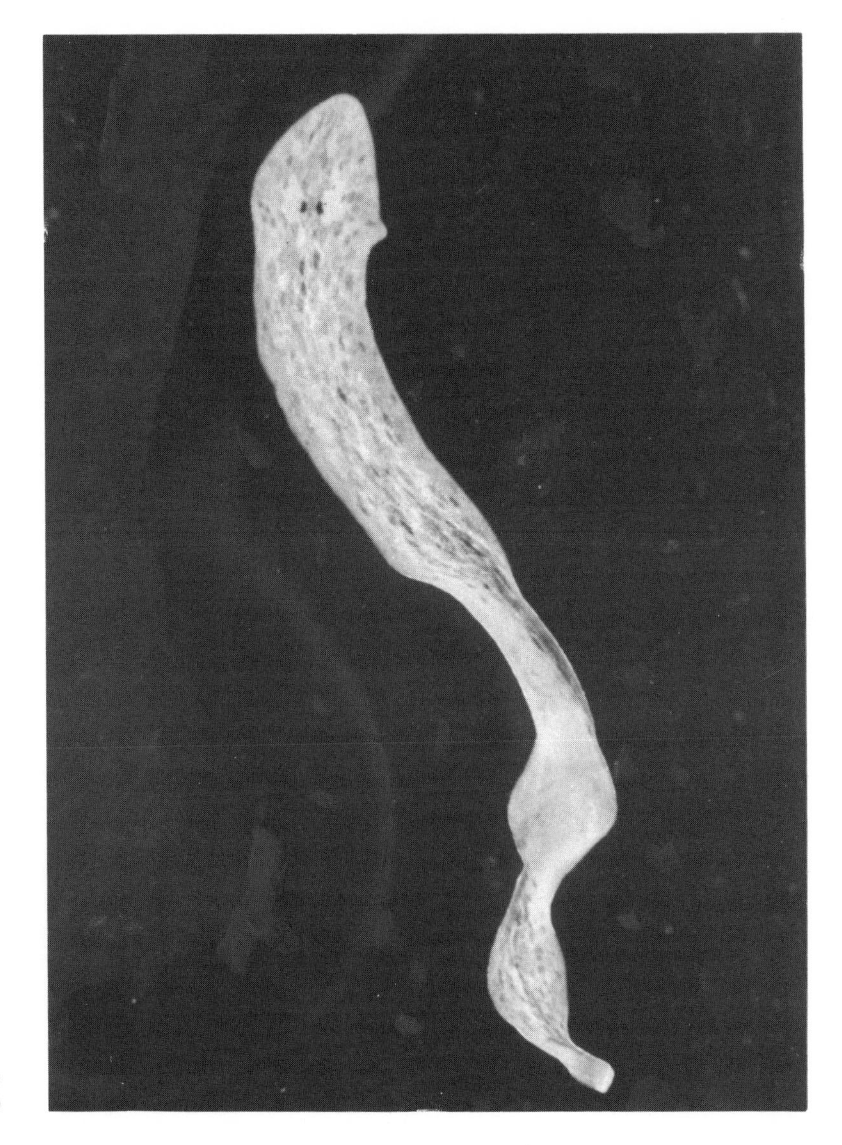

Asexual reproduction. The rear end of this planarian, already marked off by a constriction, will soon separate and regenerate to become a complete worm. Meanwhile the head end will regenerate a new tail, and there will be two worms where there was only one before. (R. Buchsbaum)

Many planarians reproduce only sexually, but some multiply by **asexual reproduction.** In this process the worm, without any evident preliminary change, constricts at a region behind the pharynx, and the posterior piece begins to behave as though it were rebelling against the domination of the anterior piece. When the whole animal is gliding quietly along, the posterior part may suddenly grip the bottom and hold on, while the anterior piece struggles to move forward. After several hours of this "tug of war" the anterior piece finally breaks loose and moves off by itself. Both pieces regenerate the missing parts and become complete worms. Species that have this habit often go for long periods without sexual reproduction, and in fact some of them rarely develop sex organs.

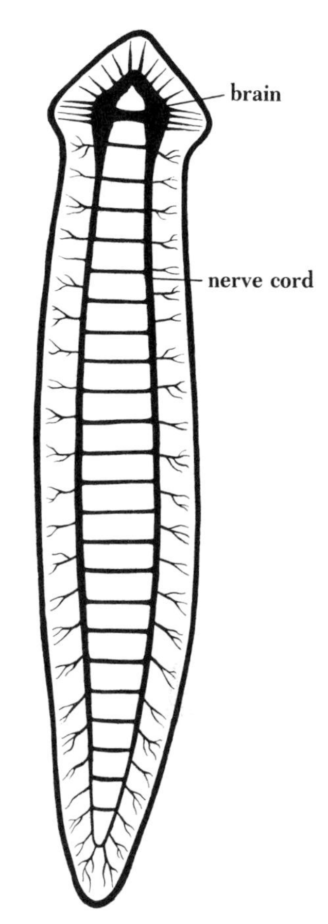

Nervous system of a planarian

Flatworms have a **central nervous system,** the kind of nervous system possessed and further centralized by most of the more complex animals. In the head of a planarian there is a large **ganglion** (plural: *ganglia*), a concentration of nervous tissue into a discrete mass. From this bilobed head ganglion, called the **brain,** two strandlike concentrations of nerve cells, the **nerve cords,** run backward through the mesenchyme near the ventral surface. From these ventral nerve cords numerous side branches are given off to the body margins. The two cords are connected with each other by many cross-strands like the rungs in a ladder, so this type of system has been called the "ladder type" of central nervous system. The brain and the two cords constitute the central nervous system, a kind of "main highway" for nerve impulses going from one end of the body to the other. The brain is not necessary for the muscular coordination involved in locomotion, for a planarian deprived of its brain can still move along in coordinated fashion. The brain serves chiefly to initiate behavior and also as a sensory relay that receives stimuli from the sense organs and sends them on to the rest of the body. The result is a much more closely knit behavior than is possible with the diffuse, noncentralized nerve net of a hydra, which lacks definite pathways and a coordinating center.

> In addition to the central nervous system, nerve nets occur locally in planarians and in almost all more complex animals. In humans, for example, a well-developed nerve net (connected with the central nervous system) serves the wall of the intestine.

Conditions in the external world are conveyed to the nervous system by **sensory cells,** slender elongated cells that lie, with their pointed ends projecting from the body surface, between the epithelial cells. Different ones are specialized to receive the stimuli of touch, water currents, and chemicals. Sensory cells are

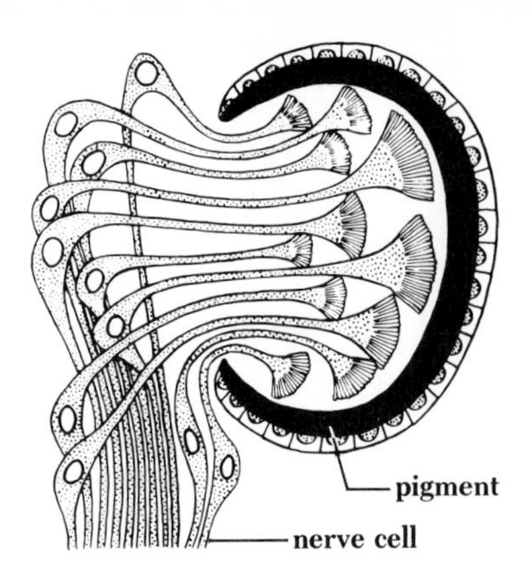

Section through the **eye** of a planarian. The eye is sensitive only to light coming toward the open end of the pigment cup. The light-sensitive nerve cells run to the brain. (Combined from R. Hesse and from Röhlich and Török.)

distributed all over the body surface, but in addition are concentrated in the head, especially in the **sensory lobes,** pointed projections on each side of the head. If the sensory lobes are cut off, the worm has difficulty finding food. The two **eyes** are sense organs specialized for light reception. Each consists of a bowl of black pigment filled with special sensory cells whose ends continue as nerves that enter the brain. The pigment shades the sensory cells from light in all directions but one, and so enables the animal to respond to the direction of the light. Unlike other regions of the epidermis, that immediately above the eye is unpigmented, and thus allows light to pass through the sensory cells. Planarians whose eyes have been removed still react to light but more slowly and less exactly than normal worms. This indicates that there must be some light-sensitive cells over the general body surface.

By virtue of abundant sensory cells, specialized sense organs, and a centralized nervous system, planarians show a more varied **behavior** and much more rapid responses than do hydras. Planarians avoid light and are generally found in dark places, under stones or leaves of water plants. If placed in a dish exposed to light, they immediately turn and move toward the darkest part of the dish. They respond positively to contact and tend to keep the under surface of the body in contact with other objects. They also react to chemical substances in the water and quickly respond to the presence of food by turning and moving directly toward it. That is why a piece of raw meat placed in a spring

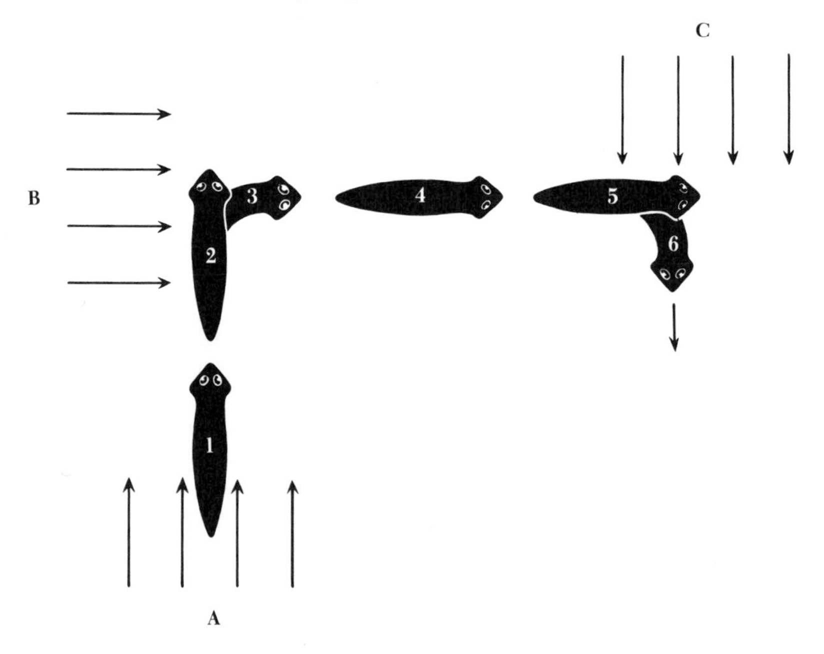

Orientation to light. 1–6: Successive positions of a planarian. In **1**, the animal was moving away from light coming from source **A.** When it reached position **2**, the light was turned off at **A** and on at **B.** The worm turned and moved away from the light. At position **5**, light **B** was turned off and **C** turned on; the worm again oriented away from the light. (Based on W. H. Taliaferro.)

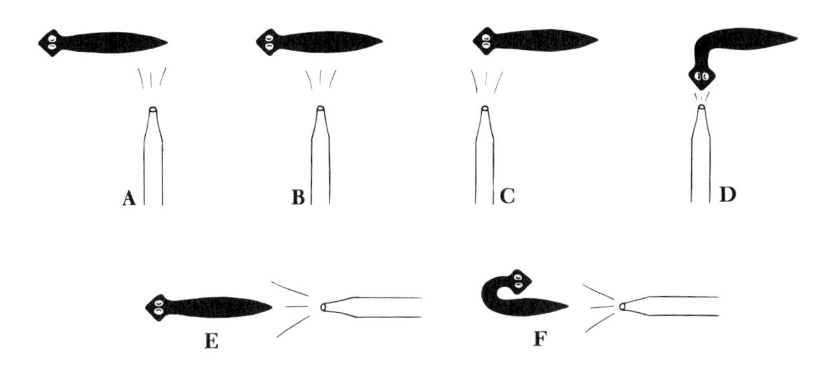

Reactions to water currents produced by a pipet. **A, B:** The current strikes the middle or rear of the body and there is no response. **C, D:** The current strikes the sensory lobe on the side of the head and the worm turns toward the current. **E, F:** The current from the rear passes along the sides of the body to the sensory lobes and the worm turns around toward the current. In nature these reactions orient the worm upstream. (Modified from I. Doflein.)

inhabited by planarians attracts hordes of worms, which glide upstream toward the food, guided by the meat juices in the current of water. Planarians react to water currents, and some species regularly move upstream against a current. They also respond to the agitation of the water produced by the animals upon which they prey.

The flatworms, as illustrated by the planarians, differ from the two-layered animals in a number of important characters that are possessed by most of the more complex animals. Flatworms have specialized anterior and posterior ends and dorsal and ventral surfaces. They have a definite head, with a concentration of sense organs, and a central nervous system. And they have an extensively developed third layer of cells, the mesoderm. Either by itself, or in combination with ectoderm or endoderm, the mesoderm gives rise to organs and organ systems.

The Shapes of Animals

Although animals range in size from microscopic protozoans to massive whales, there are only three basic styles in animal shapes: spherical, radial, and bilateral. Some version of these three kinds of symmetry is characteristic of almost all multicellular animals and many protozoans. However, forms that have a discrete shape but none of the basic symmetries are seen in various protozoans, such as some ciliates and dinoflagellates; such organisms are said to be asymmetrical.

A spherical shape is assumed by any isolated small quantity of fluid because of the physical forces acting upon it. Small bits of protoplasm or single cells assume this spherical form unless a specific structure, such as a stiff surface layer or skeleton, enables them to maintain some other shape. An ameba must ex-

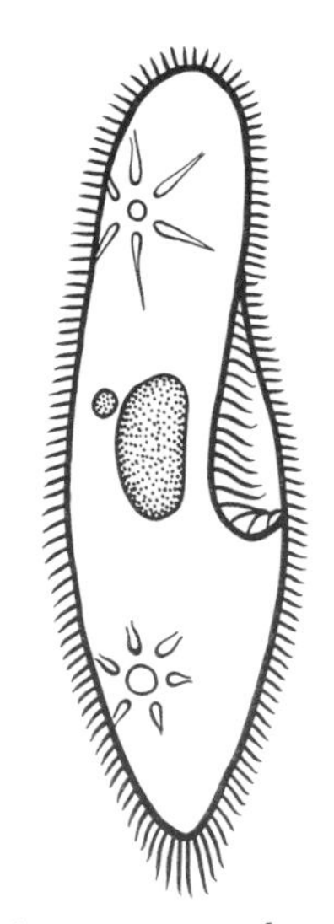

Asymmetry, a ciliate.

pend energy in extending pseudopods and becomes spherical when at rest. This type of shape, called **spherical symmetry,** is characterized by the arrangement of structures with reference to the center of a sphere. Since all radii are alike, a spherical animal can be divided into two similar pieces by a cut in any direction through the center. There is no front or rear, no top or bottom, no right or left sides, no ends—at least no permanent ones. Such a form would be a disadvantage in directed locomotion and in fact is most typical of free-floating organisms that do not move under their own power, either toward food or away from predators. On the other hand, such organisms can respond to the presence of food or other features of the external environment in any direction. Spherical symmetry is rare and is found among adult animals only in protozoans such as radiolarians that float near the surface of the oceans and feed by means of pseudopods that radiate out in all directions.

Spherical symmetry, a radiolarian.

In spherical protozoans that live attached to solid surfaces, such as some foraminiferans and heliozoans, the symmetry may be modified by a stalk that attaches the body to the substrate or by an opening in the shell through which pseudopods emerge. Such modified spherical symmetry can be called **bipolar symmetry,** as the sphere has differentiated ends, an attached end bearing the stalk or opening, and a free end opposite to this structure. (The planet Earth, although approximately spherical, shows bipolar symmetry; it cannot be divided into two similar pieces by all cuts passing through the center, but only by cuts passing through the center and both poles—on the lines of longitude—or by a cut passing through the equator—if the north and south poles are considered equivalent.)

Bipolar symmetry, a stalked protozoan.

If an animal becomes differentiated, not around a central point, but around an axis that runs between two ends, it is said to exhibit **radial symmetry.** The polyps and medusas of cnidarians are radially symmetrical; all radii are alike at any particular level, but there is a differentiation between levels along an axis from the mouth end to the end opposite the mouth. A radial animal may be cut into two similar pieces by a lengthwise cut (but not a crosswise cut) through the center in any direction, for the animal is alike all around the circumference; it has no differentiated sides. Lacking a definite side to go first in locomotion does not prevent radial animals from moving in a directed fashion. Many medusas and even some sea anemones execute directed movements. But the majority of cnidarians either drift with water currents much of the time or live a sedentary or sessile life. A sessile animal is not threatened from below, and the basal end is specialized only for attachment. The circle of tenta-

Radial symmetry, a sea anemone.

cles extending from the exposed mouth end is prepared to meet the environment from above and from all sides. Radial symmetry is seen in some protozoans and sponges but is most characteristic of cnidarians.

In many cnidarians such as sea anemones and corals, the radial symmetry is modified by an elongation of the mouth with grooves at both ends and by internal structures, so that all radii are no longer alike. The animal can now be divided into two similar halves only by cutting it lengthwise in two particular planes, one running through the long axis of the mouth and the other at right angles to this axis. Such a modification of a basically radial symmetry is called **biradial symmetry.** Only one whole phylum of animals, the phylum Ctenophora, is characterized by biradial symmetry. In ctenophores the biradiality begins as early as the eight-celled embryonic stage and continues throughout development.

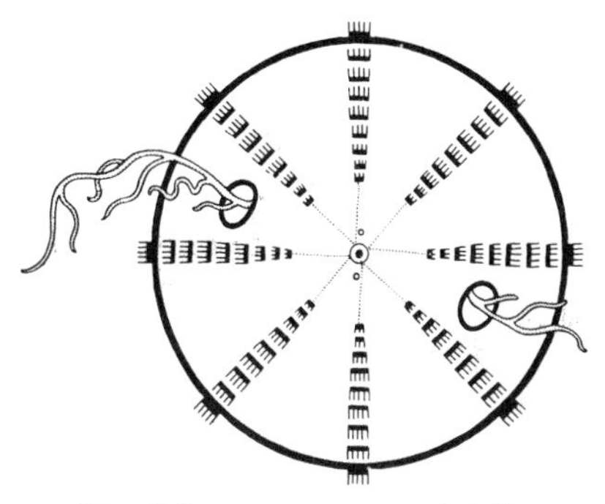

Biradial symmetry, a comb jelly.

In a planarian there is a definite front (anterior) end that bears the sense organs and that always ventures first into a new environment, and a rear (posterior) end that merely follows along. Such an animal is open to attack from the rear and from the sides, as compared with a polyp or medusa, which can detect predators and ward off their attacks on all sides. However, the concentration of sense organs in the front end may enable the animal to detect danger ahead and so better avoid it. Also, specialization of anterior and posterior ends is related to more active and agile seeking out of food than in radial animals, which usually depend upon chance encounters with prey.

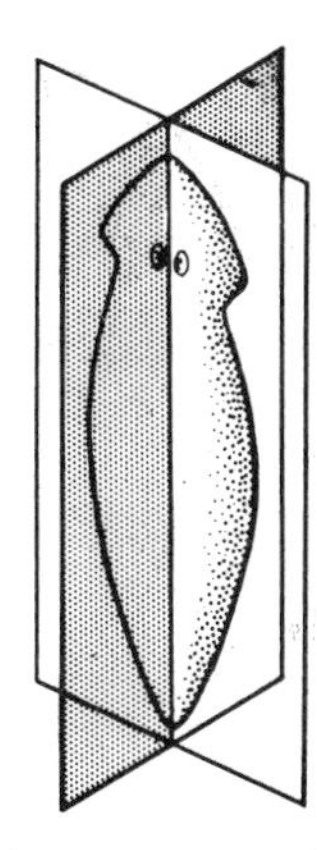

Bilateral symmetry, a planarian.

The specialization of the head end is accompanied by a differentiation of the upper (dorsal) and lower (ventral) surfaces of the body. The undersurface of a planarian bears the mouth and most of the cilia and differs from the upper exposed surface. This type of body form, in which there is a difference between front and rear ends and upper and lower surfaces, is called **bilateral symmetry.** The term "bilateral" means *two sides* and refers not to these ends or surfaces but to the fact that in these animals the body structures are arranged symmetrically on either side of a central plane that runs from the middle of the head end to the middle of the tail end. The paired eyes and sensory lobes of planarians (like the paired structures of humans) occur at equal distances on either side of this plane. Single organs are generally located in the midline and are bisected by the plane. Unlike radial animals, which lack definite sides, bilateral animals have right and left sides, and they can be cut into two similar pieces by only one particular cut—along the plane that runs down the middle of the body from head to tail and from back to belly. The two resulting pieces are equivalent but are mirror images of each other.

Bilateral symmetry is often imperfect in some degree. For example, in humans, the right arm is commonly larger and stronger than the left, and in the brain there is usually a well-developed speech center on the left side but not on the right. Much more marked asymmetries occur in some other bilateral animals, such as the coiling of snail viscera and shell into a spiral.

Many bilateral animals lead sessile or sedentary lives, and among these, the differentiation of the head and its sense organs tends to be reduced; some have even evolved a modified radial symmetry. Such **secondary radial symmetry** is seen in particular in sea stars, sea urchins, and their relatives (and some of these have become secondarily bilateral!).

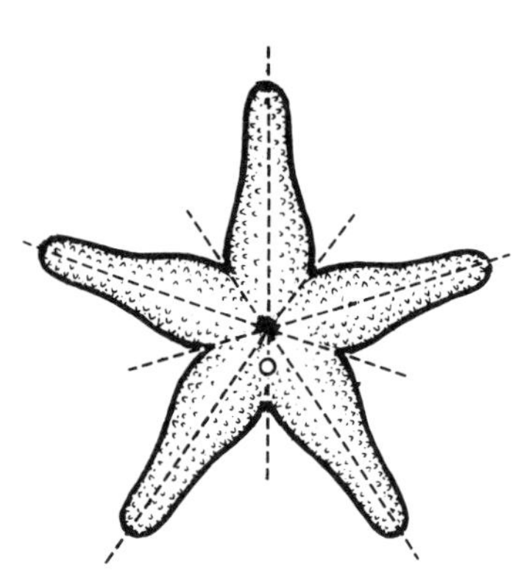

Secondary radial symmetry, a sea star.

Approaches to bilateral symmetry have evolved independently among members of some groups in which it is not the characteristic form. For example, some ciliates are bilaterally symmetrical. In many sea anemones and in soft corals the elongated mouth and gullet have a groove at only one end. Such an animal, though still radial in basic organization and in behavior, can be divided into two similar halves by only one particular cut, passing lengthwise through the long axis of the mouth.

Beginning with the flatworms, all more complex animals (unless secondarily modified) are bilaterally symmetrical. The bilateral body form lends itself readily to streamlining in an anterior-posterior direction and, with a head to direct movements, characterizes many animals whose success depends on being able to move fast. Yet no one kind of symmetry is "better" or "more advanced" than any other, and the advantages and disadvantages of each must be seen in the context of an animal's habits and habitat. An example of unconventional symmetry is the octopus, which is basically a bilaterally symmetrical animal and has in its head what are among the most elaborate eyes and brain of any invertebrate. But it is streamlined in such a way that the dorsal surface goes first and the head trails behind during fast swimming, and its agile arms are arranged radially around the mouth, enabling it to move over the bottom in any direction. Its symmetry is not a fixed or arbitrary character of its group but a functional part of its equipment for a particular way of life.

Opposite: **Octopus** (a mollusc; see chapter 15) combines elements of bilateral and radial symmetry. (R. Buchsbaum)

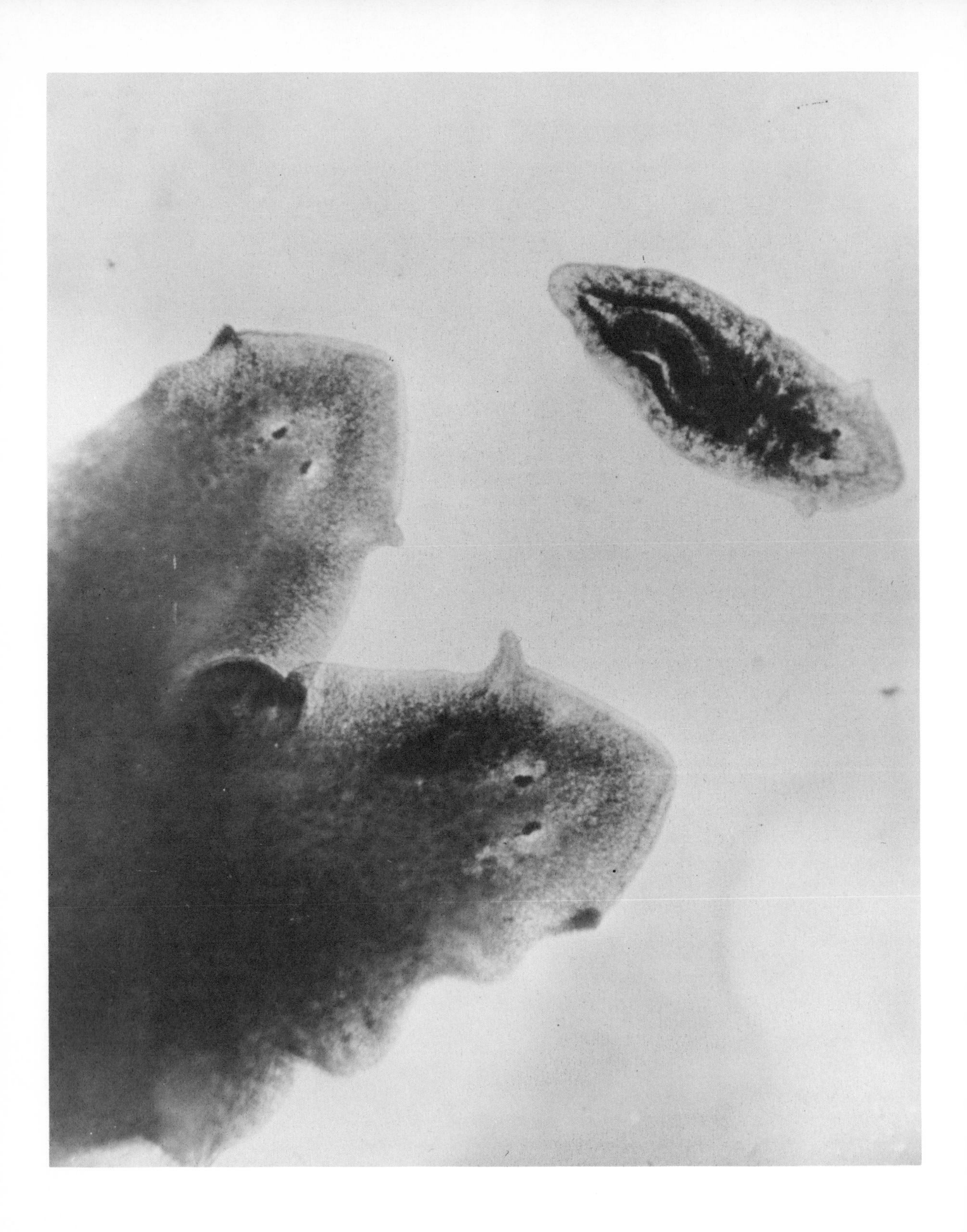

Chapter 10

New Parts from Old

Regeneration, or the ability to renew tissues and repair damage, at least to some degree, is universal among animals. In humans, the epidermal cells of the whole body surface are continually sloughed and replaced; wounds of considerable size heal and broken bones grow together again, but a lost finger or toe cannot be regenerated. Lizards, on the other hand, can pinch off and replace the tail (although the new one lacks vertebras) and salamanders can regrow an entire limb (but frogs cannot). Among many invertebrates, the power of repair is much greater. An earthworm can replace its head, a sea star its arms, and a lobster a leg or an antenna, but in general, those invertebrates with relatively complex organ systems show a correspondingly low capacity for regeneration. Those organized on a simpler level can usually regenerate to a greater extent.

Many **protozoans** have a notable capacity for regeneration. Any piece that is not too small will re-form a complete and perfect organism if it contains a nucleus; nonnucleate pieces fail to regenerate. This is not surprising when we recall that in a normal method of reproduction in protozoans the nucleus divides in two and the cell separates into two parts, each containing a nucleus and each capable of growing into a complete individual.

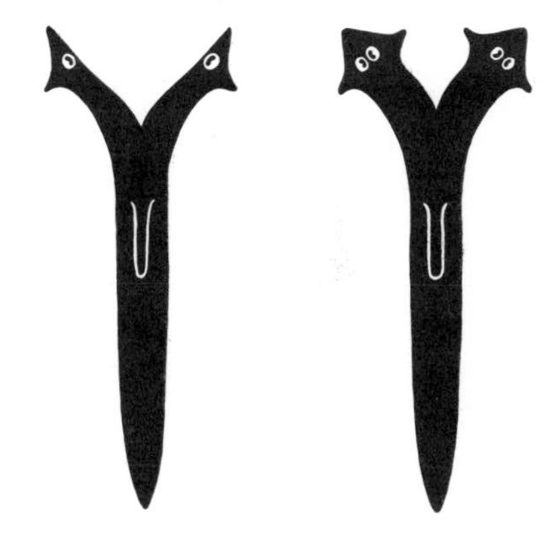

Two-headed planarian can be produced by making a longitudinal cut through the anterior end of a worm and then renewing the wound on successive days so that the cut edges cannot grow together again. Each half of the head will regenerate the missing parts. (Based on C. M. Child.)

Opposite: **Regeneration** makes strange bedfellows. The two-headed planarian was produced by making a cut down the middle of the anterior part of a worm. Both heads are of equal dominance over the rest of the body and often appear to try to go off in opposite directions. Some monsters do split apart, and each part regenerates into a complete worm. The very tiny planarian is of the same species as the two-headed monster and is not a baby. It is small (only 3.5 mm long) and disproportionately wide because it has regenerated from a small transverse slice of a normal-sized worm. In such recently regenerated worms, the pigmentation is light and the internal organs are more easily visible. *Dugesia dorotocephala.* (R. Buchsbaum)

Among the loosely organized **sponges** we saw a very marked ability to regenerate; separate cells from finely macerated sponges aggregate in small masses and develop into complete sponges.

Cnidarians also regenerate very well. Disaggregated hydra cells, if packed into a clump, will regenerate one or more complete hydras (the number depending on how many cells are in the clump). Pieces of a hydra grow into small but complete hydras.

Similarly, some **planarians** will regenerate complete worms from almost any piece. They are easily kept in the laboratory, have been the subject of many experiments on regeneration, and illustrate certain characteristics of regeneration that apply to other animals as well.

In the first place, any piece cut from a planarian usually retains the same **polarity** it had while part of the whole animal; that is, a head regenerates from the cut end that faced anteriorly in the whole animal, and a tail regenerates from the cut end that faced posteriorly. Likewise, a piece cut from a hydra will regenerate a mouth and tentacles at the end that was closer to the mouth region and will regenerate a basal disk at the end that was nearer the base. This anteroposterior or mouth-base differentiation is present even in relatively small portions and throughout the entire animal.

> That polarity of a hydra does not depend on a structural or functional "top" and "bottom" of individual cells was shown in an experiment. The mouth region and base of a hydra were cut off. The remaining column was cut into 10 pieces, which were

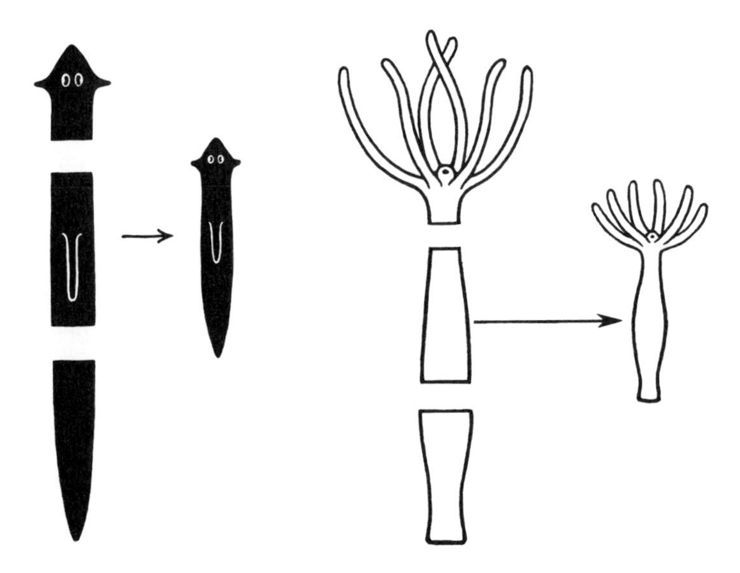

Polarity of a regenerating piece of a planarian is retained—a head grows from the anterior end and a tail grows from the posterior end. A piece of a hydra also regenerates according to its original polarity. (Based on C. M. Child.)

then grafted back together with each piece turned upside-down but replaced in its original position along the column. Now every cell in the animal was physically turned around in polarity, but the tissue that was originally nearest the mouth end regenerated a mouth and tentacles and the tissue that was originally nearest the base regenerated a basal disk. This experiment has been interpreted as indicating that polarity does not depend on the orientation of individual cells, but on differences at the tissue level.

Another generalization drawn from experiments is that the **competence** for head regeneration is greatest near the anterior end and decreases toward the posterior end. Pieces from the anterior regions of a planarian regenerate faster and form bigger and more normal heads than do pieces from posterior regions, and there is a gradual change in these respects along the antero-posterior axis. In some species of planarians, only pieces from anterior regions are able to form a head, while those farther back effect repair but do not regenerate a head.

The head of a planarian is dominant over the rest of the body, in that it *induces* the formation of more posterior structures and *inhibits* the formation of extra heads. In general, any level controls the level posterior to it, and this series of controls maintains the normal form of the worm. One way in which this **dominance** can be demonstrated is by means of grafting. If a small bit of the head region of a planarian (donor) is grafted into another individual (host) at a posterior level, it will not only grow out into a head but will induce the adjacent host tissue to form, for exam-

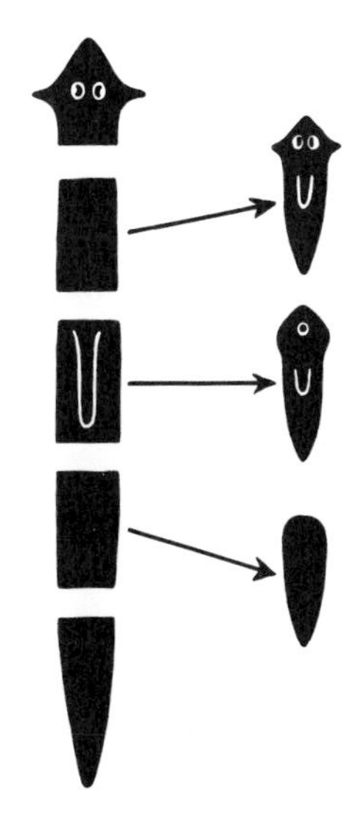

Competence for regeneration of the head decreases from the anterior to the posterior end. (Based on C. M. Child.)

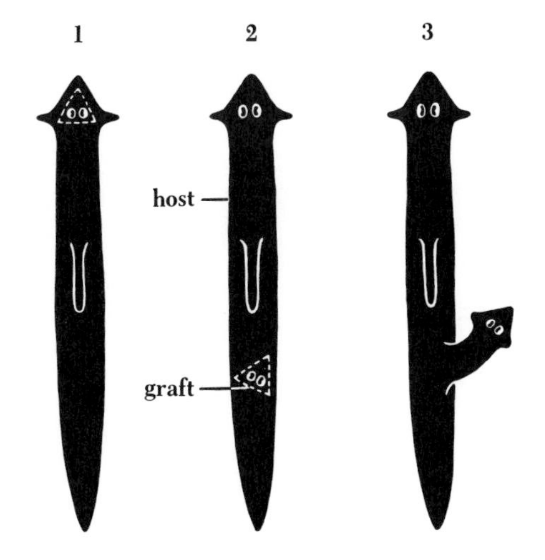

Grafting. 1: A small piece, indicated by broken lines, is cut out of the head of the donor. **2:** The graft is placed in a wound made in the posterior region of the host. **3:** The graft has grown into a small head. (Based on F. V. Santos.)

ple, a new pharynx. If a head piece is grafted into a planarian and then the host head is cut off, the grafted head may inhibit the anterior cut surface from regenerating a head (as it ordinarily would) and cause it to form a tail instead. In other words, grafts of head pieces tend to reorganize the adjacent tissues into a whole worm in relation to themselves. Grafts from tail regions do not have these effects but are generally absorbed.

The dominance of the head over the rest of the body is limited by distance. The fate of a head graft therefore may depend on its location. If placed close behind the host head, the grafted head will usually be inhibited, and will fail to live or grow. If placed far enough behind the head, the graft survives. This has suggested that dominance may depend on substances that are continuously secreted in the head, diffuse posteriorly through the body, and are gradually destroyed. If the animal grows to a sufficient length, its rear part appears to get beyond the range of dominance of the head, as in asexual reproduction when the rear part starts to act behaviorally isolated and finally constricts off as a separate animal. A reduction of the control of the head over the body is an important factor in asexual division, as shown by

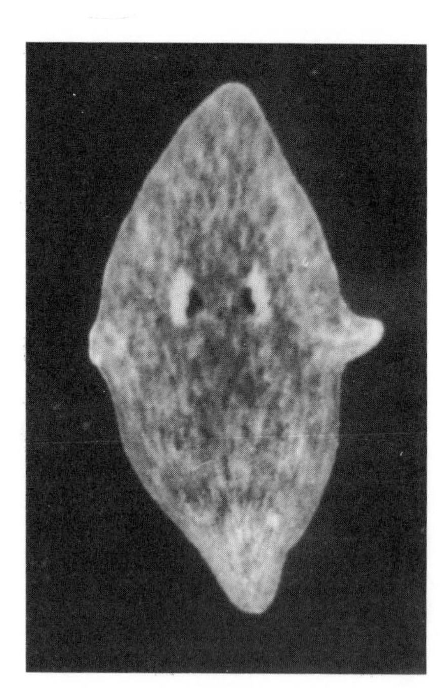

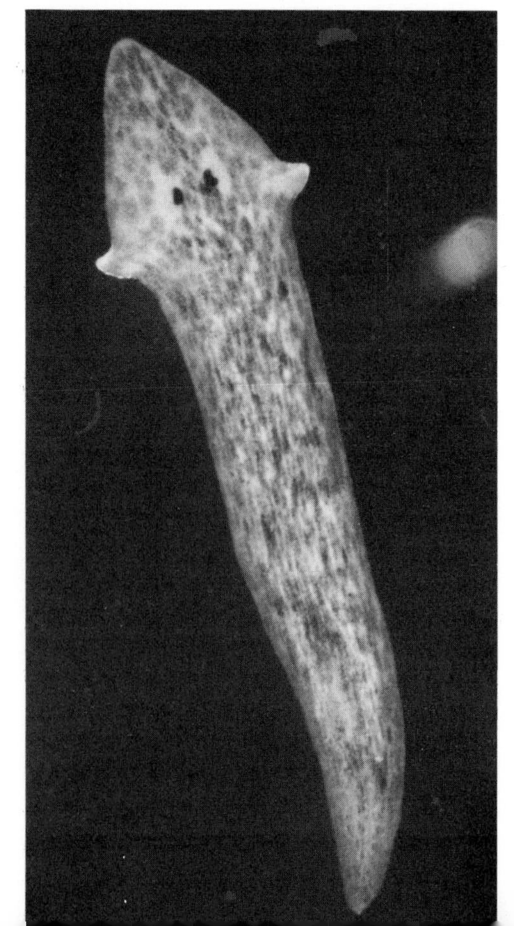

Regulation of shape is part of the process of regeneration. *Left:* A head piece has just begun to regenerate a posterior tip. This oddly shaped worm can move actively about its dish, but cannot yet feed. *Right:* As the digestive tract and other posterior parts are regenerated, the proportions of the worm slowly become more normal. How the body assesses and then adjusts its own proportions is a question that applies also to normal development from embryo to adult. (R. Buchsbaum)

the fact that separation of the rear part can be speeded by cutting off the head.

All these facts indicate that there is an **axial gradient,** a gradation related to regenerative ability, in some essential substances or activities or structures along the anteroposterior axis of a planarian. Similar experiments on hydras and hydroids have yielded similar results, and in these radially symmetrical animals, the gradient extends from the mouth end to the base of the stalk. The basis of the axial gradient is not known, but most of the observations and experimental results can be accounted for by *concentration gradients of specific inducers and inhibitors,* such as were mentioned above in connection with dominance. In planarians, experiments indicate that the brain is a source of such substances, for a suspension of ground-up brain tissue placed in the same dish with a regenerating animal will have the same effect as an intact or grafted brain. In hydras, two centers of dominance have been identified, one at each end. The mouth region produces an inducer and an inhibitor that influence regeneration only of the head end, and the base of the stalk region produces an inducer and an inhibitor that influence re-

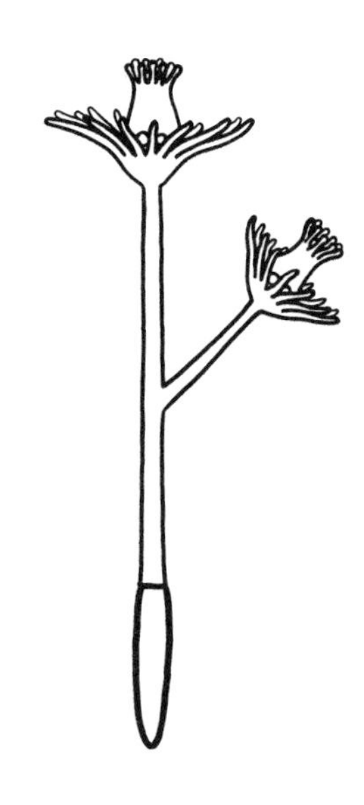

Graft-induced polyp growing from the side of the stem of a host polyp. In this hydroid *(Corymorpha),* a bit of stem removed from one polyp and grafted into the stem of another polyp will induce development of a new polyp. Fragments of stem taken near the hydranth (mouth and tentacles) of the donor give rise to a higher percentage of complete polyps than do fragments cut from more basal regions of the donor. If the host hydranth is removed before the fragment is grafted into the host stem, the percentage of complete polyps formed is greater than if the host hydranth remains present. Also, the side polyp becomes larger if the host hydranth is removed. The stem of the side polyp grows longer the farther away it is from the dominant host hydranth. This hydroid thus shows a gradation of regenerative capacity and dominance that extends from the hydranth toward the base of the stem, comparable to the gradation that extends along the body of a planarian from head to tail. (After C. M. Child.)

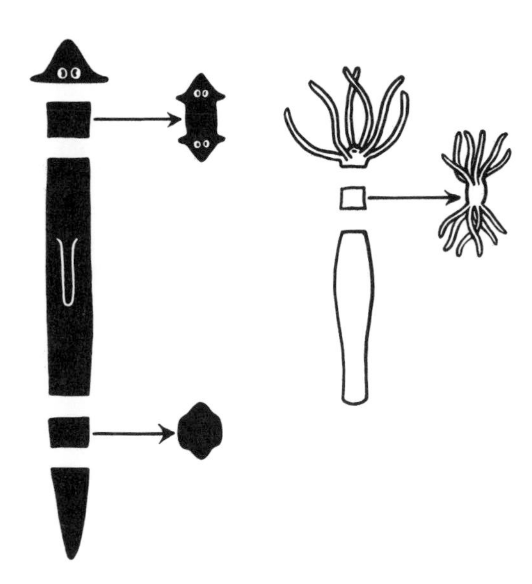

Small pieces lack polarity—they regenerate similar structures at both ends. Pieces may be cut so short that, presumably, the gradient difference between anterior and posterior cut edges is not great enough to establish polarity of regeneration.

Short pieces cut from behind the eyes of a planarian may regenerate a head at both ends, and short pieces from the posterior region will regenerate a tail at both ends. Experiments on hydras yield similar results. (Based on C. M. Child.)

Experimental reversal of polarity is easily accomplished in a hydra. **(1)** The mouth region and base are cut from a hydra, and **(2)** each is grafted back onto the opposite end (the mouth region is attached where the base used to be, and vice versa). **(3)** After several days, if a piece is cut from the middle of this animal, **(4)** a mouth region with tentacles will regenerate from the end that was originally nearer the base and a basal disk will form at the end that was originally closer to the mouth. Presumably substances diffusing from the dominant mouth and base regions, relocated at opposite ends, have somehow changed the middle region in such a way as to reverse its polarity. (Modified after Marcum, Campbell, and Romero.)

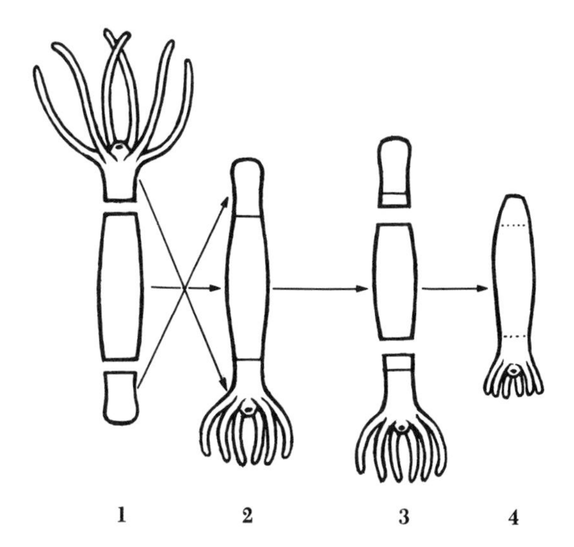

generation only of the base. When extracted and partially purified, extremely minute amounts of these sustances have dramatic effects in stimulating or inhibiting regeneration.

Just as the brain appears to be a source of controlling chemicals in planarians, the nerve cells are probably the major normal source in hydras. About 40% of all the nerve cells in a hydra occur in the mouth end, and there is also a concentration of nerve cells in the base. Clumps of disaggregated hydra cells from the mouth region regenerate faster and form more mouths and tentacles than do cells from the middle region of a hydra. Disaggregated cells from the base of a hydra regenerate basal disks first, then form mouths and tentacles later. Such differences in competence retained by disaggregated cells from different regions are presumably related to the numbers and types of cells present, perhaps especially the nerve cells.

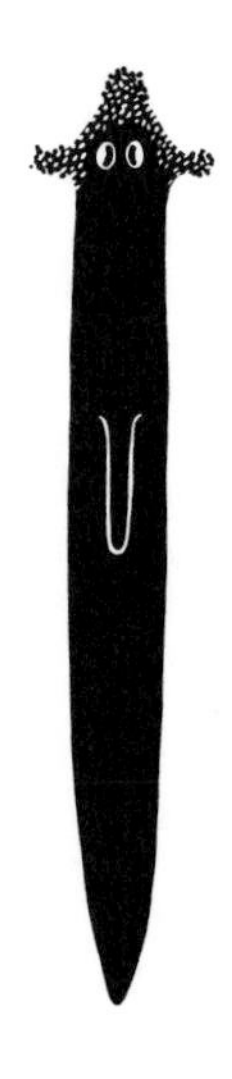

Degeneration in a poisonous solution begins at the head end. (Based on C. M. Child.)

Besides the axial gradient of regenerative ability, other features show a parallel gradation along the body. The rates of metabolic processes, such as oxygen consumption and protein synthesis, are greatest at the head end of a planarian and decrease posteriorly. If planarians are placed in a poisonous solution of lethal concentration, they begin to degenerate and eventually die; but degeneration begins at the head and continues progressively toward the tail in a regular fashion. This would happen if the parts with the highest metabolic rate are affected first and most severely by a poison, while the less active parts are more slowly affected. In a hydra, the sizes and kinds of nematocysts show a regular distribution along the body; gonads and asexual buds occur only at particular levels. It is possible that all of these phenomena are eventually to be understood in the context of the axial gradient.

Hydras and planarians have in common a prominent population of **unspecialized cells.** In hydras these are the interstitial cells, and in planarians they are mesenchyme cells called neoblasts. Both of these cell types are responsible for normal renewal of at least some types of cells in uninjured animals, but they appear to become especially active in regeneration following an injury. If the neoblasts of a planarian are selectively destroyed (by exposure of the worm to X-rays), it is unable to regenerate and eventually dies. However, if a piece of healthy tissue from another worm is grafted into an irradiated worm, neoblasts migrate from the graft into the irradiated tissue, effect regeneration, and eventually replace all of the irradiated cells.

If the interstitial cells of hydras are destroyed (by exposure to chemicals such as colchicine or hydroxyurea), the polyps soon lose all their nerve cells, which are normally replaced continually from interstitial cells. (All nematocysts and gland cells are similarly lost, and the animals must be force-fed by hand in the laboratory.) Surprisingly, although they now consist only of epithelio-muscular cells, the animals can still bud and regenerate and show normal polarity. Even experiments testing induction of structures or reversal of polarity proceed much as in normal polyps. Under these circumstances, it has been found that the epithelio-muscular cells produce the chemical inducers and inhibitors that control body form, budding, and regeneration.

One may inquire how gradients get started in animals. It seems likely that they arise early in development by the action of external factors. The position of the egg in the ovary is one such external condition if the egg is attached by one end to the ovary and is free at the other end. It is known for a good many eggs that this position is correlated with the polarity of the egg, that is, which end of the egg will become the mouth or anterior end of the animal.

It has been found that the results of many experiments on regenerating adult flatworms parallel those on developing embryos. Such parallel results suggest that gradients of some kind are an important factor in embryonic development and furnish an underlying pattern that controls the orderly development of normal form and proportion. Thus we might think of the several kinds of symmetry as resulting from differences in the number of gradients that act in development. In a spherical animal, all radii are alike and there is no one main axis of differentiation, though there are differences between the interior of the organism and its exposed surface. In a radial animal, the main axis is from mouth to base. And in a bilateral animal, there are, besides the major anteroposterior axis, two minor axes of differentiation (ventrodorsal and mediolateral) for which there is also evidence of

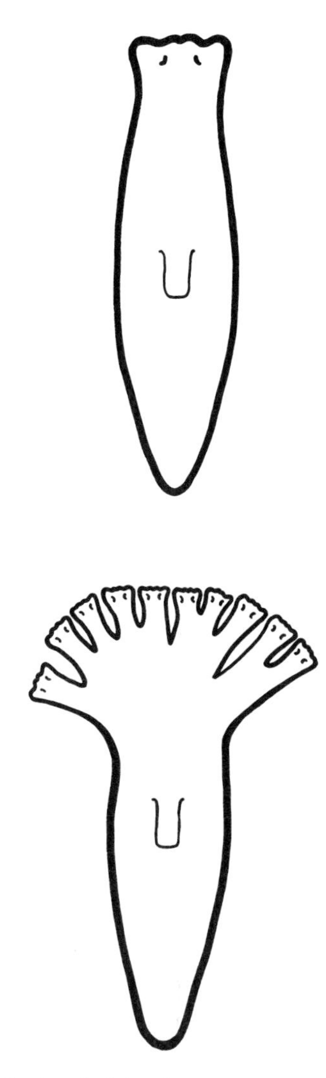

The head of this flatworm was cut repeatedly, and the cut edges were not allowed to grow together again. The result is a "monster" with ten heads. (After J. Lus.)

associated gradients. Experiments on regenerating invertebrates have helped us to understand and have been fruitful in suggesting experimental means of approach to the general problems of animal form, growth, and development.

Left: Dugesia dorotocephala, in which the black pigment is distributed over the dorsal surface in small granules against a brownish background. The ventral surface is gray. *Center: Dugesia tigrina,* which has the dorsal surface finely spotted with yellowish-orange pigmented areas against a tan background. The ventral surface is white with gray pigmented areas. In the grafting experiment host and donor were placed on a tray of ice to slow down their movements. Then with a small knife a piece of the head, including the two eyes, was cut from *D. tigrina* and placed in a hole, made just behind the eyes, in the head of *D. dorotocephala. Right:* 175 days after the operation the tissues of host and donor are completely fused. The grafted eyes lie behind those of the host. The only evidence of any effect produced by the graft on the host was the growth, in the tissues of the host, of a small projection on either side of the grafted eyes. (J. A. Miller)

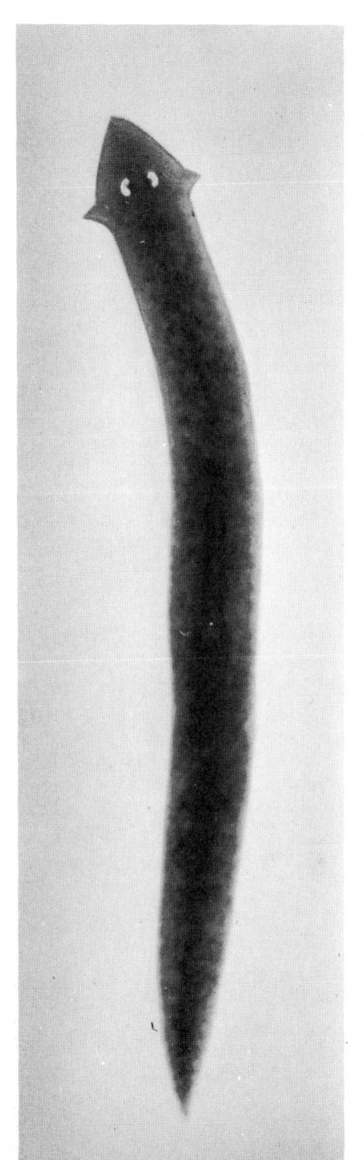

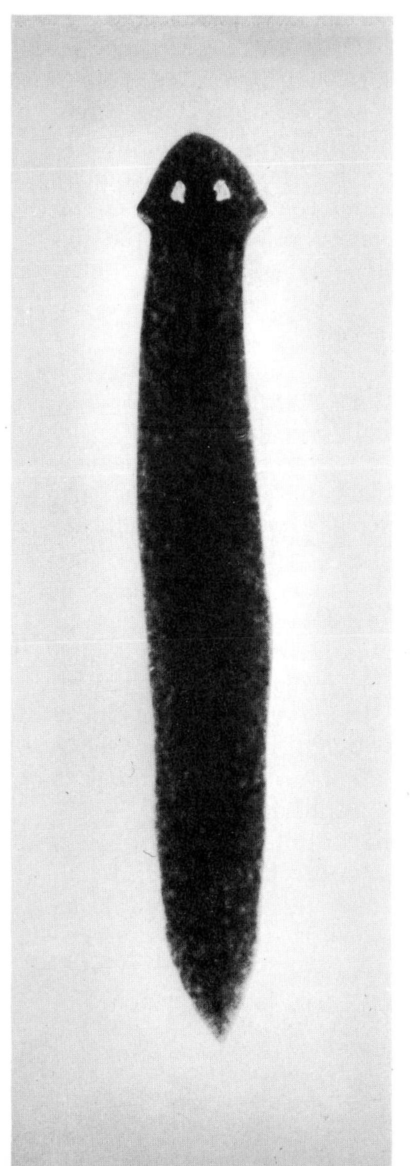

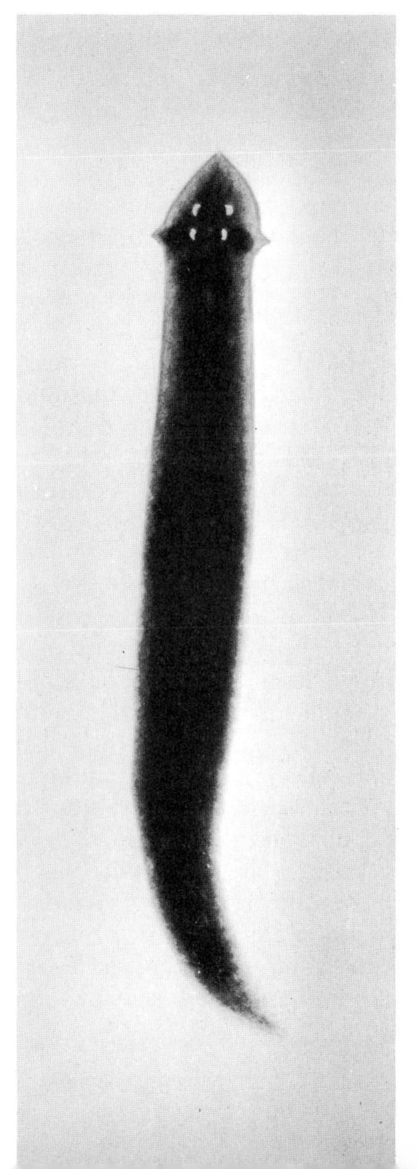

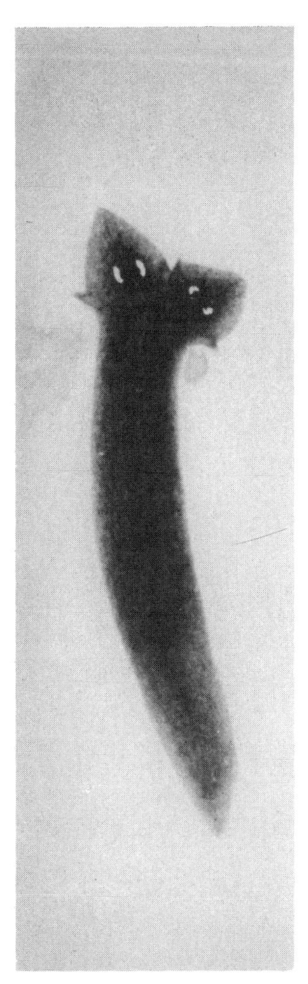
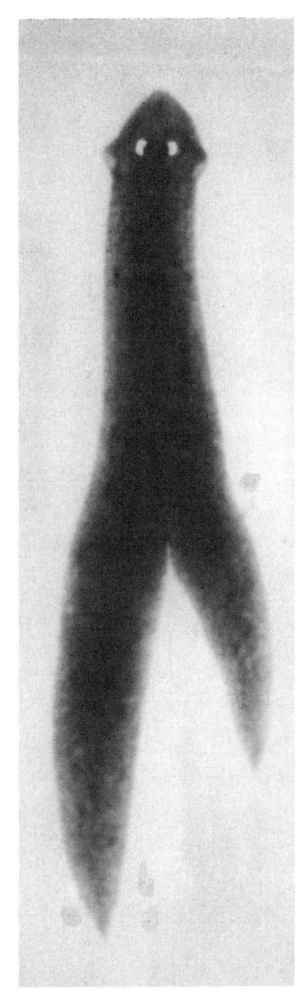
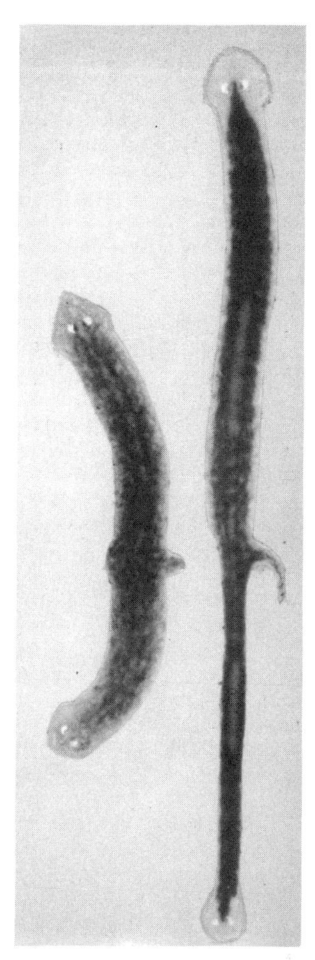

Grafting between species is particularly easy with certain species of planarians. Grafts take well, the chemical stimulatory and inhibitory substances produced by one species have the same effect on the tissues of other species, and the tissues of host and graft can be distinguished by their different patterns of pigmentation.

Left: An animal with two well-developed heads produced by a procedure similar to the one on the preceding page. In this case, however, the graft did not unite on all sides with the tissues of the host but has induced the formation of a complete head having the pointed head shape, pointed sensory lobes, and pigmented pattern of the host *(Dugesia dorotocephala)* rather than that of the donor *(D. tigrina)*, which has a more rounded head with blunter lobes and a different pigment pattern.

Center: A two-tailed worm was obtained by grafting a head piece from *D. tigrina* into the tail region of *D. dorotocephala*. The graft grew out as a complete head with *D. tigrina* characters. It induced the formation of a pharynx in the host tail. Then the host head and pharyngeal region were removed; and the cut surface, which we might expect to regenerate a new head, produced instead a pharynx and tail, presumably because of dominance by the graft head. After this photograph was made, the two tails fused along their inner borders, resulting in a double posterior region with two pharynxes. When this animal later divided asexually, the new animals so produced had four eyes, two pharynxes, and a double digestive tract.

Second from right: A graft was placed in the tail region and grew out as a head; then the posterior part of the host tore away, leaving only the graft attached to the anterior piece of the host. *Right:* The same worm, here shown greatly extended, has been fed blood to make the pharynxes stand out. The posterior pharynx, which has been induced in host tissue by the graft, takes a direction related to the graft head and opposite that of the original host pharynx. (J. A. Miller)

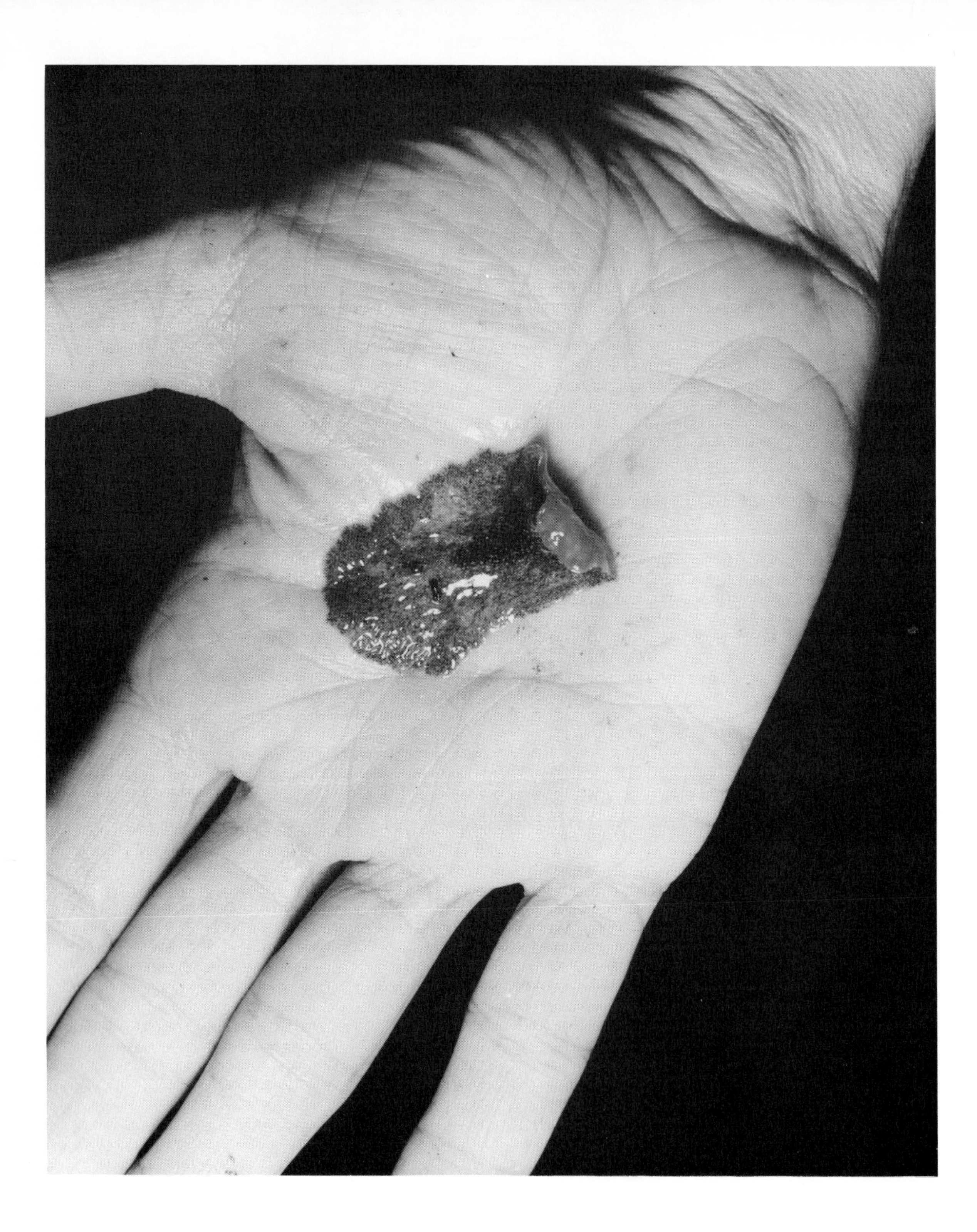

Chapter 11

Messmates and Parasites

Living at the expense of one's neighbor is an old habit among animals. Practically all animals harbor one or more kinds of parasite, and most of these are themselves hosts to still smaller parasites. The total bulk of the parasites residing in one host is necessarily less than that of the free-living animal that provides the food and lodging for so many unwelcome guests. But from the standpoint of numbers of organisms, the animal kingdom has many more parasitic than free-living individuals.

Nearly every phylum has its parasitic members, and some phyla have more than their share. In the phylum Platyhelminthes, the majority of species are flukes and tapeworms, which belong to entirely parasitic groups. But the beginnings of parasitism are to be found in the group of mostly free-living flatworms, the turbellarians.

Opposite: **Flatworm** held on outstretched palm is a marine polyclad. Wafer thin, this active animal has a large surface area in proportion to body mass. Misaki, Japan. (R. Buchsbaum)

Free-living Flatworms

Like the planarians, most of the other free-living flatworms are covered externally with cilia, the beating of which creates in the water the turbulence that suggested the name of the class **Turbellaria.** Although freshwater planarians are typical in their carnivorous or scavenging feeding habits, flame-bulb excretory system, and complex hermaphroditic reproductive system, most turbellarians are marine. Several groups of turbellarians can be roughly recognized by the form of the digestive cavity.

The group to which planarians belong is distinguished by a digestive cavity that has three main branches. Such worms are appropriately called **triclads** ("three-branched") and are large- to moderate-sized turbellarians. Besides the freshwater forms, there are also marine and land triclads.

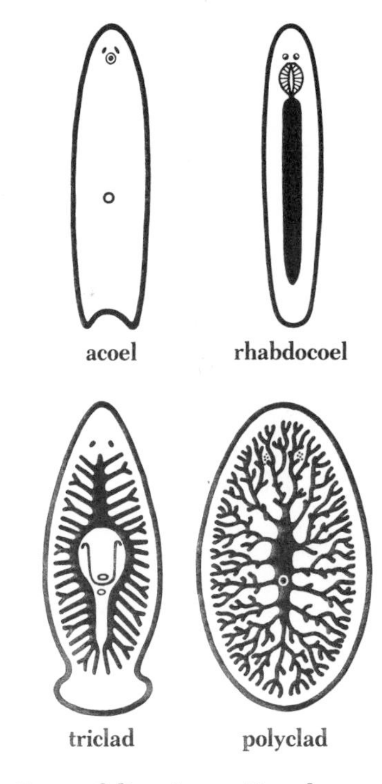

Types of digestive cavities of turbellarians.

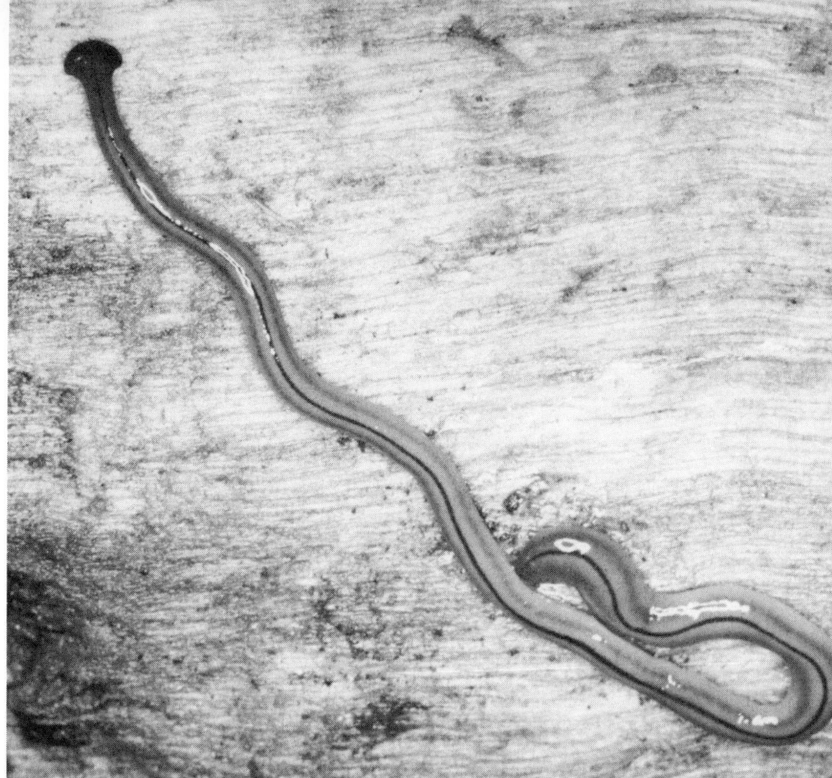

Land planarian, found under a flowerpot in a Pittsburgh, Pennsylvania garden, is far from its native home in the moist forests of tropical southeast Asia. This species, *Bipalium kewense,* has been widely distributed by plant shipments. It has a distinctive halfmoon-shaped head and dark stripes on a light background. This land triclad does not reproduce sexually in temperate climates, but maintains itself by fragmentation in gardens of California and southern states of the United States, where it has become well established. (R. Buchsbaum)

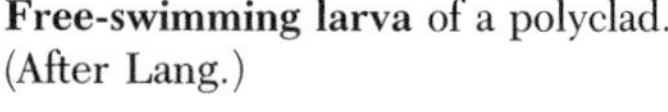

Free-swimming larva of a polyclad. (After Lang.)

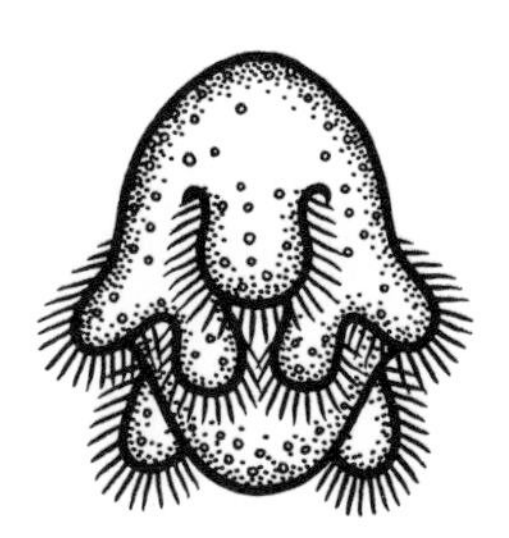

Polyclads, so named for their many-branched digestive cavity, are almost exclusively marine. They are thin, leaflike animals—sometimes almost as broad as they are long—and are among the largest turbellarians, individuals of some species reaching lengths of 15 centimeters. Many have free-swimming larvas with 8 ciliated lobes.

Various groups of smaller turbellarians with a straight, unbranched digestive cavity are referred to collectively as *rhabdocoels* ("rodlike cavity") and others with a lobed or diverticulated cavity as *alloeocoels* ("cavity of another sort"). Rhabdocoel and alloeocoel groupings have marine, freshwater, and terrestrial forms.

Among the simplest turbellarians are tiny marine worms (usually under a few millimeters in length) that have a mouth but no digestive cavity, and hence are called **acoels** ("without a cavity"). Food is swallowed into a mass of loosely packed cells and there digested. Acoels also lack an excretory system. Acoels, polyclads, and most alloeocoels have multiple pairs of nerve cords; in freshwater planarians and most rhabdocoels these have been reduced to a single ventral pair.

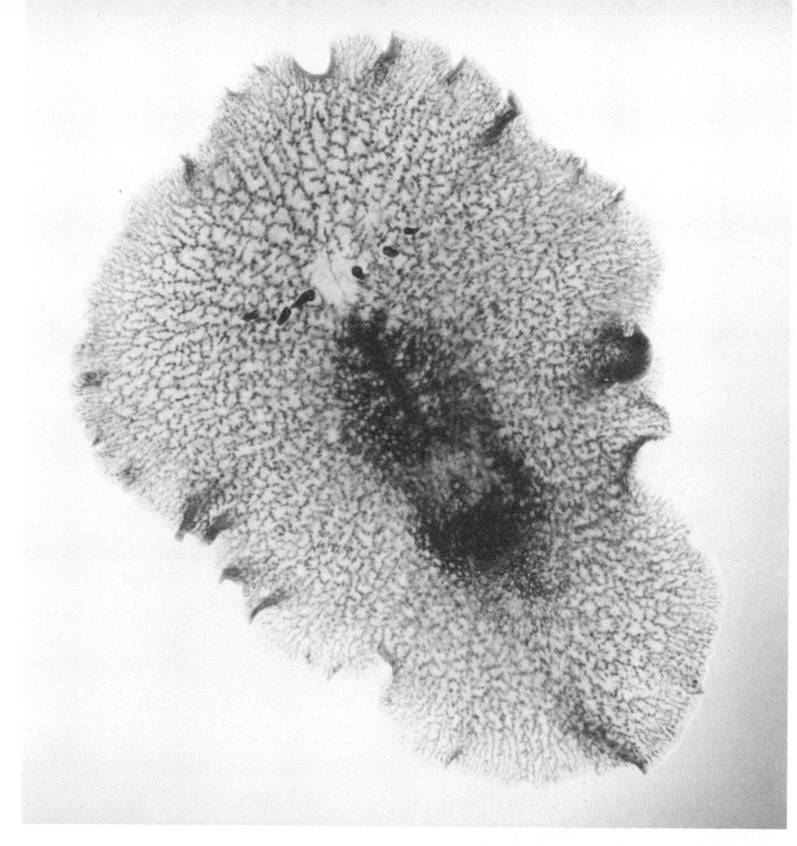

Marine polyclad moving along on a moist glass plate, its path smoothed by a bed of secreted mucus. This is the same worm shown at the beginning of this chapter. Illuminated from below it reveals the extensive branching of the gastrovascular cavity that both digests and distributes food. Near the anterior end the bilobed brain is seen as a light area, flanked on each side by three dark sensory tentacles. Misaki, Japan. (R. Buchsbaum)

Polyclad *(Pseudoceros)*, pale yellow with black stripes, found crawling on rocks or feeding on colonies of tunicates. Length, 2 to 3 cm (about 1 inch). The head, to the left, bears sensory projections. Bermuda. (R. Buchsbaum)

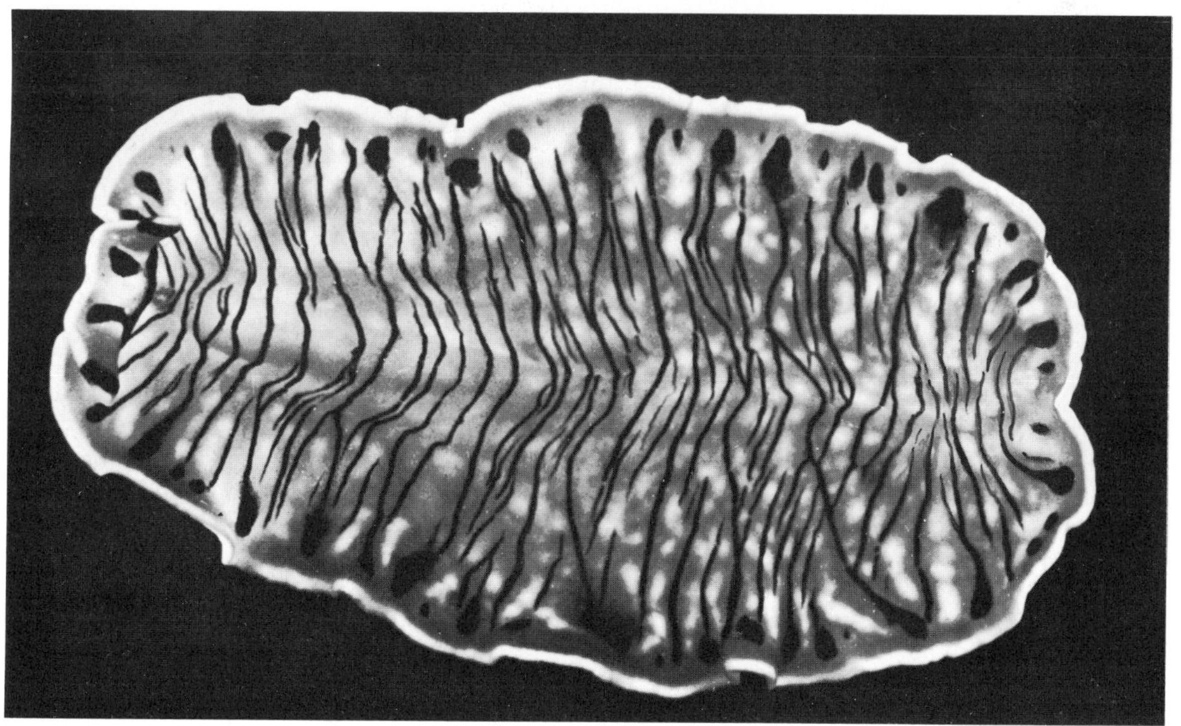

Parasitism is only one of several kinds of relationship that can occur between organisms, and it usually evolves from one of the other kinds. Two organisms of different species are often found living together in a constant association of a type called **commensalism.** The smaller one, the *commensal,* derives benefit from the relationship while the larger one, the *host,* is apparently neither benefited nor injured. Such an association is illustrated by commensal triclads of the genus *Bdelloura,* which live attached to the gills or appendages of horseshoe-crabs (see chap. 18). The worms receive free transportation, shelter, and scraps of food from their host.

Sometimes a commensal incidentally benefits the host. The host species may then undergo changes in its behavior and physiology, making conditions increasingly favorable for the commensal. Any such association of mutual benefit is called **mutualism.** Examples already mentioned are the intestinal flagellates of termites, and the unicellular algas of protozoans and corals. A well-known case of mutualism is that of a tiny green acoel, *Convoluta roscoffensis.* The young worms are colorless when first hatched, but they promptly encounter and ingest certain green flagellates *(Tetraselmis convolutae),* which conveniently cling to the egg cases from which the young worms emerge. Once inside the worms, the flagellates lose their flagella, outer cell wall, and eyespot, but continue to carry on photosynthesis, producing organic nutrients and oxygen. From the worm they receive a sheltered place in the sunlight and a steady supply of carbon dioxide and of nitrogen- and phosphorus-containing compounds (by-products of the host's metabolism). In their natural habitat, the young worms feed like other acoels, but if kept without food in the laboratory, they are able to live and grow on the nutrients derived from the flagellates, as long as they have sufficient light. Adult worms do not feed and are completely dependent for their nourishment on their green guests. Late in life the worms begin to digest the flagellates in their tissues, the numbers of green cells gradually decline, and the worms die.

Commensalism usually evolves, not in the direction of mutualism, but toward **parasitism.** A commensal that at first takes only shelter, and then scraps of food, may finally begin to feed on the tissues of the host body and to inflict harm. However, a

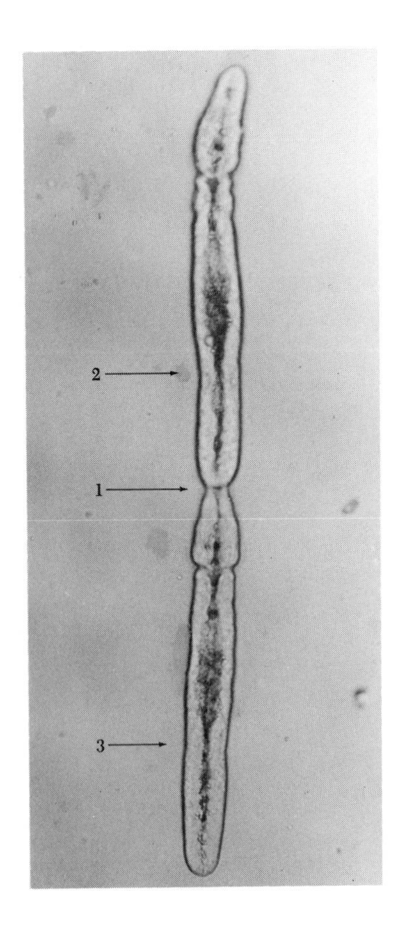

Fissioning turbellarian. This worm *(Catenula)* consists of two distinct subindividuals, each with a tapering head lobe containing the brain and a statocyst. The first fission plane is indicated by arrow 1. The next two fission planes are already marked by constrictions of the rhabdocoel-type digestive cavity and small constrictions of the worm at arrows 2 and 3. Lake Pymatuning, Pennsylvania. (R. Buchsbaum)

Commensal triclad, *Bdelloura,* is marine and clings to the gills of horseshoe-crabs. It shares the food of the host but does no apparent harm. The adhesive margin of the worm is expanded into a broad sucker at the posterior end. Florida. (R. Buchsbaum)

Green acoels, *Convoluta roscoffensis,* occur along certain areas of the Channel coast of France, where they gather by the millions at low tide on the surface of sand beaches in patches that look like streaks of dark green paint. From these concentrations, great numbers of worms may easily be scooped up in a glass vessel to be photographed and studied in the laboratory. Since the French word for "worm" *(ver)* sounds the same as the word for "green" *(vert)* and the word for "glass" *(verre),* French descriptions of this process can be confusing to English ears. Roscoff, Brittany, France. (R. Buchsbaum)

parasite and its host may evolve together until there is little damage to the host, and the parasite may even finally prove to be of some service to the host; then the parasitism has become a mutualism. Thus the three kinds of relationships may be only different stages in the process of living together, and it is not always possible to draw a sharp line between them. The term **symbiosis** ("living together") is useful as a collective word for all three types of relationships or for designating an association that is not well enough known to be accurately classified as one of the three.

It is sometimes to the advantage of a parasite to minimize damage to the host. This would be the case, for example, if prolonging the life of the individual host increases the life span and reproductive potential of the parasite. A parasite would also profit by reducing damage if it thereby avoided fully activating the host's defenses. These defenses, from the host's point of view, may even constitute the worst part of the disease, and it is then to the host's advantage to tolerate the parasite. Evidence that such adjustments do occur comes from instances in which a parasite invades an animal that is not the usual host, and both parasite and host suffer unhappy results. Of the flatworms parasitic in humans, the tapeworms seem the most benign. The flukes are often less so.

Parasitic Flatworms

The three classes of parasitic flatworms differ from free-living turbellarians in the loss of external cilia (except in larvas) and in the development of an unusual kind of epidermis. The special epidermis of a parasitic flatworm helps to protect it against digestion by the host and against the host's immune defenses. The surface is also specially modified for digestion and absorption of nutrients from the host.

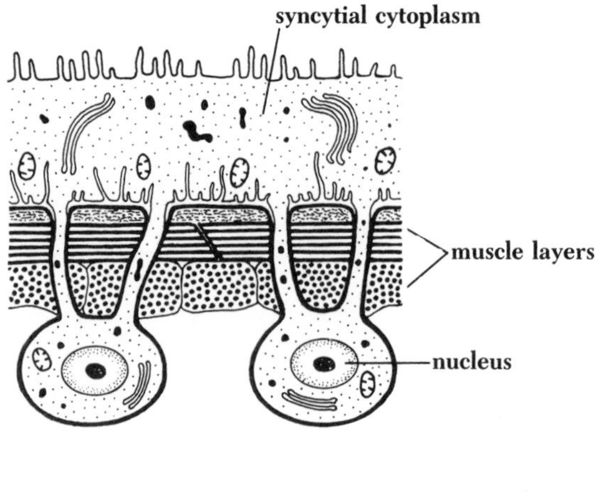

syncytial cytoplasm

muscle layers

nucleus

Modified epidermis of parasitic flatworms is a nonciliated syncytium, a continuous layer of cytoplasm not divided into separate cells. Its surface is specialized for digesting and absorbing host nutrients, and the surface area is increased by microscopic projections or invaginations. The outer layer contains no nuclei, but it is connected by narrow cytoplasmic channels to cell bodies that contain nuclei and synthetic organelles and that lie below the muscle layers. This arrangement is thought to protect important cell components from damage by the host, while the surface cytoplasm is constantly renewed or repaired. Such a surface is present in all three parasitic classes of flatworms. (Modified after K. M. Lyons.)

Monogenean, *Acanthocotyle,* an external parasite, moving about on the surface of a marine fish, its host. The worm holds on by means of the large posterior sucker. (After Monticelli.)

The members of the class **Monogenea** are few in number and include no human parasites. Most live as *external parasites* attached to the skin or gills of fishes, feeding on mucus and epithelial tissue or on blood. Hanging on to the outside of a fast-moving fish is no easy matter, and these worms frequently have numerous hooks and an enormously developed sucker (or group of suckers) at the posterior end, and sometimes one or more smaller suckers around the mouth as well. The mouth leads through a muscular pharynx into a two-forked digestive cavity. There is an excretory system of flame bulbs and canals, and much space is devoted to a complex hermaphroditic reproductive system. The nervous system and sense organs are usually well developed, but the lensed eyes present in the ciliated larva are lost or reduced in the adult. As they move about on the surface of their hosts, the worms stray occasionally into the cavities that communicate with the exterior: the mouth, nasal passages, urinary bladder, and rectum. Thus it is not surprising to find that many fishes and other aquatic vertebrates harbor monogeneans that have become adapted to live in these cavities, where the danger of being swept off is much less. Most monogeneans have simple life histories with only one host, and any worm that is accidentally separated from its host faces an even shorter and simpler life history.

Hooked sucker of *Gyrodactylus* provides a firm hold on the fish host. This parasitic monogenean is less than a millimeter long and usually does little harm to fishes, but if many worms are present, the host may weaken and die. *Gyrodactylus* is a serious pest in fish ponds and hatcheries where fishes are crowded. (Modified after O. Fuhrmann.)

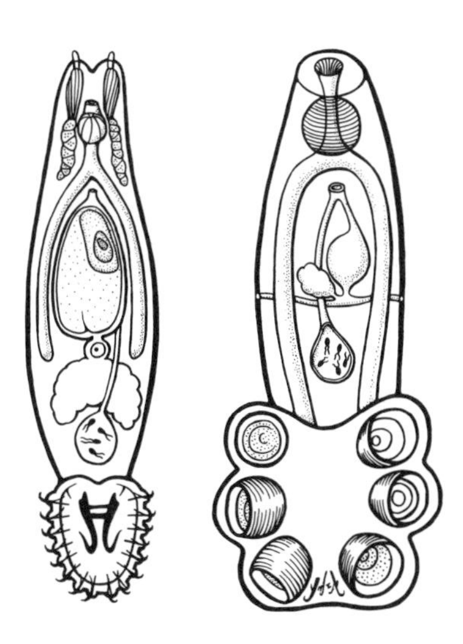

Multiple suckers, as in *Polystoma,* mark monogeneans that feed on host blood. This worm came from the mouth cavity of a turtle. Another species, which lives in the bladder of frogs, times its reproduction by the annual hormone cycles of the host. The adult worms and frogs spawn eggs in synchrony each year, and the hatching worm larvas find a ready supply of new tadpole hosts. (Modified after H. W. Stunkard.)

Flukes

People who are always talking about the "good old days" when life was simpler should be sympathetic with flukes (class **Trematoda**). Flukes probably had ancestors with simple life histories involving a single host, as in monogeneans, but modern flukes lead complicated lives with two or more hosts and with many asexual stages.

Flukes live as *internal parasites*, embedded in the tissues or clinging to the lining of cavities far from the surface. They have little trouble holding on securely, and their hooks or suckers are not as elaborate as those of external parasites. But the problem of getting their offspring established in a new host is a much more difficult one, and they require not just one host but two or three successive ones. Their chances of completing these complex life histories are improved by a tremendous increase in the number of potential offspring, which they accomplish in two ways. First, they produce *great numbers of eggs*, which they store in a long coiled tube, the uterus, not found in free-living flatworms such as planarians. Second, the young stages multiply by *asexual reproduction*.

Human blood flukes *(Schistosoma mansoni)*. (Partly after Faust and Hoffman.)

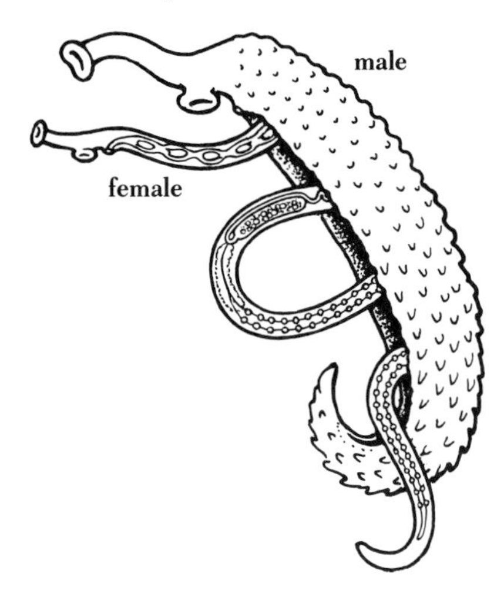

Of the flukes that parasitize humans, the most important are the blood flukes called **schistosomes**, which constitute a world health problem second only to malaria. They affect an estimated 200 million people, mostly in tropical or subtropical Asia, Africa, northeastern South America, and the Caribbean, including Puerto Rico. Schistosomes are slender and elongate flukes that differ from most in that they are not hermaphroditic but occur as separate males and females. The sides of the male fold over to form a groove in which the longer and more slender female is held. In *Schistosoma mansoni*, one of the three important species in humans, the worms live in the small veins of the large intestine, clinging to the walls of the vessels by means of suckers and feeding on blood.

The female lays her eggs in the small blood vessels of the intestinal wall, close to the cavity of the intestine. Some of the eggs reach the cavity and are carried out in the feces. If the feces were removed by a modern sewage system or deposited in a dry place, that would be the end of the young schistosomes. But in many of the places where this parasite flourishes, sanitation is casual, sewage treatment is absent or inadequate, and schistosome eggs readily reach water, where they hatch. A ciliated larva, the **miracidium**, emerges and swims about. The miracidium will perish in under 24 hours unless it encounters a snail of a certain kind. It burrows into the soft body of the snail, loses its ciliated outer layer, and turns into a saclike form called a **spo-**

Life history of a human blood fluke *(Schistosoma mansoni).* **1:** Egg shed from human host in feces. **2:** Egg hatches in freshwater. **3:** Free-swimming miracidium. **4:** Sporocyst in snail host. **5:** Asexual multiplication of sporocysts. **6:** Cercaria escapes into water. **7:** Adult flukes in human host. The species of snail identifies this scene as Puerto Rico. (Based partly on Faust and Hoffman.)

rocyst, which absorbs nutrients from the snail and proceeds to multiply by asexual reproduction. Within the sporocyst develop more sporocysts, and within these develop forms called **cercarias.** A single miracidium may give rise over a few months to over 200,000 cercarias. The cercarias resemble the adults in having two suckers and a forked digestive tube, but differ in possessing a tail. They make their way out of the snail and swim about near the surface of the water. Like the miracidia, they have only a few hours to encounter a suitable host. People who spend time wading, washing, or swimming in freshwater streams or cultivating in irrigated fields are susceptible targets. On contact with human skin, the cercaria attaches and, aided by digestive secretions, bores its way through the skin into a blood vessel. Eventually the young fluke makes its way to the blood vessels of the large intestine, where it feeds and grows into an adult worm, finally mating with another that entered at the same time or with one already established from a previous infection.

The presence of schistosomes in humans causes a disease (schistosomiasis) characterized by a rash, fever, body pains, and a cough in the early stages, dysentery later on. Victims may live for many years but gradually become weak and emaciated and eventually many succumb to other diseases because of their weakened condition. The chronic effects of the parasites are due to those eggs that do not escape into the intestinal lumen but

Egg of a human blood fluke *(Schistosoma mansoni).* The ill effects of schistosomes are not due to the presence of the adults but are caused mainly by eggs that fail to escape from the human host, instead lodging in various tissues throughout the body.

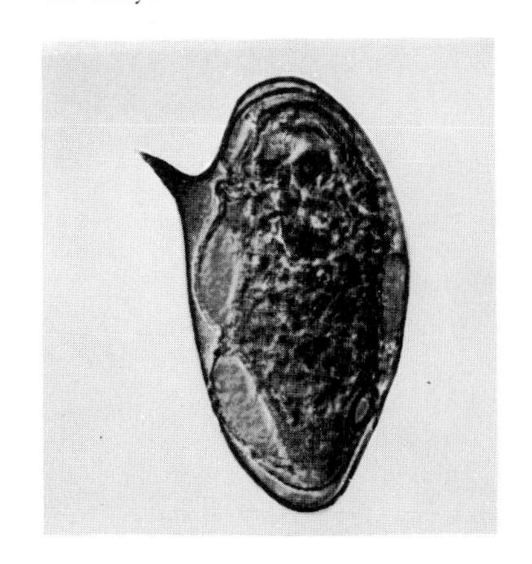

accumulate in the intestinal wall or are carried in the blood to other tissues throughout the body; each egg becomes a center of inflammation and scar tissue. Treatment with drugs is partially successful, but of little use if the patient is at once exposed to reinfection.

Control measures for schistosomiasis might reasonably begin with sanitary disposal of human feces and education to reduce contact with infected water, but these are beyond the limited means of most of the countries in which the disease is widespread. Where infection is restricted to small areas, poisoning can help to reduce the numbers of snails (though it sometimes causes other problems), and drainage of snail habitats is also effective. But these are usually losing battles, as progress in the development of dams and irrigation systems for agriculture is now favoring the populations of snails and increasing schistosomiasis in most areas.

Schistosoma mansoni occurs in parts of Africa, the Middle East, and the Western Hemisphere. Of the other two important species of human blood flukes, one affects most of the population in Egypt and is also common in other parts of Africa; the worms live in the small veins of the urinary tract, and the eggs are released into the bladder and escape with the urine. The most common species in Asia lives in the blood vessels of the small intestine; its control has been difficult, as the use of human feces as fertilizer is an important part of the traditional economy in many areas. This has made possible intensive cultivation of the same soil for thousands of years by enormous populations of people (without using nonrenewable resources to produce artificial fertilizers, as is done in the United States, for example). If farmers would conserve human manure for a few weeks before using it in the fields, all the schistosome eggs would have died; however, further treatment might be necessary to destroy other disease organisms.

When a parasite lives for part of its life history in one kind of animal and spends another part of its life history in another kind of animal, the host that harbors the sexually mature stage is said to be the **final host,** and the one that harbors the young stages is called the **intermediate host.** In the case of blood flukes, humans and some other mammals are the final hosts, and certain snails are the intermediate hosts.

A similar type of life history is shown by the **sheep liver fluke,** *Fasciola hepatica*, which inhabits the bile passages of the liver in sheep and cattle, and inflicts severe, often fatal damage—with important economic consequences for stock raisers. It also infects a variety of wild herbivores and occasionally humans. The cercarias of this fluke, after emerging from the snail, encyst on grasses or other vegetation and, as in most flukes except schistosomes, are eaten by the final host.

A fluke that lives in humans and illustrates a life history involving *two intermediate hosts* is the **Chinese liver fluke,** *Clonorchis sinensis*, of China, Korea, Japan, and parts of southeast Asia. The adult is 1 to 2 centimeters long and has two suckers, one at the anterior end and one a short distance behind this. The fluke is hermaphroditic, and the fertilized eggs pass from the host's liver into the intestine and out with the feces. If the feces get into freshwater, as they commonly do, the eggs do not hatch into free-swimming miracidia, as in most flukes, but remain on

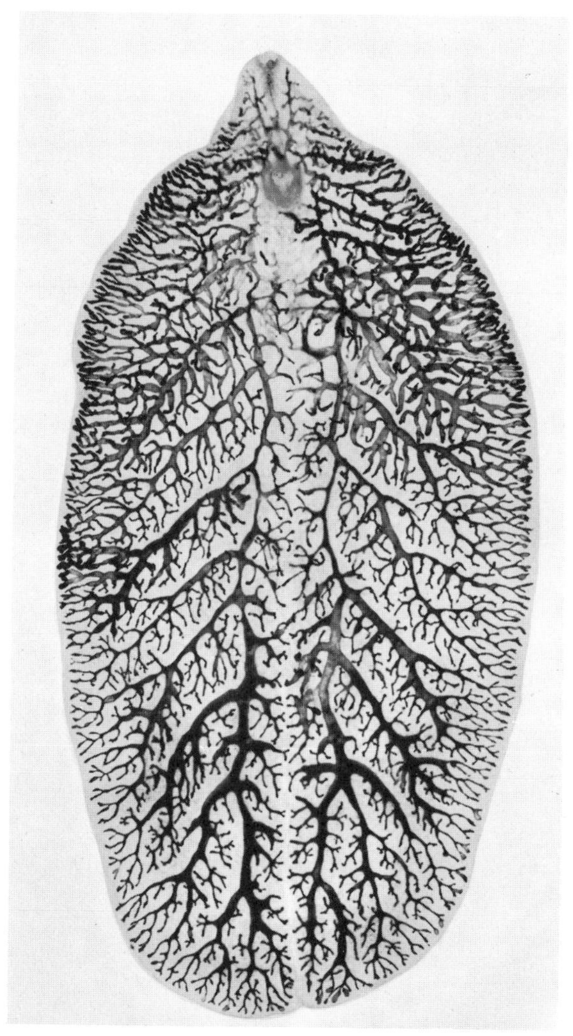

Sheep liver fluke *(Fasciola hepatica)* stained to show the highly branched digestive cavity. Length, 2 cm (about 4/5 inch). (Stained preparation.)

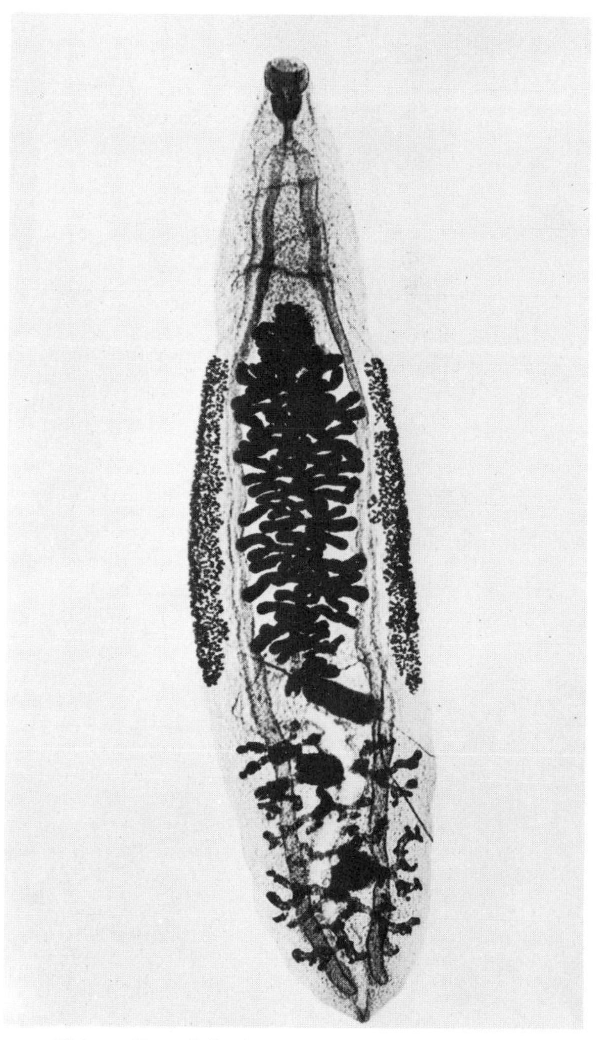

Chinese liver fluke *(Clonorchis sinensis)* showing the two-branched digestive cavity and the reproductive system. Length, 1 cm (about 2/5 inch). (Stained preparation.)

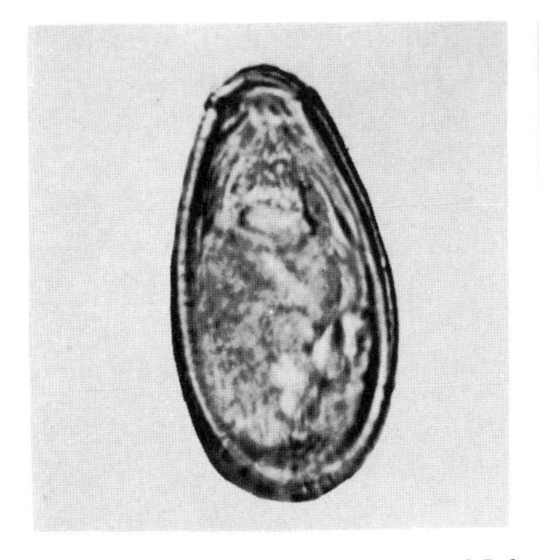

Egg of a liver fluke. Diagnosis of fluke infections is often made by examining the patient's feces and urine for eggs.

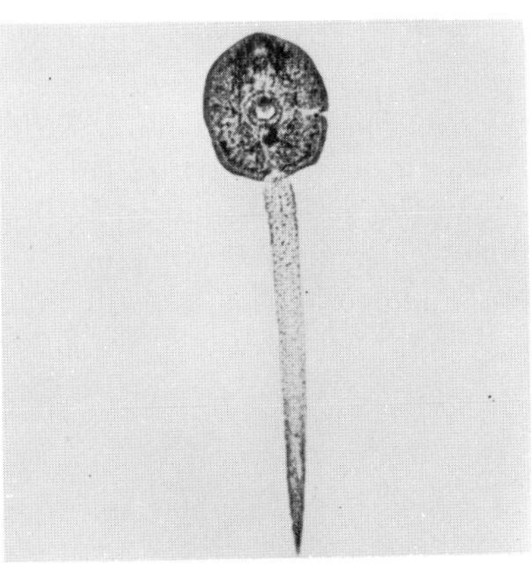

Cercaria of an unidentified species of fluke, showing ventral sucker and tail. (Stained preparation.)

Life history of the Chinese liver fluke *(Clonorchis sinensis,* or *Opisthorchis sinensis).* (Based on E. C. Faust.)

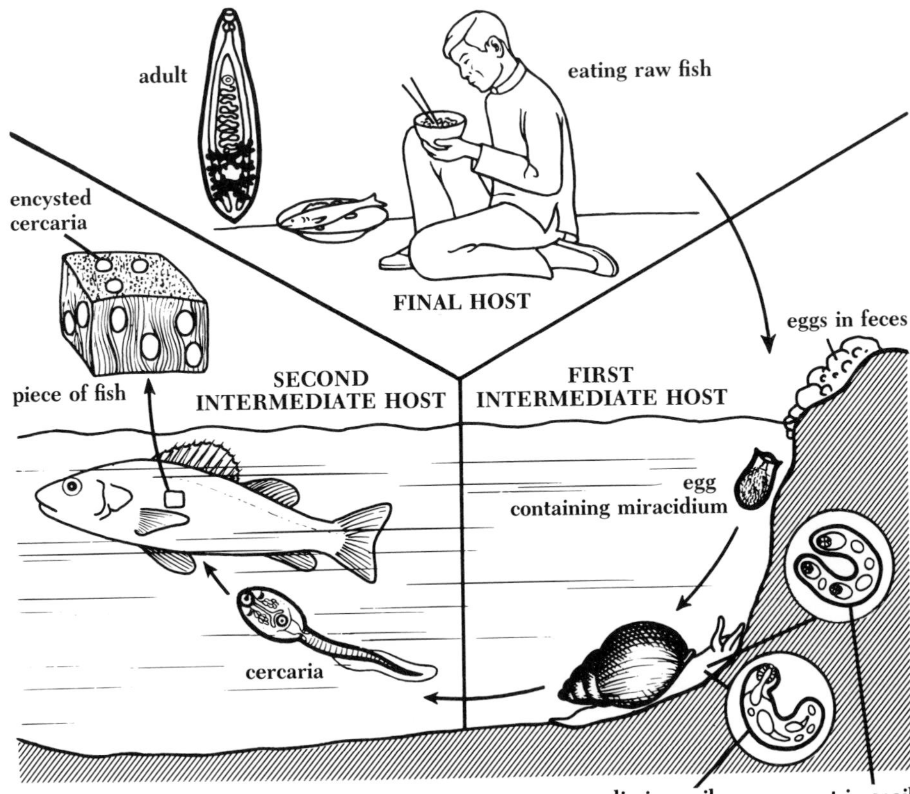

the bottom and are eaten by snails. Within the digestive tract of the snail the egg opens, and the miracidium emerges and makes its way through the wall of the digestive tract into the tissues of the snail. There it becomes transformed into a sporocyst, which produces another asexual form, the **redia.** Redias are more active than sporocysts and feed on host tissues. Each redia subsequently produces many cercarias, which escape from the snail and swim about. The cercarias encyst, not on grass like those of the sheep fluke, but in a fish, which thus serves as the second intermediate host. They burrow through the skin of the fish, lose the tail, and secrete about themselves a protective capsule. The fish responds by forming an outer capsule around each one produced by a parasite. There they remain until the fish is eaten by a suitable final host, a human (or other mammal). In the human stomach the cysts are digested out of the flesh, and in the intestine the capsule is weakened and the young fluke emerges. It makes its way up the bile duct and into the smaller bile passages of the liver, where it attaches by its suckers and feeds on blood. These flukes may persist for many years, causing serious anemia and disease of the liver from blocking of the bile passages. Flukes have been implicated in liver cancer.

Control of the Chinese liver fluke should be a relatively simple matter, for it is only necessary to cook freshwater fish thoroughly to destroy the encysted cercarias. Yet in certain regions in the south of China from 75% to 100% of the human population is affected. Not only is the cost of the fuel necessary to cook the fish an economic problem, but both rich and poor alike delight in eating raw fish, a custom long enjoyed and difficult to change.

From a parasite's point of view, it may be advantageous not to waste energy in seeking out a host, but instead rely on the host to provide the means of transfer. Parasites commonly lose active means of transportation and depend upon more or less *passive transfer* to new hosts by the activity of the hosts themselves. (The sure spread of venereal parasites among humans testifies to the reliability of this method.) In the flukes and other parasites, passive transfer is frequently achieved through the food habits of the host. Thus the sheep liver fluke is rare in humans because humans do not usually eat grass, though they do sometimes become infected by eating watercress upon which cercarias have encysted. Humans frequently get the Chinese liver fluke, however, wherever they habitually eat uncooked freshwater fish containing encysted cercarias.

Having once smuggled itself into a suitable host or succession of hosts, the problem of an internal parasite is how to get its offspring out again. The easiest way of leaving a host is with the outgoing feces, just as the easiest way of entering is by way of the mouth. But once free in the outside world, any one offspring has a slim chance of completing the life history and producing offspring of its own. The probability that an egg will reach a suitable spot for hatching is remote, and the probability that the miracidium will, within a short time, find or be eaten by a suitable snail (and not be eaten by something else) is even more remote. But if even a single miracidium manages to enter a snail, it can multiply within its intermediate host by asexual means and so compensate for the enormous loss of potential individuals by the random distribution of the eggs. The hordes of cercarias that result from such asexual multiplication again face low odds of survival, and only a few will reach new final hosts. Having to find a mate at that point would still further reduce the odds for reproductive success, and it is the option of hermaphroditic worms to be able to mate with any other individual they meet (not just the opposite sex), or even to self-fertilize, that finally tips the balance in their favor.

In the face of high losses, only parasites that produce enormous numbers of young can expect to continue their lineage, either on the individual or species level. It is not surprising, therefore, that most parasites seem to live only to reproduce, the reproductive organs occupying most of the animal's body. Looked at another way, parasites have little else to do, once they are safely established within a host; protected against the hazards of the outside world and relieved of the responsibilities of finding food, they need only protect themselves against the defenses of the host, and are otherwise free to devote most of their energies to reproduction.

Parasitism often results in weakness or disease of the host, but the *effects of parasitism on the parasite* are even more marked. Sometimes a parasite is barely recogizable as a member of its phylum, because it becomes so modified by its peculiar environment that it usually loses many of the structures characteristic of its free-living relatives and acquires others. In flukes we saw a loss of cilia, modification of the epidermis, and development of suckers. Tapeworms have still further broadened the definition of the flatworm body plan.

Tapeworms

The tapeworms (class **Cestoidea**) are mostly long, flat, ribbonlike animals that live as adults in the intestine of vertebrates, and few vertebrates are free of them. The young stages occur in invertebrates and vertebrates.

The most common large tapeworm of humans is the "beef tapeworm," *Taenia saginata* (or *Taeniarhynchus saginatus*). It attaches to the wall of the intestine by means of four suckers on the minute knoblike anterior end, or **scolex.** Behind the scolex is a short neck, or growing region, from which a series of body **sections** (proglottids) are constantly budded off. The sections closest to the neck are the youngest ones; those farthest away, the most mature. Thus the body widens gradually along its length, and the sections are in all stages of development.

The body is covered externally by a **syncytial epidermis** similar to that found in monogeneans and flukes. Unlike the flukes, which feed actively on the tissues of their host and do their own digesting, tapeworms have no mouth and *no trace of a digestive system.* They live in the intestine of their host, where digested food is readily available; there they seem simply to "soak up" their nourishment—truly the laziest way of living. But their apparent inactivity is misleading, for the body surface is a site of constant activity, perpetually busy with the active uptake of food and the never-ending battle against attack by host digestive enzymes and immune defenses.

Beneath the epidermis are longitudinal and circular muscles, by means of which the worm slowly and continually crawls forward, against the constant flow of materials. This behavior, together with the suckered scolex and the nonslip surface, helps to maintain the worm's place in the intestine. The nervous system is like that of turbellarians and flukes, but less well developed. From a small concentration of nervous tissue in the scolex, two longitudinal nerve cords run backward through the body, with cross-connections between the sections. Between the nerve cords, and parallel with them, run two longitudinal excretory canals, connected with each other by a crosswise canal near the posterior border of each body section. The canals receive fluid from branched tubules ending in flame bulbs.

The **reproductive system** lies embedded in the mesenchyme and is so highly developed in mature sections that the tapeworm is sometimes described as nothing but a bag of reproductive organs, a complete set of which, both male and female, develop at some time in every section.

> The male system starts to grow first. It consists of numerous small **testes,** scattered throughout the mesenchyme and connected by many fine tubes with a single large convoluted **sperm duct,** the end of which is modified as a muscular organ for the transfer of sperms. The sperm duct opens into the **genital chamber,** which connects with the outside through a **genital pore.** Running parallel with the sperm duct and also opening into the genital chamber is the **vagina,** a female duct that receives sperms. Cross-fertilization can take place between the sections

Section of a beef tapeworm *(Taenia saginata)*.

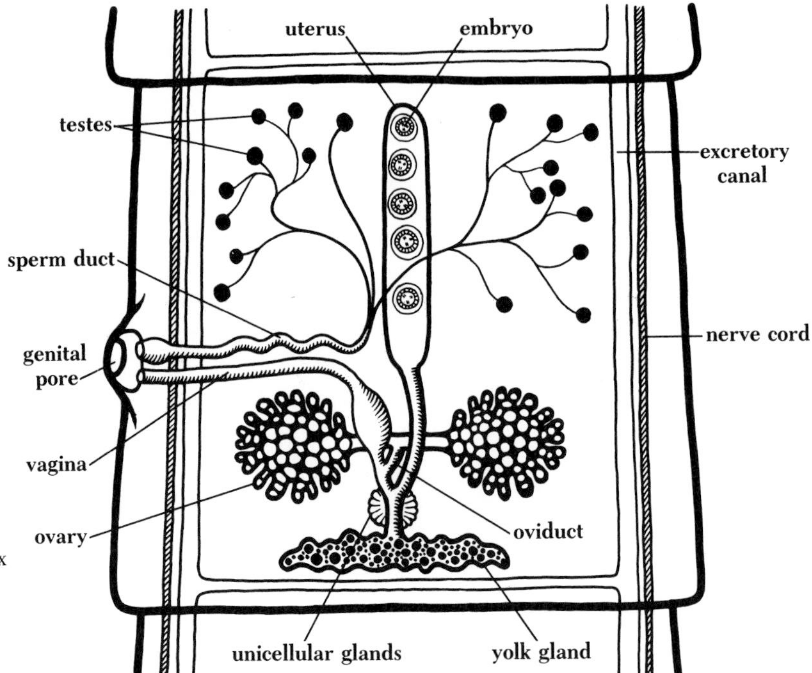

1. **Scolex of a tapeworm** *(Taenia pisiformis* from a dog, showing suckers and hooks. Behind the scolex are young developing sections. Diameter of scolex, 1 mm (1/25 inch).

2. **An immature section** with male organs developed. The female organs are only beginning to appear.

1

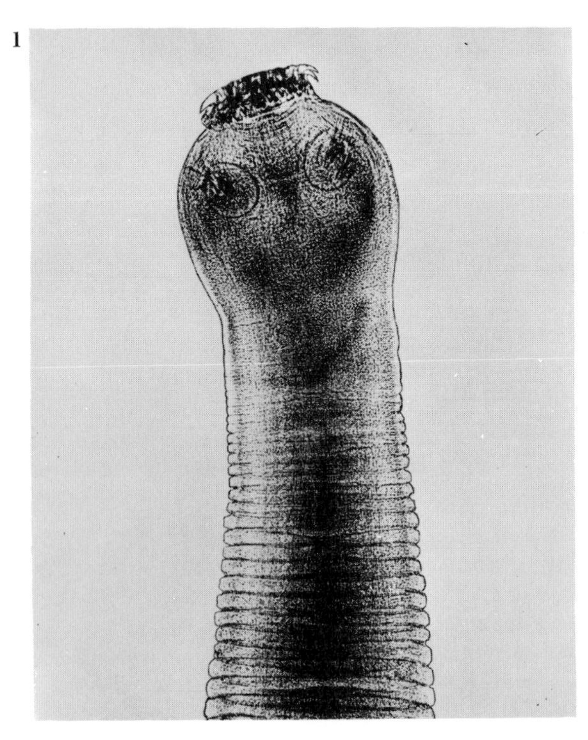

2

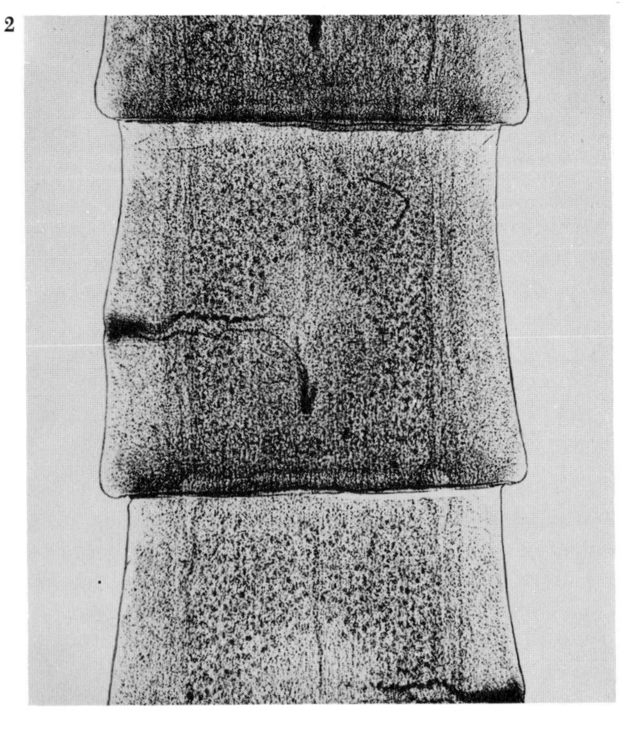

of different worms when two or more are present in the same host, or self-fertilization can occur, either within the same section or by transfer of sperms from one section of the worm to a more mature section farther down the length of the same worm. This is possible when the animal is folded back on itself for part of its length. The sperms enter through the vagina and are stored in an enlarged portion of its inner end. As each section matures, the male system degenerates while the female system develops. Eggs are produced in an **ovary** and pass into the **oviduct,** where they are fertilized.

Each fertilized egg is joined by a yolk cell from the yolk gland, and the two are then covered with a thin membrane secreted by unicellular glands. Additional protective coverings are contributed by the cells within the egg. The eggs pass forward into the **uterus,** which at first is a single sac (as shown in the diagram) but later develops numerous side branches. Eventually all the female organs degenerate except the uterus, which becomes enormously distended with developing embryos. At this stage the sections, each containing many thousands of young embryos, detach from the worm and pass out with the host's feces or may even actively crawl out under their own power.

Transfer between hosts is entirely passive and depends on certain eating habits of the intermediate host (usually referred to as a cow, although most beef cattle are male) and the final human

3. A mature section showing both sets of reproductive organs well developed. To identify the organs, see diagram in text.

4. A ripe section containing the branched and enormously enlarged uterus filled with embryos. Width, 5 mm (about 1/5 inch).

3

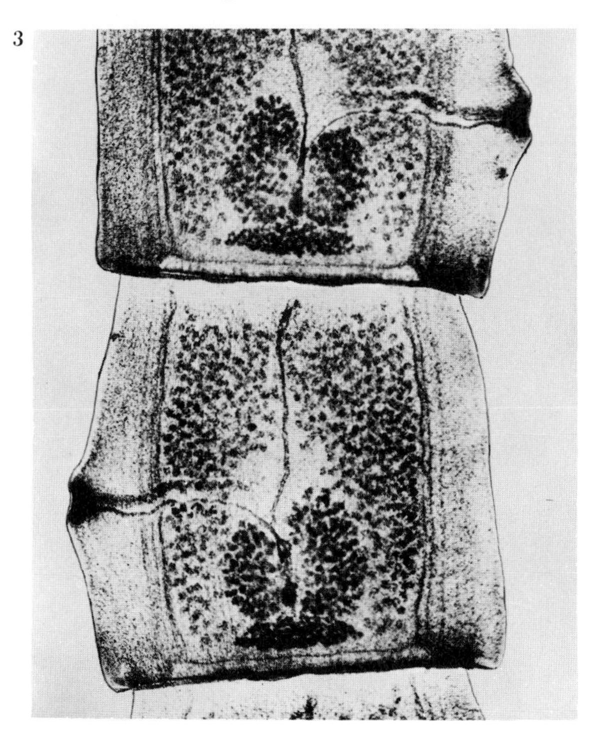

4

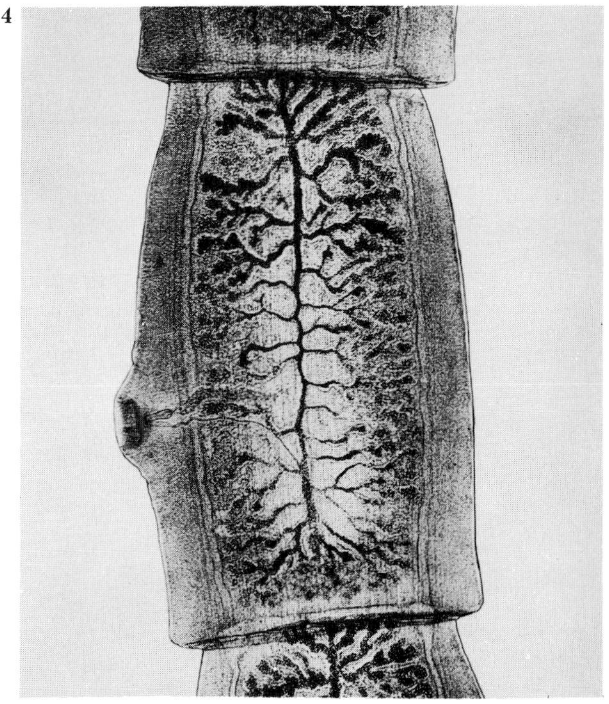

host. The cow eats human feces or vegetation on which human feces have been deposited. In the intestine of the cow the egg coverings are digested off and the **six-hooked larva** is released. The larva bores its way through the wall of the intestine, enters a blood vessel, and is carried in the blood stream to a muscle. There it remains and grows into a sac, or **bladder,** from the inner

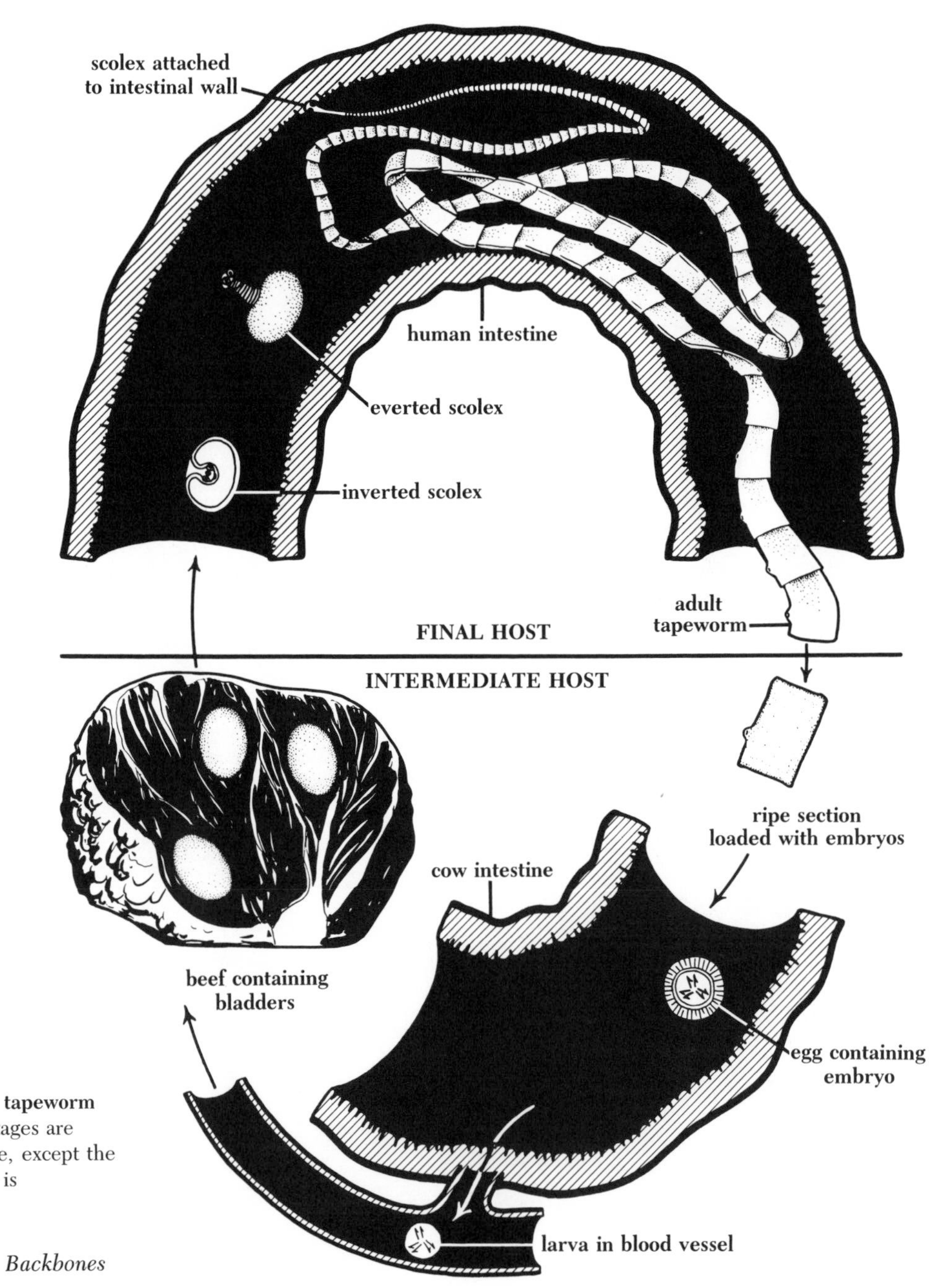

Life history of the beef tapeworm *(Taenia saginata).* All stages are shown about natural size, except the six-hooked larva, which is microscopic.

wall of which is developed the inverted scolex of the future adult tapeworm. When a human eats raw or rare beef, the enclosed bladder is digested open; the scolex everts and attaches to the intestinal wall by means of its suckers. Nourished by an abundant food supply, it begins to grow and produce eggs. Beef tapeworms commonly reach 10 meters (over 30 feet) in length and have 1,000 to 2,000 sections, each containing about 80,000 embryos when mature.

The bladders of the beef tapeworm occur most frequently in the muscles of the jaw and of the heart; these are the part of the cow usually examined by meat inspectors. In the United States, meat inspection has greatly reduced the occurrence of this once common parasite; but not all meat is federally inspected, and studies have shown that about one out of four infected cattle goes undetected. The bladders are about a centimeter long but can readily be overlooked. It is safest to avoid eating beef that is not cooked thoroughly. In many parts of Africa, Asia, and South America where sanitation is poor and beef may be prepared by roasting big pieces over an open fire, a large proportion of the human population is infected. In India, the Muslim population has a relatively high incidence of beef tapeworm, whereas Hindus, who follow religious restrictions against beef, are free of infection.

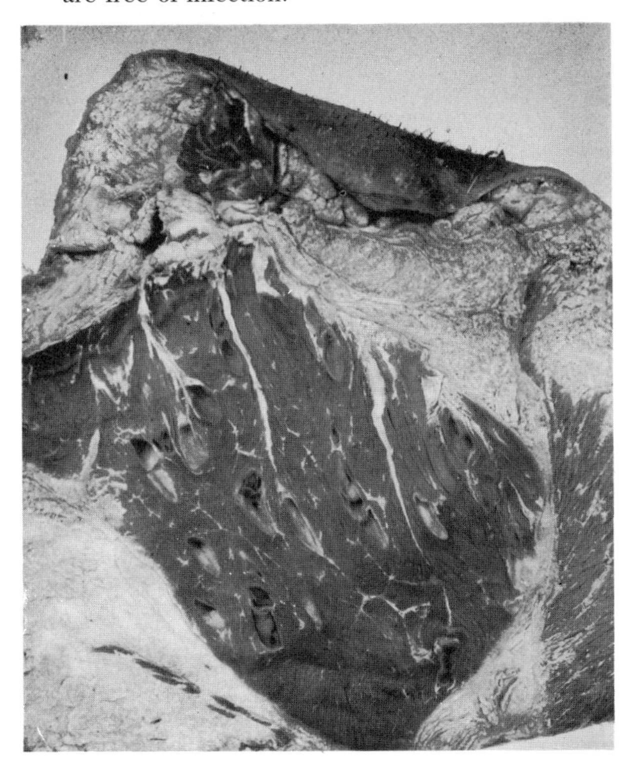

Measly beef, or beef containing bladders of the beef tapeworm, is now rarely found on the market, thanks to federal meat inspection. Meat that is cooked until it has lost its red color is safe. (Courtesy, Army Medical Museum)

Likewise, the **pork tapeworm** *(Taenia solium)* is rare among Muslims and Jews, who avoid the meat of the hog, and common in Christians in parts of Europe, Africa, the Middle East, and Mexico where pork is eaten without thorough cooking. The pork tapeworm resembles the beef tapeworm closely and has a similar life history, except that the bladders usually develop in pigs. It is especially dangerous, however, because the bladders can develop in humans. If these settle in muscles, no great harm results. Sometimes, however, they lodge and grow in the eyeball, interfering with vision. Certain cases of convulsive seizures and other neurological disturbances are due to bladders in the brain. Treatment is by surgical removal of the bladders, if possible.

A very thin person is frequently accused by his friends of harboring a tapeworm, and it is true that parasitized individuals are sometimes emaciated, partly because of loss of appetite. The worm's toxic excretory products may cause dizziness, nausea, and other symptoms. Also, the mere bulk of the worm, especially when folded back on itself many times, may block the intestine and cause distress. Or a person may harbor a tapeworm for many years—even throughout life—with ever noticing any symptoms. The presence of a tapeworm can be detected by the appearance of white sections loaded with embryos in the feces. The only way to get rid of the parasite is to take by mouth some drug that kills the worm or causes it to detach from the intestinal wall, whereupon the whole worm is evacuated with the feces.

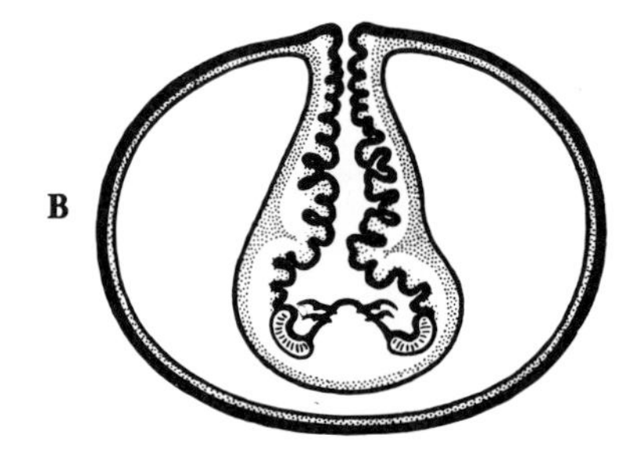

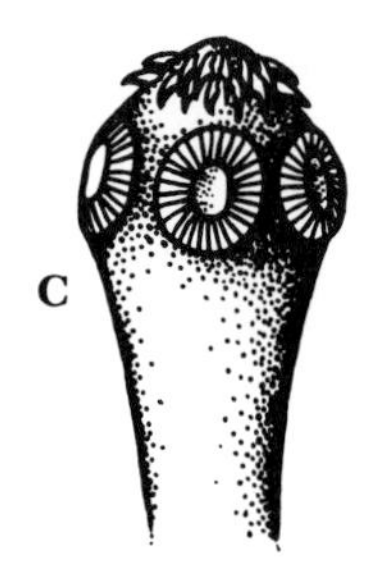

Pork tapeworm *(Taenia solium).* **A:** Six-hooked embryo. **B:** Bladderworm with inverted scolex. **C:** Everted adult scolex showing hooks and suckers. (After various sources.)

Many tapeworms have more than one intermediate host. The **broad fish tapeworm** (*Diphyllobothrium latum,* or *Dibothriocephalus latus)* is a large tapeworm with a life history requiring that the eggs reach freshwater, that the free-swimming larvas be eaten by copepods (small crustaceans), and that the copepods be eaten by fishes. Humans get the parasite when they eat raw, undercooked, or inadequately smoked freshwater fish.

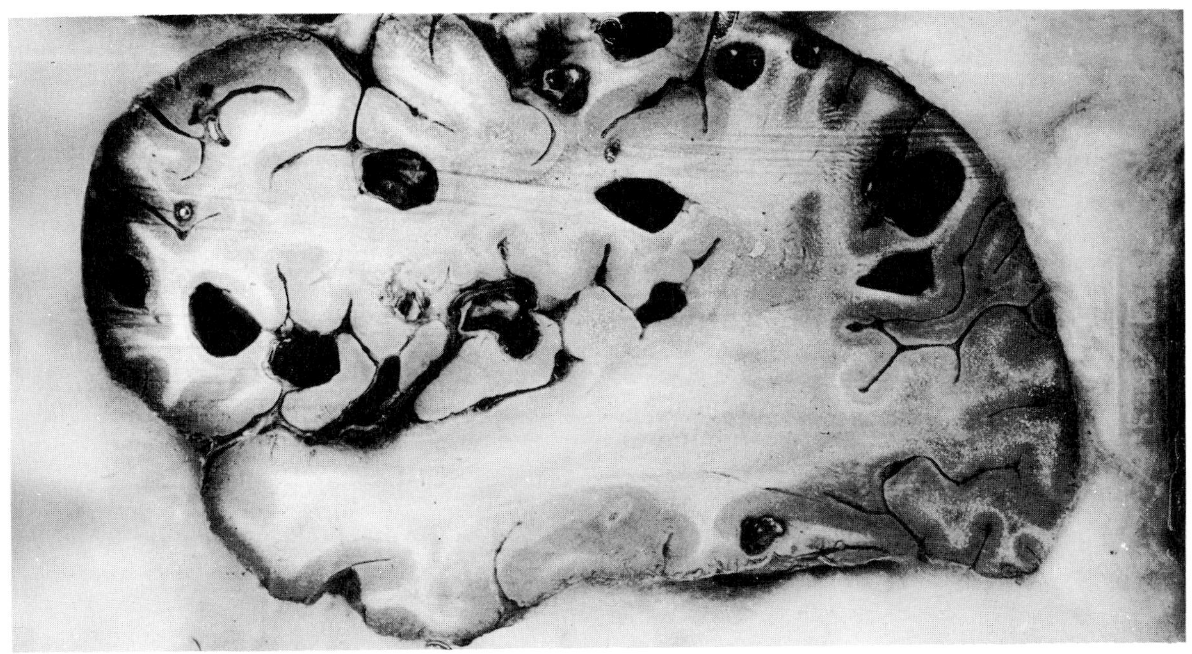

Bladderworms in brain of a woman, 34 years old, a resident of the Chicago region, who was brought to the hospital with a history of convulsive seizures. These became more frequent until three days before her death, when convulsions set in every half-hour. The brain (shown here in longitudinal section) contains 100 to 150 bladders. (Specimen in Pathology Museum, University of Chicago.)

The broad fish tapeworm occurs in many places all over the world and has been known for centuries in the Scandinavian and Baltic regions of Europe, where in some localities nearly all of the people are infected. Brought by Europeans to North America, it became established in the Great Lakes region; shipping of fish to other parts of the United States and Canada has spread this parasite. The symptoms suffered by most people infected are mild and similar to those associated with other tapeworms. However, some people infected with this tapeworm develop megaloblastic anemia, a serious disease caused by a deficiency of vitamin B_{12}, essential to normal digestion and absorption of nutrients and to production of red blood cells. The parasite takes up such great quantities of this vitamin that the human host, especially if subnormal in ability to absorb the vitamin, gets almost none.

Sometimes a human is the intermediate host for a tapeworm that lives its adult life in some other mammal. *Echinococcus granulosus* is a minute tapeworm (with only three or four sections) that lives as an adult in the intestine of dogs and occurs only as a young stage in humans. Human infection results from drinking contaminated water or from allowing dogs to lick the face and hands. Because dogs clean the anal region by licking, the tongue of an infected dog is likely to carry tapeworm eggs. The young larva develops into a hollow bladder. From the inner walls of this grow smaller bladders, and within each of these are produced numerous scolexes. The whole structure is known as a hydatid cyst and may grow to the size of an orange or even a watermelon. When such a cyst develops in the brain, the results are extremely serious. From the point of view of the parasite, development of the cysts in humans is unfortunate because humans are rarely eaten by dogs, and the cysts cannot reach the final host. The most common intermediate hosts are sheep and cattle. The parasite is common in the great stock-raising regions of the world, including parts of the United States.

No one has ever seen a free-living animal evolve into a parasite. Parasitism may sometimes evolve from harmless commensalism

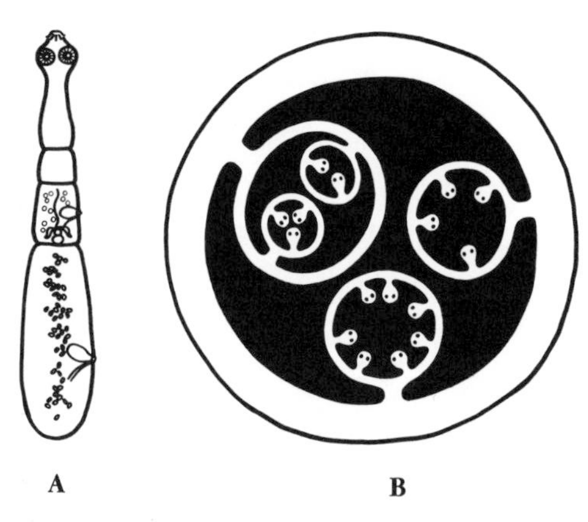

Humans are the intermediate host for *Echinococcus granulosus.* **A:** Adult, 3 to 6 mm (1/8 to 1/4 inch) long, lives in dogs. **B:** Hydatid cyst from human liver. (Modified after Leuckart.)

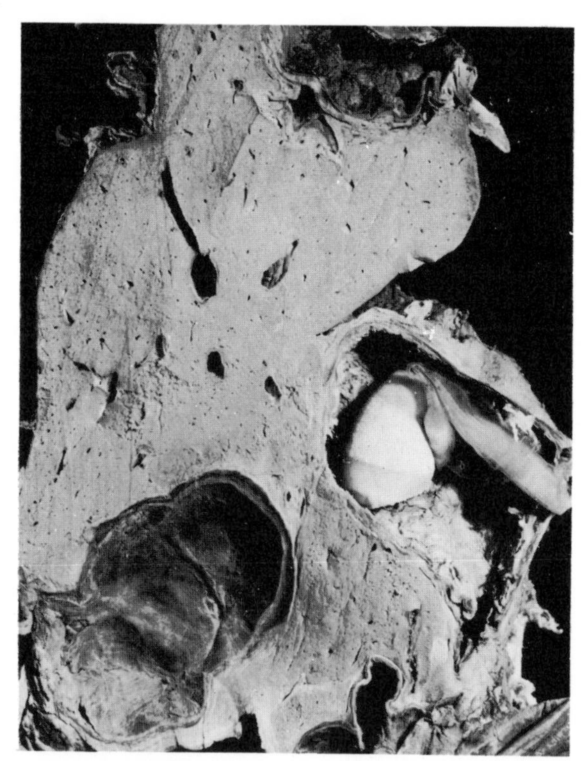

Hydatid cysts of *Echinococcus* revealed by cutting open the liver of a person who accidentally swallowed the mature sections or eggs. The adult parasites live in the intestine of dogs. These cysts were 2 to 5 cm across (4/5 to 2 inches). (Specimen in Pathology Museum, University of Chicago.)

and sometimes develop by other routes. External parasites are little changed from their free-living relatives except for the development of hooks or suckers for holding on. Internal parasites usually show more marked structural adaptations to their special environment. The sensory and muscular systems, so important for free-living activities, may become reduced. Many parasites lose the free-swimming young stages and depend entirely upon passive transfer from host to host. In some intestinal parasites digestive organs are reduced and in others are lost altogether. On the other hand, the reproductive systems of parasites are highly developed and most of the energy of these animals is directed toward one main activity: the production of tremendous numbers of eggs to offset the losses incurred in the hazardous transfer from one host to another. Asexual reproduction within the body of an intermediate host is another method for increasing the number of young, and therefore the chance that a reasonable number will reach a final host.

The young stages of many parasites live embedded in such tissues as muscle, brain, and so on. But the adult, which produces the eggs, must live in or near some cavity that has direct access to the outside. The digestive tract and its associated organs such as lungs and liver are most frequently occupied, as they present the easiest and consequently most popular highway for the entrance and exit of parasites, particularly for those that depend upon passive transfer.

The relation of a parasite to its host requires not only marked adaptation on the part of the parasite but often also an adjustment on the part of the host. The host may secrete a capsule about a young stage embedded in its muscles; this helps to confine the activities of the parasite. Or the host may develop a resistance to the toxic substances given off by the adult.

The parasite-host relationship is usually specific. Some parasites can live in a variety of closely related hosts, but many can develop in only one particular species. Some can grow in species other than their normal hosts; but when they do so, there is a lack of mutual adjustment and the host or parasite may suffer abnormal damage.

There is a tendency among most people to look upon parasitism as an aberrant way of life and upon parasites as being somehow immoral or at least less worthy of respect that their free-living relatives. But since there are more parasites than free-living individuals, a parasitic existence must be considered as one of several normal ways of life. Who can say that a parasite, the very existence of which may depend upon doing as little harm as possible to its host, is a less noble creature than a voracious carnivore that kills its victim outright?

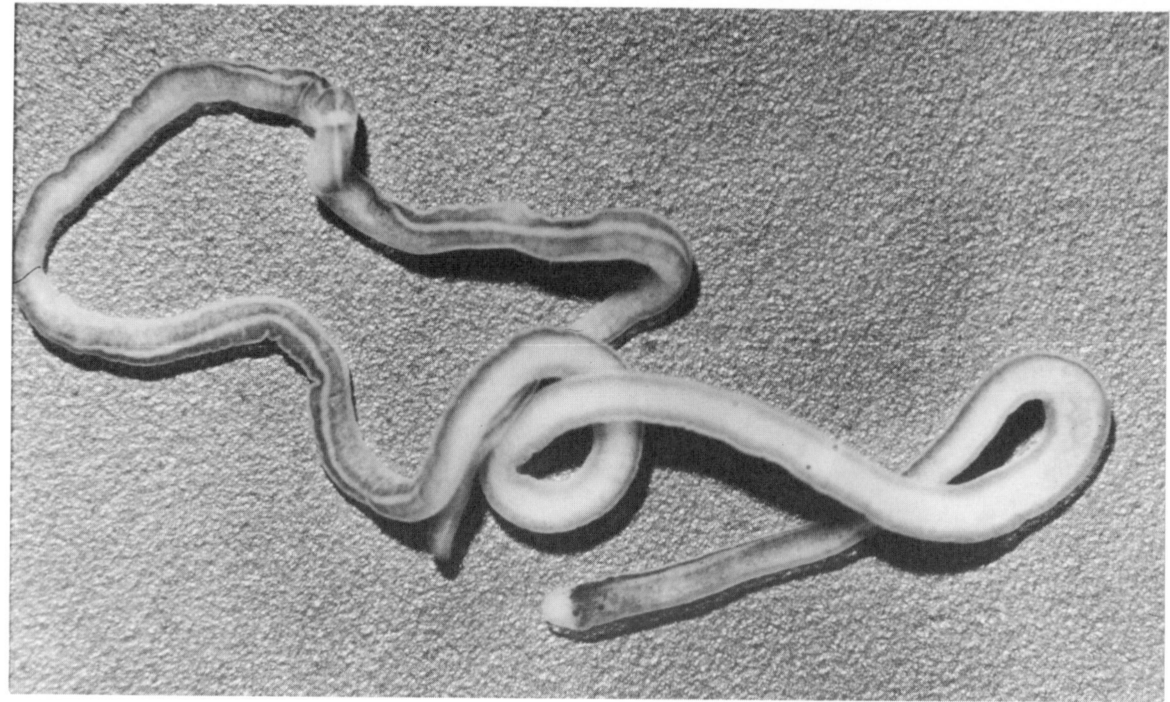

Chapter 12

One-Way Traffic: Proboscis Worms

Proboscis worms are common along seashores under stones and among seaweeds; some float in the open ocean; and a few live in freshwater or damp soil. Their elongated flattened bodies range in length from a few millimeters to many meters, and they are often colored a vivid red, orange, or green, with contrasting patterns of stripes and bars. Their most distinctive character is the **proboscis,** a long muscular tube that can be thrown out to grasp prey. The phylum name NEMERTEA was borrowed from a mythical Greek sea nymph, Nemertes, said to have unerring aim.

Like planarians, nemerteans are bilaterally symmetrical. The anterior end is not distinctly marked off as a head but often bears numerous simple eyes and specialized sensory cells. The proboscis worms are not a very large group; they are not usually noticed by visitors to the seacoast; nor do they have special economic or medical importance. They are described here because they are among the simplest animals to possess two important features of construction not found in flatworms but present with few exceptions in all more complex animals.

First, the **digestive system** has two separate openings instead of one. The mouth is near the anterior end, and most of the long, straight digestive tract is an intestine that often has side branches. The intestine extends the length of the body, terminating at the posterior end in an opening called the **anus.** The mouth serves exclusively for taking in food and the anus for the exit of undigested materials. This one-way traffic has certain ad-

Opposite, above: **Rocky shore nemertean,** *Tubulanus polymorphus,* common on California and Oregon coasts, may reach 1 to 3 m in length. Its bright red-orange color is highly visible, as are the patterns of bars or stripes on other nemerteans. Many of these worms contain distasteful or toxic substances that, together with their memorable colors or patterns, may teach predators to avoid them. (R. Buchsbaum)

Opposite, below: **Sandy shore nemertean.** Florida. (R. Buchsbaum)

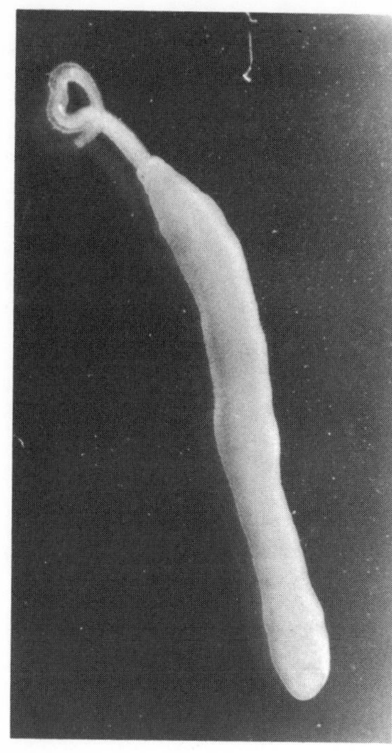

Freshwater nemertean, *Prostoma,* is a minority member of this almost entirely marine group. The long proboscis is here partly everted. Pennsylvania. (R. Buchsbaum)

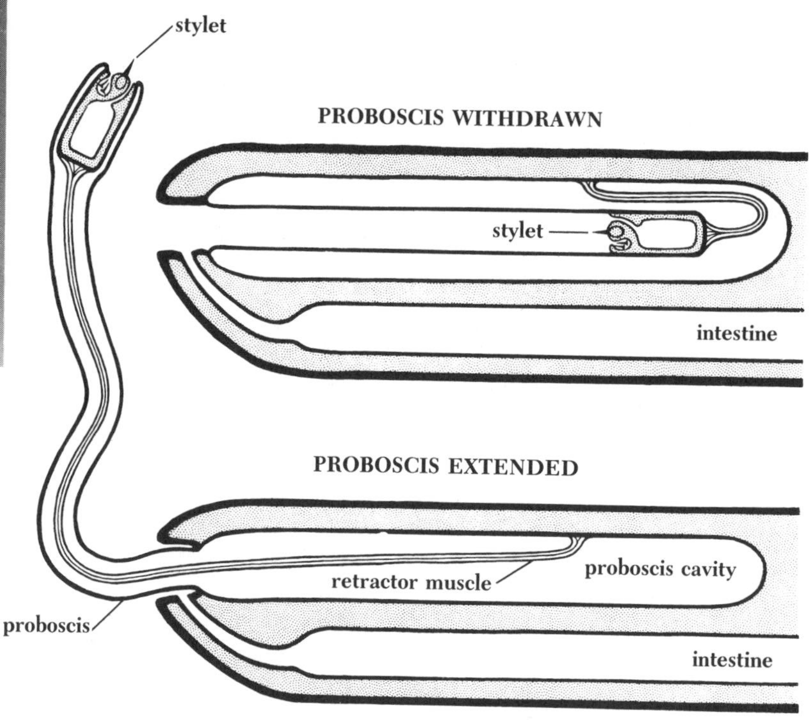

Diagram of the anterior end of a nemertean showing the **proboscis** withdrawn and extended. The proboscis, which is often as long or longer than the body, lies inside a muscular sheath just above the digestive tract. When a likely victim is encountered, the proboscis is extended quickly and wrapped around the prey, which is entangled with sticky mucoid secretions. In some nemerteans the proboscis bears one or more sharp stylets; these pierce the body of the prey, which is paralyzed by toxic secretions introduced into the wounds. When a stylet is lost, it is replaced from adjacent sacs of extra stylets. During development, the proboscis and digestive tract form as completely independent systems, and some nemerteans retain separate proboscis and mouth openings throughout life; in others, as shown here, there is only a single anterior opening in the adult. (Based on W. R. Coe.)

Worm eats worm. A nemertean *(Paranemertes)* has captured a segmented worm. The long white proboscis of the nemertean is wrapped around the prey, which will soon be swallowed. (S. A. Stricker)

vantages over the general traffic jam in the gastrovascular cavity of cnidarians and flatworms, where newly ingested food becomes mixed with partly digested food and indigestible residues. In a one-way system such as is found in nemerteans and in virtually all other animals more complex than flatworms, the food passes along a continuous digestive tract that can be differentiated into various parts specialized for separate, sequential functions.

The second important structural innovation found in the nemerteans is a new system, the **circulatory system,** which takes over much of the vascular function of the old gastrovascular cavity. The circulatory system distributes food, oxygen, and other substances to the tissues and carries off metabolic wastes. The circulatory system of nemerteans usually consists of three principal **blood vessels,** muscular tubes that lie in the mesenchyme, one on each side of and one just above the intestine. They connect with each other by transverse vessels and sinuses. The **blood** is generally colorless and contains cells. In some species the cells are red from hemoglobin, the same substance that col-

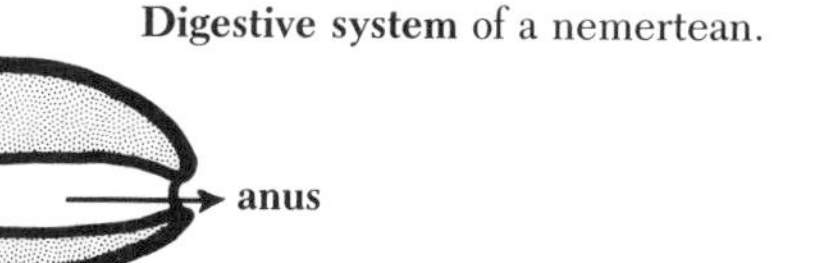

Digestive system of a nemertean.

ors human blood red. Hemoglobin is present in most vertebrates, many invertebrates (including some flatworms, which have no circulatory system), and even in a few protozoans and plants. It combines readily with oxygen, allowing blood and other tissues to contain more oxygen than would simply occur in solution. The circulation of nemerteans is primitive in some respects. There is no special pumping organ, or heart, to circulate the blood. The larger vessels are contractile, but blood flow follows no consistent direction and depends mostly on pressure from muscular waves that pass along the body wall. Also, the blood vessels are not finely branched; therefore materials must move longer distances to and from cells by the slow process of diffusion.

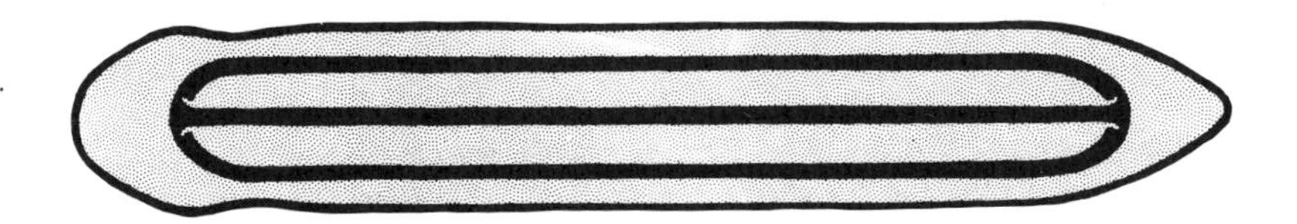

Circulatory system of a nemertean. Only the three main blood vessels are shown.

The **general body structure** of nemerteans, aside from the two new features described above, is very similar to that of planarians and includes all of the same organ systems. The animal is covered completely with a ciliated epithelium that contains many gland cells. Under the epithelium are thick muscular layers, circular and longitudinal, by which the highly contractile animal can execute agile movements of all sorts. Some muscle cells are usually associated with the digestive tract, but food is moved

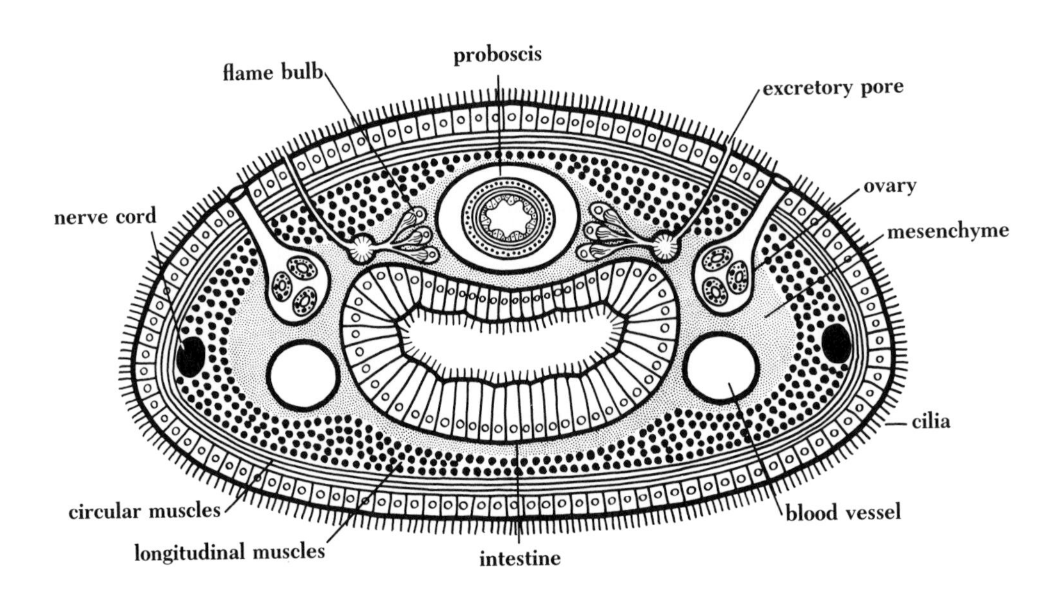

General body structure of a nemertean shown in cross-section. (Based on W. R. Coe.)

along chiefly by means of muscular contractions in the body wall. These pass down the animal from front to rear and force the food along as they press on the intestine. As already mentioned, muscular waves also assist the flow of blood. Between the gut and the body wall there is a layer of mesenchyme cells. Imbedded in the mesenchyme is the circulatory system, already described, and also the excretory system, consisting of canals with side branches ending in flame bulbs, much as in flatworms. Wastes removed from the mesenchyme and blood pass into the canals, which open on the surface by pores. The nervous system is also similar to that of flatworms, but the brain is more massive and forms a ring around the proboscis apparatus; longitudinal cords run the length of the worm on each side.

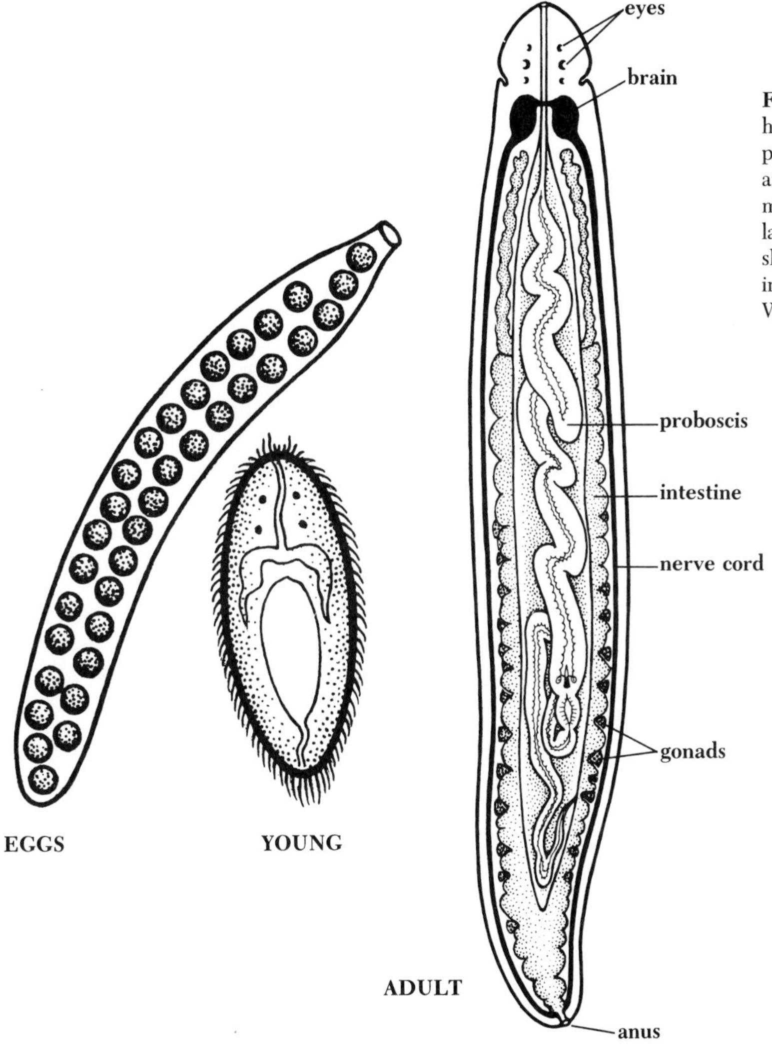

Freshwater nemertean, *Prostoma,* has 3 pairs of simple eyes and a proboscis longer than its body; it is among the few hermaphroditic members of the phylum. Eggs are laid and fertilized in a mucoid sheath, and they develop directly into wormlike young. (Modified after W. R. Coe.)

Eggs and sperms are produced, usually by separate female or male individuals, in little sacs that lie in the mesenchyme between the branches of the intestine. Each gonad opens directly to the outside through its own pore on the surface. There is none of the complicated sexual apparatus seen in flatworms, the gametes being simply shed to the outside, and in this respect the nemerteans are less complex than the flatworms. In many nemerteans, the fertilized egg develops directly into a little worm, but in some marine nemerteans there is a ciliated larva shaped like a helmet with earflaps and known as a **pilidium.** It has a ventral mouth but no anus, and at the end opposite the mouth is a sense organ topped by a tuft of long flagella. During a rather complicated metamorphosis, the pilidium develops a whole new outer epithelium beneath the original larval skin, which is shed and often eaten by the young worm. This sort of development may be compared to the transformation of caterpillars into butterflies, but its presence in animals as simple as nemerteans is a puzzle.

Pilidium larva. (Partly after C. Wilson.)

Stout-bodied nemertean, *Amphiporus bimaculatus,* of the west coast of North America, is 15 cm (6 in) long. When disturbed, it may swim by undulations of the body. (R. Buchsbaum)

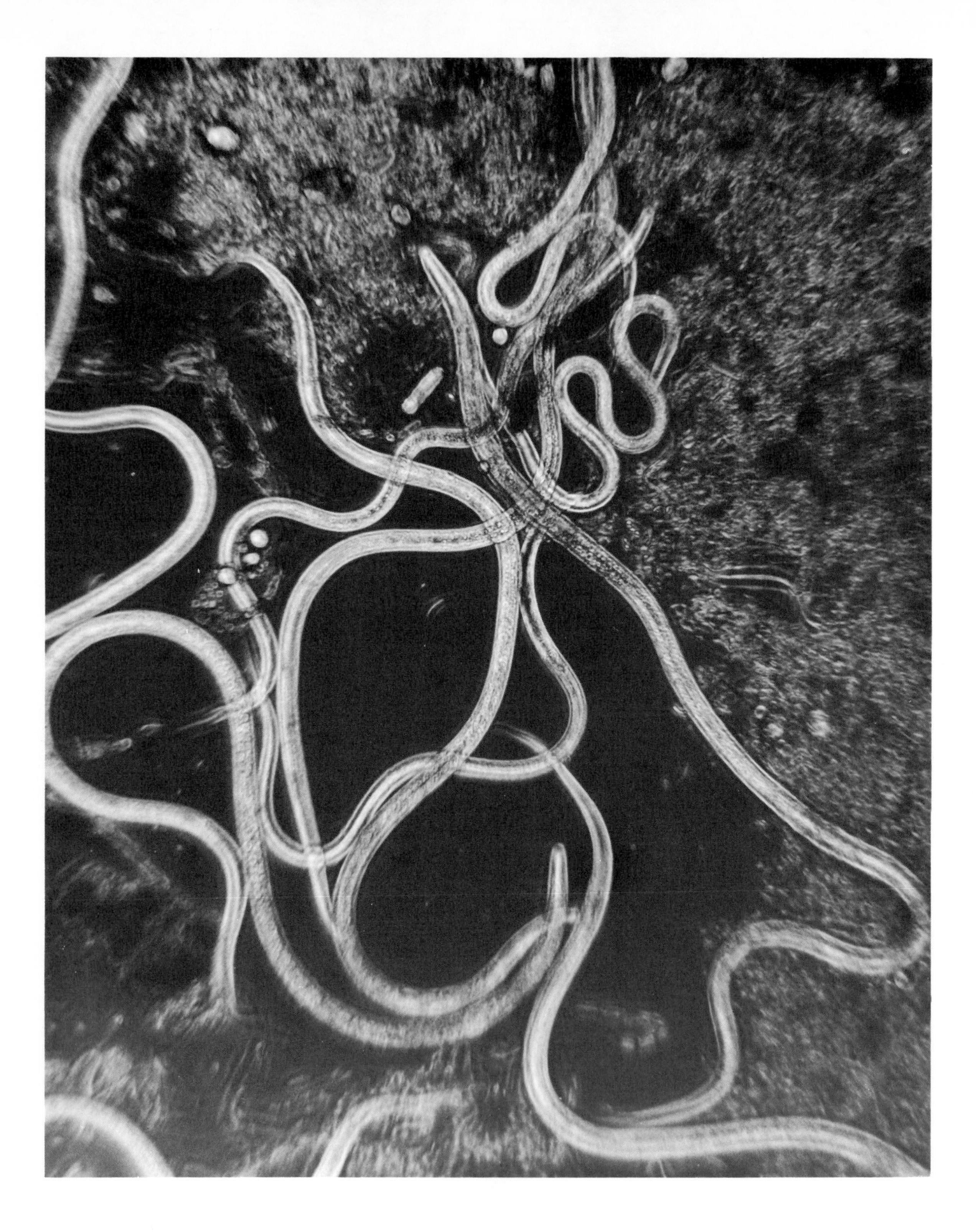

Chapter 13

Roundworms

Most people are host at some time or other to the cylindrical worms called "roundworms." Many different species of parasitic roundworms have been found in humans, but relatively few are common. Of these, some are harmless and do not even make their presence known, while others cause mild to very serious diseases. Besides these of immediate human interest, there are many economically important species that parasitize domestic stock and crops, as well as parasites of wild animals and plants, and an even larger number of free-living types. Free-living kinds feed on bacteria or on other small organisms in all moist soils and in the bottom sediments of oceans and freshwaters. Among both free-living and parasitic types are ones with activities beneficial to humans, such as recycling of soil nutrients and destruction of various agricultural pests. And one free-living soil species (*Caenorhabditis elegans*) has become a well-established laboratory animal; its genetic and developmental biology are known in more detail than for almost any other organism on earth and promise to contribute much to unanswered questions about the biology of all organisms, including humans.

Many roundworms look like animated bits of fine sewing thread; and from the Greek word for thread, *nēma*, has been derived the technical name of the group, phylum **Nematoda.** This is by no means a small or obscure phylum. More than 15,000 species have so far been described, and the actual number has been estimated at close to half a million. Nematodes are so abundant that a square meter of garden soil teems with millions of them. When we see a sick dog, our first guess is that it has more roundworms than it can tolerate. And even in the best-regulated cities, roundworms occur in the drinking water.

Opposite: **Vinegar eels** are nematodes. They used to be easily collected from any jar of vinegar, by taking a bit of the fine sediment in the bottom. There they lived together with the bacteria responsible for making vinegar from fermented apple juice, wine, or other alcoholic liquids, by converting the alcohol to sour acetic acid. These worms are harmless, except to the vinegar bacteria on which they feed, but one may look for them in vain in most modern distilled, filtered, or pasteurized vinegars. *Turbatrix aceti*, about 2 mm long. (R. Buchsbaum)

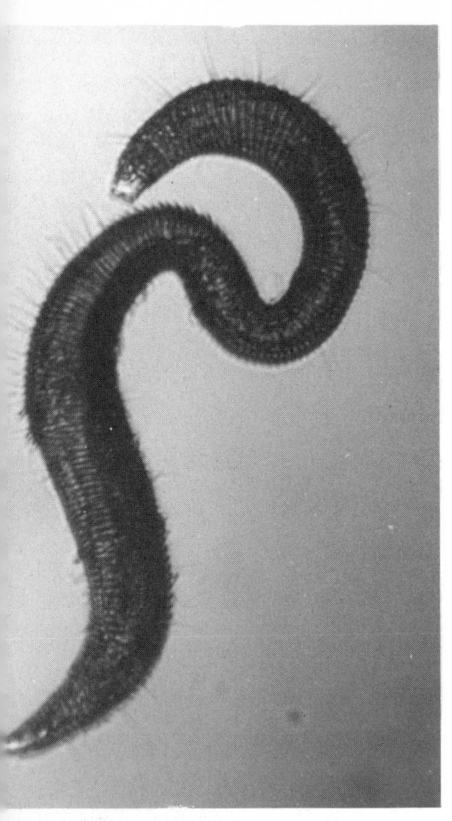

Marine nematode, its surface decorated with rings and bristles that help it to grip the bottom sediment through which it moves. Bimini, West Indies. (R. Buchsbaum)

Some roundworms occupy very limited niches. On the other hand, many species of nematodes are cosmopolitan, and members of one species sometimes occupy a remarkably broad range of habitats. The group as a whole lives anywhere that other animals can live, even in such unlikely places as hot springs or ice. Any collection of earth or of aquatic debris from an ocean, a lake, a pond, or a stream, if examined with a lens, will reveal the tiny colorless worms thrashing about in a way so characteristic of roundworms that it immediately identifies them. The widespread occurrence of this group inspired a leading student of nematodes to write:

> If all the matter in the universe except the nematodes were swept away, our would would still be dimly recognizable, and if, as disembodied spirits, we could then investigate it, we should find its mountains, hills, vales, rivers, lakes, and oceans represented by a film of nematodes. The location of towns would be decipherable, since for every massing of human beings there would be a corresponding massing of certain nematodes. Trees would still stand in ghostly rows representing our streets and highways. The location of the various plants and animals would still be decipherable, and, had we sufficient knowledge, in many cases even their species could be determined by an examination of their erstwhile nematode parasites. [N. A. Cobb]

Nematodes are so remarkably alike in external appearance and internal structure that the following description, based mostly on *Ascaris lumbricoides*, a common human parasite, roughly fits almost any other roundworm. The elongated cylindrical body tapers at both ends. Curiously enough, it is entirely devoid of cilia, outside and in, except for modified nonmotile cilia in sense organs. The body is covered with a thick, tough **cuticle,** which is molted 4 times during the nematode's life. However, molting is not necessary for growth (as it is in insects, for example); the cuticle grows with the rest of the body between molts and after the final molt to the adult stage. The cuticle is secreted by the underlying **epidermis,** which, in adult ascarids, is a **syncytium,** its many nuclei not separated by cell membranes. Under this is a **longitudinal muscle layer,** divided into four lengthwise bands by four thickenings or ridges of the syncytium. The muscles consist of large lengthwise cells with bulbous cytoplasmic expansions that project into the interior. The lack of circular muscles, and the nervously controlled synchrony of both dorsal or both ventral bands, permit bending of the body only in the dorsoventral plane. On a flat surface, the worms wriggle along on their sides. When the worms are free in the water, the whiplike movements of the body result in erratic locomotion; but when they are in the soil, or among the tissues or intestinal contents of a host, the solid particles provide surfaces against which to push, and the worms move along with speed and agility.

The **mouth** is at the anterior tip, encircled by **sense organs** in the form of small protuberances and pits, and it leads into a muscular **pharynx** by means of which the worm sucks in food. The **intestine** is made up of only one layer of cells and is not encircled by a muscle layer; the food is moved along mostly by the ingestion of more food and by the general movements of the body. The digestive tract opens through the **anus** near the posterior tip. Between the digestive tract and the body wall is a fluid-filled **body cavity** kept under pressure by contractions of the body wall muscles and by the elastic cuticle. Against this continuous pressure, food must be pumped into the intestine by the muscular pharynx and prevented from escaping at the rear by a strong sphincter or valve. When the anus is opened, the jet of intestinal contents from a large ascarid may shoot half a meter.

Cross-section of a female ascarid.

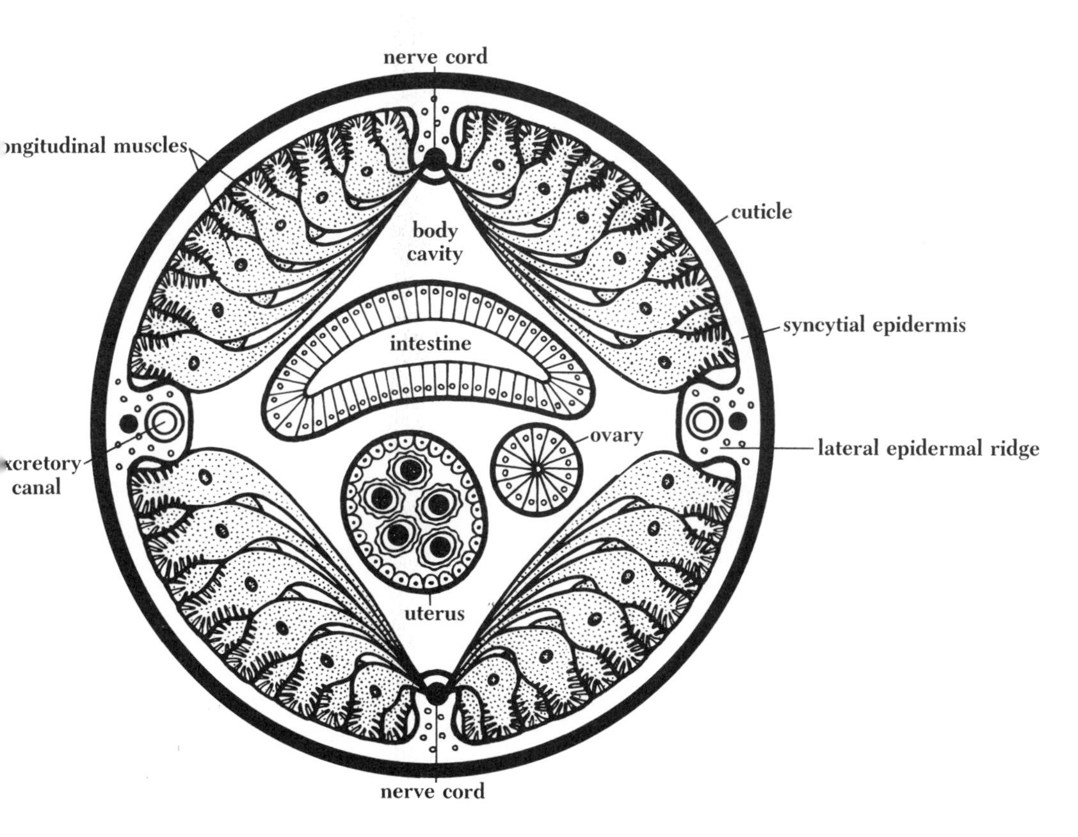

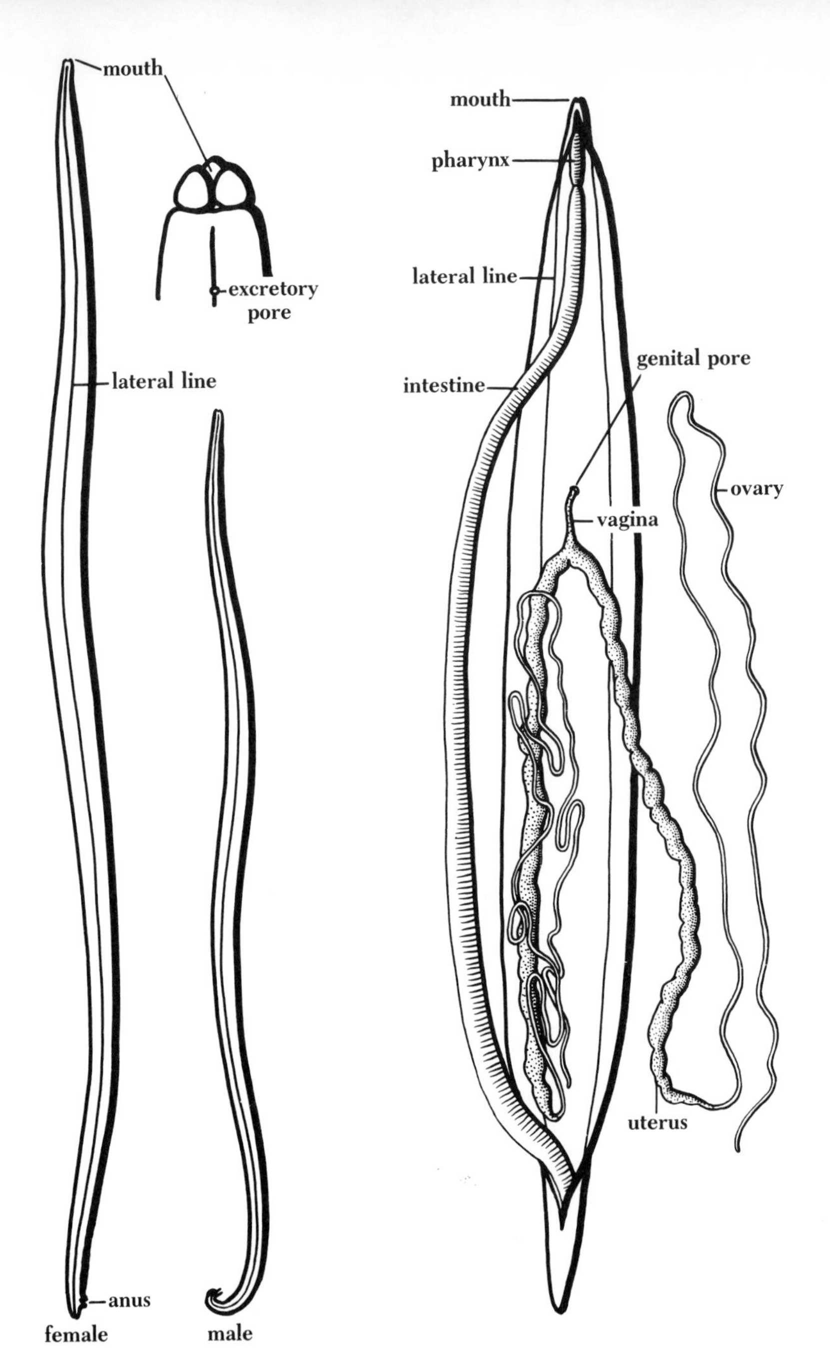

Left: **Ascarids,** natural size, taken from the intestine of a pig.

Right: **Dissection of a female ascarid.** The animal has been slit open along the mid-dorsal line, revealing the internal organs.

The fluid in the body cavity aids in distributing nutrients and oxygen, and nematodes have *no special respiratory or circulatory systems.* Regulation of the body cavity fluid is probably a major role of the **excretory system,** which consists of two canals that run in the lateral epidermal ridges and unite near the anterior end to form a single tube, opening ventrally through a pore.

The **nervous system** consists of a nerve ring and associated ganglia, around the pharynx, from which longitudinal trunks run backward in the epidermal ridges. The large ventral nerve cord also has a series of ganglia. Nematodes are peculiar in that no nerves branch out to the muscles; instead, long processes from the muscles extend to the midline and make contact with the dorsal and ventral nerve cords (as shown in the cross-sectional diagram).

The **reproductive system** lies in the space between the intestine and the body wall. The sexes are separate in ascarids, as in most nematodes, and the males are smaller than the females. The tubular reproductive system coils back and forth in the body cavity. In a female, the genital opening is in the middle part of the worm on the ventral side. In a male, the sperm duct joins the digestive tract to open near the posterior end. The sperms are transferred to the vagina of the female with the aid of a pair of horny spicules.

In the **development** of nematodes, the pattern and number of cell divisions is amazingly constant for any species. The number of resulting cells is small and fixed, and by the time the young worm hatches, cell division is largely complete. Most growth is by cell enlargement. So it has been possible to trace the fate of individual cells as they differentiate to form the various organs of the worm. Such studies have revealed that each cell is the product of a given series of divisions, always develops in the same way, and occupies the same position in each worm. This rigid pattern of development is in sharp contrast to the more flexible pattern of most other animals (in which the fates of cells depend to a greater extent on their position within the embryo and the kinds of cells around them). Nematodes are therefore particularly convenient animals in which to investigate the effects of a local damage or of a single genetic mutation on a group of cells, and to try to determine the consequences for the development of structure and behavior of all stages of the worm.

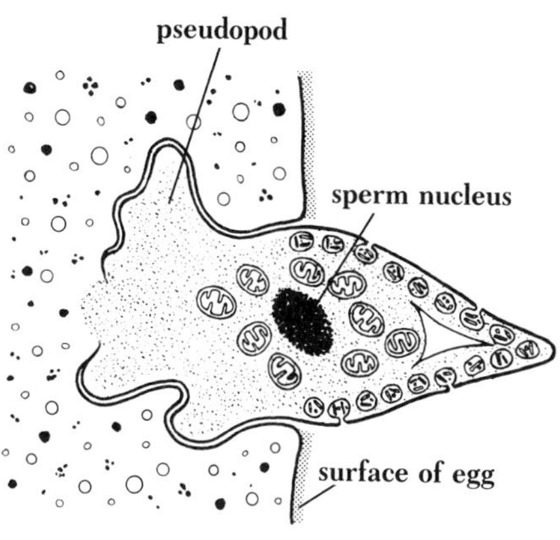

Ameboid sperm of *Ascaris lumbricoides*, a species parasitic in humans. Nematodes have no motile cilia or flagella, and their sperms move along within the female reproductive tract by means of pseudopods. This sperm is beginning to fuse with the large egg, only a small portion of which is shown. (Based mostly on W. E. Foor.)

The events of development and the genetic basis of their control are being actively studied in a tiny free-living soil nematode, *Caenorhabditis elegans*. Only 1 millimeter long when mature, these worms can be raised by the hundreds of thousands in small laboratory dishes, taking only 12 hours from fertilization of the egg to hatching of the juvenile worm. In that time successive cell divisions produce 671 cells, of which 113 are programmed to die, leaving 558 in the worm that hatches. When the juvenile reaches maturity it will have exactly 959 cells (not counting any eggs or sperms in its reproductive system). Such a number is adequate for studying the development of complex organ systems, but not so many as to make it impossi-

Caenorhabditis elegans, an adult and a young stage, photographed under the microscope on the surface of an agar plate. Great numbers of these tiny nematodes are reared in laboratories, grown on agar plates and fed on bacteria, by researchers studying the genetic control of their structure and development. (Specimens, courtesy R. Edgar and M. Kusch. Photo, R. Buchsbaum.)

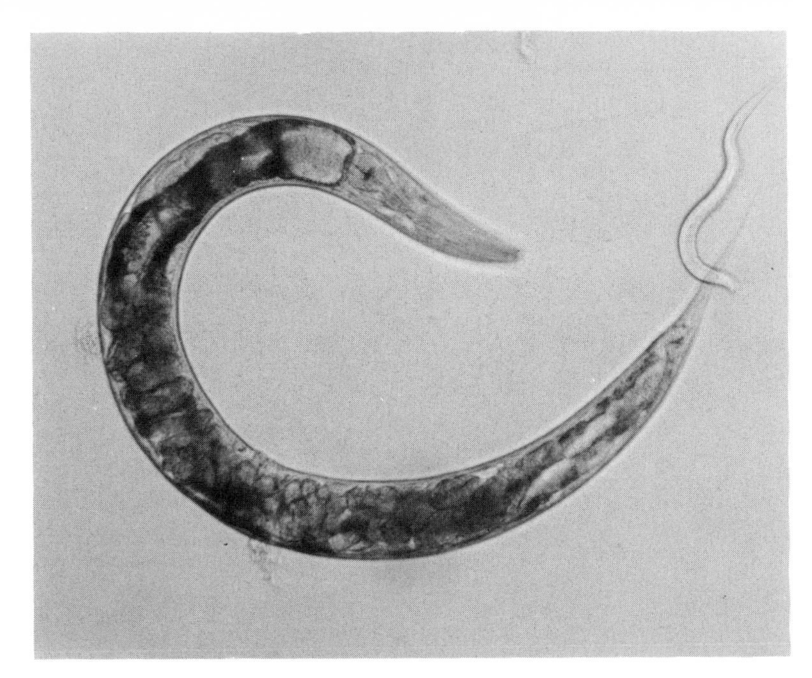

ble to follow the divisions of each cell. Detailed studies of both normal and abnormal development in worms of known heredity may eventually help us to understand how heredity controls development in all organisms, including humans.

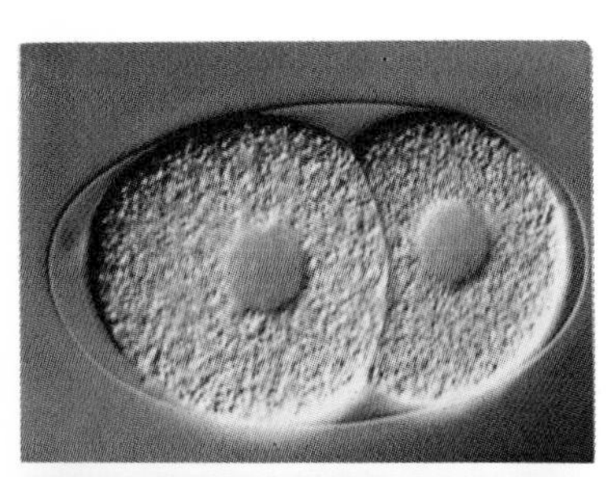

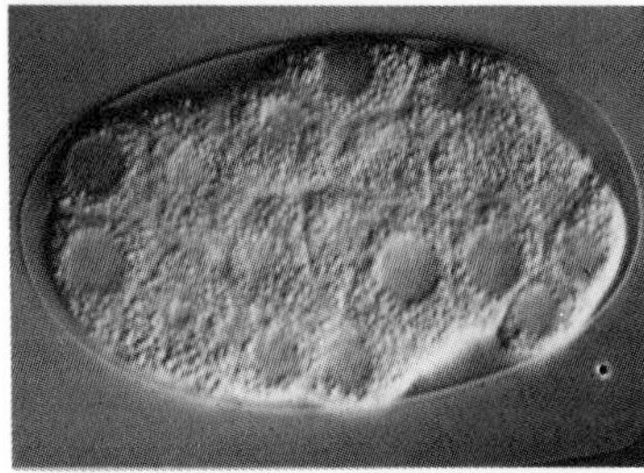

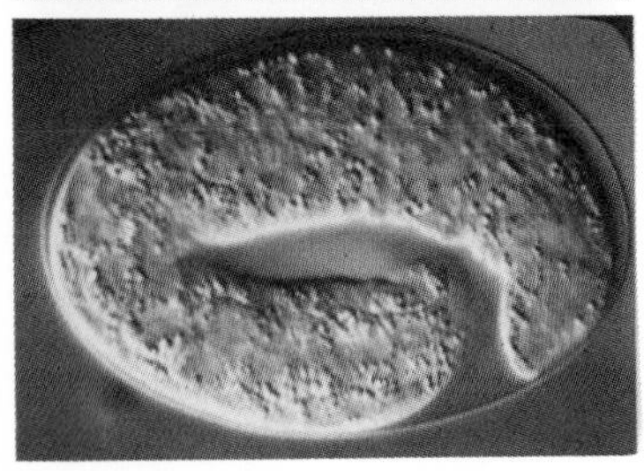

Development of a nematode, *Caenorhabditis elegans*, through several stages. *Top*, 2-celled stage; *middle*, 28-celled stage; *bottom*, embryo has attained a wormlike form. (E. Schierenberg)

The widespread importance of **plant-parasitic nematodes** went unappreciated until recent decades, perhaps partly because the worms tend to be small and less visibly destructive than the insects or the snails and slugs that usually come to mind as garden and agricultural pests. Only after the testing of the first pesticides effective against nematodes was the great impact of these worms revealed. Now it is known that many established agricultural practices such as crop rotation, cultivation, addition of organic matter, and control of soil moisture and acidity are effective at least partly because they reduce populations of harmful nematodes. Recent estimates, however, still hold nematodes responsible for reducing crop yields by about 10% in the United States and provoking distribution of over 50 million kilograms of pesticides over hundreds of thousands of acres in a single year. Economic losses amount to several billion dollars annually.

On the other hand, nematodes sometimes benefit agriculture by parasitizing insect pests; and researchers, growers, and companies are now cooperating to test and promote the use of **nematodes for biological control.** For example, one species of nematode attacks both larvas and adults of a woodwasp that

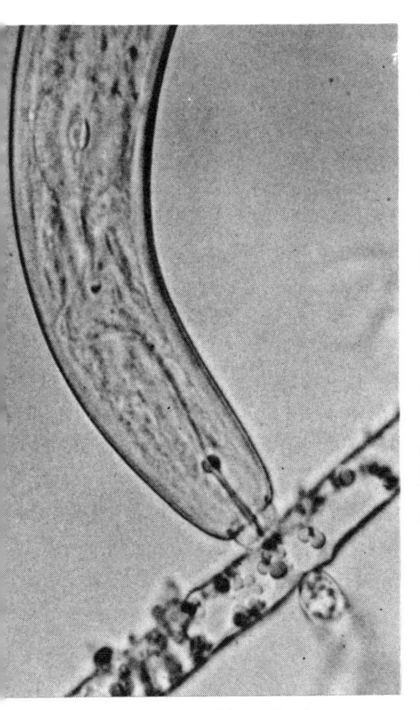

Hollow stylet is used by all plant-parasitic nematodes to pierce the cell walls and to feed on the plant juices and cell contents. This nematode (*Aphelenchoides blastophthorus*, diam. about 25 μm) is piercing a fungal filament. Various nematodes feed on fungi, algas, mosses, and ferns, as well as on flowering plants. Some of these nematodes attack the plant from the outside; others invade the plant tissues. (C. C. Doncaster)

Cyst nematodes are important parasites of many plants. The young worms penetrate the roots, feed, and grow. The "cysts" such as those shown here are the swollen bodies of the adult females, filled with developing eggs. When the females die, the cysts drop off into the soil, and the embryos remain alive for 10 years or more, ready to hatch when stimulated by materials released from the roots of host plants. Clover-cyst nematode, *Heterodera trifolii*, cysts about 0.8 mm long. (C. C. Doncaster)

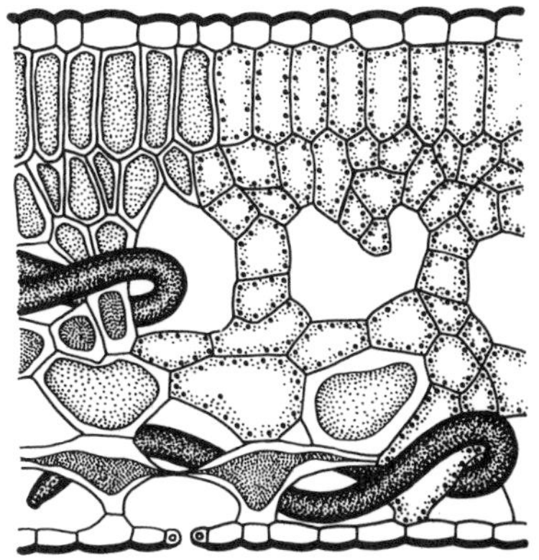

Nematodes parasitic in plants cause untold damage. They suck the contents of the plant cells, causing wilting and withering of the leaves, stunting the plant, and sometimes inducing the growth of galls. The worms here *(Aphelenchus)* are wriggling along in the tissue spaces of a dahlia leaf, shown in section. They have already injured the cells at the left. (After H. Weber.)

damages pine trees. The eggs of the wasps are damaged, but the adults are not killed and themselves spread the nematodes from tree to tree. Certain other nematode parasites enter the body cavity of insect hosts, introducing specific bacteria. When the bacteria have infected and killed the insect, the nematode feeds on the bacteria and perhaps also on the broken-down insect tissues. These nematodes have proved promising as biological controls against a number of important caterpillar and beetle pests. Still others are being investigated for use against snails that harbor schistosomes (see flukes, chapter 11).

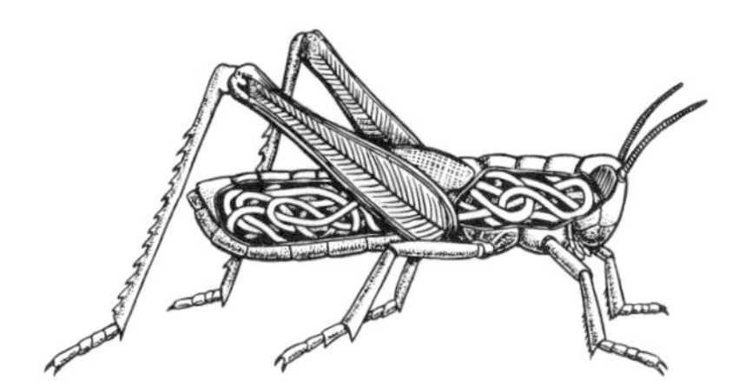

Mermithid nematode, *Agamermis*, lives as a juvenile in insects, mostly grasshoppers. Nourished by the host, it grows and stores up food. Finally it emerges from the doomed host and becomes a free-living adult that reproduces but does not feed. Most juvenile mermithids parasitize insects, but many find their nourishment inside other invertebrates such as crustaceans, spiders, or snails. As adults they live free in soil or water. (After J. R. Christie.)

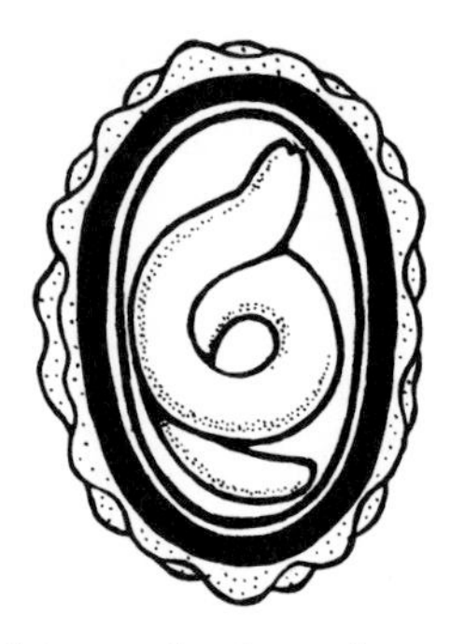

Egg of *Ascaris lumbricoides* is 50 μm to 80 μm in longest diameter, brownish yellow, and thick-shelled, with prominent projections on the outer layer. This egg contains a young developing worm, protected against many unfavorable environmental conditions. This is the infective stage.

Of the parasitic nematodes, one of the largest is an **ascarid,** *Ascaris lumbricoides,* which inhabits the human intestine. The adult females are up to 40 centimeters long. The males are smaller, seldom over 30 centimeters, and can be distinguished by the curved posterior end. The oval eggs are easily recognized, in microscopic examination, by their warty shells. A female may lay 200,000 eggs a day. The eggs pass out in the feces and develop inside their shells into little worms that, if swallowed, hatch in the human small intestine. With so many eggs it would seem that everybody would be infected (and in some areas of Asia they are); but the developing worms encounter many environmental hazards, the chief of which are human sanitary practices. On hatching, the little worms do not remain in the intestine but take a sort of "tour" through the body. They burrow through the intestine into the blood or lymph vessels and are carried about to various organs. In the lungs they are filtered out by the capillaries and bore through the lung tissue into the bronchial (air) tubes. They ascend into the throat, and are then swallowed back into the stomach, finally again reaching the intestine. In the intestine, the worms feed upon the partly digested food of their host and grow rapidly to adult size. They resist digestion

Human ascarids, *Ascaris lumbricoides.*

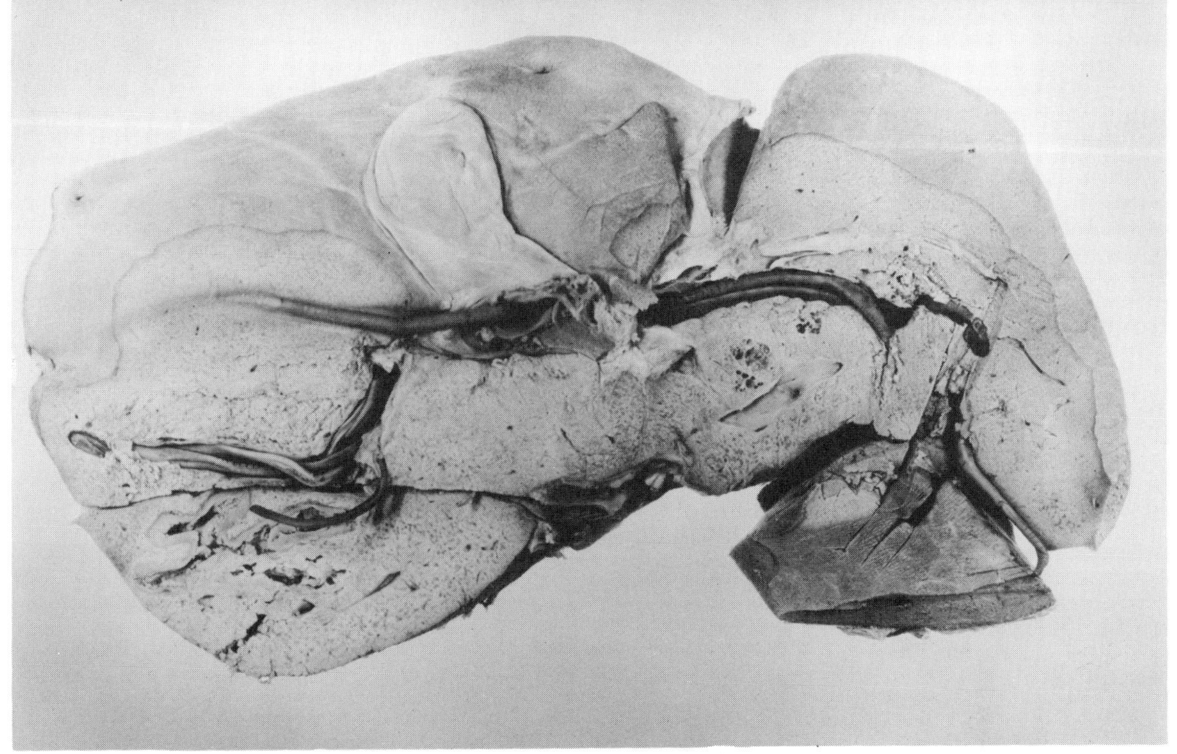

Wandering ascarids may cause death as they migrate into various body tissues, instead of remaining in the cavity of the intestine, where they usually live and do relatively little harm. This is a human liver cut away to show ascarid nematodes that have entered through the bile duct. (U.S. Army Medical Museum)

themselves by secretions that counteract the action of the host's enzymes, but the worms die in 9 to 12 months and are then digested or expelled by the host.

The greatest damage to the host is done during the migrations of the young; their mysterious detour probably represents a vestige of an ancestral life history involving an intermediate host. The adult worms in the intestine seem to be relatively harmless unless they occur in large numbers. Up to 5,000 worms have been found in one host, but in a small child even a hundred worms may block the intestine completely and cause death. Especially if irritated by antiworm medications, they may wander into the liver, the appendix, the stomach, and even up the esophagus and out through the nose, to the horror of the startled host.

Infection occurs chiefly in human communities with bad sanitary habits, and people who are themselves careful may become infected, for example, in a restaurant that serves inadequately washed salad vegetables or strawberries grown in soil contaminated with human feces.

In addition, humans may become infected with dog and cat ascarids, three species of *Toxocara*. Again, it is the migrating young stages that pose the greatest dangers. They occur primarily in children, who usually become infected by the habit of eating dirt.

Merchant of worming medicine at the Sunday Market in Bangkok, Thailand, displayed dozens of jars of worms, many large ascarids conspicuous among them. He presented the masses of worms as evidence of the effectiveness of the medicines that had expelled them from their human hosts and to spur potential buyers to take action against such unwelcome guests by purchasing his wares. (R. Buchsbaum)

Horse ascarids, *Parascaris equorum*, are the only ascarids of horses, asses, and mules. Because they are common and readily obtained, a historic role was played by horse ascarids (then known as *Ascaris megalocephala*) in the development of our understanding of cell structure, of the fundamental events of fertilization, and of the contributions of egg and sperm to the heredity of offspring. Ascarids infect almost every foal because horses feed on the ground, readily ingesting the ascarid eggs in the feces of their mothers. The eggs are resistant and persist on pastures and paddocks for months or years, so prevention consists of giving medication and in cleaning up the feces. The life history of these intestinal worms is similar to that of human ascarids and so are the complications they cause in the horse. Serious infections run to more than 1,000 worms, and almost a bucketful of worms can be removed from one horse. Young horses usually develop immunity as they age and may eliminate many of their worms spontaneously by the time they are a year old. The worms seen here, up to 18 cm long, are from one of the batches expelled over a period of days by a 3-year-old that had received medication. Santa Cruz, California. (R. Buchsbaum)

Much more serious than big ascarids are tiny **hookworms** that occur in close to one-fourth of the world's population—about 900 million people. Progress achieved in developing countries is canceled out by human population growth, so that hookworm disease remains more prevalent than any affliction except the common cold. The mouth cavity of a hookworm contains teeth or plates by which the worm grasps a bit of the intestinal lining of the host and holds on while it sucks in blood and tissue fluids. The eggs pass out in the feces and hatch into young worms that live in the soil for some time, feeding and growing. After they have attained a certain size and have stored up food, they cease to feed and are capable of infecting a human host. They invade by burrowing through the human skin, and infection most often occurs from the habit of going barefoot in localities where the soil contains human feces. After entering the skin, the worms pursue the same course as described for ascarids, eventually reaching the intestine.

Hookworms, natural size.

Hookworm disease occurs mostly in tropical and subtropical regions of the world where adequate moisture and favorable temperatures permit development of hookworms in the soil. Two species are largely responsible for human infections, *Ancylostoma duodenale* and *Necator americanus*. The latter causes 90% of all cases; it appears to have been introduced from Africa

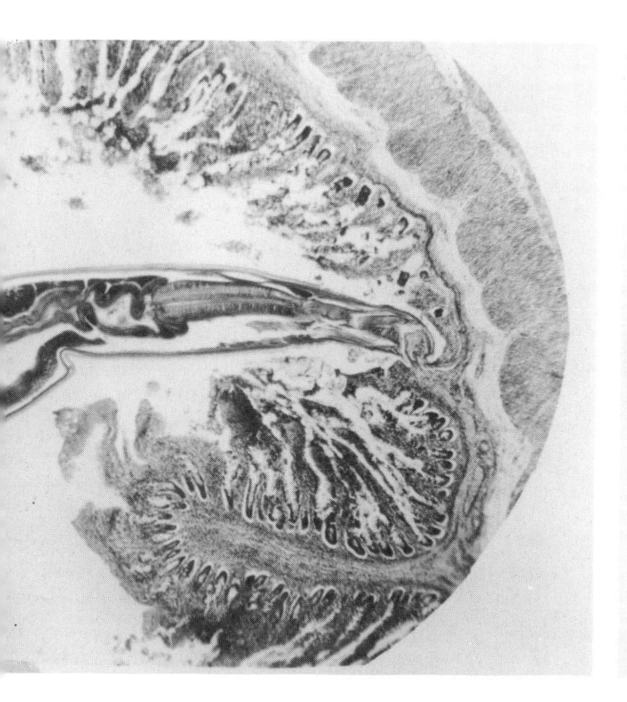

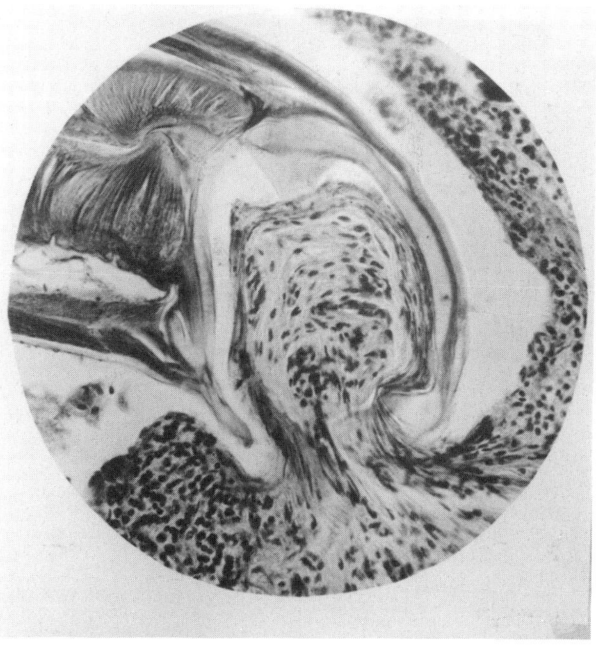

Section through hookworm biting intestinal wall. *Necator americanus* holds on by sharp cutting ridges and feeds on blood and tissue fluids. *Left:* The anterior portion of the worm and the layers of the intestine are seen. *Right:* Closeup of the same section shows the head of the worm with a bit of intestinal lining in its mouth. Stained preparation. (U.S. Army Medical Museum)

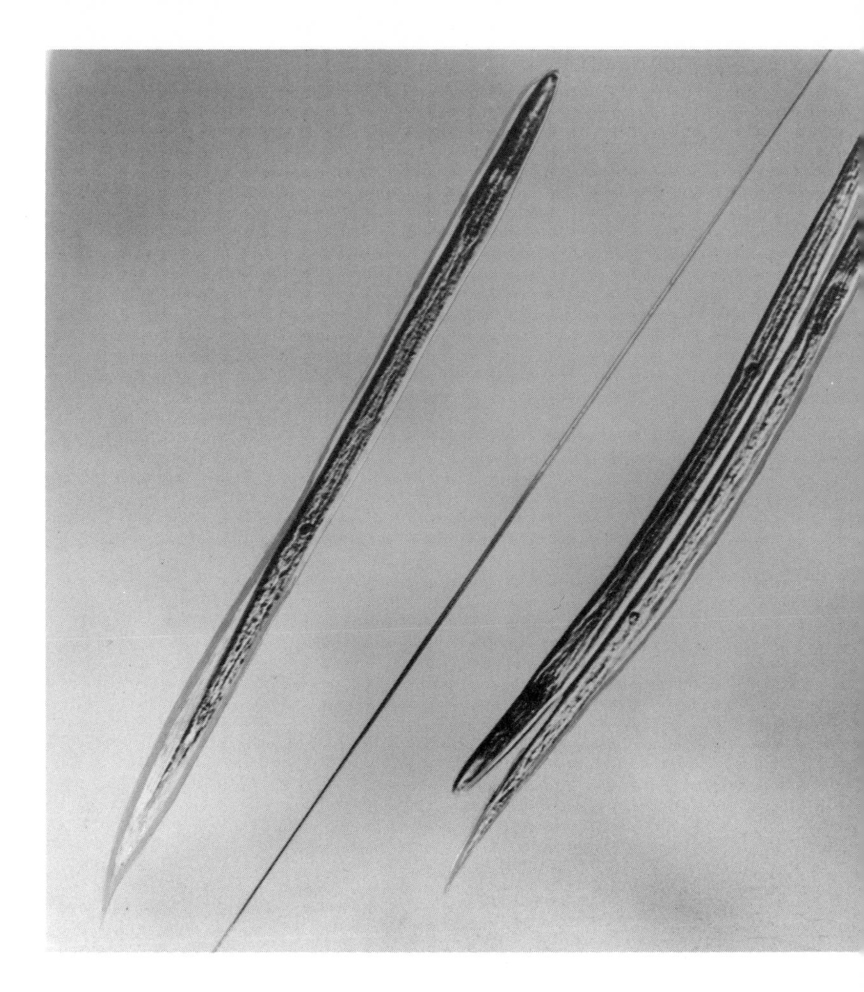

Juvenile hookworms from soil. This is the infective stage. Length, 0.5 mm. (A. C. Lonert)

into North America, and in the United States it is largely confined to rural parts of the southeastern states.

The symptoms of the disease are anemia, diarrhea, and general lack of energy, leading to a retardation of physical and mental development, so that an infected child of 15 years of age may appear to be only 10 years old. Simple treatment with drugs eliminates most of the worms, but in addition to treatment it is necessary to prevent new worms from entering. Wearing shoes and otherwise avoiding contact of the skin with contaminated soil is one precaution. A second is the sanitary disposal of feces that contain eggs.

The **trichina worm** *(Trichinella spiralis)* is a much-dreaded parasite of humans and is common in other mammals as well. Humans usually become infected by eating insufficiently cooked pork that contains encysted juvenile worms but occasionally contract infection from other kinds of meat—bear or wild boar meat, for example. The worms become sexually mature in the wall of the human intestine. The eggs hatch within the reproductive system of the female worm, which thus gives birth directly to young worms. These young gain access to the blood and lymph vessels of the human intestine and are carried about through the body. They leave the vessels and burrow into the muscles—usually those of the eyes, tongue, and jaw and of the diaphragm, ribs, and chest. In the muscles the juvenile worms increase about 10 times in size, to a length of 1 millimeter, and then encyst, curling up and becoming enclosed in a calcified wall. They develop no further and eventually die, unless the flesh (containing the cysts) is eaten by a suitable host. Pigs obtain the worms by eating the flesh of other animals, sometimes rats but usually fragments of slaughtered pigs included in uncooked garbage. In the body of a pig the worms go through the same history as described for a human infection.

The adult worms do no harm, and after a few months disappear from the intestine. The greatest injury occurs during the migration of the young worms, when half a billion or more of them may simultaneously bore through the body. At this time there are excruciating muscular pains, muscular disturbance and weakness, fever, anemia, and swellings of various parts of the body. It is during this stage of the disease that death occasionally occurs, usually from respiratory or heart failure. The victim usually survives this period, the young worms become encysted, and the symptoms subside, though the muscles may be permanently damaged. In less heavily infected cases the symptoms may be mild and are likely to be diagnosed as "intestinal trou-

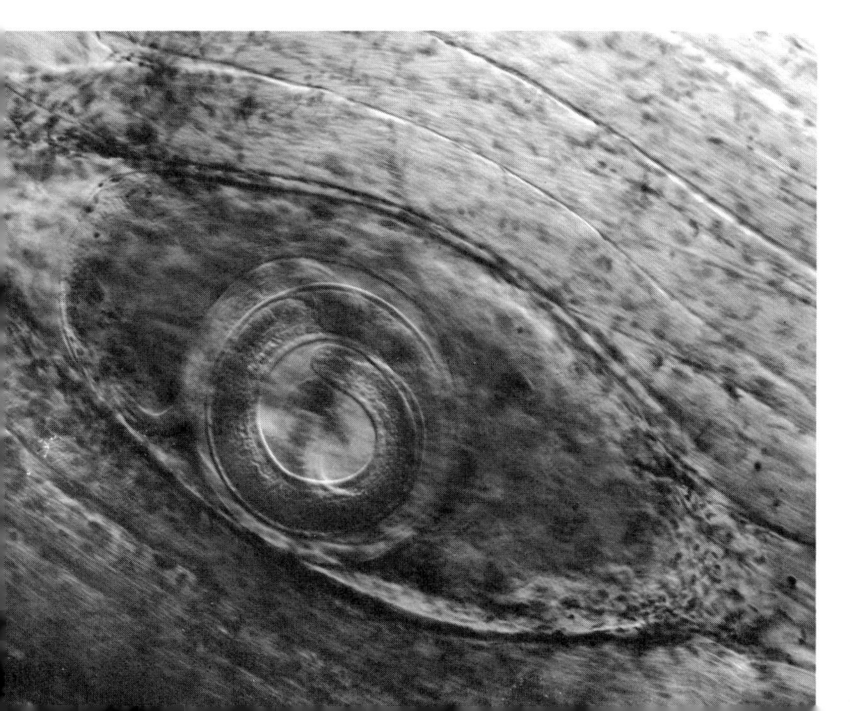

Trichina cyst in pig muscle. This is the infective stage that, if eaten by a human or other host, will hatch out, mature, and reproduce; the juvenile worms cause the serious disease known as trichinosis. The cyst itself does no harm, and the worm within eventually dies. Length of cyst, 0.5 mm. Stained preparation.
(P. S. Tice)

ble." Serious cases are sometimes diagnosed as typhoid or other fevers. Thus the actual occurrence of this disease may be higher than is supposed. However, the rate of infection in the U.S. population (as revealed by autopsies) has declined since the introduction of laws against feeding uncooked garbage to pigs.

At the present time the U.S. government does not inspect pork for the occurrence of encysted trichina worms, since such inspection requires microscopic examination, and light infections could be readily overlooked anyway. Inadequate inspection is worse than none, because it gives a false sense of security to the consumer. The absolute prevention of trichinosis lies with the consumer, who has only to cook all pork and pork sausage thoroughly. If large pieces are to be roasted, it is especially important to determine that heat has penetrated to the center; large public barbecue picnics are often a source of epidemic trichinosis. An ounce of heavily infected pork sausage may contain 100,000 encysted worms. If half of these are females, and each gives rise to 1,500 juveniles, the host would have to cope with 1.5 million juveniles, enough to cause death. All market animals are parasitized in some way or other, and it is understood that consumers will prepare food in such as way as to safeguard themselves against infection.

Young filarial nematode, a stage called a microfilaria. This microfilaria in the blood of its host is encased in a transparent sheath, really the inner lining of the egg from which it hatched. Three red blood cells are shown for scale. (Modified after E. C. Faust.)

The **filarial nematodes** differ from the other parasitic types already described in that their life histories require an arthropod intermediate host in addition to the final vertebrate host. The several kinds of filarial diseases are widespread, especially in the tropics, and are estimated to affect at least 270 million people. Some of the most notorious types of filarias are *Wuchereria bancrofti,* which causes a condition known as elephantiasis and which is spread by mosquitoes; *Onchocerca volvulus,* the agent of "river blindness," spread by blackflies; and *Loa loa,* the eye worm, spread by deer flies.

But most people in the United States are more likely to encounter filarias as **heartworms** in their pet dogs. The heartworm, *Dirofilaria immitis,* is transmitted by the bite of various mosquitoes and is widespread around the world in dogs. An infected dog contains in its blood young stages called microfilarias. These are sucked up by a hungry mosquito and develop within the insect host to an infective stage that is passed on to the next dog the mosquito bites. The tiny worms migrate through the dog's tissues to its heart, where they grow to adult length, the males about 15 centimeters, the females up to 35 centimeters. The adult worms irritate the heart lining and obstruct blood flow to the lungs, and the microfilarias circulating in the blood cause damage to various organs. A small number of worms may be tolerated, but it is wise to have dogs checked regularly for heartworms before symptoms appear (coughing, weight loss, weak-

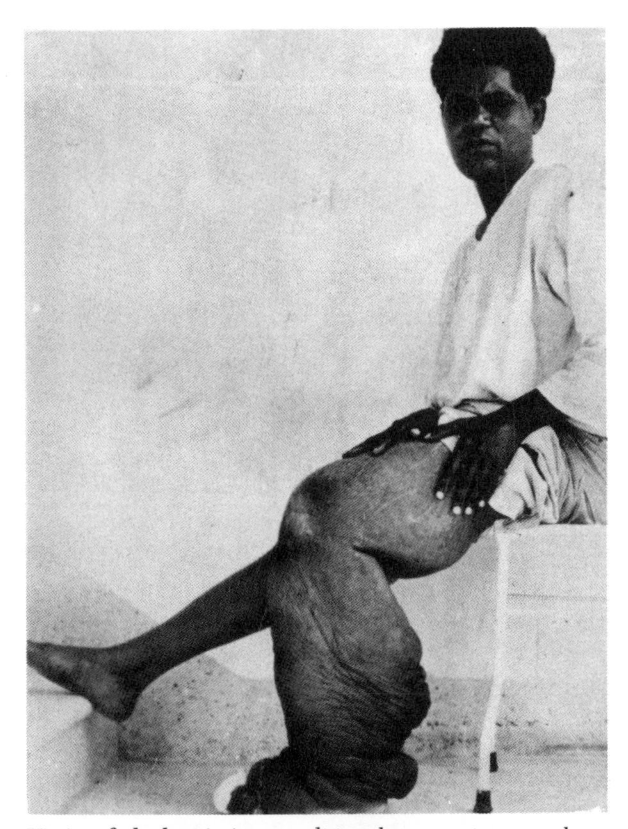

Victim of elephantiasis, a condition that sometimes results from infection with filarial nematodes, *Wuchereria bancrofti.* The adult worms look like coiled strings as they lie in the lymph glands or ducts of an infected person. The female worm is 8 to 10 cm long, and the male about half this length. The female gives birth to small juveniles, known as microfilarias, which get into the blood vessels and develop no further unless sucked up by a mosquito of the right species. Within the mosquito, the juvenile worms continue development and migrate to the biting apparatus. When the mosquito bites another person, the worms are transmitted. The chief consequence of filarial infection is the blocking of the lymph channels, which results in immense swelling and growth of affected parts, usually the limbs or, in males, the scrotum. Puerto Rico. (O'Connor and Hulse)

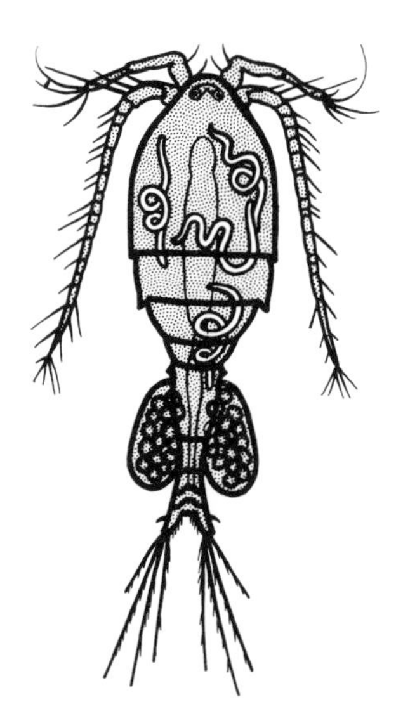

Young guinea worms, *Dracunculus medinensis,* in a tiny crustacean (freshwater copepod; see chapter 17). In tropical areas where these nematode worms are common (India, north and central Africa, the Middle East), a human who drinks unfiltered water and swallows infected copepods acquires the worms. They penetrate into many tissues, grow and mate, and finally come to lie in large blisters just under the host's skin, a stage associated with nausea, diarrhea, and other symptoms. Serious infections of the painful blisters with bacteria may cause loss of a limb or even death. (After Martini.)

ness, heavy breathing), as by this time the damage done by the worms is not readily repaired. Heartworms can also occur in domestic cats and in wild carnivores such as foxes, wolves, and coyotes. Human infection is uncommon, and the microfilarias rarely survive to reach the heart and mature.

Horsehair Worms

Horsehairs that have fallen into water, according to an old belief, may transform into worms called, naturally, horsehair worms. These are the members of the small phylum **NEMATOMORPHA** ("form of a thread"). It is not difficult to understand how the erroneous notion of their source got its start, for these slender worms look not unlike the hairs of horses' tails; though somewhat thicker, they come in about the same range of colors and lengths, and they are often found in watering troughs, as well as in ponds, streams, and other bodies of freshwater, or even in moist soil. Also, it seemed necessary to explain why one should see no trace of them on one day and then find numbers of these worms in the same place on the next day. We now know that this sudden appearance of the worms is due to the fact that the young worms

Horsehair worm in a typical complex of coils. This habit suggests the "Gordian knot" of Greek mythology and gives members of the group their other common name, gordian worms. California. (R. Buchsbaum)

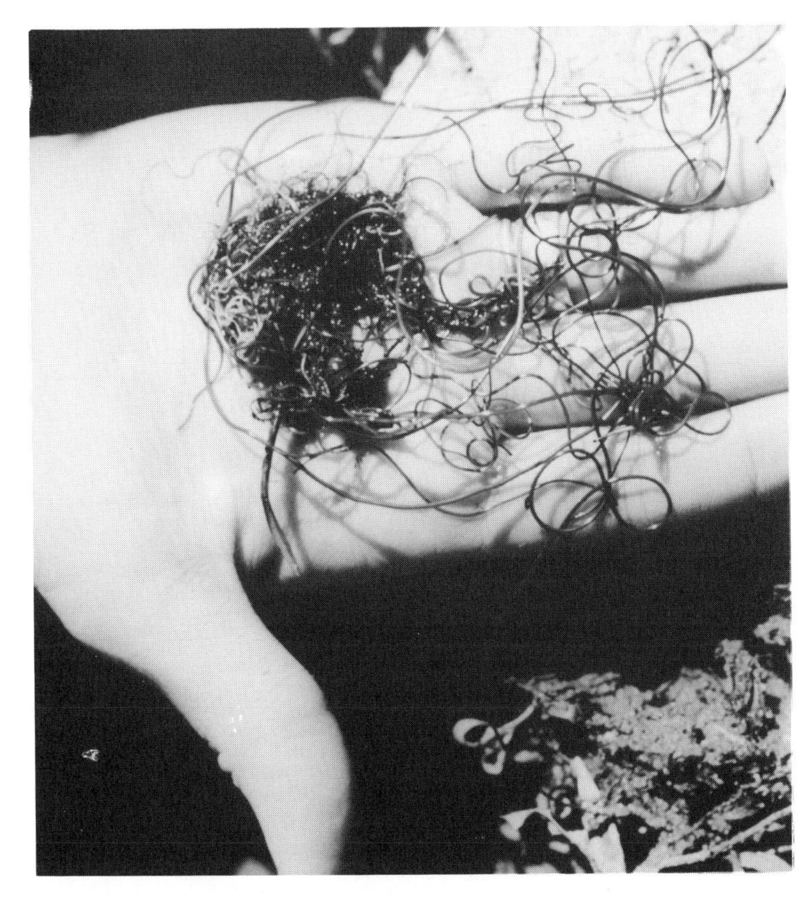

Tangled masses of adult horsehair worms are often found in streams. *Paragordius varius.* Illinois. (C. J. Swanson)

develop as parasites in arthropods, usually insects, and the full-grown worms emerge when the terrestrial host approaches fresh-water. Observations suggest that the worm, when ready to emerge, influences its insect host to seek water, but no one knows how the parasite does this, or how it senses when the insect has reached a moist spot suitable for the worm to emerge. Emergence of the worms is quickly followed by the death of the host.

Horsehair worms resemble nematodes in structure, and their life history is much like that of one group of nematodes that also parasitize insects as juveniles and emerge from their hosts as adults. Adult horsehair worms are described as free-living but they do not feed and depend entirely on nutrients obtained during their parasitic phase; these nutrient stores may sustain them for months. The males and females mate, and the females lay eggs in long strings that are wound around water plants. The larvas that hatch, if ingested by a suitable host, use the spiny proboscis to make their way through the host's intestine into the body cavity. There they grow by absorbing nutrients through the body surface; the gut of horsehair worms is rudimentary.

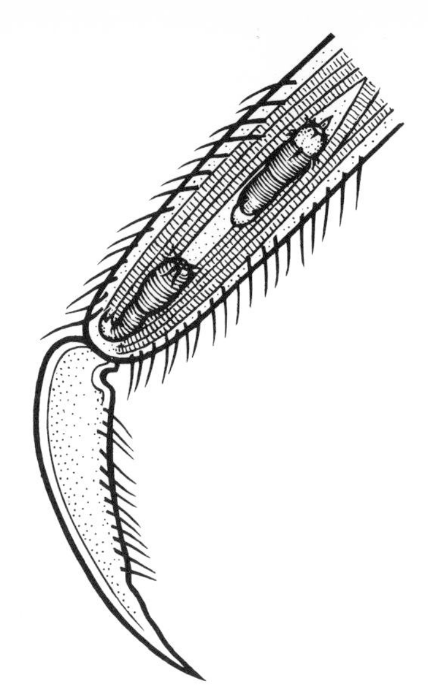

Larval horsehair worms in an appendage of a mayfly larva. The proboscis is everted in one, withdrawn in the other. In this host, the larvas will encyst and will develop further only if the mayfly is eaten by a suitable host. (After G. Meissner.)

Spiny-Headed Worms

A small group of elongate worms that live as adults in the intestine of vertebrates comprise the phylum **Acanthocephala.** The

Spiny-headed worm, or acanthocephalan

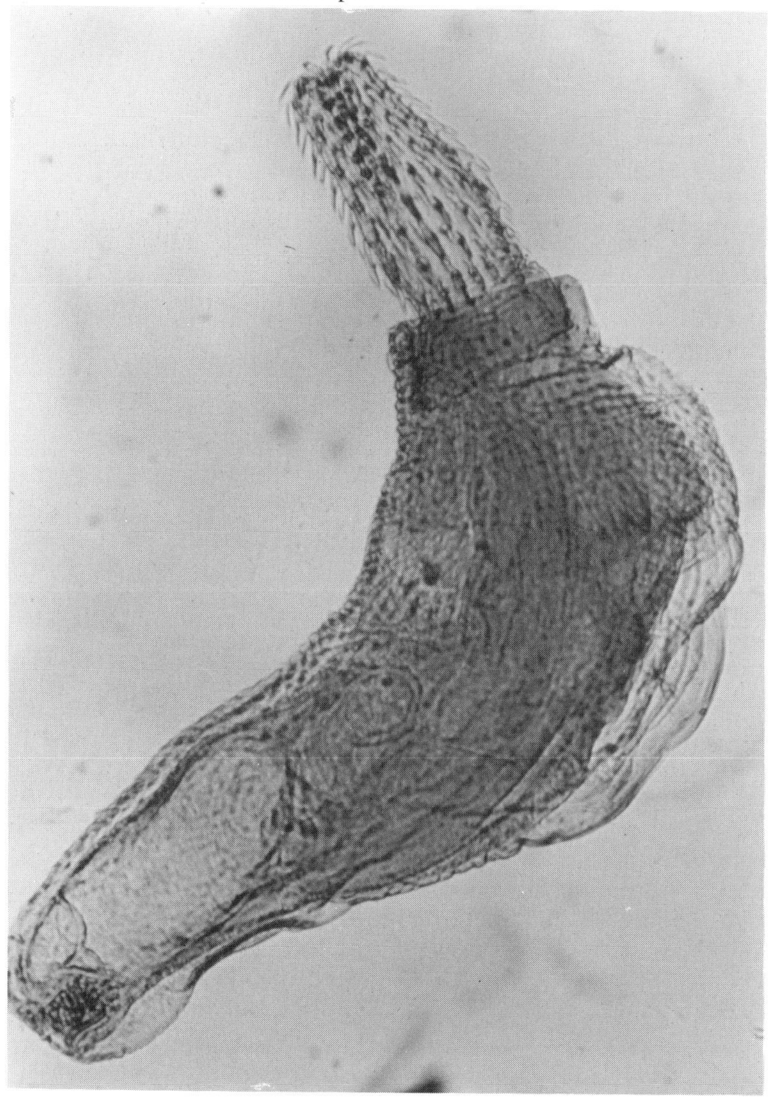

phylum name means "spiny-headed" and refers to their most characteristic structure, an anterior retractile proboscis armed with rows of stout, recurved hooks. Behind the proboscis is a short neck region and then the trunk, which is a somewhat flattened cylinder. By means of the burrlike proboscis the worm clings to the intestinal lining of its host, absorbing nourishment through the syncytial surface of its body, as does a tapeworm. There is no trace of a digestive tract.

Males and females mate, and the eggs begin their development within the body of the female. When shed, the eggs already contain a hooked larva. They pass out with the host's feces and hatch only when eaten by an intermediate host. For worms parasitic in aquatic vertebrates, the intermediate host is a small crustacean or aquatic insect; for land vertebrates, it is a land insect. The larva hatches in the gut of the intermediate host, bores through into the body cavity, and there grows into a young worm. Further development depends on the intermediate host being eaten by a suitable vertebrate. An acanthocephalan species common in rats and another that lives in pigs are both occasionally found in humans. Rats become infected by eating cockroaches, which sometimes form their chief article of diet; pigs acquire the parasites by eating certain grubs (beetle larvas), which they find as they root about in the soil. Humans presumably become infected when they unwittingly swallow cockroaches or beetles that get into foodstuffs. Though most species are small, the acanthocephalan of pigs is a huge worm, up to 65 centimeters long, with a pinkish wrinkled body. These parasites do not cause serious disease unless many are present in one host.

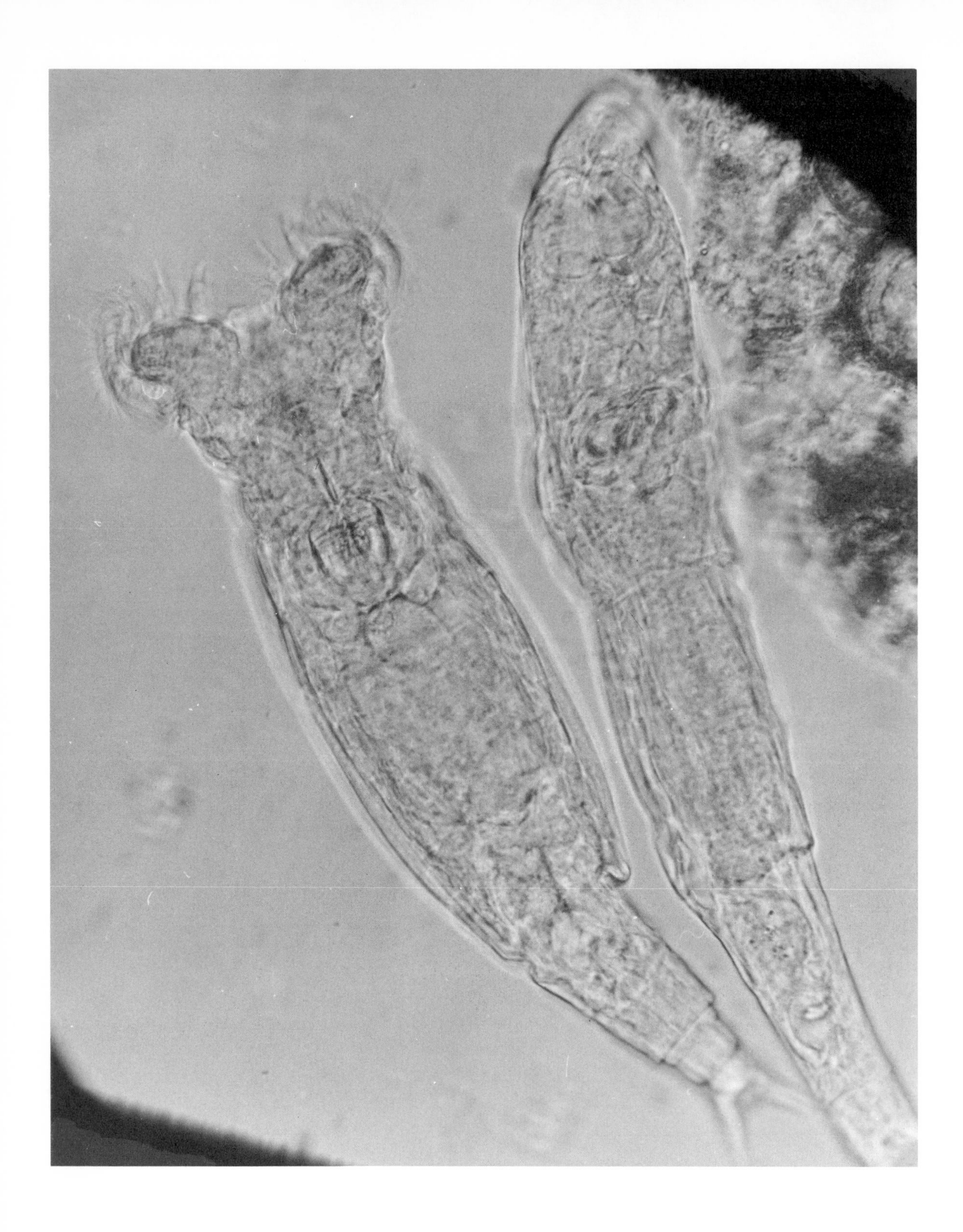

Chapter 14

Lesser Lights

The animal kingdom is divided into more than 30 phyla. The exact number depends on how many body plans the classifier thinks are different enough to be distinguished as phyla. Some of the phyla are more important than others—in terms of their numbers of species or their impact on people—but some of those usually considered less important are no less fascinating. Five of these "lesser lights" are described in this chapter. They are included here for one or all of the following reasons: they have a small number of species or of individuals; the members are of microscopic size; they are neither pests nor important sources of food or disease for humans; and they illustrate no principle of theoretical importance that is not as well shown by other phyla.

Rotifers

When examining a drop of pond water under the microscope, one is almost certain to find rotifers, animals about the size of protozoans but with a grade of structure a little more complicated than that of flatworms. Although abundant, these tiny animals are little known except to zoologists and amateur microscopists, who seldom fail to be captivated by their great variety of shapes (many of them truly fantastic) and by their rapid and often seemingly incessant motions.

The commonest rotifers can be recognized at once by the presence at the anterior end of a **crown of cilia,** which serves as the chief organ of locomotion and also as the means of bringing food to the mouth. In these forms the beating of the cilia, which are arranged around the edge of a pair of disk-shaped lobes, gives the appearance of revolving wheels—hence the name of the phylum, ROTIFERA, which means "wheel-bearers."

Opposite: **Rotifers** in a drop of pond-water are each about half a millimeter long. The crown of cilia is extended in the rotifer on the left, withdrawn in the one on the right. *Philodina.* Pennsylvania. (R. Buchsbaum)

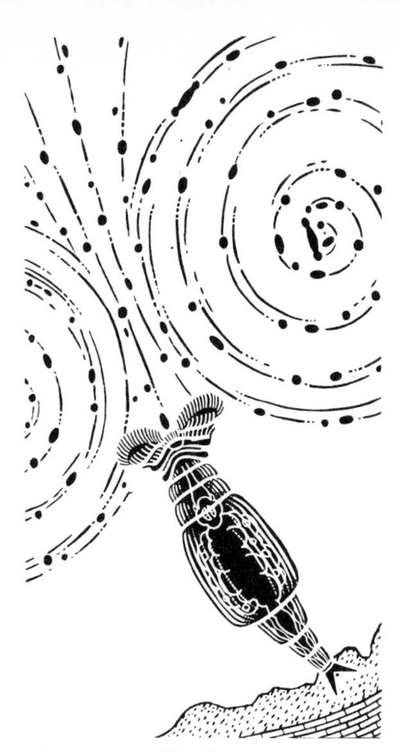

The pattern of **feeding currents** created by the beating cilia of a rotifer may be studied by tracing the paths of moving particles. This drawing was based on a film made by attaching a movie camera to a microscope focused on a common freshwater rotifer.

Rotifers vary in shape from wormlike bottom dwellers, or flowerlike attached types, to rotund forms that float near the surface; but all are bilaterally symmetrical. In many species the body is elongated and is roughly distinguishable into three regions: a **head** that bears the mouth and cilia; a main central portion called the **trunk;** and a tapering tail portion quaintly named the **foot.** At the end of the foot are the "toes," pointed projections from which open cement glands that secrete a sticky material used to anchor the rotifer during feeding. The toes aid in a second method of locomotion in which the animal proceeds in inchworm fashion. It stretches out, takes hold at the front end, releases the toes, and contracts the body; then it fastens the toes again, stretches, and so on. The whole body is enclosed in a transparent, flexible, dense layer that is cuticle-like but lies just *beneath* the surface of the epidermis. In the species most often seen, it is divided into sections that can be telescoped one into the other when the animal contracts.

When **feeding,** most rotifers remain attached to a bit of debris, and the rapid beating of the cilia draws a current of water toward the mouth. Protozoans and microscopic algal cells are swept through the mouth into a **muscular pharynx,** which contains a chewing or grinding apparatus consisting of little hard **jaws** operated by muscles. In predatory rotifers the elongate jaws can be extended through the mouth and used, like a forceps, for grasping prey. The pharynx leads into a straight digestive tract that opens by an anus at the junction of trunk and foot.

In **general structure** rotifers show similarities to flatworms and nemerteans and to nematodes. Like nematodes, rotifers have a fluid-filled body cavity between body wall and digestive tract. The stiffened surface layer serves as a place of attachment for the muscles by which the animal moves. Muscular activity is coordinated by a simple nervous system that centers about a head ganglion, or brain, in the anterior end, from which run two main ventral nerve cords. Many rotifers have simple eyes and sensory projections. The excretory system, which serves mainly to regulate water content, is like the flame-bulb systems of flatworms and nemerteans. There are no respiratory or circulatory systems in rotifers, and this is what one would expect of such minute organisms. Substances are moved in the fluid of the body cavity or simply diffuse the microscopic distance from the gut to muscle and other tissues.

The **jaws of the pharynx** are the most distinctive structures of rotifers and are used by taxonomists to distinguish one species from another. (After Harring and Myers.)

Most tissues of rotifers are not divided into distinct cells but, like some of the tissues of flatworms and nematodes, consist of **syncytia.** Cell membranes are present in embryonic stages but later disappear. It is also a striking fact that the number of cells of the late embryo, or the number of nuclei of the adult, is con-

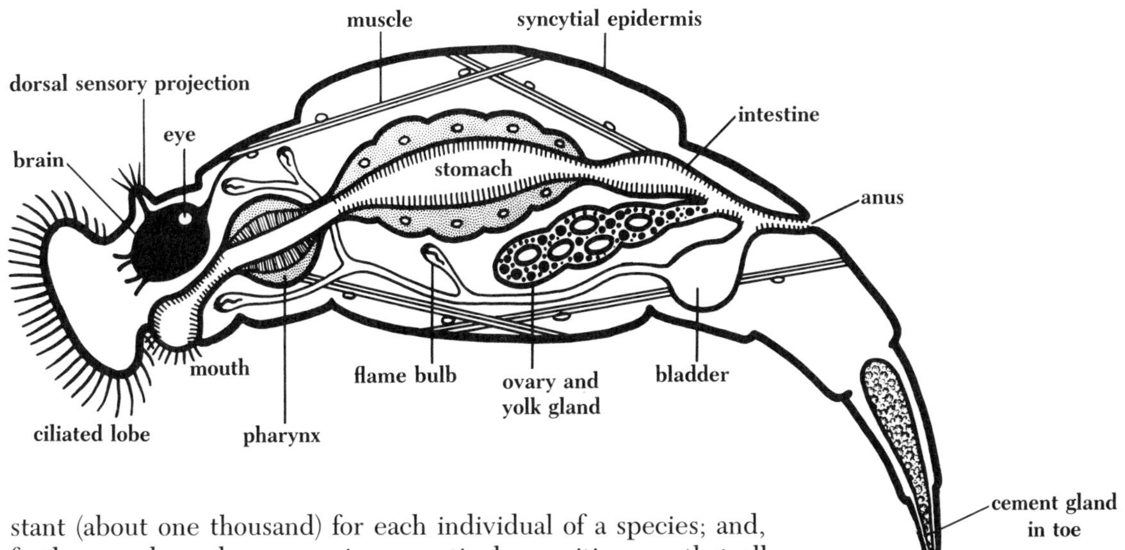

stant (about one thousand) for each individual of a species; and, further, each nucleus occupies a particular position, so that all the nuclei of a rotifer can be numbered and mapped. Such cell constancy also appears to some degree in nematodes and in some other phyla.

During most of the year, in the typical life history, females give rise to other females by way of eggs that develop without being fertilized, a process called **parthenogenesis,** which occurs also in some other phyla. During brief periods of the year, some rotifers engage in **sexual reproduction.** As the sexual season approaches, certain of the females lay eggs that are smaller and

Rotifer, showing general structure. Only a few of the nuclei are shown. In some rotifers the terminal portion of the excretory tube is enlarged into a bladder that pulsates, ejecting its contents into the most posterior part of the intestine. (Combined from several sources.)

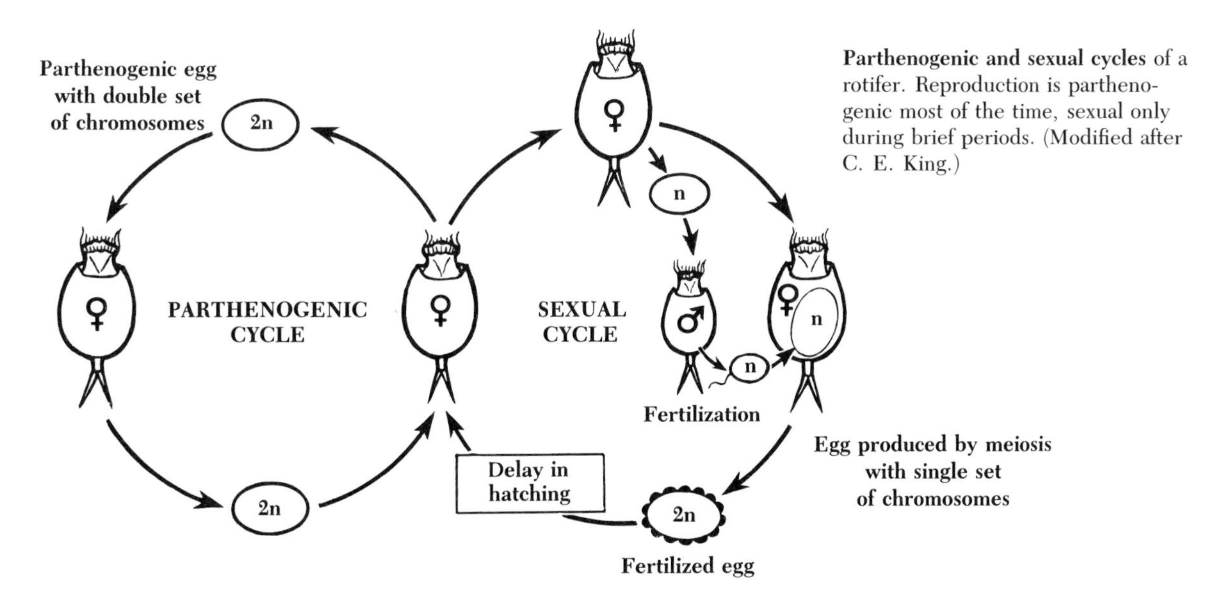

Parthenogenic and sexual cycles of a rotifer. Reproduction is parthenogenic most of the time, sexual only during brief periods. (Modified after C. E. King.)

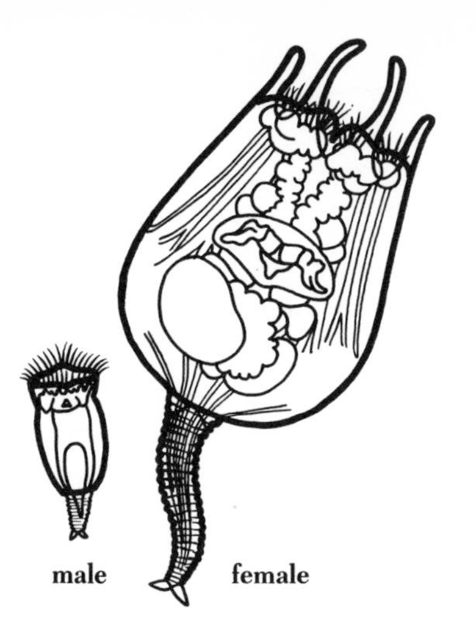

Sexual dimorphism. Male rotifers are smaller and often simplified. (After Hudson and Gosse.)

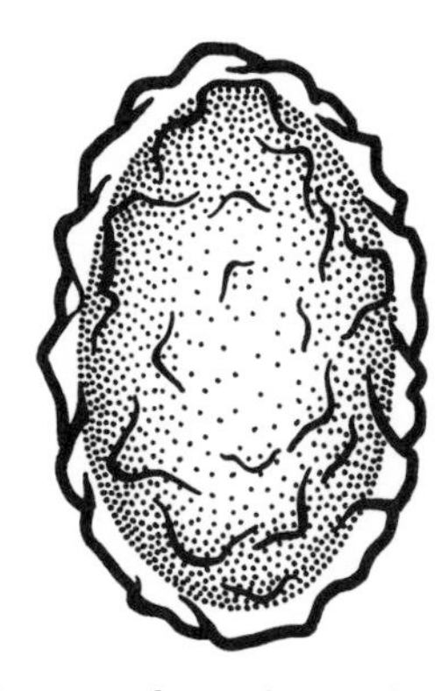

Resting egg of a rotifer is a fertilized female-producing egg with a hard, thick shell that protects the egg during unfavorable conditions. (After H. Miller.)

differ in other ways from the usual female-producing eggs. If not fertilized, these smaller eggs hatch into males. Males are smaller than females (in some species only one-tenth the size) and sometimes entirely lack the digestive and excretory systems. Such individuals can live for only a few days. The males and females mate, after which fertilized eggs are laid. These are distinguished from the parthenogenic ones by a hard thick shell, often ornamented. The fertilized eggs can withstand drying, freezing, and other unfavorable conditions, and after a resting period (commonly the local dry season or winter) they hatch into females. In most species of rotifers males have never been seen and perhaps they do not occur.

Some rotifers can withstand **drying** to a degree even greater than that seen in many protozoans and nematodes. In an almost completely dried state, rotifers with a normal lifespan of only a few days or weeks may survive for years. As soon as they are again immersed in water, they swim about and feed actively. Because of this capacity, rotifers can live in places that are only temporarily wet, such as roof gutters, cemetery urns, rock crevices, among moss, and similar habitats. When the water evaporates, the animal contracts to a minimum volume and loses most of its water content. Sometimes the animal itself dies but its contained eggs survive until moisture returns. There are some marine rotifers, but the group is much more abundant in freshwater. Because of their small size and their capacity for withstanding temporary drying, certain rotifers have been distributed the world over, chiefly by wind and by birds. If environmental conditions are similar, a lake in Africa is likely to contain some of the same species of rotifer as a lake in North America.

Gastrotrichs

Almost any aquatic debris that contains rotifers will also contain a few members of the small phylum **Gastrotricha.** These minute many-celled animals are about the size of rotifers and resemble them in many details of structure. They have no crown of cilia but swim by means of tracts of cilia on the ventral surface. The body is encased in a cuticle, often clothed with scales or bristles, and gastrotrichs are likely to be confused with ciliated protozoans.

The digestive system is a straight tube with a muscular sucking pharynx that most closely resembles that of nematodes. In the species commonly seen in freshwater, the tail end of the body is forked; and at the tip of each fork is the opening of a cement gland, which serves the same function as in rotifers.

Nearly one-half of the known gastrotrichs live in the ocean; these are hermaphroditic. The rest live in freshwater, and almost all are females that apparently reproduce by parthenogenesis; a few forms may be hermaphroditic. No males have ever been seen.

Gastrotrich from a Pennsylvania lake. (R. Buchsbaum)

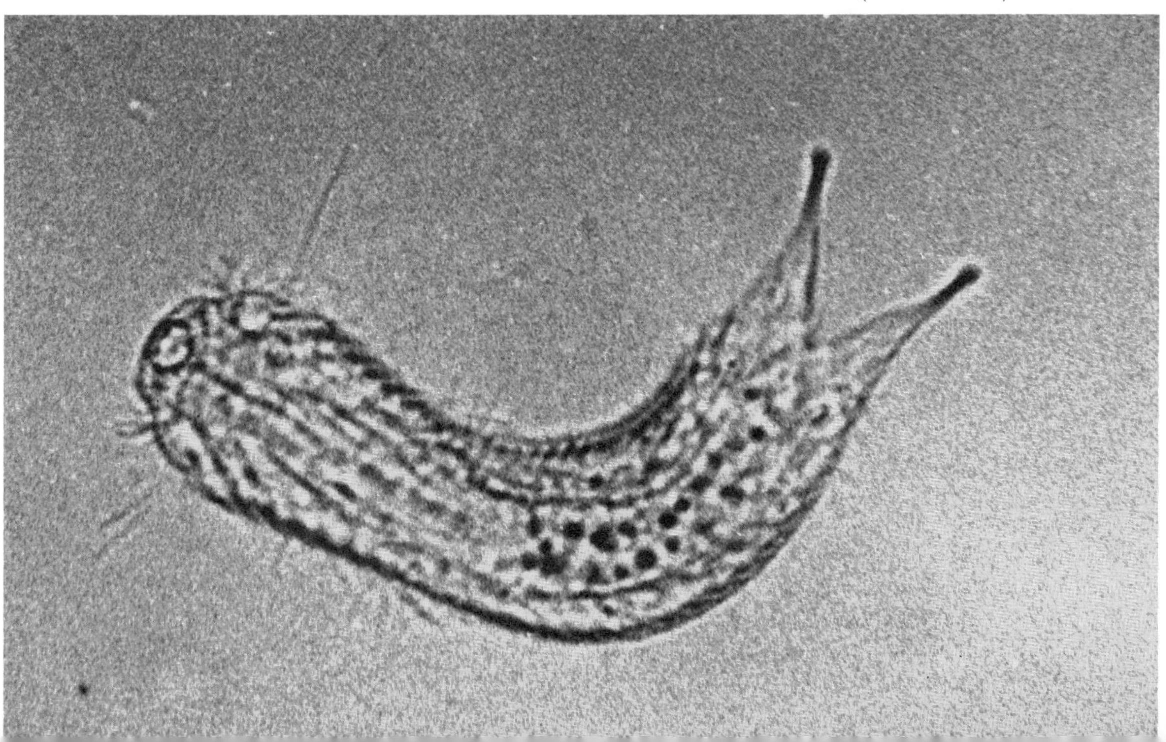

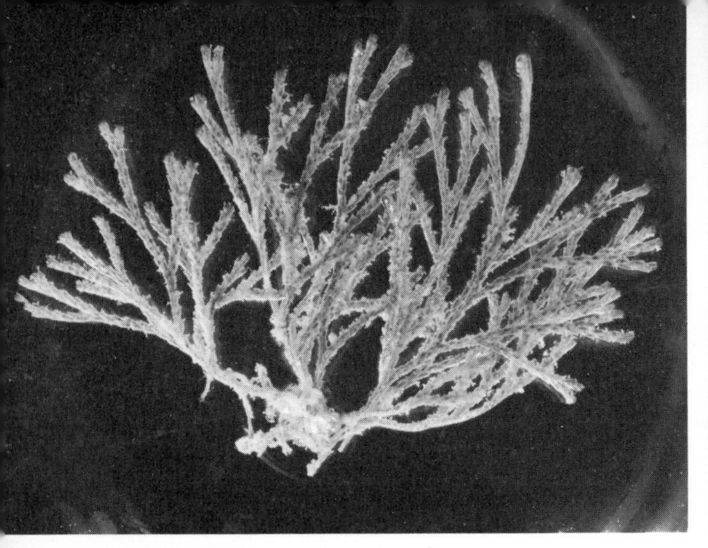

Branching bryozoan colony, *Bugula,* looks superficially like a hydroid colony or a bit of delicate seaweed. *Left,* colony about 25 mm across, from Beaufort, North Carolina. Closeup of a similar colony, *right,* shows individual members with their graceful circlets of tentacles extended. Panacea, N.W. Florida. (R. Buchsbaum)

Bryozoans

Some of the small and more delicate "seaweeds" admired by visitors to the seacoast are not seaweeds at all but branching colonies of members of the phylum **Bryozoa,** a name that means "moss animals" and refers to their plantlike appearance. Some colonies are shrublike and hang from blades of kelp or from under ledges of rock; others grow as flat encrusting sheets on seaweeds and boulders; and some freshwater bryozoans grow as gelatinous masses around stems and twigs that have fallen into the water.

At first glance a bryozoan colony resembles a hydroid colony, for each of the tiny individuals has a delicate ring of hollow **tentacles** surrounding the mouth. The tentacles are borne on a circular or horseshoe-shaped ridge, the **lophophore.** Unlike hydroids, bryozoans are **suspension feeders:** cilia on the tentacles create water currents that carry microscopic organisms to the mouth. Moreover, each individual has a U-shaped digestive tract, and undigested food remains are voided through an anus that opens just outside the ring of tentacles.

Each individual of a colony is enclosed by an outer chitinous boxlike **case,** sometimes calcified, into which the delicate feeding tentacles can be withdrawn. The animal protrudes the tentacles "cautiously," and they spread out in the water, setting up strong feeding currents; but at the slightest disturbance the tentacles are withdrawn in a flash. The case is secreted by the epidermis. The epidermis and a layer of mesodermal cells constitute a thin body wall. Between the body wall and the digestive tract is a fluid-filled space, completely lined with mesoderm. Such a body

withdrawn

extended

retractor muscle

Two members of a bryozoan colony, one withdrawn and one extended. (Modified after Delage and Hérouard.)

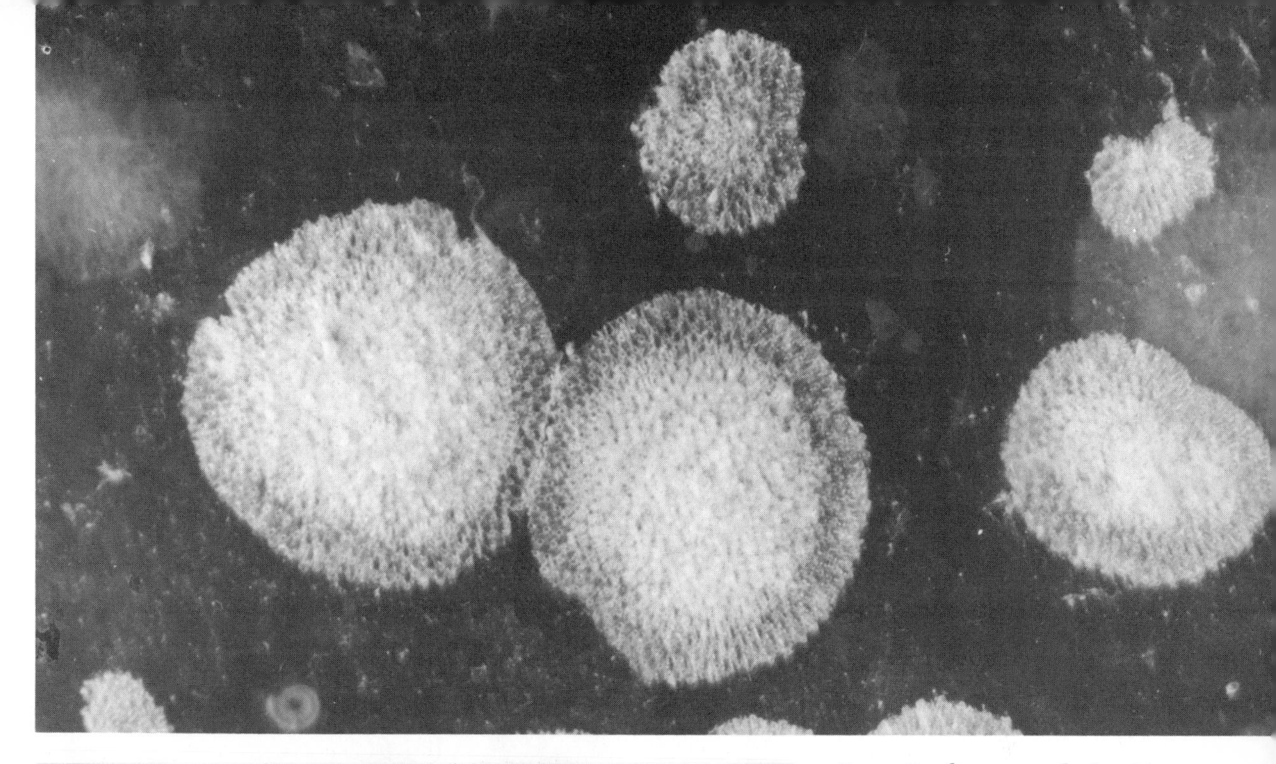

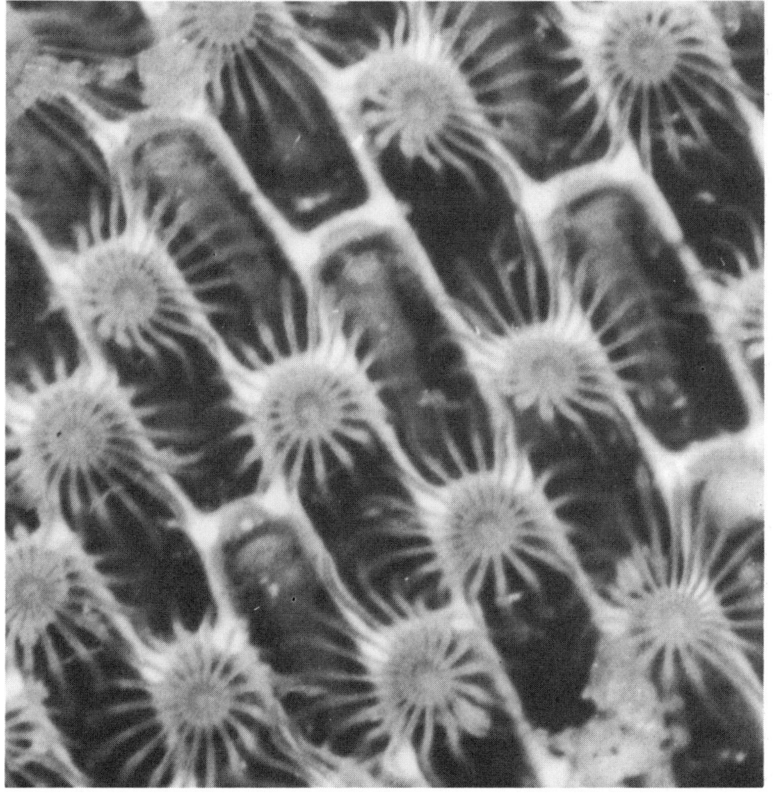

Encrusting bryozoan colonies, *Membranipora,* form lacy circles on seaweed, *above.* Maine. The precise geometry of the colony is evident, *left,* in the boxlike cases from which each individual member extends a circlet of tentacles. North Sea.
(R. Buchsbaum)

cavity lined with mesoderm is called a **coelom** ("hollow"). The coelom serves as a *hydrostatic skeleton,* against which muscles around the body wall can contract to push the tentacles out of the case. A thin retractor muscle, stretched between the ridge of the lophophore and the base of the surrounding case, can pull the tentacles back inside with remarkable rapidity (this is among the fastest muscles known).

Like some hydroids, bryozoans may display **polymorphism.** In some of the shrublike forms, individuals at the base of the colony are specialized for attachment and have no feeding tentacles. In most species there are specialized individuals called *ovicells* that hold and nourish the embryos. But the most highly specialized individuals resemble a bird's head and so are called

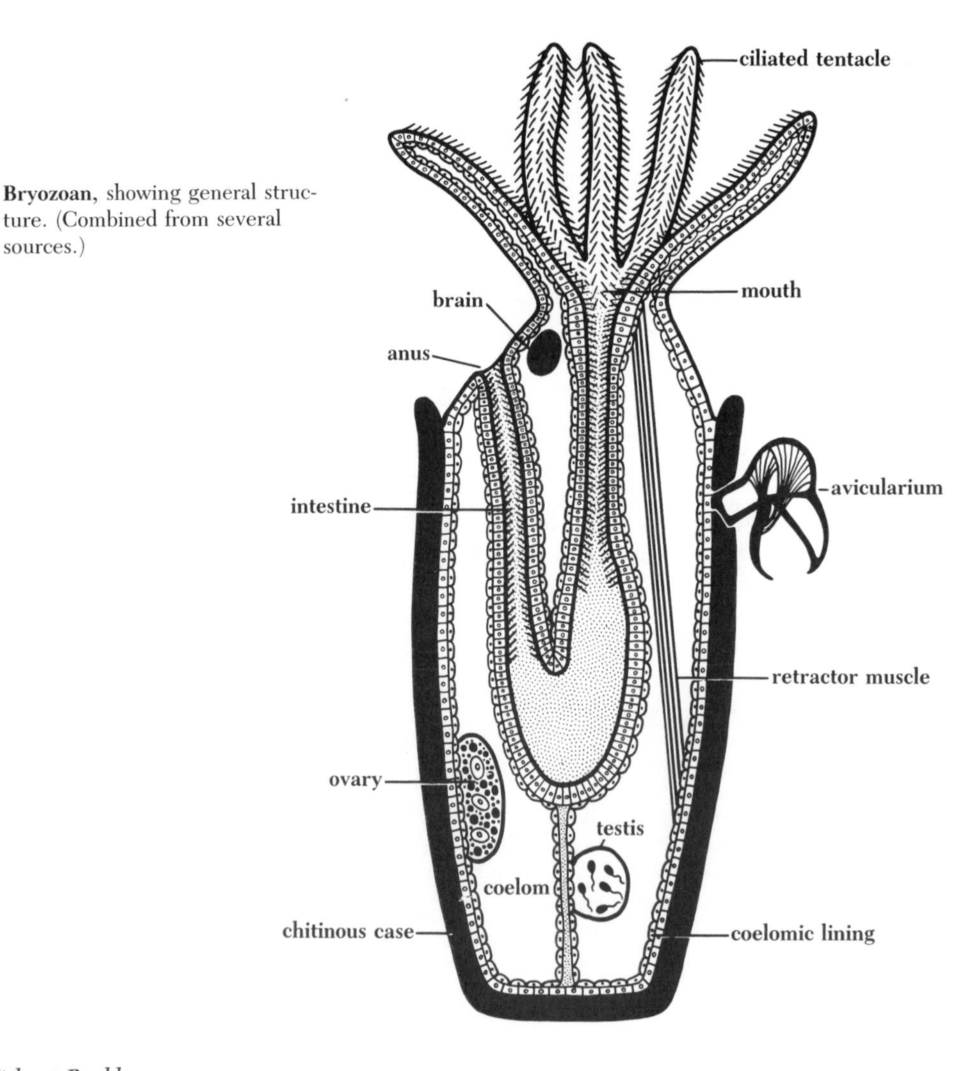

Bryozoan, showing general structure. (Combined from several sources.)

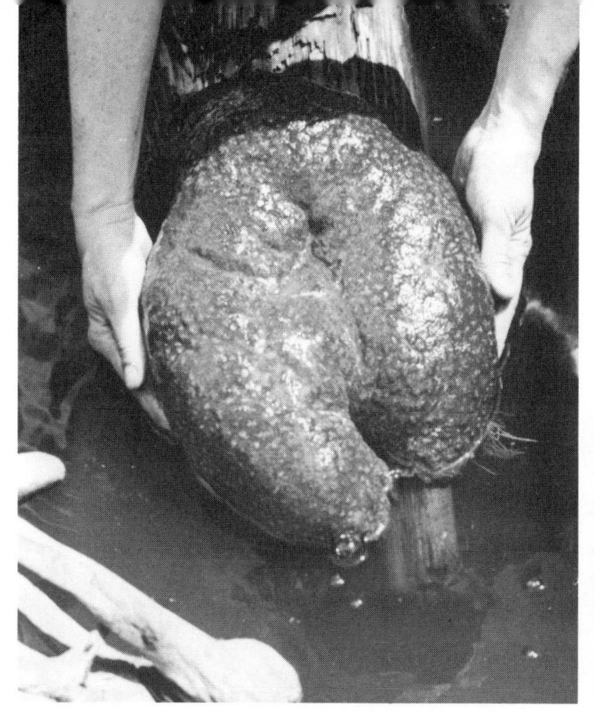

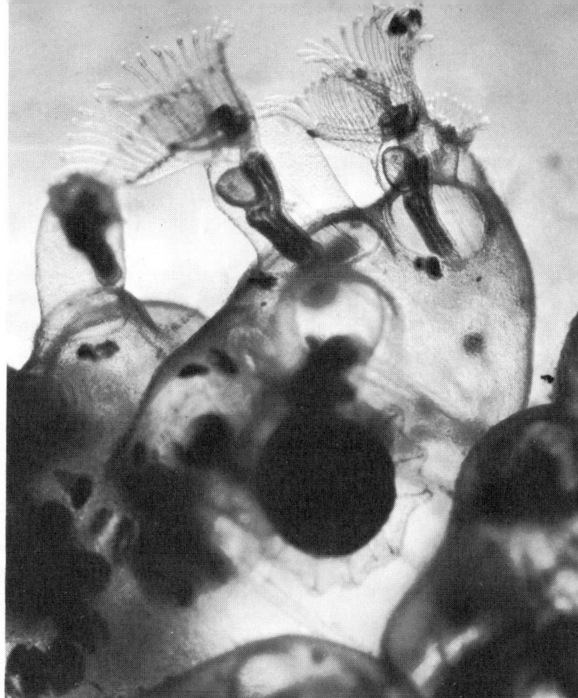

Gelatinous bryozoan colony, *Pectinatella magnifica,* from a Pennsylvania lake, *left.* These freshwater bryozoans commonly encrust sticks and stones in quiet waters, but large floating colonies occasionally clog the screens of public water systems or hydroelectric plants in such quantities that it is a full-time job to clear them. Closeup of the colony, *right,* shows several individuals, each with a statoblast at its base. (R. Buchsbaum)

avicularia. Each avicularium has a pair of jaws, operated by muscles, that can snap shut on small animals wandering over the colony. Presumably, these bizarre-looking jawed individuals prevent other animals from settling upon, and interfering with, the feeding activities of the colony.

Growth of bryozoan colonies is by budding of new individuals, so while all the individuals remain small, the colony can become quite large, with hundreds or thousands of feeding members. Each communicates with its neighbors through small holes in the body wall and case. New buds, until they develop tentacles and a digestive tract, are nourished by their neighbors. In addition to budding new individuals, established members at times undergo a special form of **rejuvenation.** The tentacles and digestive tract degenerate and form a compact mass known as a *brown body.* These parts then regenerate, and sometimes the brown body is eliminated through the new anus. Because bryozoans lack an excretory system, brown body formation may be a means to dispose of worn-out tissues and certain wastes.

Freshwater bryozoans often live in ponds and streams that

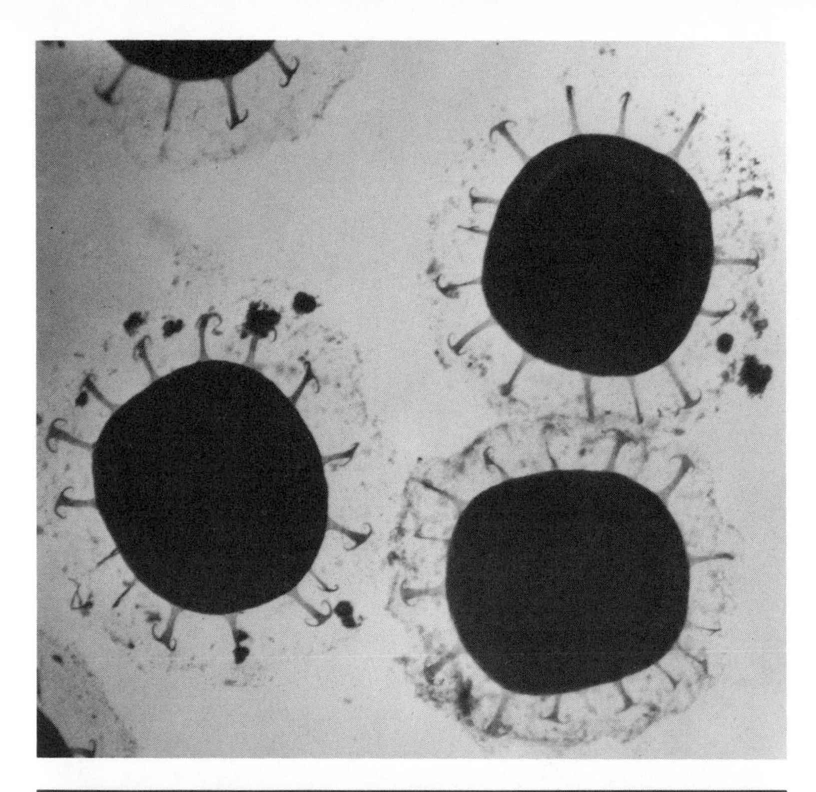

Statoblasts of *Pectinatella*. The colony dies at temperatures below 16°C (60°F), so only the statoblasts survive the winter. (R. Buchsbaum)

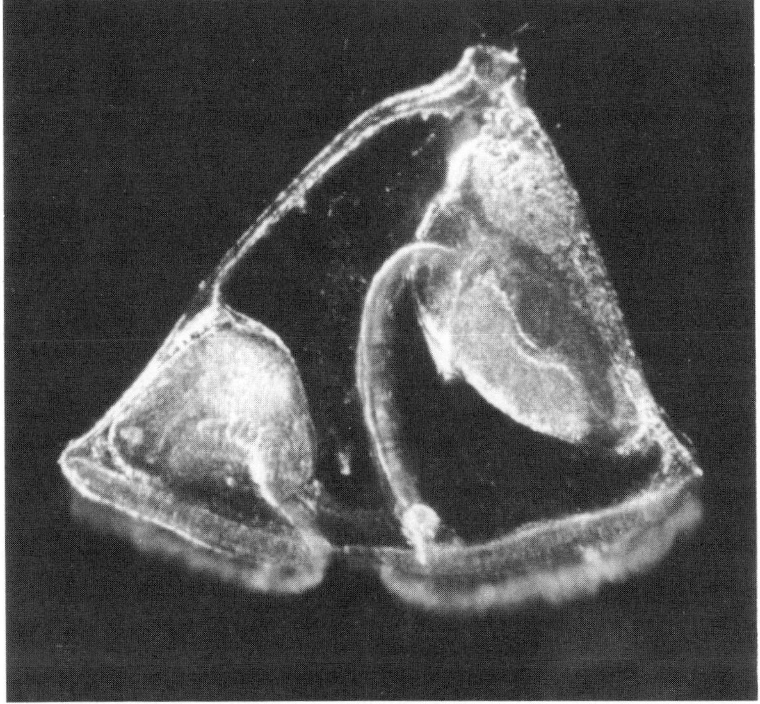

Cyphonautes larva of a marine bryozoan. Plymouth, England. (D. P. Wilson)

may freeze or dry up. These species form special buds, known as **statoblasts,** each consisting of a mass of cells surrounded by a protective covering. Like the gemmules of sponges, statoblasts can survive desiccation and extreme temperatures, to develop into new individuals with the return of favorable conditions.

Most bryozoans are **hermaphrodites,** with ovaries arising from the coelomic lining of the body wall and testes forming on the strand of tissue fastening the stomach to the body wall. Eggs and sperms are shed into the coelom, and in at least some species, self-fertilization occurs. In most species embryos develop in ovicells on the colony, and are released as little ciliated, spherical, nonfeeding larvas that swim only a few hours before settling and forming the first feeding individual of a new colony. In a few species that live on kelp, which can periodically disappear, the eggs are shed directly into the sea where they develop into free-swimming, feeding larvas, called **cyphonautes larvas** ("crooked sailors"); these larvas can live in the plankton for months before finding suitable kelp to settle upon.

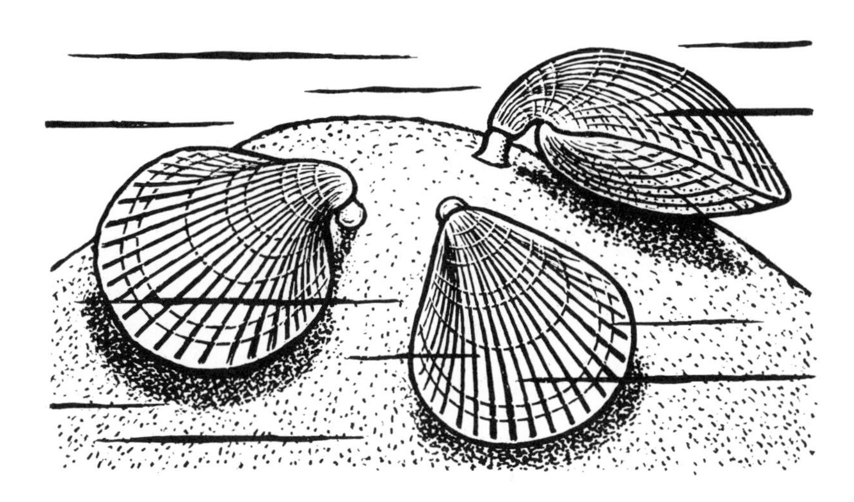

Brachiopods, attached by the stalk to rock, typically lie "upside-down," with the ventral valve uppermost and the dorsal valve below.

Brachiopods

One of the early investigators who pried open the shell of a brachiopod and looked inside, thought that the two spirally coiled ridges within were "arms" by which the animal moved and that they corresponded to the foot of a clam. From this mistaken notion came the name of the phylum **BRACHIOPODA,** which means "arm-footed." But the two shell valves of a clam are right and left, while those of a brachiopod are on the dorsal and ven-

Brachiopod, showing general structure. (Modified after Delage and Hérouard.)

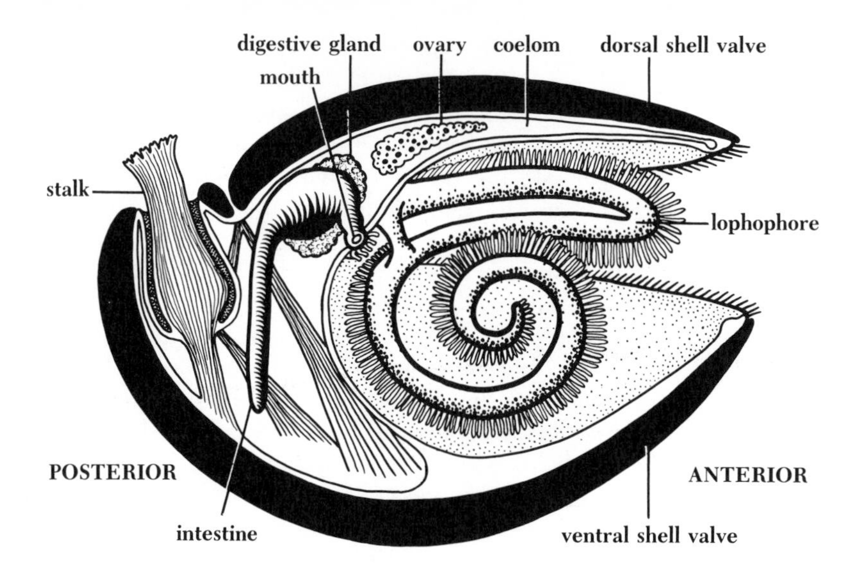

tral surfaces. The gape of the brachiopod shell is at the anterior end, and the hinge is at the posterior end. The shell valves can be opened and closed by means of muscles. The posterior end of the body extends beyond the shell as a stout, tough stalk. Some brachiopods attach with the stalk to a hard surface such as a rock, while others use it to anchor in sand or mud.

Brachiopods at first glance resemble clams, but closer inspection reveals a more basic alignment with bryozoans. If we could imagine a solitary bryozoan growing to 50 times its size, it might turn out something like a brachiopod. The calcareous or chitinous shell, like the case of a bryozoan, is secreted by the epidermis and encloses most of the body. Within the shell are rows of hollow tentacles arranged on a **lophophore** that is greatly enlarged and usually forms spirally coiled arms supported by a calcareous skeleton. The beat of cilia on the tentacles sets up currents of water over the arms, and microscopic organisms are swept into a central groove down the arms and into the mouth. The stomach is supplied with large digestive glands that fill much of the mesoderm-lined coelom. In most species there is a short intestine but no anus; undigested food is eliminated through the mouth.

The water currents across the tentacles also maintain a steady supply of oxygen to the animal and, as in bryozoans, oxygen can be transported to the internal tissues in the fluid of the coelom. In addition, brachiopods have a simple circulatory system (with a small contractile "heart") that supplements the role of the coelom in transport of materials around the body; and there is also a system of excretory tubes that void wastes from the coelom to

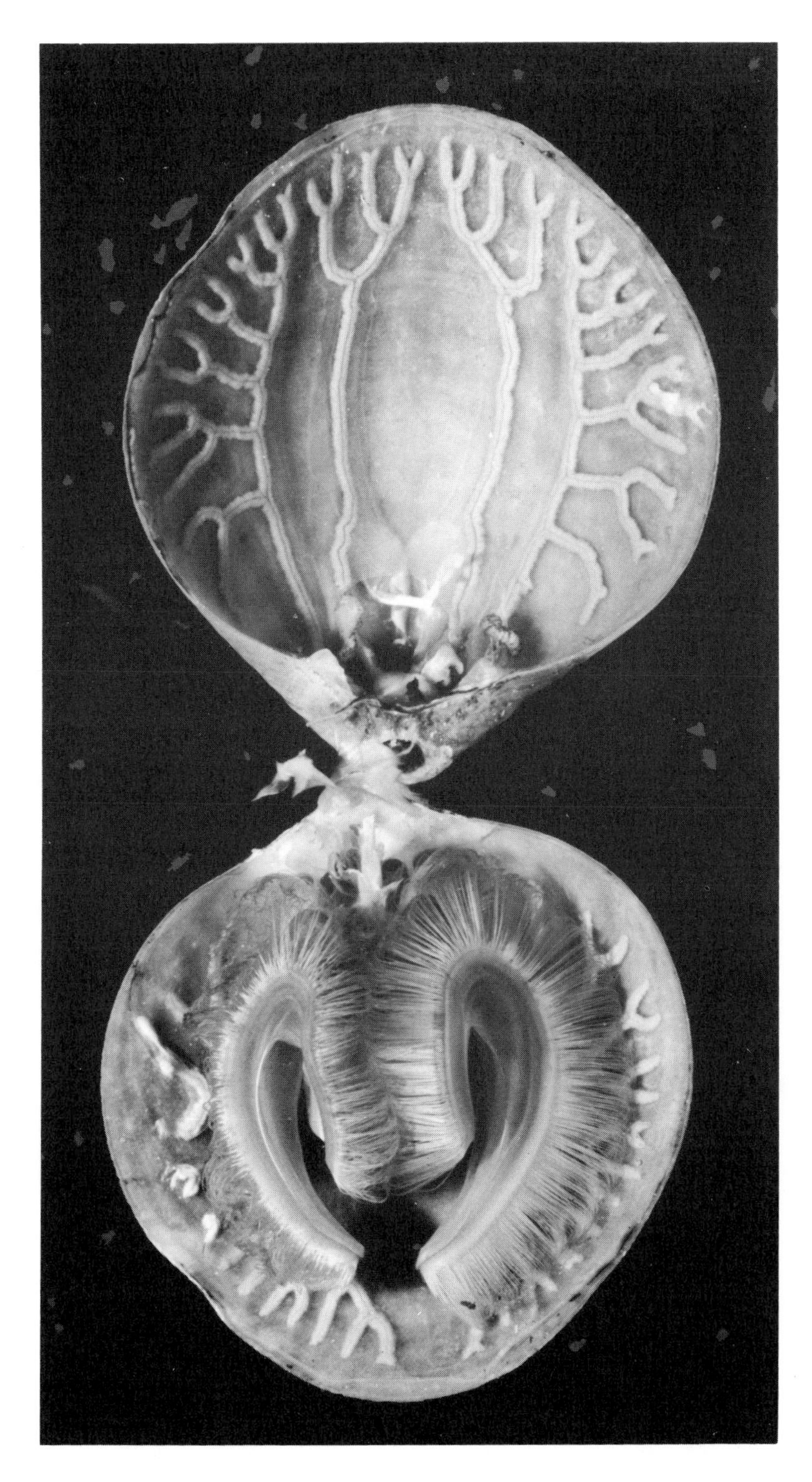

Brachiopod laid open to reveal the internal structure. The ventral shell valve, which normally lies uppermost, is at the top of this photo; coelomic canals are conspicuous in the tissue lining the shell. Calcareous extensions of the dorsal shell valve (at the bottom of the photo) support the two coiled ridges of the lophophore, bordered by delicate hollow tentacles. *Magellania*. (R. Buchsbaum)

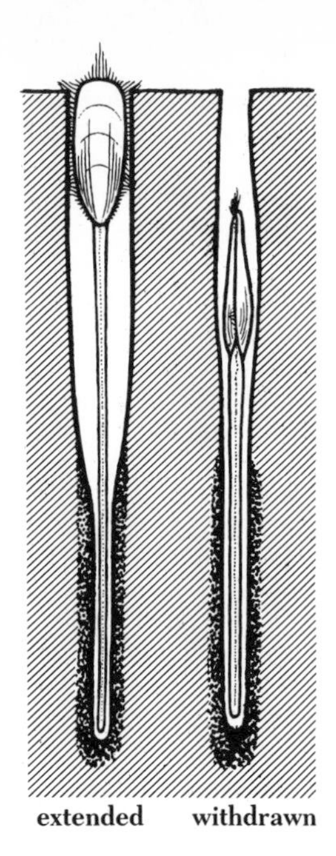

Lingula lives in vertical burrows in the sand, attached to the bottom by the long stalk. (Modified after François.)

the outside. Both circulatory and excretory systems, absent in bryozoans, are presumably related to the larger size of brachiopods.

The sexes are usually separate, and the gonads are outgrowths from the lining of the coelom. Eggs or sperms are shed into the coelom and then expelled from the body through the excretory tubes. The ciliated, nonfeeding larva resembles those of bryozoans and lives for only a short time before settling and metamorphosing into the adult form.

There are two distinct groups of brachiopods. In one group, of which *Lingula* is an example, the two shell valves are chitinous, somewhat rectangular in shape, and of equal size. They are held together only by muscles and connective tissue; as there is no hinge, these brachiopods are called *inarticulates*. The muscular stalk is usually very long, passes out posteriorly between the valves, and anchors the animal in its burrow or enables it to move short distances in the sand. As in many other animals that live in burrows or tubes, the digestive tract is recurved so that the anus opens, in the burrow opening, near the mouth. In the other, more abundant and more varied group of brachiopods, called the *articulates*, the shell valves are calcareous and hinged together by a tooth-and-socket arrangement. The ventral valve is larger than the dorsal, and at its posterior end is a kind of up-

Lingula. In Japan and the South Pacific people eat these brachiopods, which are often abundant in shallow waters. Thailand. (R. Buchsbaum)

turned "beak" through which the short stalk passes to attach to a hard substrate. In this group the dorsal valve bears two calcareous coiled projections that support the lophophore, and there is no anus.

Brachiopods are all marine but not very abundant in most areas of today's seas. There are only a few hundred living species known. But there was a time in past geological ages when brachiopods were extremely diverse and abundant worldwide. At that time brachiopods played an important ecological role, feeding on plankton and providing space on their shells for many other organisms; they appear to have occupied the niche filled today by clams and oysters.

Modern species of *Lingula* are almost identical in structure with species that lived in ancient seas over 500 million years ago. This is a record of evolutionary stability, and *Lingula* has the "honor" of being the oldest-known animal genus. Many fossil brachiopods are distinctive and easy to recognize, so they are of great value to geologists in dating rock layers.

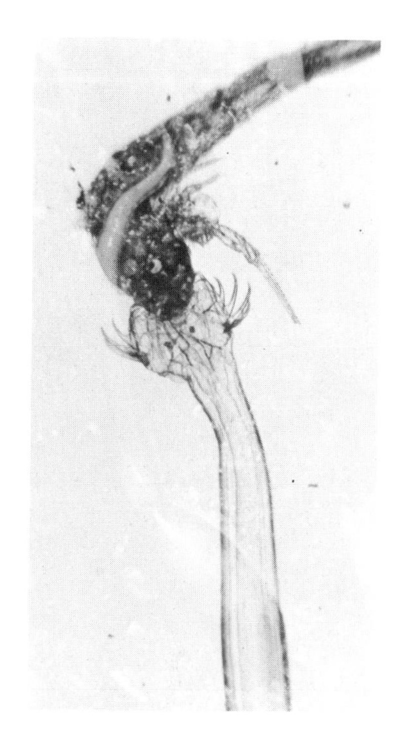

Arrow worm, with bristles spread on either side of the head, eating a small crustacean. Western Samoa. (K. J. Marschall)

Arrow Worms

In the surface layers of the oceans we find transparent, slender animals, usually less than 10 centimeters long, that look like cellophane arrows as they dart after their prey. Arrow worms are members of the phylum **CHAETOGNATHA** (KEET-og-natha), a name that means "bristle-jawed" and refers to the curved bristles, on either side of the mouth, that aid in catching prey. Arrow worms detect their prey by means of sensory cilia that are stimulated by vibrations in the water. Although the phylum includes relatively few species, at certain seasons arrow worms occur in incredible numbers. At such times they form a large part of the food of fishes, but not before they themselves have devoured great numbers of young fishes, as well as any other small members of the plankton that they can catch. As arrow worms are very sensitive to temperature and salinity, they are good indicators of water and fishing conditions. Oceanographers have used certain species to trace and identify water masses. The torpedo-shaped body is divided by transverse partitions into head, trunk, and tail and has finlike projections, which probably serve as stabilizers. Arrow worms are hermaphroditic; and the large coelomic spaces, divided into right and left halves by longitudinal partitions, contain ovaries in the trunk and testes in the tail. The body plan is so different from that of other groups that it is difficult to say what evolutionary relationships arrow worms have to other invertebrates.

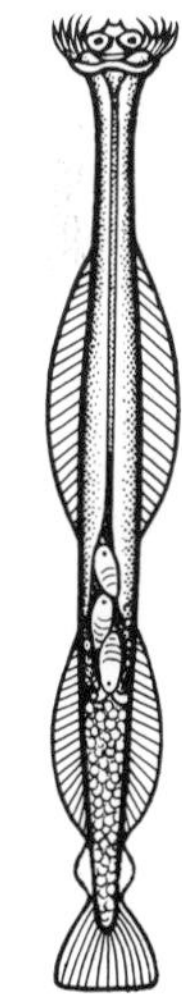

Arrow worm, *Sagitta*, that has eaten 3 small crustaceans. (After D. A. Parry.)

Chapter 15

Soft-Bodied Animals

If there were competitions among invertebrates for size, speed, and intelligence, most of the gold and silver medals would go to the squids and octopuses. But it is not these flashy prizewinners that make the phylum **MOLLUSCA** the second largest in the animal kingdom, with more than 100,000 described species. That honor has been won for the phylum mostly by the slow and steady snails, with some help from the even slower clams and oysters. The name Mollusca means "soft-bodied," and the tender succulent flesh of molluscs, more than that of any other invertebrates, is widely enjoyed by humans. But many molluscs are better known for the hard shells that these slow-moving, vulnerable animals secrete as protection against potential predators. Ironically, it is for the beauty and value of these shells that many molluscs are most ardently hunted by humans, in some cases nearly to extinction.

Despite the differences in the external appearance of a snail, a clam, and a squid, their body plan is fundamentally the same and is distinct from those of all of the other invertebrate groups. The characteristic features of a mollusc are much modified, and some are even lost, in a highly specialized animal like a sedentary clam or a speedy squid. They are less changed—from what we think was the condition of the primitive molluscan ancestors—in a chiton (class **Polyplacophora**).

Opposite: **Chitons** are typical molluscs. The exposed (dorsal) surface is protected by a row of *eight shell plates,* a stable character found in fossil chitons that are at least 400 million years old. The two chitons shown here have eroded shells; the outer pigmented organic layer is mostly gone; but the thick calcareous portion of the plates is still sturdy. Encircling the overlapping plates is the mantle edge, studded with calcareous spines but flexible enough to fit tightly against rock of uneven contour. This reduces water loss during low tide and also the chance of being dislodged by predators or surf. Chitons usually feed at night or during high tides. During the day many chitons shelter under boulders or in rock hollows to which they return after each feeding foray. Japan. (R. Buchsbaum)

Chitons (KY-tons) are sluggish animals that browse on the algas growing on rocky seashores. When disturbed, they clamp down on the rock so tenaciously with their powerful muscles that it takes much persistence to pry them loose.

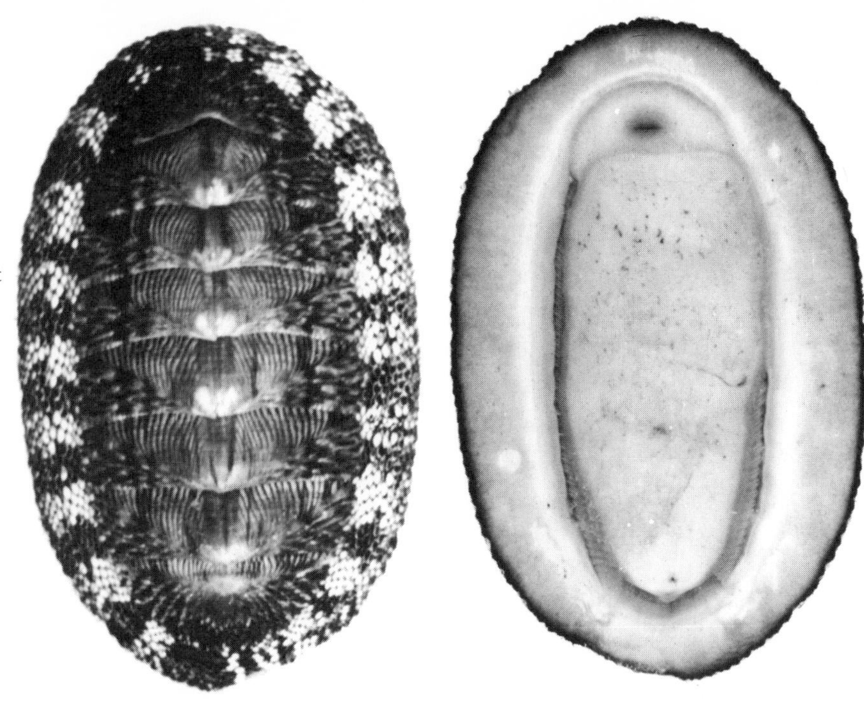

A chiton gives us some idea of the kind of animal from which snails, clams, and squids evolved. The upper surface *(left)* is protected by eight overlapping shell plates. The under surface *(right)* is occupied mostly by the large fleshy foot, in front of which is the small head bearing the mouth. Surrounding the shells in the dorsal view and the foot in the ventral view is the mantle edge. Bermuda. (R. Buchsbaum)

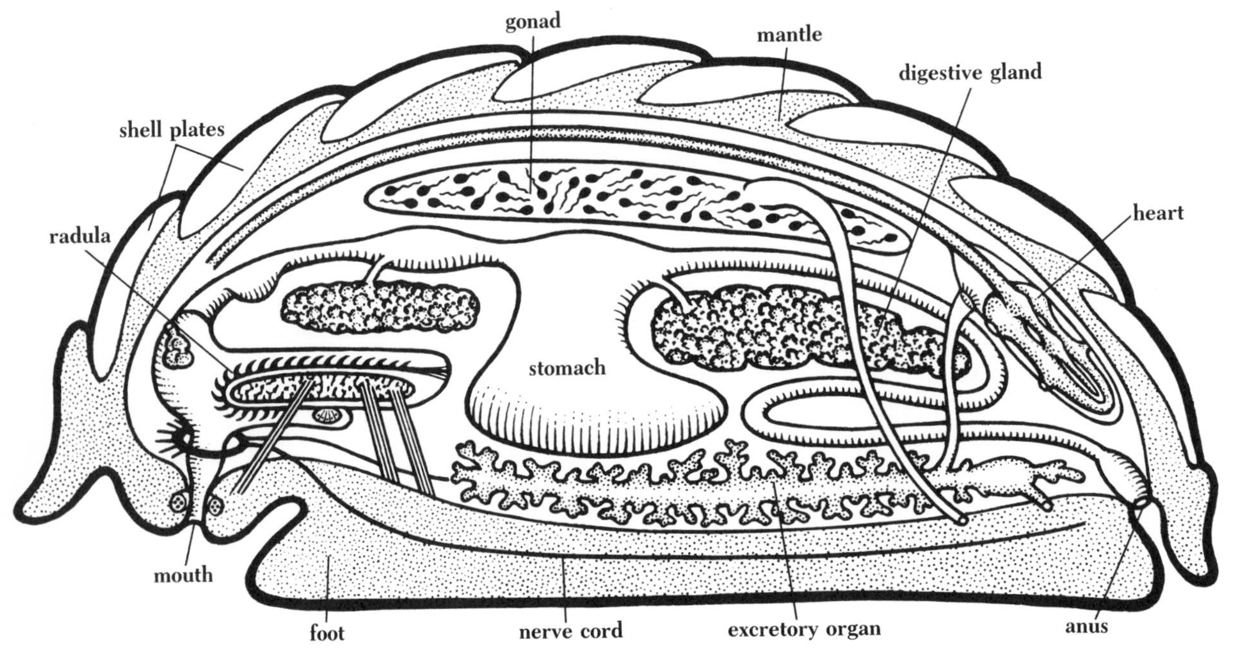

Chiton illustrates the principal molluscan features in a fairly typical form.

The body is bilaterally symmetrical. At the anterior end is a small and inconspicuous head. The ventral surface is largely taken up by a broad, flat, muscular creeping **foot,** supplied with mucus glands. Most of the soft internal organs are contained in the **visceral mass,** which lies dorsal to the foot and is completely covered by the **mantle,** much as a roof covers a house. A fold of mantle tissue extends down around the foot, and the space under the "eaves" is the **mantle cavity.** On its upper surface the mantle secretes a **calcareous shell,** which in chitons consists of eight separate plates, overlapping from front to rear like shingles on a roof. Between the mantle edge and the foot, hanging from the ceiling of the mantle cavity on each side, is a row of **gills,** thin-walled featherlike structures that serve in respiratory exchange. Respiratory currents flow between mantle edge and foot and past all the gills, from front to rear.

The **digestive system** is a tube extending from the mouth, in the head, to the anus, at the posterior end of the animal. The mouth cavity leads into a muscular chamber, the pharynx. Where these join, the **radula** protrudes from its tubular sheath. Unique to molluscs, the radula is a chitinous ribbon covered with many rows of hard recurved teeth, supported on two cartilaginous structures and operated by a complicated array of muscles. When feeding, a chiton protrudes the radular apparatus through the mouth; and as the teeth of the radula move over the surface of rocks or seaweeds, they rasp off small fragments of food. Behind the pharynx the esophagus opens into the stomach, from which a long intestine runs to the anus.

Rolling up is one of the few defenses chitons have when dislodged from the rock substrate that normally protects the more vulnerable undersurface. *On the left,* the mantle edges are pulled back to reveal the row of gills in the mantle cavity on each side of the foot. This is a middle-sized gumboot chiton, *Cryptochiton stelleri.* Common along the U.S. West Coast, these chitons are the world's largest and may reach a length of 35 cm. The eight shell plates are not visible on the dorsal surface but are completely overgrown by the mantle. (R. Buchsbaum)

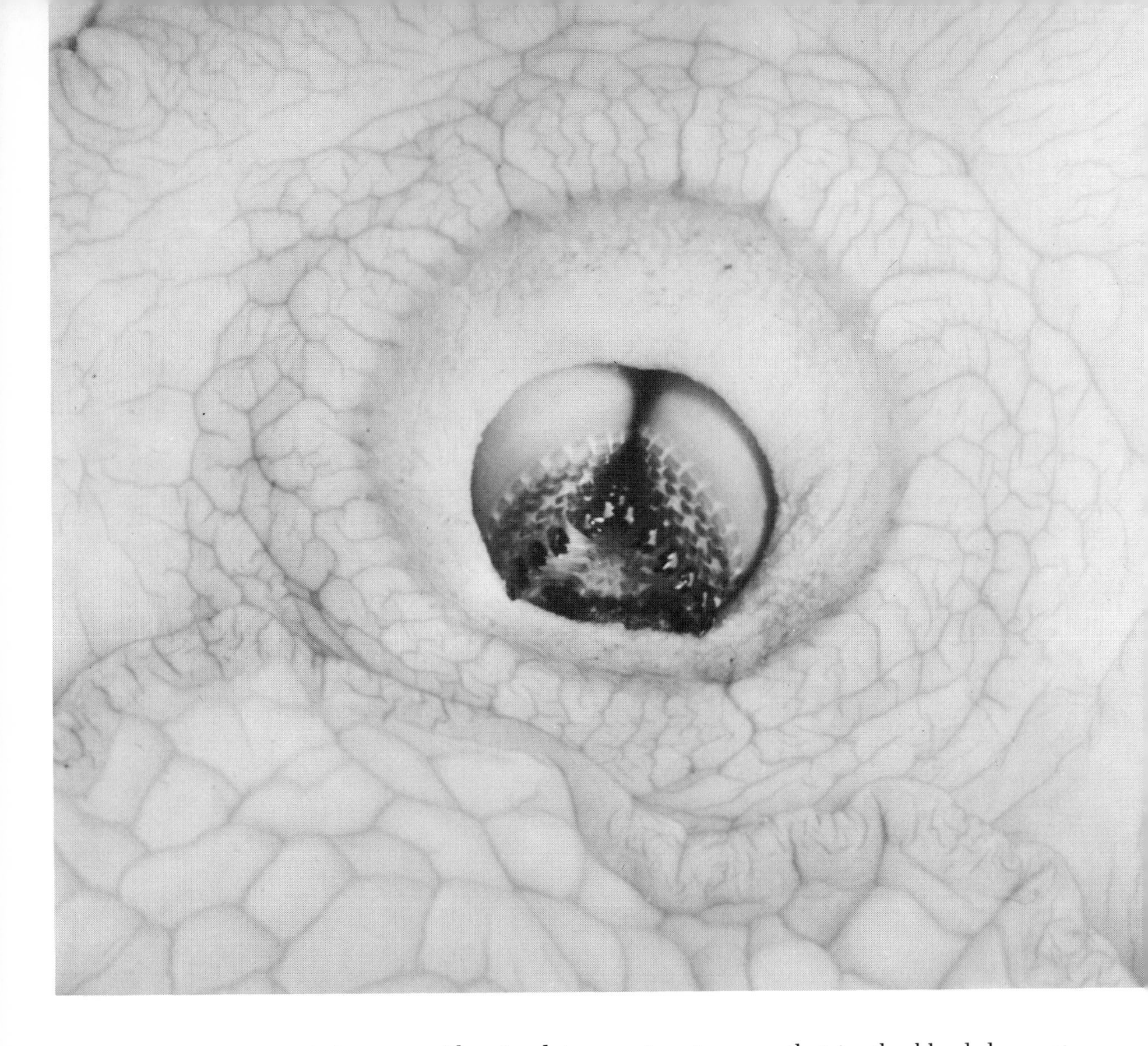

Closeup of chiton mouth reveals the radula in action as it scrapes algal film from the glass of an aquarium. At the very front, a transverse row of teeth show the mineralized (dark) crowns that make chiton teeth hard enough to scrape grooves in rock. Farther back the radula is folded, enclosing the teeth and preventing injury to the lining of the mouth cavity. *Cryptochiton.* (R. Buchsbaum)

The **circulatory system** is *open*, that is, the blood does not flow through a continuous system of discrete vessels but through **blood spaces** that surround the organs. However, the circulatory system includes a few tubular blood vessels and a specialized pumping organ, the **heart.** The heart lies in a cavity, the **pericardial cavity,** which is a mesoderm-lined coelom. The coelom of molluscs is not a major body cavity but consists of small spaces around the heart and also the cavities of the gonads and excretory organs.

The two **excretory organs** are connected with the pericardial cavity and lined with a glandular epithelium. They extract nitrogenous wastes from the blood and reabsorb salts and other useful

Four rows of teeth from the **radula** of a chiton *(Nuttallina)*. The radula is ribbonlike; in this chiton, about 30 mm long, the radula was about 1.2 mm wide and 12 mm long, with about 50 rows of teeth. Scanning electron micrograph. (D. Eernisse)

materials from the pericardial fluid as it passes through them. Waste fluid is discharged to the outside by way of two pores near the anus.

The **nervous system** includes an anterior ring of nervous tissue around the gut. Two pairs of longitudinal nerve cords go to the muscles of foot and mantle. It is a "ladder type" of nervous system, not very different from those of flatworms or nemerteans. Patches of sensory cells in various parts of the body enable a chiton to detect light or to sense chemicals in the water.

Eggs or sperms are shed into the sea. The fertilized egg develops into a rounded larva called a **trochophore,** which swims by means of a prominent band of cilia around its equator, just in

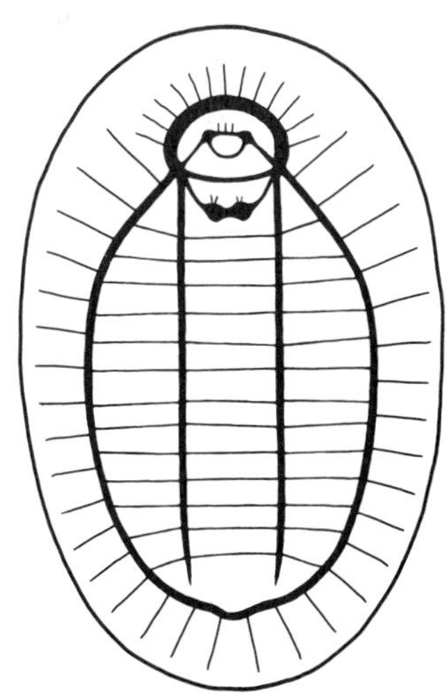

Nervous system of a chiton. (Based on several sources.)

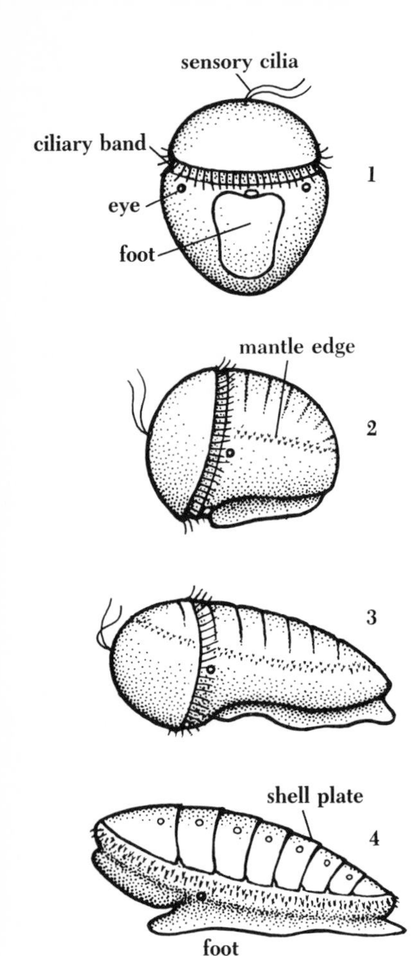

Development of a chiton.
1: Swimming trochophore shows ventral foot. **2:** In side view of swimming larva, beginnings of shell plates are already visible. **3:** Although still free-swimming, the larva has begun to elongate and flatten. **4:** Settled larva has lost the band of swimming cilia, the shell plates are clearly marked, and the adult appearance is developing. (After H. Heath.)

front of the mouth. At the anterior pole is a group of sensory cells that bear a tuft of long cilia. The trochophore has a developing digestive tract but does not feed. It swims for only a few hours or days before settling down on a rock and metamorphosing into a baby chiton. Many marine molluscs (and also annelids and related phyla to be introduced later) have a trochophore stage in their development. However, in most molluscs the trochophore is not a free-swimming larva but only a stage passed within the egg.

Though less common and familiar than their more conspicuous and more economically valuable relatives, the chitons are described here because they display the molluscan body plan in its most typical form. The body consists of three main regions: a ventral muscular foot, a dorsal mass containing the viscera, and a fleshy mantle that secretes the protective shell. Not all molluscs have a radula, but nothing like it is found anywhere else in the animal kingdom.

A hard shell is not peculiar to molluscs, but other groups with shells are mostly sessile and completely encased by their shells, like the brachiopods. Molluscs, on the other hand, have managed to combine a shell with some degree of freedom. **Clams** and their relatives are more or less completely encased by their shells and are often sessile, but many are capable of burrowing and a few can even swim. **Chitons** are free to crawl about, for the shell plates cover only the dorsal surface, and the exposed mantle and foot are extremely tough (whoever called molluscs "soft-bodied" probably wasn't thinking of chitons). Most of the **snails** have compromised; head and foot can be extended when the animal is moving or feeding, withdrawn completely into the shell when danger threatens. **Squids and octopuses** have lost the shell and depend on speed and cunning to escape from predators. How the various kinds of molluscs have adapted the same body plan to their many different ways of life is the subject of the rest of this chapter.

Gastropods

The class **Gastropoda** bears a name that means belly-footed and attempts to describe a group in which the broad foot occupies most of the underside. The group includes such common animals as snails and slugs, limpets and abalones, and is by far the largest of the classes of molluscs. Indeed, five out of six molluscs are gastropods, and most have all of the chief molluscan features already presented: foot, visceral mass, mantle and shell, and radula. Compared to chitons, gastropods have a more prominent

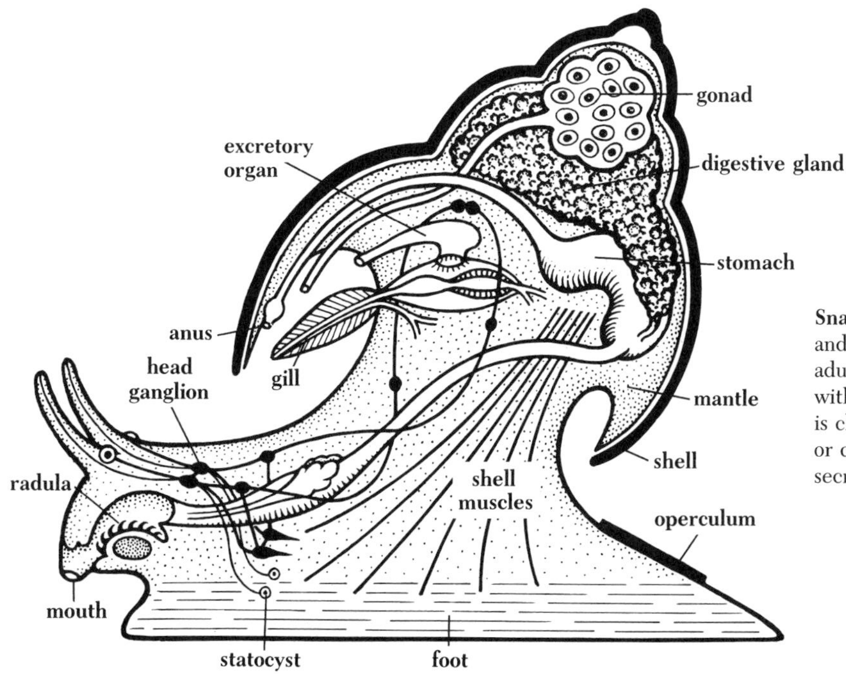

Snail illustrates the effects of coiling and torsion on the structures of an adult gastropod. When the snail withdraws into its shell, the opening is closed by the *operculum,* a horny or calcareous plate that some snails secrete on the back of the foot.

head, with eyes and sensory tentacles, and a smaller number of gills, usually one or two. Most gastropods are snails, with a **helically coiled shell and visceral mass.** The helical coiling results from asymmetrical growth and is said to provide a more compact and stable package than would a symmetrical cone-shaped shell large enough to accommodate the same amount of tissue. Since the great majority of snail shells are in the form of a right-handed coil, which leaves less room for the organs on the right side, many snails are missing the right gill, the right excretory organ, and the right auricle of the heart.

Quite apart from the lopsidedness that relates to helical coiling, gastropods are highly modified by a peculiar process called **torsion,** during which the visceral mass, mantle, and shell become reoriented at 180° with respect to the head and foot. As a result of torsion, the nervous system becomes twisted into a figure eight, and the products of the animal's digestive, excretory, and reproductive systems are all discharged on top of its own head. The advantages of such an arrangement are not clear, but the process can be observed during gastropod development and appears to result from a combination of rapid asymmetrical muscular contraction and slower asymmetrical growth.

During **development** most young gastropods go through a trochophore stage within the capsule in which the eggs are laid. They hatch out as swimming larvas called **veligers,** in which the girdle of cilia is expanded into a *velum,* in the form of large cil-

Trochophore of a gastropod. (After W. Patten.)

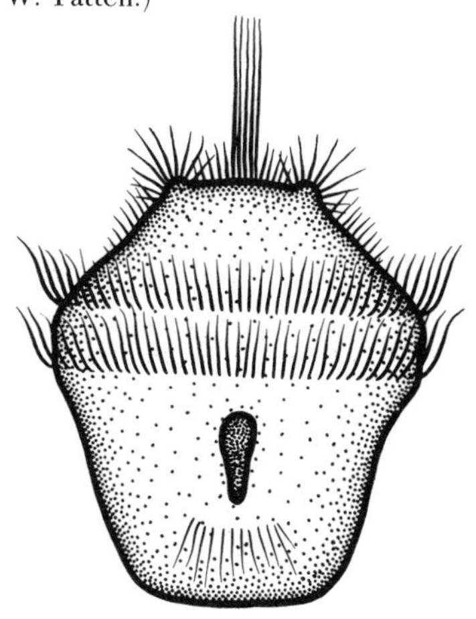

iated lobes that serve in swimming and also to bring food to the mouth. The development of the foot, shell, and adult organ systems is well begun in the veliger. At metamorphosis, after the larva settles to the bottom, the velum is lost or sometimes eaten, and the juvenile gastropod crawls away.

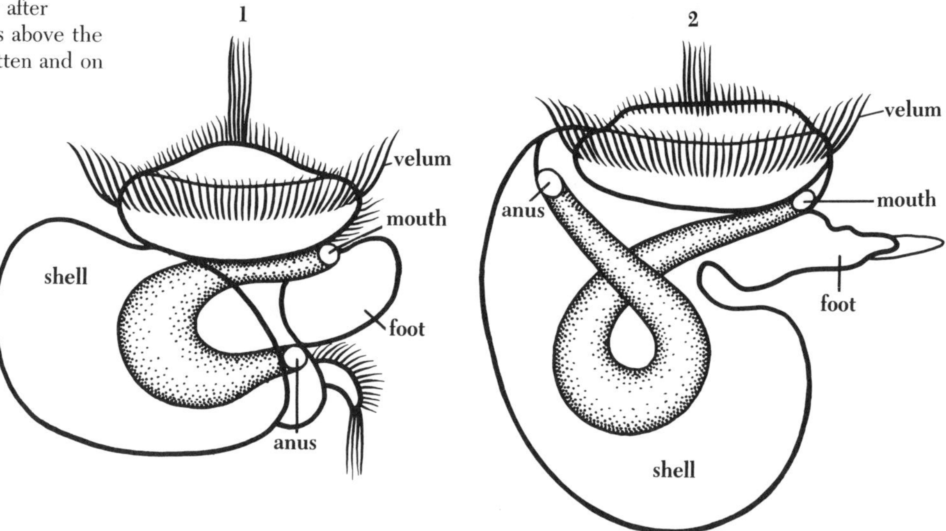

Veliger of a gastropod. **1:** Veliger before torsion. **2:** Veliger after torsion; the anus now lies above the mouth. (Based on W. Patten and on A. Robert.)

Veligers of the red abalone, *Haliotis rufescens*, swim and feed actively, the cilia of the velum sweeping phytoplankton into the mouth. Three days old at the time of this photo, these larvas developed from eggs spawned and fertilized at the Granite Canyon Laboratory, California Department of Fish and Game. (R. Buchsbaum)

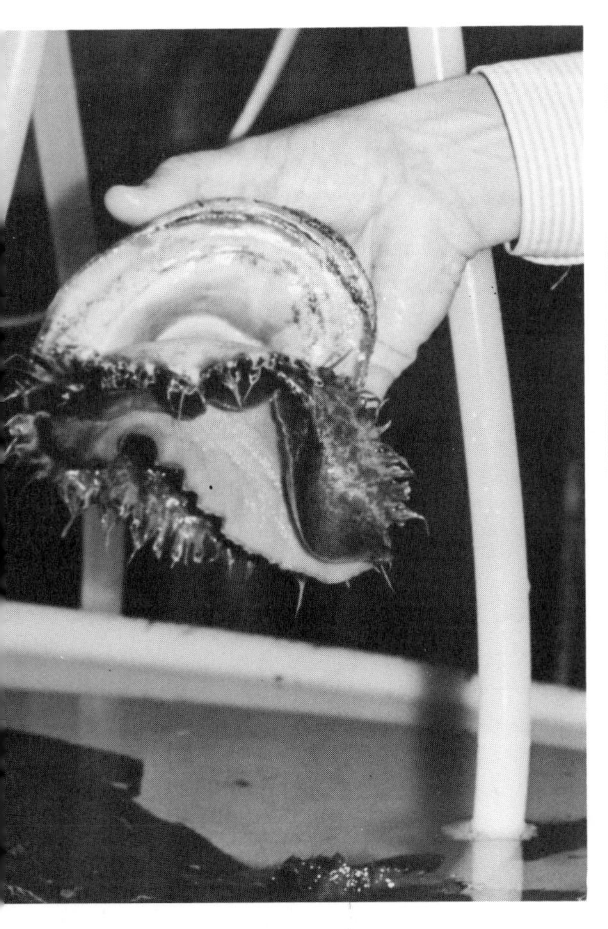

Mariculture of abalones is an alternative to collecting from wild populations, which have been greatly reduced by sport and commercial divers along the California coast, in spite of regulation. Cultured red abalones *(Haliotis rufescens)* take about a year to reach a shell length of 25 mm *(right)* and can be lab-reared to maturity, fed on kelp. But commercial mariculture of abalones is still an infant industry. Monterey Abalone Farms. (R. Buchsbaum)

Some gastropods pass through even the veliger stage within the capsule and emerge as crawling juveniles. In freshwater and land gastropods, the eggs usually develop without a recognizable veliger stage. In some cases they are not laid at all but develop within the body of the parent. A free-swimming trochophore larva occurs only in some of the relatively primitive gastropods.

As with development, gastropods display a great range of habits as adults. The relatively primitive gastropods, such as abalones and limpets, live much like chitons, as slow grazers and scrapers of plant material on rock surfaces. But others have evolved as lively predators that range into many marine habitats, feeding on a variety of invertebrates and sometimes on fishes.

For all its advantages, a shell is a handicap to active locomotion. And there is a tendency among many groups of gastropods toward reduction or even complete *loss of the shell,* accompanied by an uncoiling and an untwisting (detorsion) of the visceral

Limpets are mainly herbivorous, scraping algal material from the surface of rocks and seaweeds. Some kinds make scars in the rock, depressions that just fit their shell and to which they return after each feeding expedition. The limpets in this photo *(Lottia digitalis)* live high in the intertidal zone; they are wetted intermittently by splashing waves but submerged only by the highest tides. Oregon. (R. Buchsbaum)

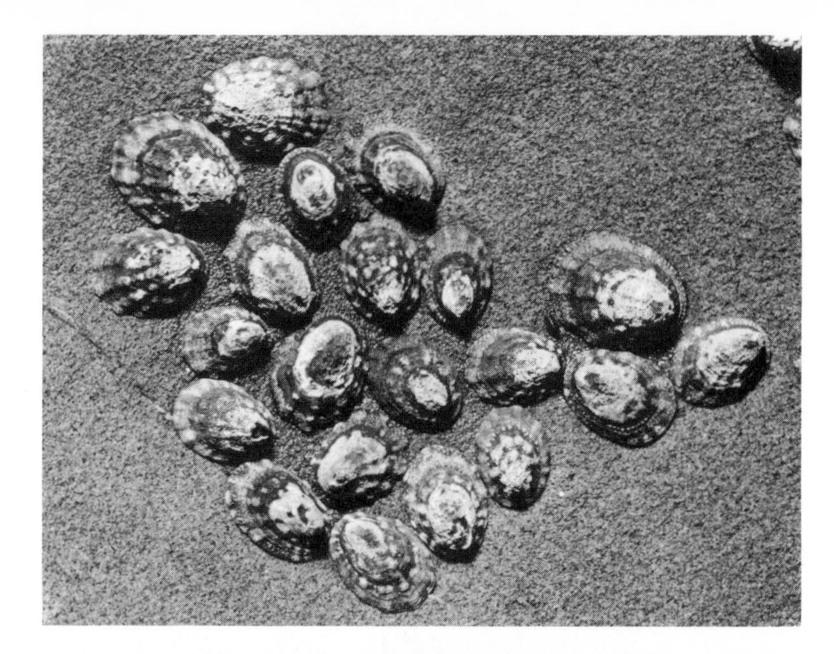

Keyhole limpet, *Diodora aspera,* has an opening for exit of the respiratory current at the apex of the conical shell. Keyhole limpets have a pair of gills, one of which can be seen protruding from under the mantle on the limpet's right. This head-on view shows the mouth and two large sensory tentacles on the head, as well as smaller ones around the edge of the mantle. Members of this species reach about 70 mm and occur from Alaska to Baja California. They are exceptional among limpets in having omnivorous habits; they feed on small animals such as bryozoans in preference to algas. (R. Buchsbaum)

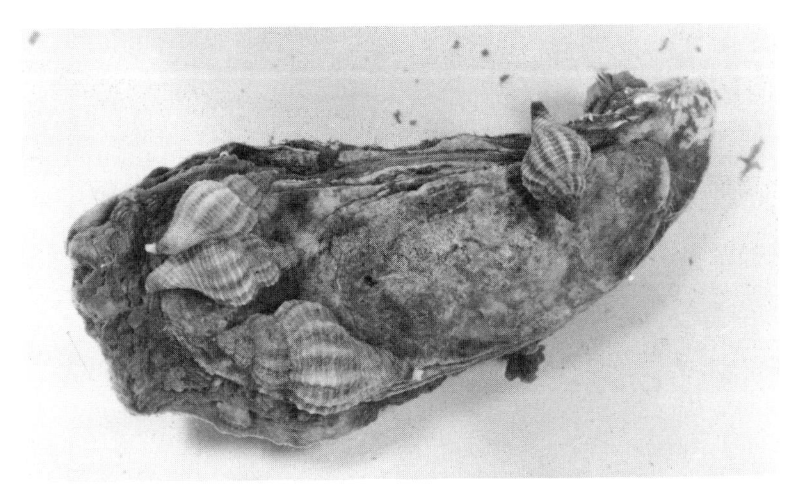

Four oyster drills *(Urosalpinx cinerea)* are here in the process of drilling holes through the thick shell of a large oyster. In drilling, the snail uses both its scraping radula and an acid secretion from its foot that weakens the shell. It is slow work and may take the snails a full week before they reach the soft flesh. Naturally they prefer young thin-shelled oysters and may eat so many of these that they are serious pests of commercial oyster beds. Beaufort, North Carolina. (R. Buchsbaum)

Snails laying egg capsules. In the winter and spring these marine snails *(Nucella lamellosa)* gather in breeding groups, to which they may return year after year. After mating the females deposit their bright yellow capsules, each containing many eggs, on the undersides of intertidal rocks. The eggs pass through a veliger stage within the capsule and hatch out as tiny snails. Both eggs and young face many predators, as well as other hazards, and only one in a hundred will survive the first year. These snails are themselves predators, feeding on other molluscs and on barnacles. Oregon. (R. Buchsbaum)

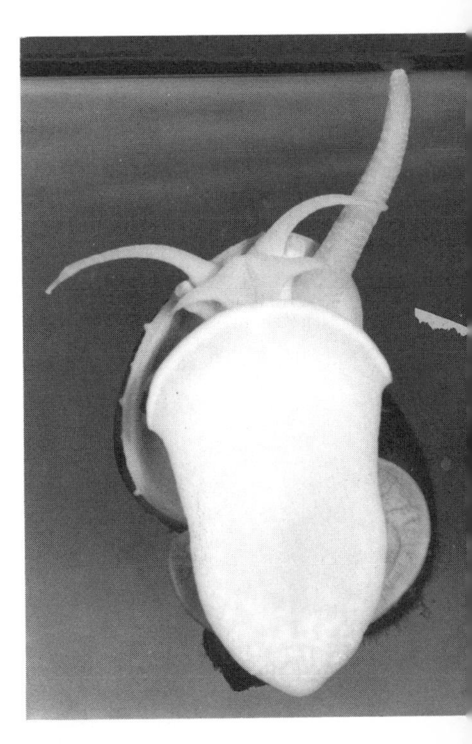

Freshwater snail taking air through its equivalent of a snorkel, a long tubular extension of the mantle. This versatile snail also has a gill that supplements its oxygen requirements between trips to the surface. Native to South American swamps, it can tolerate stagnant waters and does well in aquariums. *Ampullarius.* Length, 3 cm. (R. Buchsbaum)

mass. That these forms have descended from typical gastropods is shown in their larval development. The larvas have a coiled shell and undergo torsion, followed by loss of the shell, uncoiling of the viscera, and detorsion. The **sea slugs,** lacking the protection of a shell, have developed other defenses. Some are marvelously camouflaged to match their habitat, while others produce copious mucus and noxious substances in their tissues. Inexperienced fishes that attempt to eat such sea slugs are observed to spit them out quickly. Like many other poisonous, prickly, or otherwise noxious animals, these sea slugs are often adorned with bright, bold patterns that presumably help fishes or other visual predators to learn to avoid them. A number of sea slugs that feed on cnidarians can transfer the undischarged thread capsules to their own tissues and use them in defense.

Garden snails mating. Like other pulmonates, they are hermaphroditic and here are engaging in a mutual exchange of sperms, introduced through a genital pore on the right side of each snail just behind the head. Later both snails will lay eggs. *Helix aspersa* has been spread from Europe around the world. Freed of its natural predators and parasites, it has become a pest in congenial climates such as coastal California. In French cuisine it takes second place to the larger *Helix pomatia* but is eaten when the favored species is scarce. California garden. (R. Buchsbaum)

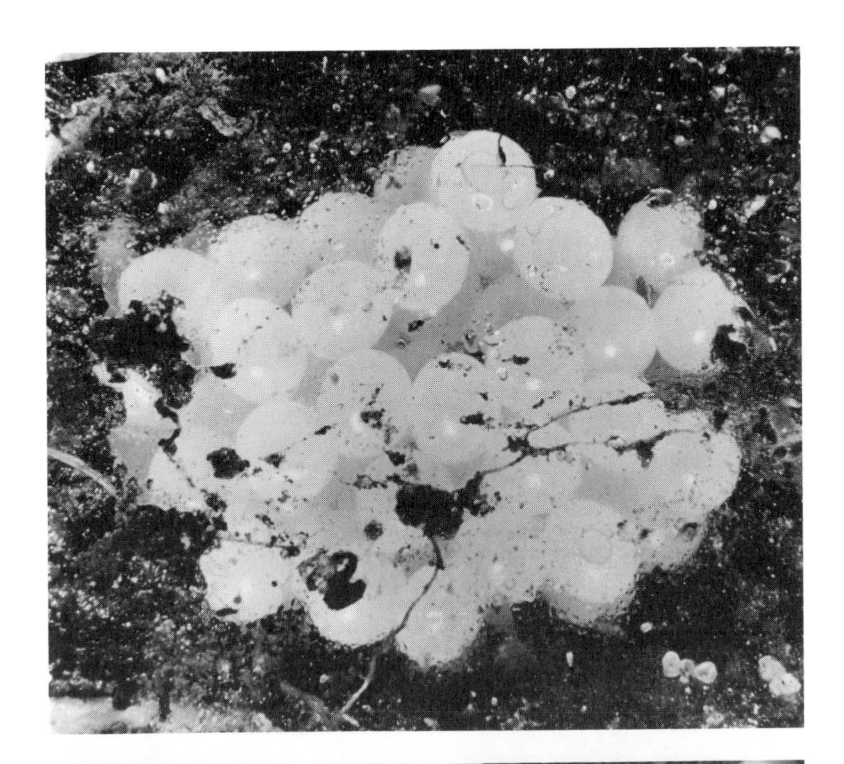

The eggs of a garden snail look like glistening pearls. They are laid in a hole that the parent digs in moist earth and hatch in a little over a week. (R. Buchsbaum)

The newly hatched snails are soft and pale. Their light-seeking behavior presumably helps them to find the exit from the nest. Soon they switch to the light-avoiding behavior that characterizes adults and keeps them in moist, protected hiding places during the day. (R. Buchsbaum)

Mucus trail laid down by a snail's foot smoothes and lubricates its path. Snails may also use mucus to attach to substrates and seal the shell opening when resting during the day, or for longer periods when the weather is dry. (R. Buchsbaum)

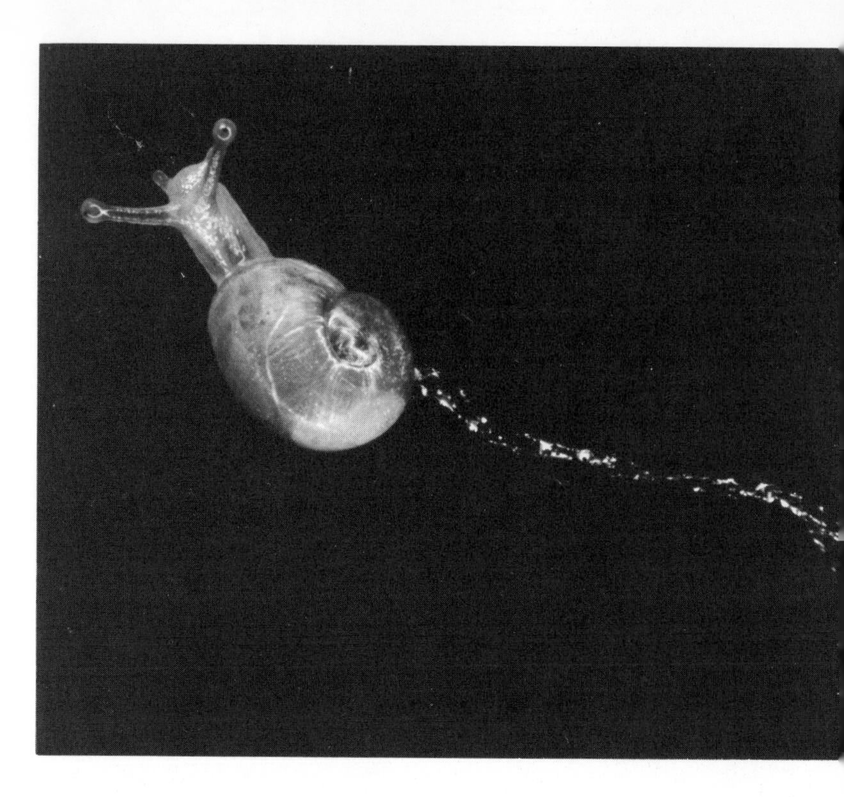

A slug is a land pulmonate that has lost the external shell, having only a thin plate embedded in the mantle. The mucus these animals secrete and on which they glide is lubricating and protective, as is demonstrated by these pictures of a slug passing unharmed over the sharp edge of a razor. (O. Croy)

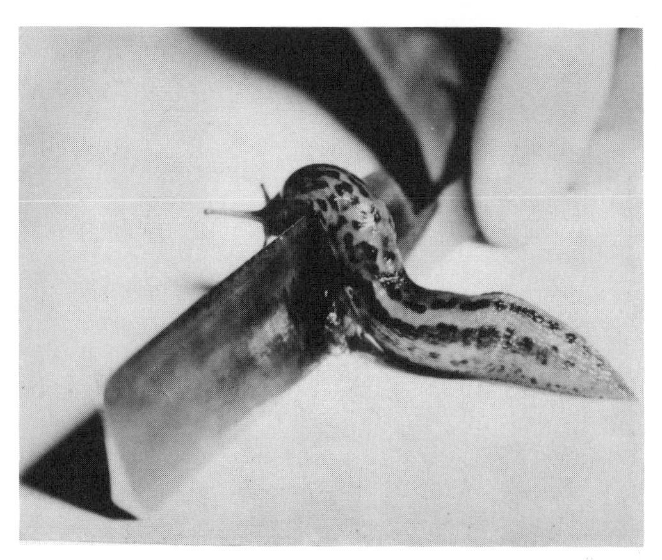

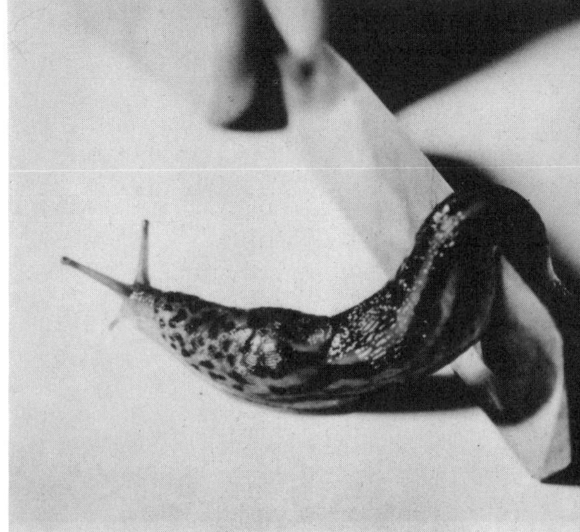

Giant African snails, *Achatina fulica,* have been introduced into southern Asia and many Pacific islands, where they have become serious agricultural pests; in addition, they are hosts to parasitic worms that cause severe, even fatal infections in humans. On reaching a new area, the snails multiply rapidly, achieving alarming densities, but later decline in numbers. The main effect of various predators, hastily introduced to control the giant snails, has been to endanger the native snails of the islands. The body of *Achatina* may reach almost 30 cm, the shell 20 cm. Oahu, Hawaii. (R. Buchsbaum)

Many gastropods, such as the familiar garden snails and slugs, have invaded land. These are called **pulmonates** because they have a modified mantle and mantle cavity that act as a lung for breathing air. Many pulmonates have gone into freshwater; but organs once lost do not generally reappear, and these aquatic snails have no gills. They must come to the surface periodically to take air into the lung.

A nudibranch ("naked gills") lacks not only a shell but also a mantle cavity. The gills that in most molluscs hang within the mantle cavity have been replaced by branched or fingerlike projections on the back of these sea slugs. Nudibranchs feed on sponges, cnidarians, and other sessile animals. These and other sea slugs have nervous systems with relatively small numbers of large cells that lend themselves to studies of physiology and behavior. *Phidiana crassicornis.* Monterey Bay, California. (R. Buchsbaum)

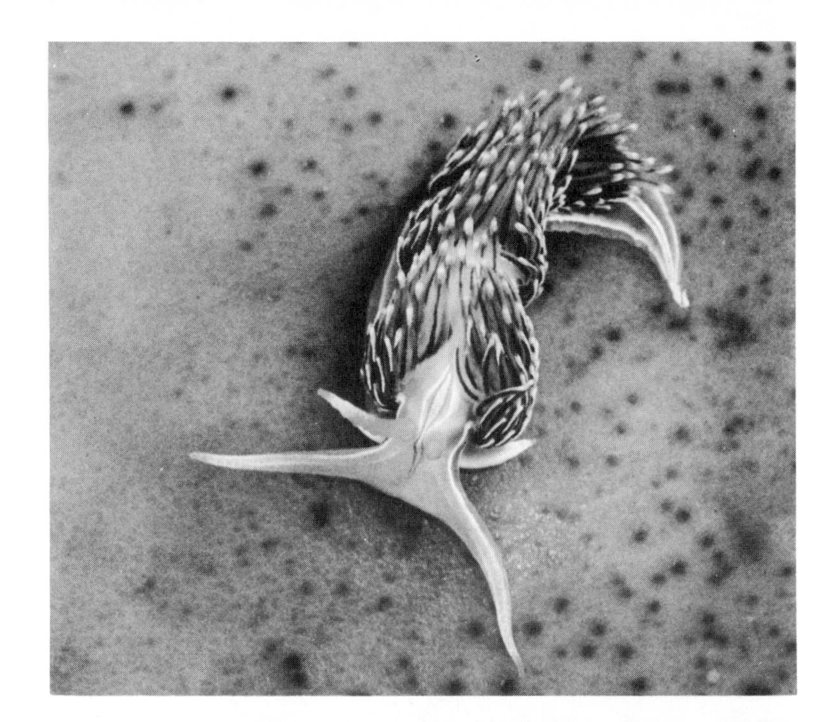

Sea hare *(Aplysia)* is aptly named. Its curled earlike tentacles and herbivorous habits do remind one of rabbits, and its soft rounded back (which conceals a thin shell) invites gentle petting. However, the animal will eject a slimy purple ink if roughly handled. Sea hares often live only a year or so, but they grow extremely fast to large sizes and may be sexually mature when only a few months old. A species of the U.S. West Coast *(Aplysia vaccaria)* is probably the largest gastropod in the world, reaching over 75 cm (about 30 in) in length and weighing up to nearly 16 kg (about 35 lb). Like other sea slugs, sea hares are hermaphrodites. Under favorable conditions they become very abundant and may sometimes be seen mating in large groups. Panacea, N.W. Florida. (R. Buchsbaum)

Tooth Shells

A small class of marine molluscs is the **Scaphopoda,** in which the shell is tubular and open at both ends. The inconspicuous head bears a number of extensible filamentous tentacles that are sen-

Four scaphopods dug up from the sand and laid on the surface. The one at the left has extended its foot and is preparing to burrow in. *Dentalium.* Roscoff, channel coast of France. (R. Buchsbaum)

Tooth shells are often used in necklaces; here two large ones flank the central pendant. Tooth shells were used as ornaments on clothing and also as "wampum" (monetary exchange) by the Amerindians of the northwest coast of North America. Otherwise scaphopods are neither useful nor harmful to humans. (Necklace and photo by C. K. Pearse.)

sory and gather small organisms and food particles. There is a large radula. The animal burrows into sand with its muscular foot and lies almost completely buried with only the narrower, posterior tip of the shell and mantle protruding above the surface into the seawater. A current of water is maintained in and out of the upper end of the shell, and the mantle serves as the respiratory organ, as the gills are lost. The larva goes through free-swimming trochophore and veliger stages.

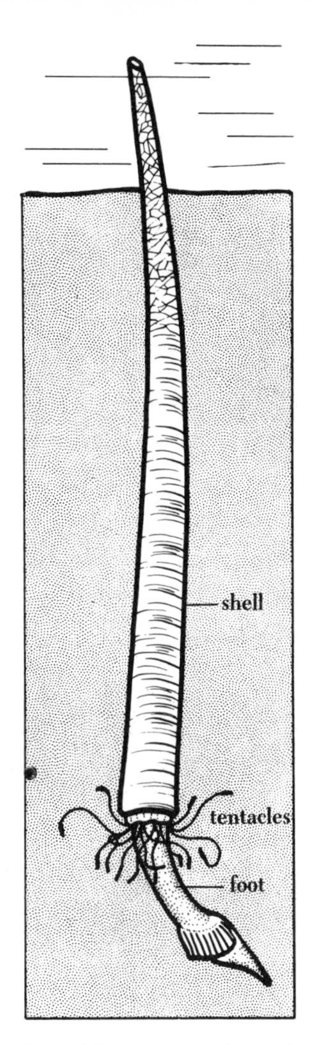

Scaphopod buried in sand. (Modified after Sars.)

Bivalves

Clams, oysters, scallops, and others of the molluscs with a shell in two parts belong to the class **Bivalvia** ("two valves"). Most bivalves are marine, but some clams are very abundant in freshwaters and the following description applies to almost any of these.

A **clam** is flattened from side to side. The two valves of the shell represent *right and left* sides, and are fastened to each other along the *dorsal* edge by an elastic horny **ligament.** The gape of the shells is *ventral.* Near the *anterior* end is an elevated

Clams are bivalves. The hardshell clams shown here are among the most popular commercial varieties, as implied by their name, *Mercenaria mercenaria.* They are being washed in running seawater to remove sand. Besides clams, which come in many shapes and sizes, other common types of bivalves are cockles, mussels, scallops, and oysters. Lewes, Delaware. (R. Buchsbaum)

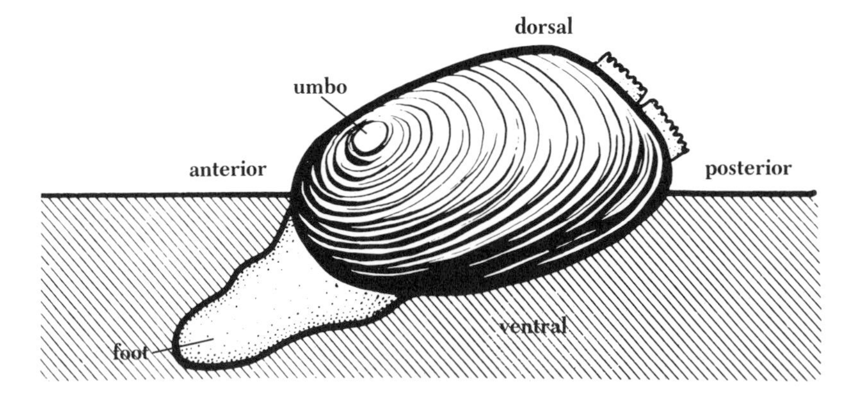

Clam, viewed from the left side. A freshwater clam lies obliquely in the sand or mud.

The concentric growth lines that mark the shell of this clam are punctuated with sharp ridges that help to anchor it firmly in the sand. *Mercenaria campechiensis.* N.W. Florida. (R. Buchsbaum)

knob, the umbo, which represents the oldest part of the shell. At the opposite or *posterior* end are the openings through which currents of water enter and leave the clam.

The **shell** is enlarged as the animal grows. The mantle secretes successive layers of shell, each projecting beyond the last one laid down, and this results in a series of concentric lines of growth that mark the external surface, a permanent record of the successive outlines of the shell.

The shell consists of two portions. A thin outer *organic layer,* continuous with the ligament, protects the *calcareous layers* that form the body of the shell. The organic layer, or periostracum, prevents erosion of the calcareous portion and discourages many boring marine animals. It often wears off with age, but by then the calcareous layers are thick and less vulnerable. In freshwater

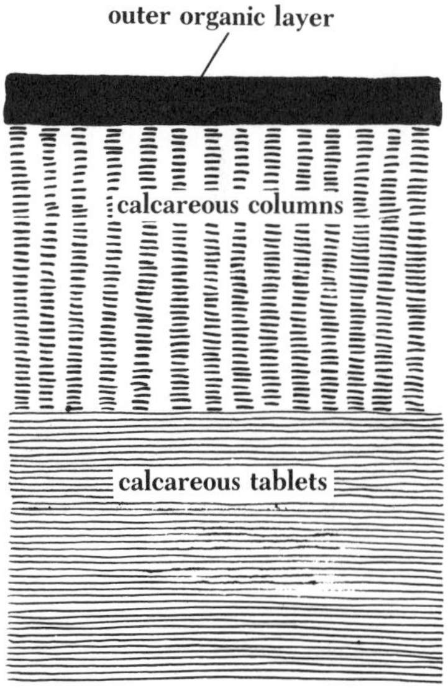

outer organic layer
calcareous columns
calcareous tablets

The shell consists of organic and calcareous layers.

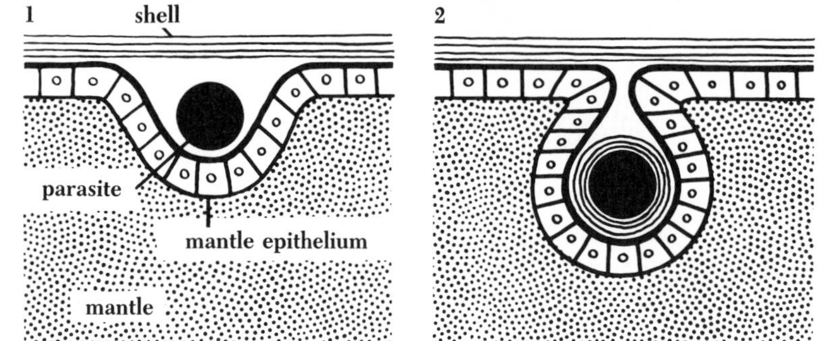

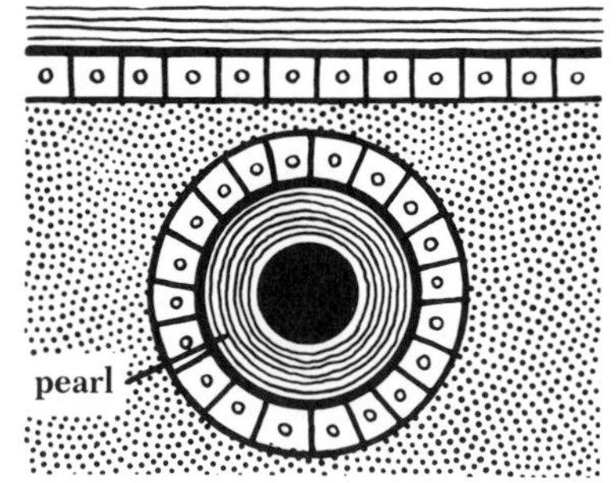

Formation of a pearl. 1: A parasite lodges between the shell and the mantle. **2:** It is almost completely enclosed in a sac formed by the mantle epithelium, which secretes thin concentric pearly layers around it. **3:** A pearl of good size has surrounded the parasite and prevented it from harming the clam. (Based on F. Haas.)

Pearl buttons are cut from the shells of freshwater clams. Millions of kilograms of shells, mostly from the Mississippi River Valley, were once used annually for this purpose. But now the damming of rivers, and also pollution, have greatly reduced populations of freshwater bivalves, and most buttons are plastic. (C. Clarke)

Freshwater pearls, although mostly irregular in shape, are sometimes used in jewelry. The most valuable pearls come from marine "pearl oysters" *(Pinctada).* Naturally occurring pearls of perfect shape are rare, and most pearls are cultured in live pearl oysters by inserting a round pellet of shell, which the oyster coats with a thin pearly layer. It takes up to seven years to produce a valuable pearl. (C. Clarke)

clams and snails and especially in land snails, a thick periostracum helps to prevent dissolution of the calcareous portion by acids in lakes or streams and in leaf litter or soil.

The calcareous portion of the shell is composed of crystals of calcium carbonate laid down within a sparse matrix of organic material. The outer calcareous layer consists largely of columnnar crystals that lie perpendicular to the surface. The inner calcareous layer consists mostly of thin tablets parallel to the surface of the shell.

The periostracum and outer calcareous layer are secreted only by the edge of the mantle, and hence show the concentric markings of discontinuous growth. The inner calcareous layer is laid down by the whole surface of the mantle and has a smooth, sometimes lustrous surface. It is this shiny inner material that coats *pearls,* secreted by the mantle as protection against some foreign body, usually a parasite such as a larval fluke.

When undisturbed, a clam lies partly buried in the sand or mud with the ligament up and the shell valves slightly agape ventrally. If stimulated to move, the animal protrudes its fleshy

(1) **A moving clam** extends its foot into the sand; (2) the tip of the food swells and acts as an anchor; (3) the muscles of the foot contract, drawing the clam forward; (4) the foot is again extended; (5) the tip is anchored; (6) and the slow process continues.

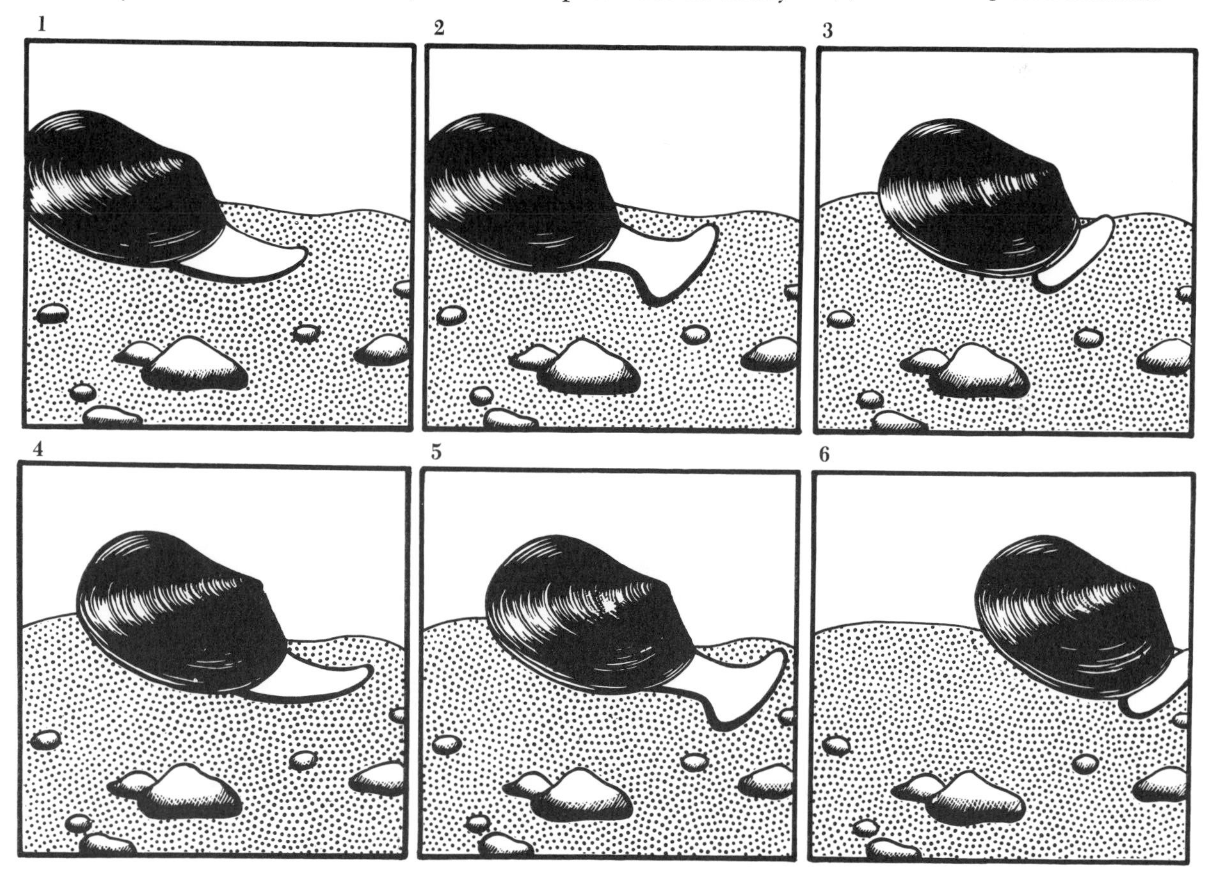

foot and burrows through the mud like an animated plowshare. First, the pointed foot is extended forward into the mud and anchored by a turning or by a swelling of the free end (owing to an influx of blood into a cavity within the foot). Then, the muscles of the foot contract, drawing the body of the clam forward. Such a slowly moving animal, with a shell that is heavy and cumbersome to carry about, could hardly run down prey. Instead, like so many other sedentary animals, clams filter-feed by drawing water through the body and straining out microscopic organisms and other nourishing organic particles. For protection a clam relies on its retiring habits and on its heavy shell. The shell valves may be held tightly closed by the contraction of two large muscles. When these relax, the shell is opened by the elasticity of the ligament.

Within the shell, the **visceral mass** lies dorsally, most of it between the two muscles that close the shell. The **mantle** covers the visceral mass and extends ventrally as two large lobes, one just beneath each shell valve. Between the mantle lobes lies the **mantle cavity.** At the posterior end the lobes form openings (in some clams, long tubes) for the entrance and exit of water. A current of water flows into the mantle cavity through the ventral or *incurrent opening* and out through the dorsal or *excurrent opening.*

Hanging freely into the mantle cavity, on either side of the foot, are a pair of thin folded **gills,** covered with cilia, which create the water current. The water is drawn through microscopic pores in the sievelike gills, leaving suspended food particles on

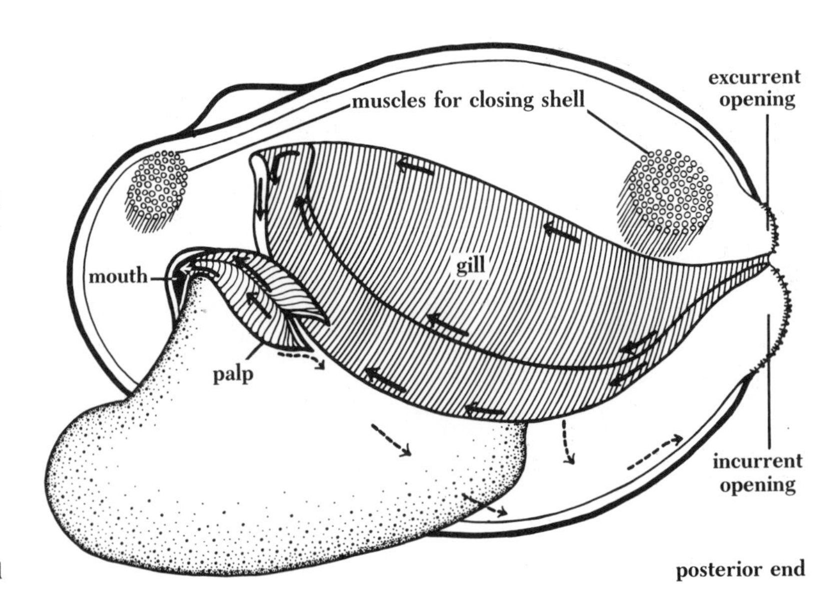

The feeding mechanism of a clam is ciliary. Food particles, caught on the surface of the gills, are carried to the mouth, as shown by the *solid arrows.* Rejected particles are removed from the gills and palps, as shown by the *dotted arrows.* (The left shell valve and mantle lobe have been removed.)

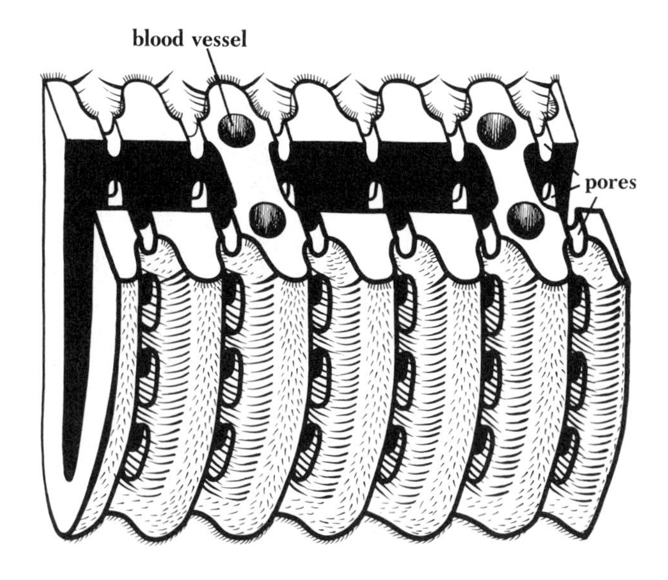

Small portion of a gill, greatly enlarged, shows the microscopic pores in the gill walls through which water flows, leaving food particles trapped on the surface. The space between the inner and outer walls of the gill fold is divided into vertical water channels by partitions (in which blood vessels run).

the gill surface. Within the gills, the water flows upward through channels and finally posteriorly and out through the excurrent opening. The water that flows through the mantle cavity supplies both food and oxygen, and the rate of flow depends on the type and size of the animal and on many other factors such as temperature, water chemistry, and even time of day. Pumping rates measured in large oysters have reached up to 40 liters per hour.

Food particles left on the surface of the gills by this steady stream of water are separated, mostly by their small size, from silt and other undesirable materials during their passage to the mouth. Heavy particles of sand or mud drop from the surface of the gill to the edge of the mantle, are carried backward by cilia on the mantle, and are expelled posteriorly. Lighter particles become entangled in mucus secreted by the gills and are carried, always by beating cilia, to the ventral edge of the gill and then forward until they meet the ciliary tracts on the **palps,** a pair of folds on each side of the mouth. Further sorting occurs here, and selected food materials are finally carried in a deep groove into the mouth. As might be expected in animals that feed only on microscopic particles, there is no radula.

Food-laden strings of mucus are drawn through the narrow esophagus into the saclike stomach by being wound around a gelatinous rotating rod, the **crystalline style,** which lies in a sac off the intestine and projects into the stomach. The food is mixed with digestive secretions set free in the stomach by the slow dissolution of the style tip as it is constantly rubbed against a shield on the stomach wall. However, as we might expect in animals

Clam with left mantle lobe, left gill, and part of the foot cut away to show the **principal organs.**

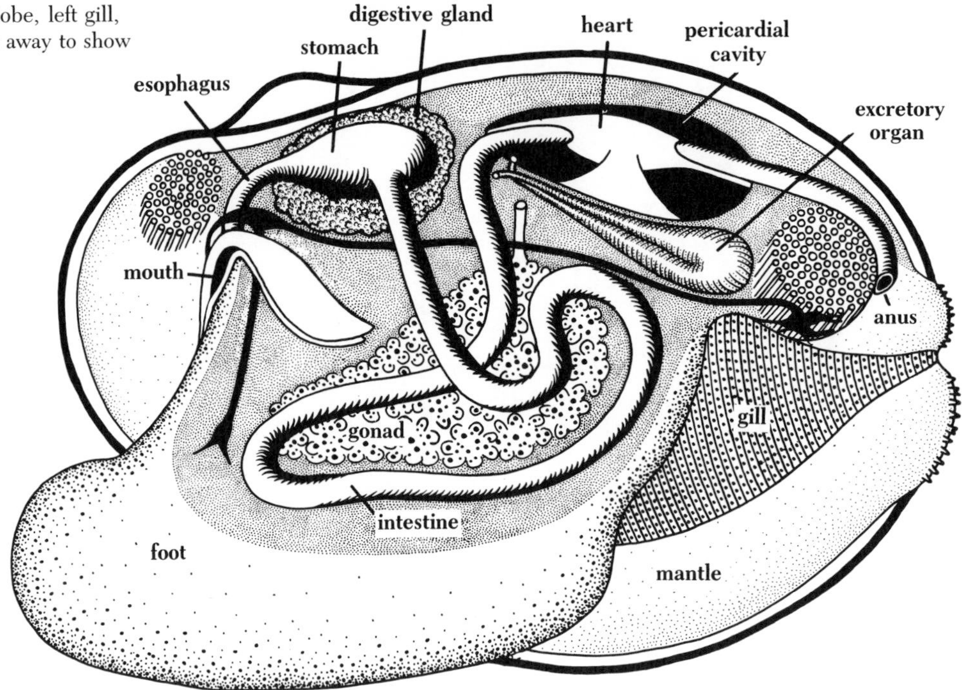

that eat only finely divided food, digestion is mostly intracellular, taking place within the cells of a large **digestive gland** that surrounds the stomach and is the main organ of digestion and absorption. From the stomach the intestine runs ventrally, makes several coils through part of the foot, then runs dorsally again, passing through the cavity that surrounds the heart and appearing to pass through the heart itself (actually the heart is wrapped around the intestine). The anus opens near the excurrent opening, and the feces are carried away in the outgoing current, along with waste fluid from a pair of tubular **excretory organs.**

The **circulatory system** consists of a three-chambered heart in a pericardial cavity, closed blood vessels, and open blood spaces. As mentioned before, shunting of blood into a large blood space in the foot helps to produce swelling of the tip of the foot during locomotion. In most bivalves the blood is colorless but contains ameboid cells. Some clams, especially those in poorly oxygenated habitats, have hemoglobin contained within cells.

The simple **nervous system** serves to coordinate the slow movements of the muscular body. There is no recognizable head, apart from the site of mouth and palps, nor should we expect more in an animal that lives with its anterior end buried in the

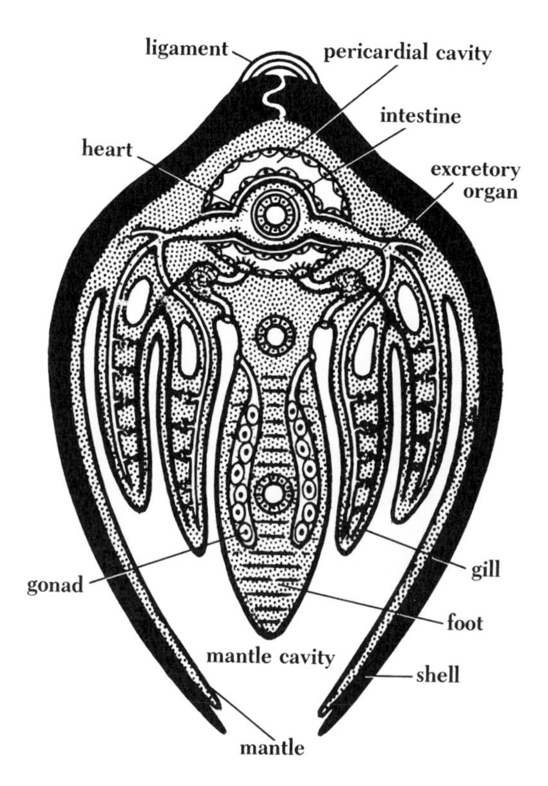

Cross-section of a clam shows the gills suspended in the mantle cavity between the foot and the mantle lobes that line the shell; the heart and surrounding pericardial cavity; several sections through loops of the intestine; and other organs.

mud. But there are a pair of head ganglia. These lie near the anterior closing muscle and from them run long nervous connectives to a pair of visceral ganglia near the posterior closing muscle and to another pair of ganglia in the foot. Near the foot ganglia are two statocysts. A patch of yellow sensory cells, thought to be sensitive to chemicals in the water, lies on the visceral ganglia, and other sensory cells are scattered in the mantle, especially on small projections along the edges and around the incurrent and excurrent openings. These cells probably respond to light and touch. When a clam is irritated, the foot and mantle edges are withdrawn, and the two valves close tightly—or, as we say, "shut up like a clam."

The **reproductive system** of bivalves consists of a pair of lobed gonads that open, near the openings of the excretory organs, so that the gametes or (if fertilization is internal) the embryos leave in the outgoing water current. The sexes are usually separate. Most marine bivalves shed sperms or eggs into the sea, where fertilization takes place. There is first a trochophore and then a veliger larva. In freshwater clams, and some marine species, only the male releases gametes into the water; the female retains her eggs in chambers within the gills, where they are fertilized by sperms that enter with the incoming water current. The embryos

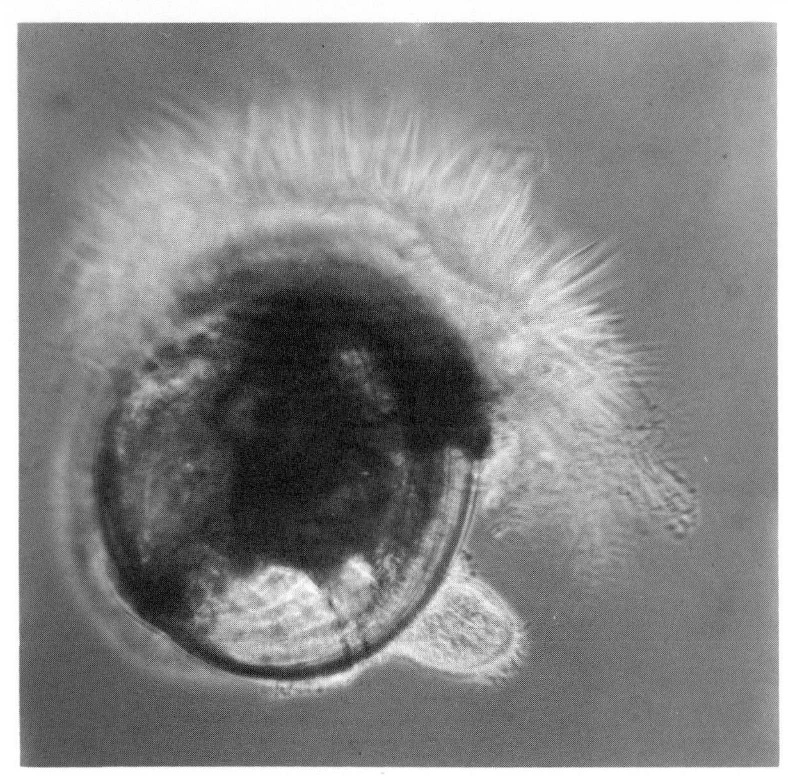

Veliger of an oyster has a bivalved shell from which protrude the lushly ciliated velum and the small rounded foot. Both velum and foot are later lost, and the adult lives cemented by one shell valve to rock or some other hard substrate. This larva was raised at the mariculture facility of International Shellfish, Moss Landing, California. (R. Buchsbaum)

Young stages in the life history of a freshwater clam. (After Lefevre and Curtis.)

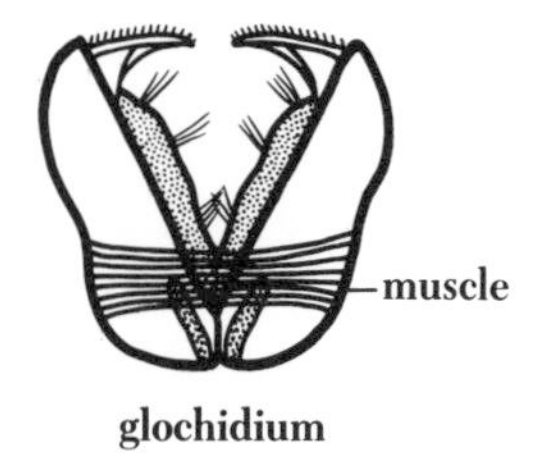

foot

juvenile clam

develop within the mother's gill chambers and are released, as veligers or as juveniles, through the excurrent opening.

In the freshwater clams on which our account is based, the zygotes develop within the gills to bivalved larvas called **glochidiums.** Huge numbers of glochidiums are produced and expelled into the water. To develop further they must, within a few days, become attached to the fins or gills of a fish. In some clams the passing of a fish stimulates the female to eject a cloud of glochidiums directly at the fish. In other species the freed glochidiums are stimulated to increased clapping of the shell valves when a fish is near. Most glochidiums fail to locate a fish and slowly sink to the bottom, where they die. But the successful ones clamp their valves tightly into the tissues of the fish host and live as parasites until they have developed into young clams. Then they drop from the fish and take up an independent adult life.

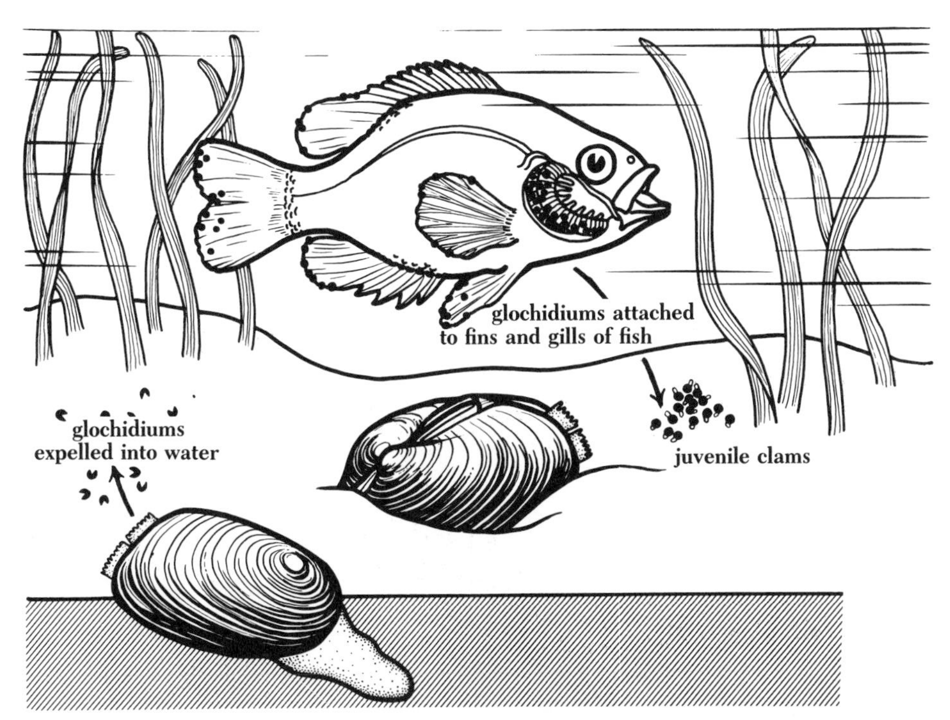

Life history of a freshwater clam. (Based on Lefevre and Curtis.)

Mussels live on rocky shores, not cemented to the rock like oysters, but attached by a bundle of strong proteinaceous threads (center of photo) that the foot secretes. Mussels are delicious but, like other bivalves, are subject to local collecting limits and may become highly poisonous when feeding on certain dinoflagellates. These toxic food organisms are sometimes so abundant in the water, especially from late spring to fall, that collecting is forbidden; such quarantines should be taken seriously, for the poisoning may be fatal to humans, though it does not seem to harm the mussels. *Mytilus edulis*. Roscoff, channel coast of France. (R. Buchsbaum)

Horse-hoof clam *(Hippopus)* lies with the ligament down. It has a reduced foot and never moves. Although it reaches more than 30 cm in length and is one of the largest of clams, it is dwarfed by some species of giant clam, *Tridacna*, which may be 150 cm long and weigh 250 kg or more. Australia. (O. Webb)

Siphons, long tubular extensions of the mantle with the incurrent and excurrent openings at their tips, allow clams to burrow down into the mud, beyond reach of many predators, while still maintaining their feeding and respiratory current. This marine bent-nosed clam uses its longer siphon to sweep the surface of the mud, sucking in food particles. *Macoma nasuta.* Monterey Bay, California. (R. Buchsbaum)

Scallops habitually rest on sand bottom but, when disturbed, they can swim about erratically by clapping the two shell valves. Along the mantle edges is a row of steely blue eyes, which show here as bright spots. The large muscle that closes the shell is the only part of a scallop that is eaten. *Aequipecten.* Woods Hole, Massachusetts. (R. Buchsbaum)

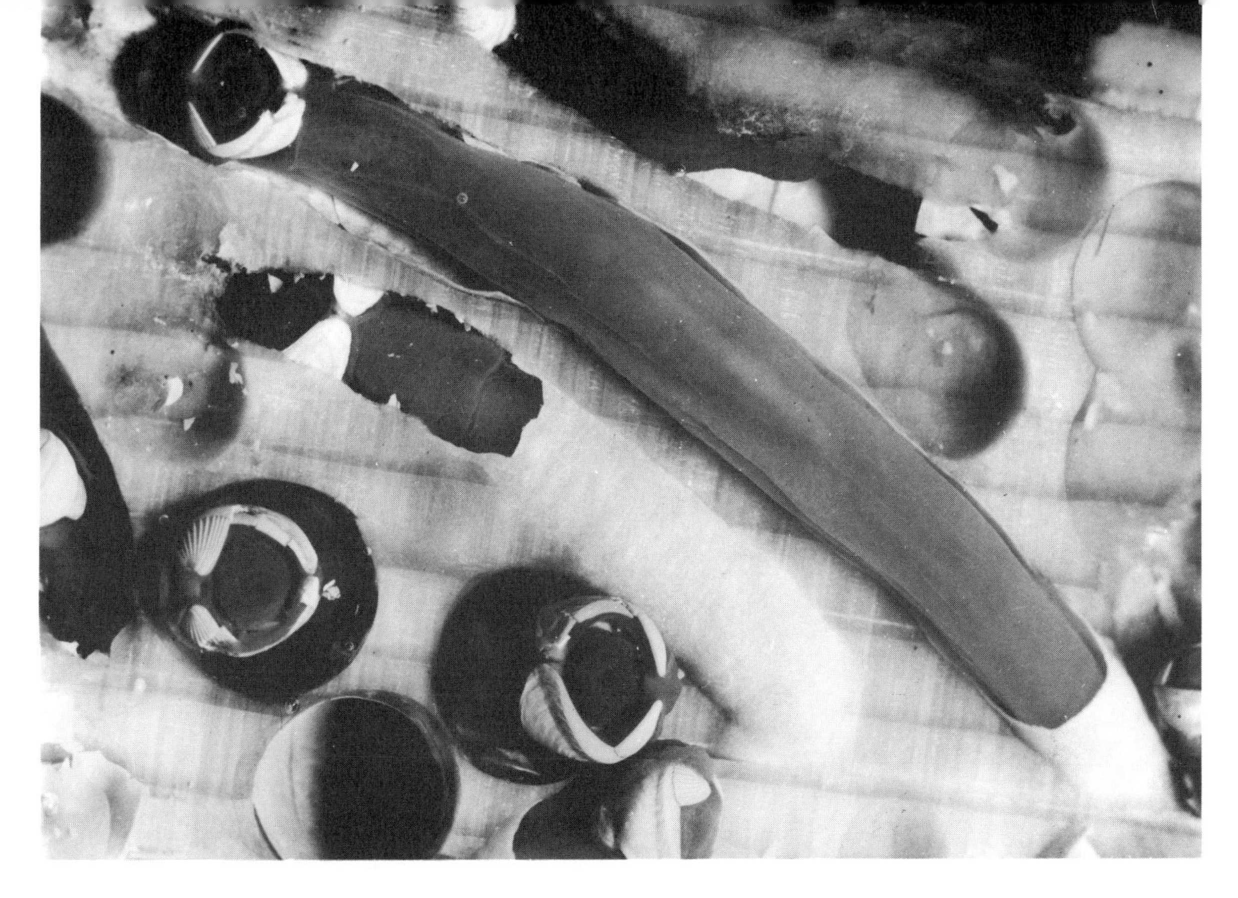

Wood-boring bivalves exposed in their burrows (about 5 mm in diameter) in a piece of wood that has been split open. Known as "shipworms," they are not worms at all, but greatly elongated clams. The two shell valves, which enclose only a small part of the anterior end of the body, have a ridged and roughened surface that rasps away the wood as they are rotated back and forth. The animals feed on wood particles, as well as on minute organisms brought in by the respiratory current. Every year shipworms do millions of dollars' worth of damage to wooden wharf pilings and ships. *Teredo*, preserved specimens. There are also rock-boring bivalves. (R. Buchsbaum)

Cephalopods

It would be hard to imagine any molluscs more different from the sedentary filter-feeding clams than are the nautiluses, squids, and octopuses—swift, predatory members of the class **Cephalopoda.** The name means "head-footed," for in these animals the foot, which is divided up into a number of "arms," is closely associated with the head. As in gastropods, the shell of cephalopods has been reduced in varying degrees. The nautiluses have a large, calcareous, external, coiled shell; the squids have only a thin uncalcified vestige of a shell embedded in the mantle; and the octopuses have virtually no shell at all.

A description of certain features of a **squid** will illustrate the ways in which it has adapted the molluscan body plan to an active, free-swimming life. Whereas most bilateral animals are elongated in an anteroposterior direction, the long axis of a squid is dorsoventral. Thus a squid usually swims with the ventral surface forward and the much elongated dorsal surface trailing; the upper surface is structurally dorsal and anterior, and the lower surface is structurally dorsal and posterior. To compare the body

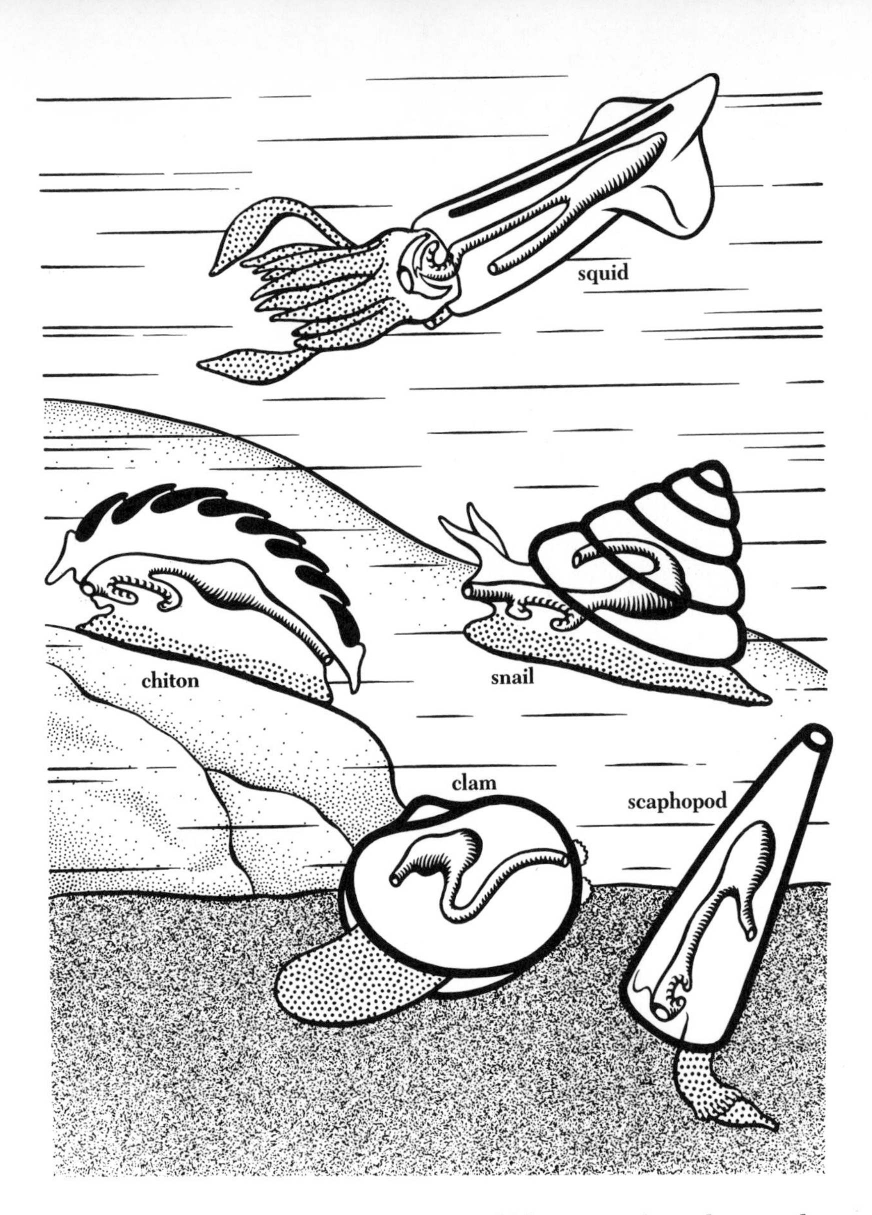

The molluscan body plan has been modified in the various groups. In this diagram, the digestive tract is shaded, the foot is stippled, and the shell is marked by a heavy black line. The squid is most readily compared to the other molluscs when oriented with the head and foot down.

with that of a snail or clam, one would have to place the squid so that the ventral region or foot was down and the tapered end of the body was up.

A squid relies for protection not on a heavy shell but chiefly on its ability to leave the scene of danger in a hurry. The **shell** is vestigial and is represented by a feather-shaped plate with a texture like that of cellophane, buried under the mantle of the upper surface. The **mantle** is thick and muscular and is the chief swimming organ. At the pointed rear end, its surface is extended into a pair of triangular folds, or *fins,* which act as stabilizers; they can be undulated to move the animal slowly and adjusted

A squid may turn darker or lighter in color by expanding or contracting the dark pigmented spots on the body; color changes occur when the animal is excited or when it swims over different substrates. Squids swim in schools and prey on crustaceans, fishes, and other animals. New Zealand. (W. Doak)

to steer. The mantle ends in a free edge, the collar, which surrounds the neck between the head and visceral mass. The **funnel,** a conical muscular tube (derived from part of the foot) projects beyond the collar on the underside of the head. When the mantle is relaxed, water enters the mantle cavity around the collar. When the mantle contracts, the collar is tightly sealed against the visceral mass and water is forced out through the funnel. When a squid is excited by the sight of prey, the mantle is contracted strongly, forcibly expelling a jet of water from the funnel. The tip of the funnel is bent backward, and the jet sends the squid quickly forward to seize its prey. If a squid is threatened, the funnel is directed forward, and the animal shoots backward like a torpedo; this is its usual behavior in escape. When attacked, a squid may emit a cloud of inky material from a special **ink sac** that opens into the funnel. The ink cloud serves either as a "smoke screen" or as a dark object that distracts the predator while the squid goes off in another direction. Discharges from the anus, gonad, and excretory organs also exit through the funnel.

Besides the funnel, the ten **sucker-bearing arms** that surround the mouth are derived from the foot of a squid. When the animal is swimming, the arms are pressed together and aid in

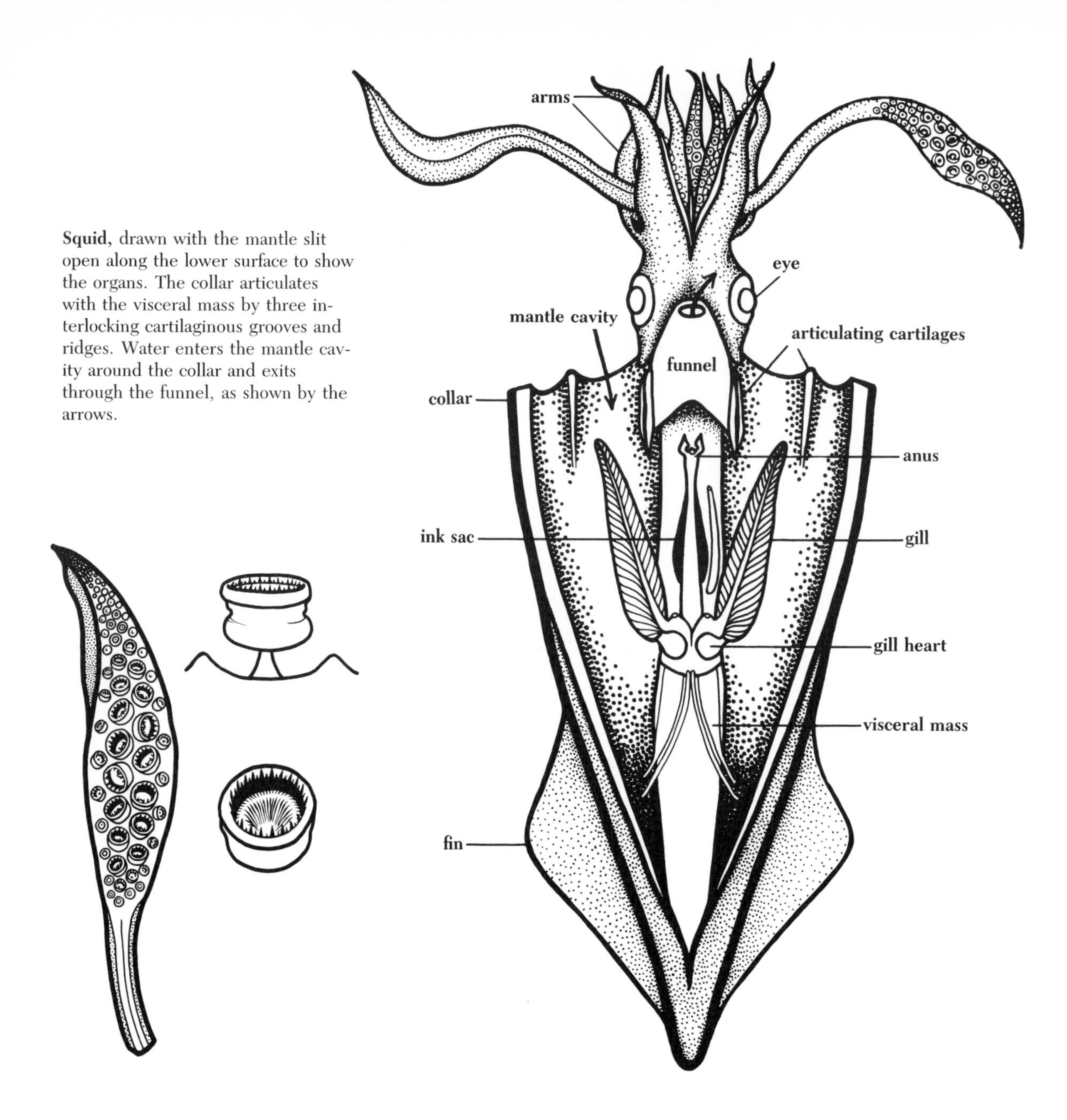

Squid, drawn with the mantle slit open along the lower surface to show the organs. The collar articulates with the visceral mass by three interlocking cartilaginous grooves and ridges. Water enters the mantle cavity around the collar and exits through the funnel, as shown by the arrows.

Suckers of a squid grip the prey. *Left:* Tip of one of the two longer arms. *Right:* each sucker is attached by a muscular stalk and lined by a toothed, horny ring.

steering. Two of the arms, longer and more slender than the rest, can be extended forward to seize the prey and draw it toward the mouth. There it is held firmly by the other arms, while two strong hard **jaws** in the mouth pierce the prey, biting out large pieces, which are then swallowed so rapidly that the small **radula** probably plays a minor role.

Squid for sale hangs in a small restaurant in the Sunday market in Bangkok, Thailand. This half-meter squid will probably be sliced up and cooked in soup. (R. Buchsbaum)

The active life of squids is made possible by much more elaborate **respiratory and circulatory systems** than those that serve the sedentary clams. On the other hand, squids require well-aerated water and could not survive in the poorly oxygenated habitats where many of their bivalved relatives thrive. A squid expends considerable energy in constant contractions and expansions of the muscular mantle that provide a vigorous circulation of water through the mantle cavity, in which lie two large gills. Unlike the open circulatory systems of other molluscs, in which the blood flows slowly and irregularly through unlined blood spaces, the circulatory system of squids is *closed*. The blood flows within discrete tubular vessels that are lined throughout with an epithelium. The tissues are permeated with networks of small, thin-walled vessels, the **capillaries,** through which gaseous exchanges take place rapidly. Respiratory exchange is facilitated by special features of the blood chemistry. In addition, there are separate pumping mechanisms for blood going through the gills and that going out to the other organs. Deoxygenated blood enters two **gill hearts,** each of which pumps blood through one gill. This gives the blood a fresh impetus, so that it passes through the gills at higher speed and pressure. Freshly oxygenated blood from the gills enter a single **systemic heart,** from which it is pumped out again to the various organs.

The **nervous system** of a squid is highly complex and centralized—in sharp contrast with that of the slow-moving clams. The several pairs of ganglia, which in clams and other molluscs are spread out over the body, are concentrated in squids and fused into a large **brain** that encircles the esophagus and lies between the eyes. This permits the different centers of nervous control to communicate rapidly and in a highly integrated fashion. Besides an olfactory organ and a pair of statocysts, the squid has two large **image-forming eyes.** They are remarkably like the eyes of humans and other vertebrates in construction but are developed in quite a different way. When two similar structures having a similar function appear in two distantly related groups, so that there is no possibility of a common ancestor that could have possessed such a structure, then those structures must have evolved independently. Thus the similarities in the eyes of squids and the eyes of humans are said to have arisen by **convergent evolution.**

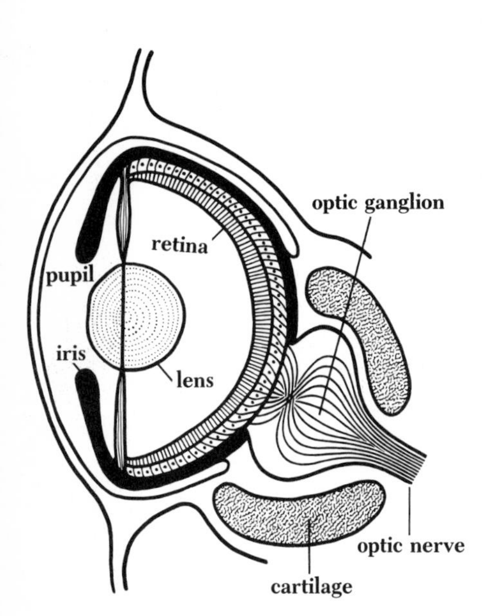

Eye of a squid, like that of a human or other vertebrate, is a "camera eye," so called because it is built on the same principle as a camera, which consists of a dark chamber to which light is admitted only through a small opening in the diaphragm; in the eye, the *pupil* opens through the *iris*. In both camera and eye, the light passes through a *lens* that focuses it on a light-sensitive film or, in the eye, *retina*.

Squids and fishes, with their highly developed eyes, also have light-producing organs that are the most complex known. Some of these have—besides the cells that produce the light—lens tissue, reflector cells, a pigment layer, a colored filter, and a fold of skin that can act like a blackout curtain or be lifted to let the light shine through. The occurrence of such organs in the two groups is another example of convergent evolution. It is not surprising, for **bioluminescence,** or the production of light by living organisms, is a widespread phenomenon found in some members of about half the phyla of animals.

Bioluminescence is rare in freshwater but common among marine organisms, including bacteria (the light of some of the luminescent squids and fishes is actually produced by symbiotic bacteria that live in the light organs). Protozoans (dinoflagellates) are responsible for most of the bright, diffuse displays of light in the ocean. Bioluminescent marine invertebrates include many cnidarians and comb jellies, scattered molluscs (besides the

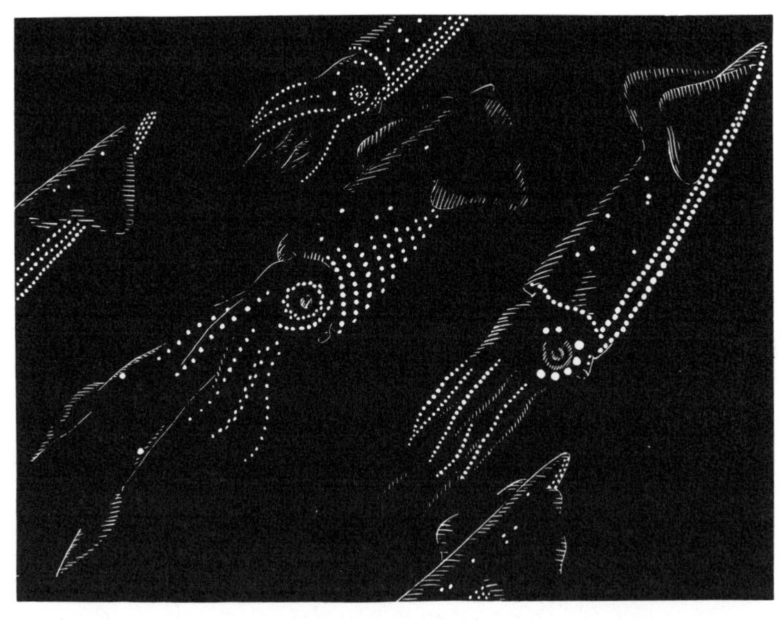

Luminescent squids are mostly deep-sea forms that live in perpetual darkness. (Based on C. Chun.)

squids) and segmented worms, a number of shrimps and other crustaceans, a few echinoderms and acorn worms, and some pelagic tunicates. On land there are luminescent centipedes and millipedes, and the familiar fireflies (which are beetles), as well as large luminescent mushrooms and smaller fungi that sometimes cause decaying wood to glow.

The adaptive value of bioluminescence certainly varies from organism to organism, and is clear in only a few cases. In squids, the light may serve to startle or confuse predators, to attract prey, or to help keep schools of the animals together. In certain marine worms and in fireflies, it is related to mating, apparently helping to identify individuals as to species and sex, and to attract mates. It is more difficult to understand luminescence by segmented worms that live all their adult lives buried and in opaque tubes, or by rock-boring clams.

The chemical basis of bioluminescence has proved easier to study. Among the first important steps was the discovery that luminescence in many organisms requires oxygen and that it involves the interaction of two different substances that can be extracted from the luminous tissues. One of these substances, a relatively stable compound, was called *luciferin* ("light-bearing"). The other substance, an enzyme easily destroyed by heat, was named *luciferase*. If separate extracts of luciferin and luciferase are mixed together in a test tube in the presence of oxygen, light is produced. The mixture continues to glow until all of the luciferin has been oxidized; then, if more luciferin is added, light production is renewed. Since most luminescent animals do not produce light continuously, but are able to do so promptly if disturbed, they must somehow be able to control the mixing of their light-generating chemicals. The luciferin and luciferase of a few organisms have been subjected to detailed chemical analysis and even synthesized in the laboratory. The substances differ in different groups of animals, although the luciferin and the luciferase of closely related species may cross-react to produce light. Active cross-reactions between the luciferin and luciferase of certain fishes and crustaceans (ostracods) have led researchers to suspect that the fish luciferin is acquired by eating the crustaceans. Some animals have modifications of the luciferin-luciferase system, and others appear to have rather different chemical mechanisms for making light.

Giant squid *(Architeuthis).* This one, stranded at Ranheim, Norway, in 1954, was 9.24 m (30 ft) in total length—still relatively small compared to the largest documented specimens, which have measured up to 18.3 m in total length (5.2 m in mantle length) and been estimated to weigh 1,000 kg or more. Such giants face few predators except sperm whales. (E. Sivertsen)

Giant squid, the largest of all invertebrates, is responsible for many of the sea-monster stories. Such stories are often embroidered to include an attack on the ship by the monster. But in one of the authentic cases—a squid encountered by the French steamship *Alecton* in the Atlantic in 1861—the unfortunate monster was vigorously and enthusiastically attacked by the ship's crew. An account in French by the ship's commander was read before the Academy of Sciences in Paris: "After several encounters which permitted only of its being struck by ten or so musket-balls, I succeeded in coming alongside it, close enough to throw a harpoon as well as a noose . . . a violent movement of the animal disengaged the harpoon; the part of the tail where the rope was wrapped around broke, and we brought on board only a fragment. . . . Officers and seamen begged me to have a boat lowered and to snare the animal again, . . . but I feared that in this close encounter the monster might throw its long arms, equipped with suckers, over the sides of the boat, capsize it, and perhaps choke several seamen. . . . I did not think that I ought to expose the lives of my men, . . . and despite the fever of excitement that accompanies such a chase, I had to abandon the mutilated animal. . . ." (Adapted from an old engraving in Figuier, based on a drawing made by one of the officers of the *Alecton*.)

Another example of convergence between squids and vertebrates is the development of **internal cartilaginous supports.** A squid has a number of internal cartilages that support muscles or form interlocking surfaces, but most interesting in this connection is the large cartilage that encloses and protects the brain, reminding us of the vertebrate brain case. Squids, perhaps more than any other invertebrates, have evolved along the same lines followed by the fast-moving predatory aquatic vertebrates: large size, streamlined shape, rapid locomotion, internal skeletal supports, elaborate respiratory and circulatory systems, large brain, and highly developed sense organs.

Cuttles, or cuttlefishes, resemble squids in structure and habits. The shell, a calcareous plate embedded in the fleshy mantle, is the "cuttlebone" given to cage birds as a source of calcium. The contents of the ink sac provide a rich brown pigment, sepia, once widely used by artists. *Sepia officinalis.* Monte Carlo. (R. Barba)

An octopus *(Octopus vulgaris)* has virtually no shell. Unlike the ten-armed squids, octopuses have only eight arms and are mostly bottom-living cephalopods. They move by pulling themselves over the rocks with their arms or swim short distances by forcibly expelling water from the funnel. The entrance to an octopus home, in a sheltered place among rocks, may sometimes be identified by the accumulated litter of crab shells, which the octopus breaks open by a pair of horny jaws and the radula. As far as is known, octopuses are the most intelligent invertebrates; their ability to learn and to make subtle visual discriminations has been intensively studied. Giant octopuses of the U.S. northwest coast may weigh 50 kg (110 lb). (R. Buchsbaum)

Octopus eggs are guarded and are kept clean and aerated by the mother. The eggs, like those of other cephalopods, contain a large amount of yolk and hatch directly into juveniles; there are no free-swimming ciliated larvas. These eggs were laid in an empty cockle shell by a small octopus *(Octopus joubini)* in Panacea, Florida. (R. Buchsbaum)

Nautilus. A few shy species of *Nautilus* in the tropical Indo-Pacific are the only living cephalopods with an external shell, although shelled cephalopods were dominant members of ancient seas, and thousands of fossil species are known (see chapter 23). Nautiluses appear to be active at night, using their 90 or so short tentacles to seize crabs and fishes; they rest during the day in coral crevices. Philippines. (N. Haven)

Preserved nautilus in section shows the animal in the largest, outermost chamber of the coiled, many-chambered shell. The chambers are filled with variable amounts of gas and liquid, permitting the nautilus to adjust its buoyancy when swimming at a range of depths. New chambers are sealed off successively as the animal grows. (American Museum of Natural History)

By comparing the squids and clams with snails and other molluscs, we see that the fundamental body plan of an animal may become so modified in adaptation to a particular way of life that many of its structures reflect the kind of life it leads more than its relationship to its more typical relatives.

Chapter 16

Segmented Worms

The members of more than half the phyla in the animal kingdom are worms—a name bestowed on almost any invertebrate that is much longer than it is wide. But for most people, and especially those fond of fishing or gardening, the worms that first come to mind are **earthworms.** As earthworms are adapted to living on so abundant and widely distributed a food as the decaying organic matter of the soil, it is not surprising that they occur in countless numbers in moist soils all over the world. Ever since Darwin made their activities the object of a careful study and concluded that "it may be doubted if there are any other animals which have played such an important part in the history of the world as these lowly organized creatures," it has been recognized that the work of earthworms in enriching and cultivating the soil is of tremendous importance to agriculture.

The most noticeable feature of an earthworm is the ringing of the body, which is not merely external but involves nearly all of the internal structures, and the name of the phylum to which earthworms belong is ANNELIDA ("ringed"). The ringed condition

Opposite; **An earthworm,** as its name suggests, lives underground. It stays in its burrow and avoids full sun, unless rudely dug up as this one was for the making of its portrait. It is beginning to dig in again at the pointed anterior end. The ringing of the body is conspicuous except in the glandular region (clitellum) which secretes the egg capsules. (R. Buchsbaum)

Repetition of parts is a major feature of the annelid body plan.

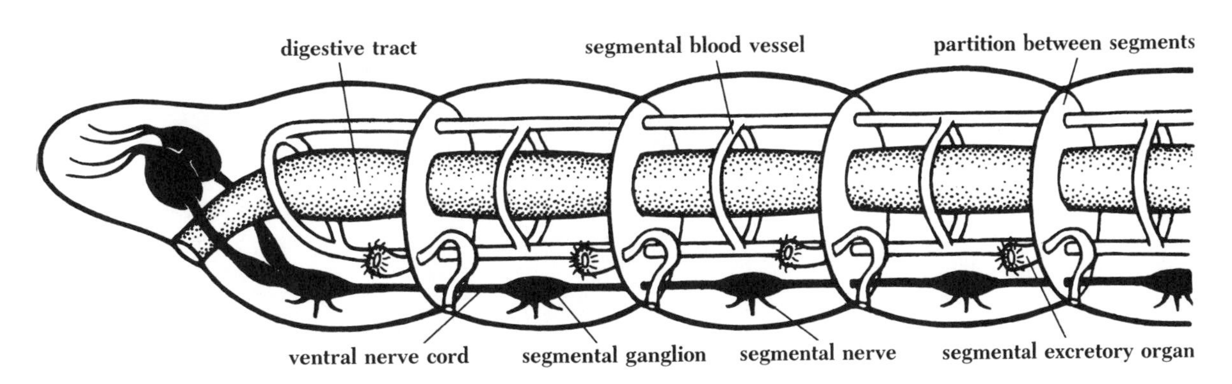

is known as **segmentation,** and each ring is called a **segment.** To describe one segment of an earthworm is to describe nearly the whole worm.

As in other burrowing animals, the body is streamlined and has no prominent sense organs on the head or any projecting appendages on the body that would interfere with easy passage through the soil. The outer covering of the body is a thin, flexible, collagenous **cuticle,** which is secreted by the underlying epidermis. Beneath the epidermis is a layer of **circular muscles,** then a layer of **longitudinal muscles.** Together these various layers constitute a definite **body wall.** The muscles run the length of the worm but are divided up by **partitions** between the segments.

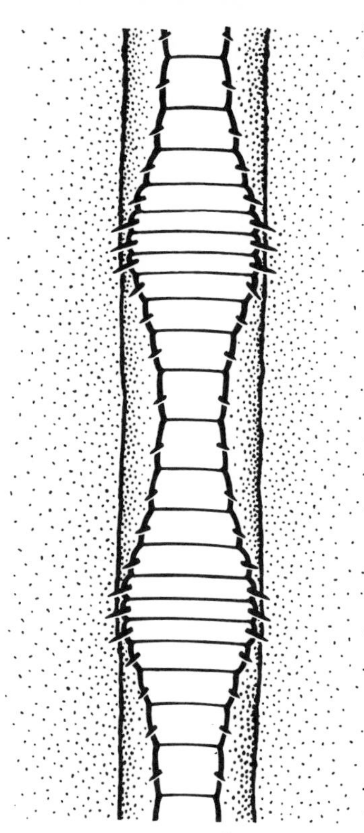

Worm moving upward in its burrow extends or retracts the bristles of each segment as muscular waves pass down the body from head to tail.

On each segment are 4 pairs of **bristles,** which protrude from four small sacs in the body wall and are extended or retracted by special muscles. The bristles are used to anchor the worm in its burrow, and the firmness of their grip has frustrated many hungry birds and eager bait collectors. But the main function of the bristles is assisting in **locomotion.** As the worm works its way along, there pass down the body successive waves of thickening and thinning (7 to 10 per minute), as the longitudinal and circular muscles contract alternately. At each place where the body bulges out at a given moment, the bristles are extended and grip the burrow walls. As the wave passes on, the extended bristles are retracted, and the segments of the bulge thin out and are moved ahead, about 2 to 3 centimeters each time.

An essential component of this system is the mesoderm-lined **coelom,** which in earthworms is a large fluid-filled body cavity between the body wall and the digestive tract. The coelomic

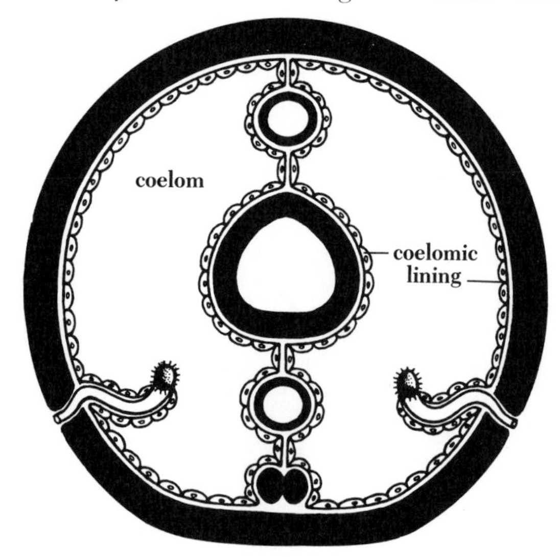

Diagrammatic cross-section of an annelid showing the **coelom and its mesodermal lining.** The coelomic lining covers the digestive tract and excretory organs; and it forms supporting mesenteries in which the major blood vessels run, above and below the digestive tract.

fluid acts as a *hydrostatic skeleton* around which the muscles contract. The partitions between the segments prevent coelomic fluid from merely sloshing from one part of the worm to another as the muscular waves pass. Instead the force of the contracting muscles is effectively localized and transmitted through hydrostatic pressure to the walls of each segment.

> It is the combination of the circular and longitudinal muscle layers, fluid-filled coelom, and segments divided by partitions that enables earthworms to sustain powerful burrowing movements. Annelids that do not burrow actively often show reduction or loss of the partitions between segments and of the coelomic cavity.

Another important advantage of the coelom is that it separates the body wall from the digestive tract, allowing the muscles of both to contract without interfering with each other. The coelomic fluid bathes all of the internal organs and thus supplements the role of the circulatory system (although it has no direct connection with that system). The coelom also plays a role in excretion and reproduction.

The coelom arises by the formation of a pair of spaces in the embryonic mesoderm of each segment of the body. These spaces enlarge, and are lined by a thin layer of mesoderm, giving rise to a paired series of coelomic sacs. The inner walls of these sacs envelop the digestive tube; and where they meet in the midline, they form a double layer of coelomic lining, the **mesentery,** which supports the gut above and below. In earthworms and in many other annelids, part of the ventral mesentery disappears during embryonic development, and right and left coelomic spaces are confluent ventrally. The anterior and posterior walls of the coelomic sacs form the partitions between the segments.

> The presence of a coelom is considered of such importance that animals are often divided into two large groups, those with a coelom and those without, categories that correspond roughly to what is meant by "higher" and "lower" invertebrates. A space between the digestive tube and body wall occurs in many of the phyla we have studied already. But in such groups as the roundworms and rotifers it has no definite mesodermal lining and therefore is not considered to be a coelom, although it serves many of the same functions. In bryozoans, brachiopods, and arrow worms the coelom is a large body cavity. In molluscs—and in the arthropods, to be described in the following chapters—it is limited to small spaces associated with some of the organs.
>
> A coelom occurs in all vertebrates, but it develops in a different way from that of annelids and probably evolved independently. In humans the coelom is divided into an abdominal cavity, a cavity surrounding the heart, and two cavities that contain

the lungs. The coelomic lining is called the *peritoneum;* and when it becomes infected, as from a ruptured appendix, the serious condition that results is known as peritonitis.

Giant earthworm from Colombia dwarfs a moderate-sized earthworm like *Lumbricus terrestris,* which measures 30 cm and has about 180 segments compared with 500 to 600 in some terrestrial giants. There are only 7 segments in some of the minute aquatic oligochetes. (R. Buchsbaum)

In loose ground earthworms may burrow simply by pushing the dirt away on all sides, but in compact earth the soil must actually be swallowed. Earthworms spend most of their time swallowing soil underground and depositing it on the surface around the mouths of their burrows in the form of the *castings* familiar to everyone. The swallowing of soil is not only a means of digging burrows, but also a mode of **feeding.** The soil passed

through the digestive tract contains organic materials of various kinds: seeds, decaying plants, the eggs or larvas of animals, and the live or dead bodies of small animals. These organic components provide nourishment, while the main bulk of the soil passes through. When leaves are abundant on the surface, the worms drag them into the burrows, and few castings are thrown up. When few leaves are taken in as food, the amount of castings increases.

The effects of the worms on the soil are many. The earth of the castings is exposed to the air, and the burrows themselves permit the penetration of air into the soil, improve drainage, and make easier the downward growth of roots. The thorough grinding of the soil in the gizzard of the worm and the sifting out of all stones bigger than those that can be swallowed is the most effective kind of soil cultivation. Leaves pulled into the ground by earthworms are only partially digested, and their remains are thoroughly mixed with the castings. This organic matter is then further broken down by microorganisms of the soil, releasing nutrients in a form available for absorption by plants. In this way earthworms have helped to produce the fertile humus that covers the land everywhere except in dry or otherwise unfavorable regions.

Thus when earthworms are present in the soil, agricultural productivity is generally higher, and greater crop yields have been achieved by introducing earthworms into soils where they were absent. Unfortunately, agriculture does not similarly benefit earthworms, which usually decrease in abundance when grassland is plowed and put under cultivation, and especially when heavy doses of some pesticides are used. Decline in earthworm populations is less when as much plant material as possible is left on the land.

The quantity of earth brought up from below and deposited on the surface in temperate regions has been estimated to be as much as a centimeter per year, if spread out uniformly, or 90 metric tons per hectare (40 tons per acre). The layers of soil are thus thoroughly mixed, seeds are covered and so enabled to germinate, and stones and other objects on the surface become buried. In this way ancient buildings have been covered and so preserved.

The **digestive system,** which plays such an important part in the good works of earthworms, is the only system uninfluenced by the segmental body plan. Stretching uninterrupted from the mouth at the head end to the anus at the posterior tip, it is differentiated into a number of regions, each with a special function. Food enters the mouth, is swallowed by the action of the muscular pharynx, and then passes through the narrow esophagus, which bears on each side the **calciferous glands.** These

Digestive system of an earthworm has several different regions at the anterior end, followed by a long uniform intestine. Two pairs of calciferous glands are in segments 11 and 12.

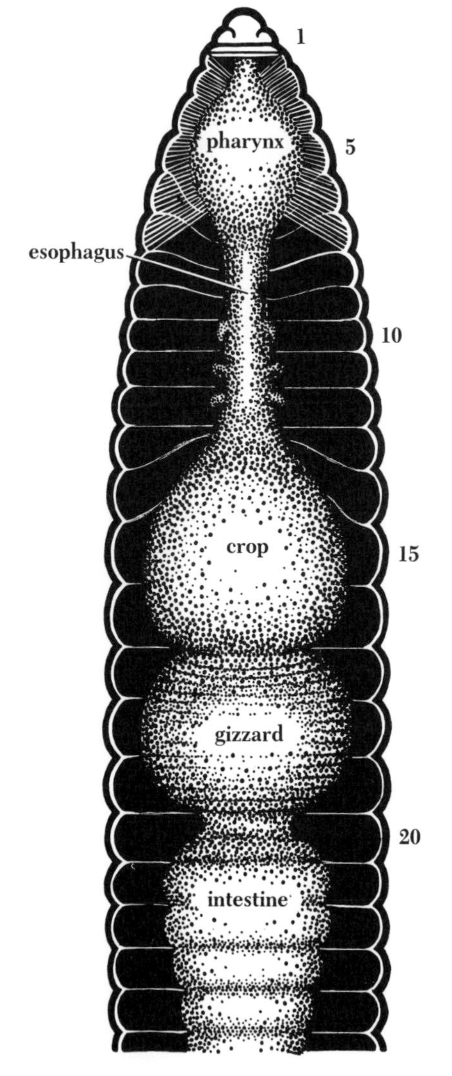

glands excrete calcium carbonate into the esophagus, ridding the worm of both excess calcium obtained from the food and carbon dioxide produced in metabolism. If the glands are removed, and the worms placed under conditions of high carbon dioxide (such as may well be normal in stuffy earthworm burrows), the body fluids become acid and high in calcium.

The esophagus leads into a large thin-walled sac, the **crop,** which apparently serves only for temporary storage, since the food undergoes little change and does not remain there very long. Behind the crop is another sac, the **gizzard,** with heavy muscular walls that (aided by mineral particles and minute stones swallowed by the worm) grind the food thoroughly. From the gizzard the food passes through the long uniform intestine, where the bulk of digestion occurs, to the anus. The roof of the intestine dips downward as a fold that increases the digestive, absorptive surface. The contractions of layered muscles in the intestinal wall produce **peristalsis,** a succession of rhythmic waves of constriction that push the food along. The intestinal wall is also abundantly supplied with blood vessels, and digested food is distributed to the rest of the body by the circulatory system.

The **circulatory system** is a *closed* system with extensively branched vessels. A median contractile **dorsal vessel,** which lies above the digestive tube and accompanies it from one end of the body to the other, is the main collecting vessel. In it the blood flows forward, propelled by rhythmic peristaltic waves. A median noncontractile **ventral vessel,** suspended beneath the digestive

Anterior circulatory system of an earthworm. Blood flows forward in the dorsal vessel and downward through the five pairs of hearts into the ventral vessel. In front of the hearts, blood flows forward in both dorsal and ventral vessels to the head, then backward in a subneural vessel. Behind the hearts, blood flows backward in the ventral vessel and out into the segmental branches.

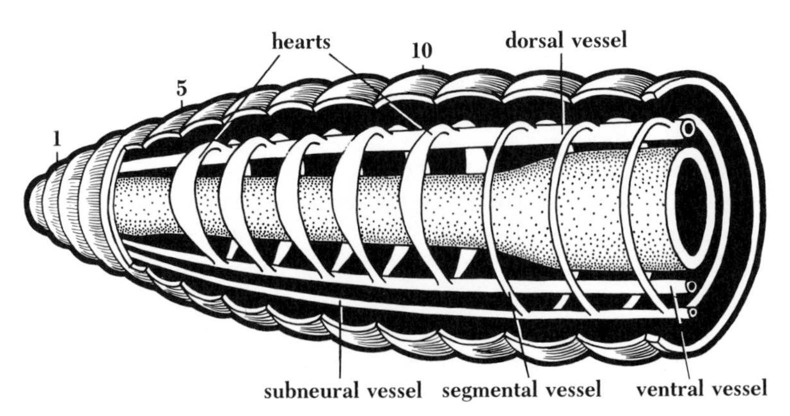

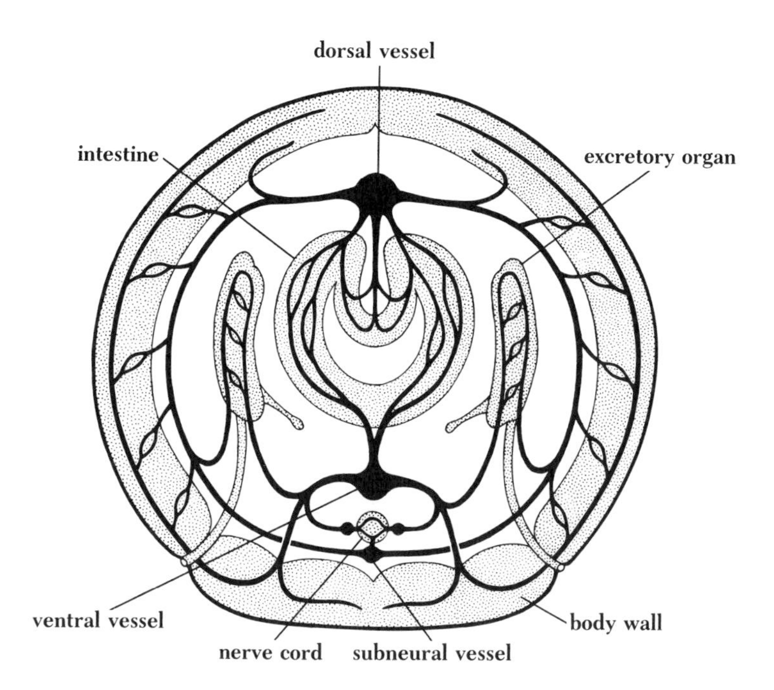

The main segmental blood vessels of an earthworm, displayed in cross-sectional view.

tube in the ventral mesentery, is the main distributing vessel. In it the blood flows backward and out into segmental branches that supply the various organs. In the region of the esophagus (segments 7 to 11) the blood flows from the dorsal vessel to the ventral vessel through five pairs of enlarged muscular transverse vessels, the **hearts,** which pump blood through the ventral vessel. In almost every segment the blood flows from the ventral to the dorsal vessel through **capillary beds** of the body wall, digestive tract, excretory organs, and nerve cord. The walls of the capillaries are composed of only a single layer of flattened epithelial cells and permit a rapid exchange of dissolved food substances, nitrogenous wastes, and respiratory gases. Their extensive ramification ensures that substances are delivered and picked up almost "at the door" of every cell and do not have to move long distances by the slow process of diffusion. Most of the oxygen carried by the blood is simply dissolved in the blood plasma, but about 40% is bound to a blood pigment, hemoglobin. As mentioned before, the coelomic fluid also helps to circulate substances.

Earthworms are terrestrial animals, but they have not really solved the problems of land life; they have merely evaded them by restricting their activities to burrowing in damp soil and by emerging only at night, when the evaporating power of the air is low. The earthworm caught by the early bird is no early worm but one that stayed out too late. Earthworms that remain above ground after sunrise risk not only daytime predators but severe damage by sunlight. Most earthworms retreat into their burrows by day, burrowing deep underground during hot, dry weather. In cold weather they plug the opening of the burrow and again retreat into its deepest part, usually enlarged into a chamber where one or several worms, rolled up together into a ball, pass the winter.

> The sun's ultraviolet rays are potentially damaging to all animals of the land surface and even those in shallow waters. Animals habitually exposed to the sun often develop skin coverings or pigments that provide at least partial protection, for example, the tanning of human skin in sunlight. But earthworms, which are rarely exposed to the sun, have a red pigment (protoporphyrin) in the body wall that happens to sensitize the tissues to damaging ultraviolet rays. Brief exposure to strong sunlight causes paralysis in some worms, and longer exposure is fatal. This may explain the death of many of the earthworms seen lying in shallow puddles after a rain. They have not been drowned by the water, as many people suppose, for earthworms can live completely submerged in water. However, during a rain the water that fills their burrows has filtered down through the soil and therefore contains very little oxygen. This forces some of the worms to come to the surface, where they are injured by light and after a time can scarcely crawl.

Animals well adapted for land life have a heavy impermeable skin or cuticular covering that prevents excessive drying, but it also prevents respiratory exchange. In such animals oxygen reaches the internal tissues by means of special respiratory devices, such as lungs. Earthworms, on the other hand, carry out **respiratory exchange** in the same way as their aquatic ancestors. That is why they can live for months completely submerged in water, yet will die if dried for a time. The cuticle of an earthworm is thin and must be kept moist so that respiratory exchange can occur by diffusion through the general body surface, which is underlain by capillary networks. Moistening of the surface is accomplished by mucus glands in the epidermis and also by the coelomic fluid that issues from *dorsal pores* located in the middorsal line in the grooves between segments.

The **excretory system** is segmentally arranged, with a pair of excretory organs in nearly every segment. Each organ consists essentially of a tube that opens at one end by a ciliated funnel into the coelom and at the other end by a ventral pore to the exterior. As coelomic fluid passes through the tube, useful substances are reabsorbed, while wastes extracted from the blood are secreted into it. The resulting urine contains ammonia, urea, and other nitrogenous wastes.

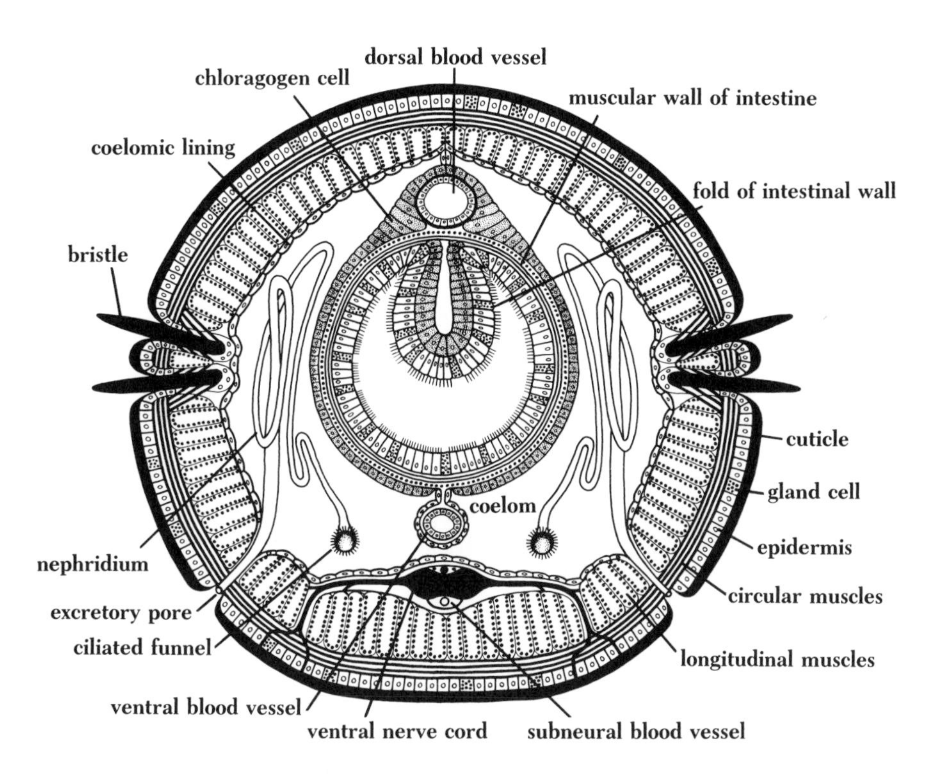

Cross-section of an earthworm. Each nephridium really occupies two segments: the ciliated funnel opens into the segment anterior to that which contains the body of the tube and the external pore. Only two of the four pairs of bristles are shown.

The excretory organs of annelids are called **nephridia.** Each nephridium really occupies two segments, because the ciliated funnel opens into the segment anterior to that which contains the body of the tube and the external pore. Although earthworms can regulate their water and salt balance to a limited extent, they are more remarkable for their ability to tolerate extreme changes in water content. An earthworm taken from moist soil has a water content of about 85% by weight. Submerged in water, the worm will gain about 15% of its original weight. Placed in dry air, it will survive a loss of 60% or more of its weight.

The nephridia are not the only means of excretion in earthworms. The calciferous glands of the esophagus have already been mentioned, and about half the total nitrogen excreted is in the mucus secreted by the epidermis. In addition, the coelomic lining surrounding the intestine and the main blood vessels is modified into special **chloragogen cells.** These cells are thought to participate in nitrogen metabolism and excretion, as well as in metabolism and storage of starches and fats. Pigmented granules accumulated by the chloragogen cells are released into the coelom, where they are engulfed by ameboid cells, along with foreign particles such as bacteria. The ameboid cells deposit their burden of solid waste in the body wall and in "brown bodies" in the coelom.

Anterior end of an earthworm, showing the **nervous system.** The digestive system is drawn as if transparent. (Modified after W. N. Hess.)

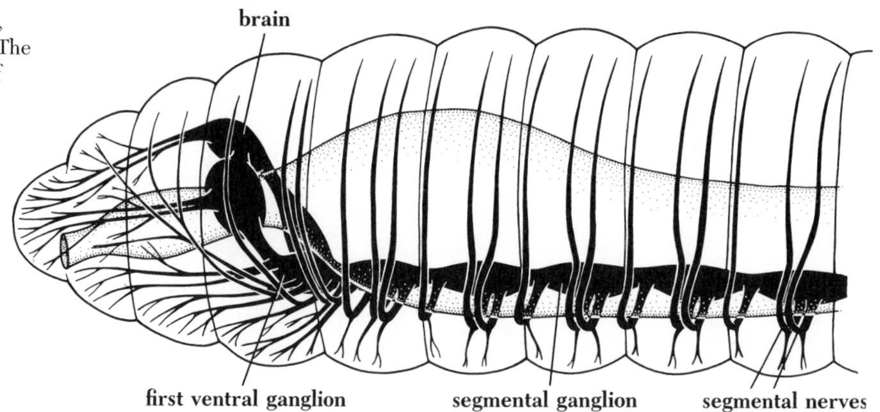

Like the circulatory and excretory systems, the **central nervous system** of an earthworm is neatly segmental. The **brain,** a bilobed dorsal ganglion lying above the pharynx, connects by two nerves with the **first ventral ganglion,** below the pharynx. These two ganglia send nerves to the sensitive anterior segments and are considered to be the "higher centers." The brain appears to direct the movements of the body in response to sensations of light and touch. And it has important inhibitory functions, for if it is removed the worms move continuously, but otherwise their behavior is affected little. Removal of the first ventral ganglion has a more obvious effect: the worms no longer eat and cannot burrow in normal fashion. From the first ventral ganglion a double **nerve cord** runs to the posterior end of the body, enlarging in each segment to a double **segmental ganglion** from which three pairs of **segmental nerves** branch off in each segment. Each ganglion receives impulses from **sensory cells** in the skin and sends impulses that control local muscles. The ganglia coordinate the impulses so that the longitudinal muscles relax while the circular muscles contract, or the opposite. Without this arrangement the two sets of muscles might only counteract each other's activities, and no movement would result.

The smooth muscular waves that pass down the body in ordinary creeping movements are not controlled by the large anterior ganglia, for almost any sizable piece of an earthworm will creep along as well as the whole worm. The coordination is achieved partly through impulses relayed from one segment to another by nerve cells in the ventral cord and partly through mechanical stimulation of successive segments. Evidence for this comes from experiments in which coordinated movements were found to continue after the nerve cord or body wall was cut, but not if both were cut.

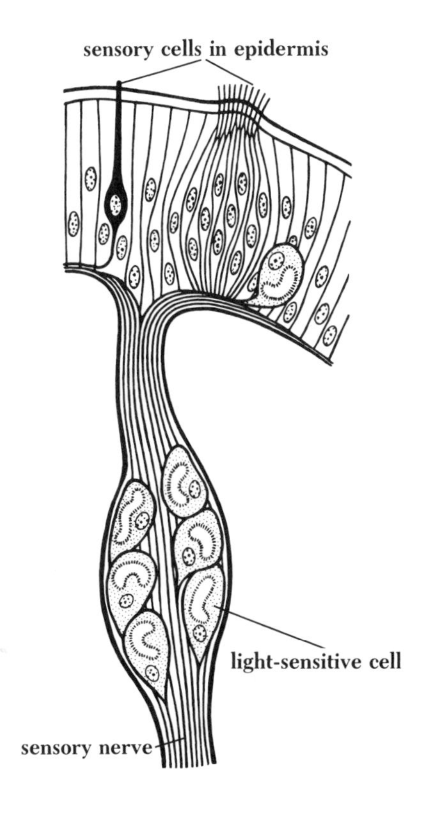

Sensory cells of an earthworm *(Lumbricus terrestris)*. The lack of prominent sensory organs on the head does not mean that earthworms are insensitive to stimuli. *Light-sensitive cells* containing clear vesicles occur in and below the epidermis. They are most abundant in those regions most frequently exposed to light, the anterior and posterior ends, and are absent from the ventral surface. Groups of 35 to 45 tall slender cells, each with a hairlike process projecting through the cuticle, may be sensitive to *touch*, changes in *temperature*, or contact with various *chemicals*—stimuli to which earthworms are known to respond. *Taste cells* occur in and near the mouth, and the worms seem to show definite food preferences—neglecting cabbage if celery is also offered, and passing up celery if carrot leaves are available. The sense of *smell* is feeble; and the worms are unresponsive to *sound*. More important for a subterranean animal is the ability to detect *vibrations* transmitted through solid objects. To these, earthworms are extremely responsive. It is said that one way to collect earthworms is to drive a stake into the ground and then move it back and forth, setting up vibrations in the ground, which cause the worms to emerge from their burrows. There are no statocysts, and it has been suggested that the sense of *gravity* may depend on certain cells in the muscle layers that respond to stretch or tension. (Modified after W. N. Hess.)

Impulses in the small nerve fibers of the ventral cord travel slowly and are subject to delays in transmission from one fiber to the next. The speed of travel of the waves of thinning and thickening in the body of a creeping worm is only about 25 millimeters per second. However, earthworms are capable of speedier action. If the anterior end, extended from the burrow, receives some strong unfavorable stimulus, the longitudinal muscles of the whole body contract, and the worm disappears into its burrow almost instantly. Such a response requires very rapid nervous transmission, and we do find certain **giant fibers** in the ventral nerve cord that pass over long distances or even throughout the length of the cord. Impulses in these fibers are subject to no appreciable delay at the divisions between segments, and speed of transmission has been measured at up to 45 meters per second, among the highest recorded for invertebrates. Transmission in the motor nerves of mammals is not much faster (up to 120 meters per second), taking into account their higher temperature, but is achieved with fibers of much smaller diameter.

The **reproductive system** of earthworms is complex and differs in several respects from that of the majority of annelids, which are marine. In the first place, earthworms are *hermaphroditic*, a condition common in animals that encounter each other only infrequently and mostly by chance. If all individuals are simultaneously male and female, then when two animals do happen to meet, reciprocal exchange of sperms can occur and

The reproductive system of an earthworm is located in the anterior end, each organ in a particular segment. The male gametes are formed in two pairs of **testes,** located in segments 10 and 11, and each pair is enclosed within a **testis sac** (drawn as if transparent). The testis sacs communicate with **sperm sacs** in which the gametes develop to maturity. Mature sperms pass back into the testis sacs, into the **sperm funnels,** and through the **sperm ducts** to the two male genital openings on the ventral surface of segment 15. Two pairs of small sacs, the **sperm receptacles,** in segments 9 and 10, open through pores to the ventral surface; during mating these receive sperms from the other worm. Eggs are formed in a pair of **ovaries** in segment 13. As they become mature, they are shed from the ovaries into two **egg funnels** on the posterior face of segment 13. Each funnel leads into an **egg sac,** in which ripe eggs are stored. The **oviducts** open by two pores on the ventral surface of segment 14. This diagram is based on *Lumbricus terrestris;* in some other species, certain of the details are different.

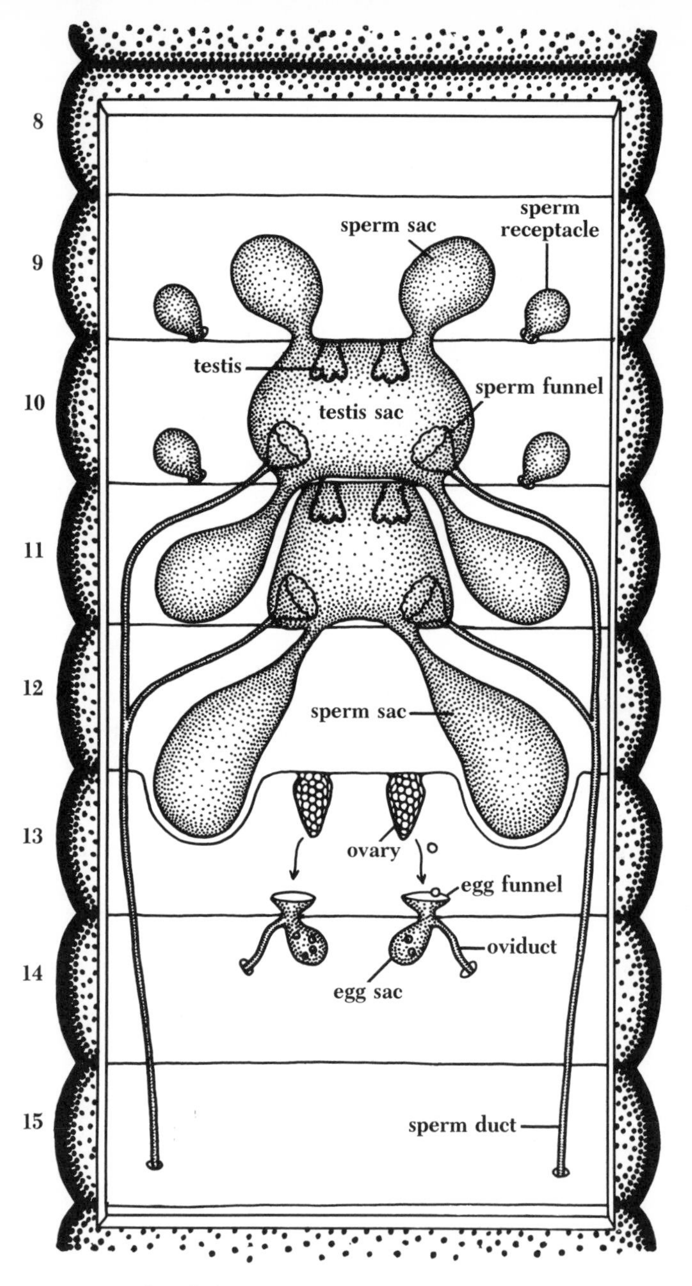

the most is made of the encounter. The complexity of the earthworm reproductive system results partly from arrangements to keep the eggs and sperms of each individual separate and prevent self-fertilization. In addition, earthworm reproduction is modified for *life on land,* where the gametes and the developing embryos must be protected against drying.

Mating earthworms appose their ventral surfaces and exchange sperms. The sperms received will fertilize the eggs at a later time. Though an earthworm has both male and female reproductive systems, it does not fertilize itself. This photo was made at night in a Chicago park. (L. Keinigsberg)

Mating is no simple process for earthworms. During the sexual season, spring to early fall, when the ground is wet following a rain, the worms may emerge at night and travel some distance over the surface before they mate. Where abundant, they merely protrude the anterior end and mate with a worm in an adjoining burrow. With their heads pointing in opposite directions, the two worms appose the ventral surfaces of their anterior ends. Mucus is secreted until each worm becomes enclosed in a mucus sheath that extends from the openings of the sperm receptacles (segments 9 and 10) to the posterior edge (segment 37) of the **clitellum,** a ring of thickened glandular epithelium that, during mating, is opposite the sperm receptacles of the partner. Sperms are expelled from the openings of the sperm ducts (segment 15) and are moved backward in longitudinal grooves (roofed over by

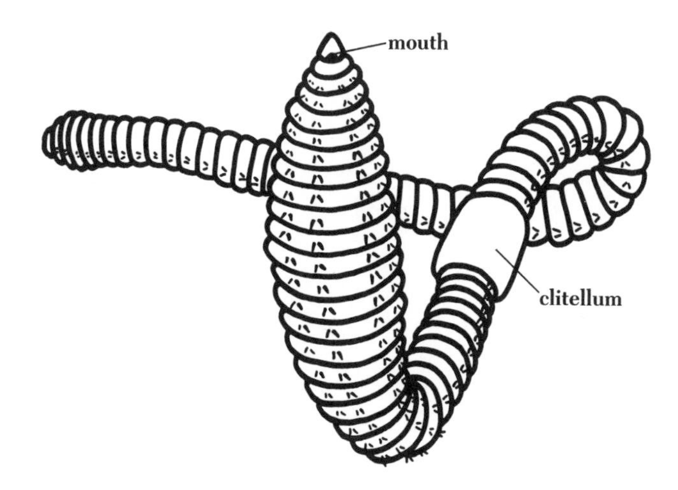

An ant's-eye view of an earthworm with the anterior end lifted to show the four **rows of bristles.** The thickening of the epidermis in the region of the **clitellum** obscures the bristles and external segmentation of that region, segments 31 or 32 to 37 in *Lumbricus terrestris.* The position of the clitellum is definite for each species of earthworm.

the mucus sheath) until they reach the region of the clitellum and are passed into the sperm receptacles of the other worm. Then the worms separate; egg laying and fertilization take place later.

At the start of **egg laying,** the gland cells of the clitellum produce a ring of secretion. The ring quickly becomes somewhat hardened, and the worm begins to wriggle backwards out of it. As the ring passes the openings of the oviducts (segment 14), it receives several ripe eggs; and then, as it passes the more anterior openings of the sperm receptacles (segments 9 and 10), it receives sperms that were deposited there previously by another worm during mating. Fertilization of the eggs takes place within the ring, which finally slips past the anterior tip of the worm and closes at both ends to form a sealed **egg capsule** (sometimes called a "cocoon"). Within the capsule, buried in the soil, the

Egg capsules of a giant earthworm from Colombia are of a size that befits the worms that deposit them below the soil surface. Capsules laid by *Lumbricus terrestris* are lemon-shaped and less than a centimeter long. (R. Buchsbaum)

zygotes develop directly into young worms, which finally escape and begin to face the hazards of the outside world.

Besides some of the physical dangers already discussed (especially drying), earthworms suffer from diverse parasites and predators. The animals that eat earthworms range from frogs, turtles, snakes, and a variety of birds, to moles, badgers, foxes, and (occasionally) people, notably the Aborigines of Australia and the Maoris of New Zealand. And how many people who would not think of eating an earthworm have nevertheless been guilty of involuntary earthworm slaughter with a sharp garden spade? Perhaps not as many as one might think, for earthworms can **regenerate** both head and tail, at least within limits. The extent of regeneration depends on the species, as well as on the position of the cut. So if you are lucky, that earthworm which has been on your conscience may now be a fully regenerated worm (or even two), obligingly improving the soil of your garden as it burrows through the earth.

Polychetes

The majority of annelids are not earthworms but marine worms of the class **Polychaeta,** named for their bunches of *many bristles,* which are borne on appendages called **parapods** ("side

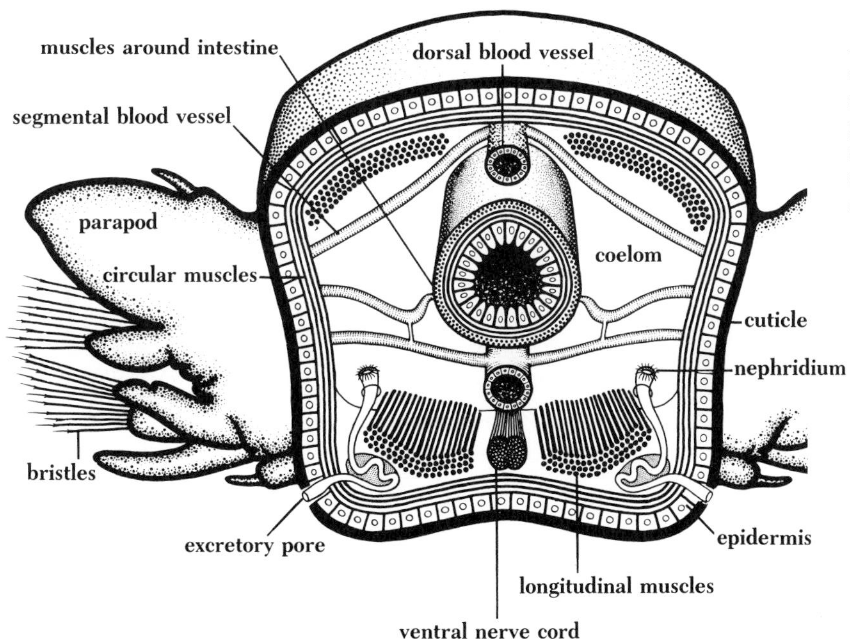

A three-dimentional diagrammatic **cross-section through a polychete,** a nereid. Special muscles move the **parapods.** The longitudinal muscles do not form a continuous layer but occur in four large blocks. (The coelomic lining is omitted.)

feet"), fleshy lobes that project from the sides of each segment. Polychetes are among the most common animals of seashores, and the various kinds live and feed in many different ways. Some are free-ranging predators or herbivores or scavengers, crawling over the bottom or swimming through the water in search of food. Others burrow in sand or mud, feeding on organisms and organic material in and on the substrate, or sit quietly in permanent tubes, filter feeding. The parapods act as little legs or as paddles to assist the strong muscular undulations of the body in locomotion or in circulating water through a burrow, and they also provide a large surface for respiratory exchange.

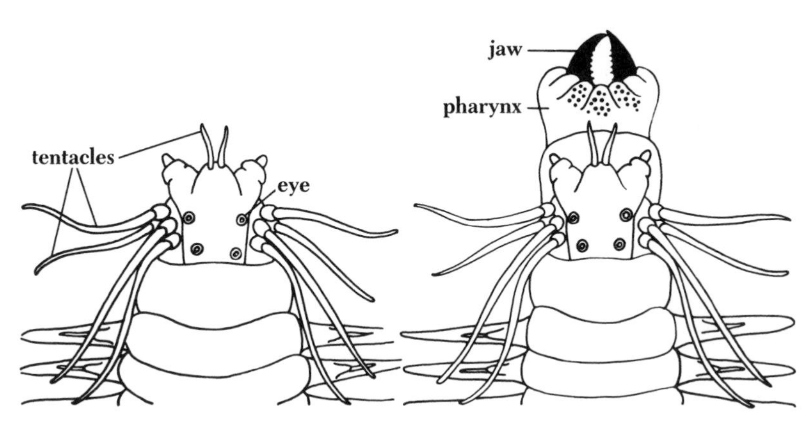

Head of a nereid, *left,* shows concentration of eyes and sensory projections. *Right,* the pharynx and jaws are everted.

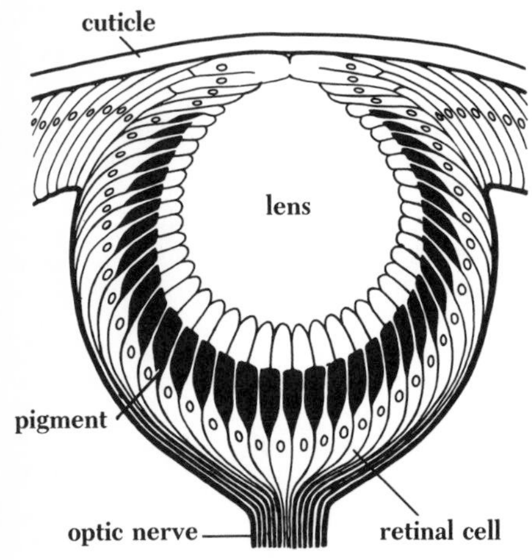

Eye of a nereid has a gelatinous lens that concentrates light on the inner rodlike ends of a layer of pigmented light-sensitive cells (retina). The outer ends of the retinal cells are continued as nerve fibers that run in the optic nerve to the brain. (Combined from various sources.)

Active **crawling or swimming polychetes** typically have a well-differentiated head with conspicuous sensory structures: complex eyes and a variety of sensory tentacles; an eversible pharynx bearing chitinous jaws or teeth, with which they seize food; and great numbers of similar segments with prominent parapods. These worms may spend much of their time in temporary burrows that they have dug or in ready-made crevices, but they can leave these shelters to seek food or mates. Many are graceful swimmers and move with equal agility over substrates.

Sedentary polychetes, those that live in permanent burrows or tubes that they do not leave, tend to be more like earthworms in having an inconspicuous head without prominent sensory structures or jaws; and typically they have fewer segments with smaller parapods, compared to more active polychetes. Some burrow in rich muds and feed by passing large quantities of substrate through the digestive system to obtain the organic matter mixed with it, or collect particles more selectively from the surface of the substrate with long ciliated tentacles. Others live in tubes that stick up into the water; from the top of the tube these

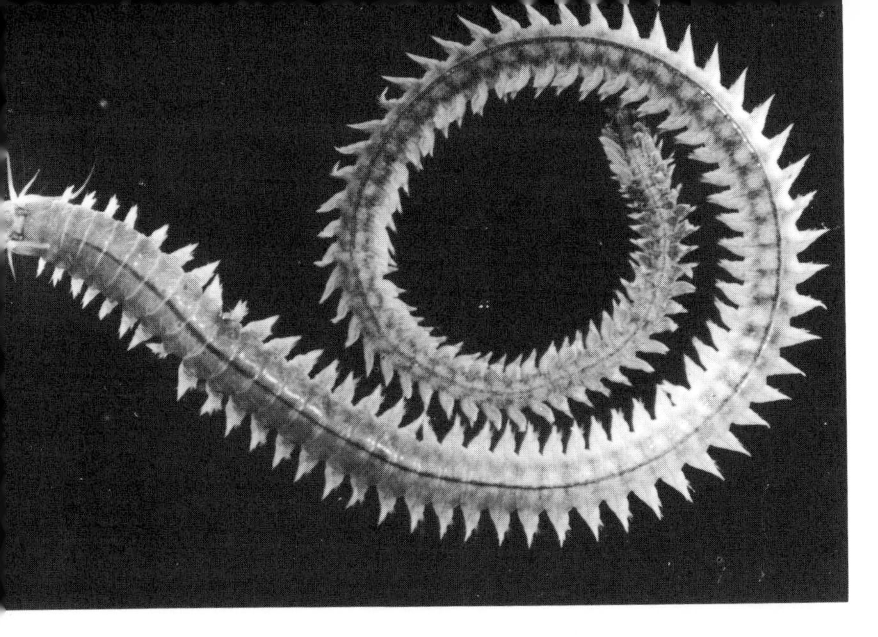

Nereid is an active crawling and swimming polychete with eyes and sensory tentacles on the head, prominent parapods along the many similar segments of the long trunk, and additional sensory projections at the posterior end. Nereids feed mostly at night and hide by day. To the people who use them as fishing bait, nereids are known in the United States as clam worms, from their habit of taking shelter in empty clam shells, in Great Britain as ragworms, presumably from the ragged outline produced by the large, flapping parapods. (K. B. Sandved)

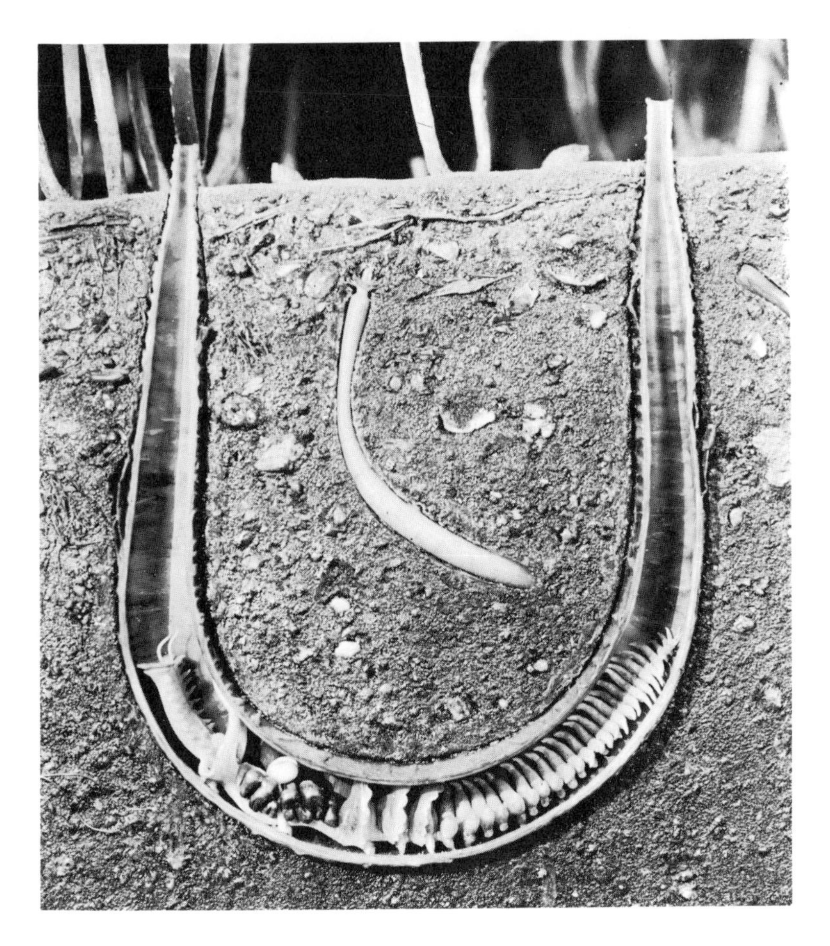

Tube-dwelling polychete, *Chaetopterus*, has marked differentiation of successive regions of the body. Some parapods are enlarged and fanlike and propel a water current through the burrow. Others secrete mucus that traps food particles, which are then rolled up in mucus and carried, by ciliary action, back to the mouth at the anterior end. There the mucus-wrapped food pellets are swallowed. *Chaetopterus* emits a luminous secretion, but no one knows how or whether this has any adaptive value for a polychete that does not leave its tube. (Photo of a model, American Museum of Natural History.)

Featherduster worm protrudes its crown of feeding and respiratory tentacles from an erect tube built in a rock crevice or among stones. The tentacles are supported by flexible internal skeletal rods and bear light-sensitive spots. The shadow of a hand passing over the crown of an extended worm, *left,* will cause it to pop back into the tube with lightning speed, *right.* Monterey Bay, California. (R. Buchsbaum)

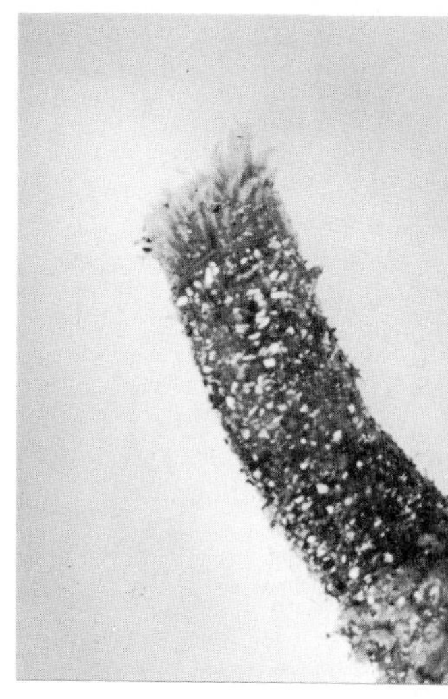

Coral worm protrudes the two spiraled halves of its feeding and respiratory crown from its calcareous tube, embedded in the hard coral of a tropical reef. Each half of the tentacular crown is cone-shaped and colorful, suggesting the common name Christmas-tree worm. A calcareous stopper seals the tube when the crown is withdrawn. *Spirobranchus giganteus.* Puerto Rico. (K. B. Sandved)

worms protrude an often brilliantly colored crown of ciliated tentacles with which they collect small organisms and food particles from the water.

Reproduction in polychetes presents many contrasts to the process we saw in earthworms. The reproductive system of polychetes is simple in structure, and the sexes are usually separate. The gametes arise from certain cells in the coelomic lining of many or most of the segments and develop free in the coelom. Most polychetes shed eggs and sperms into the seawater, sometimes through the ducts of the excretory organs or through short genital ducts in each segment, sometimes by rupture of the body wall. These outwardly simple events are regulated by a complicated system of hormones and are often accompanied by dramatic changes in structure and behavior during the reproductive season. For example, the body wall of a nonreproductive polychete would not ordinarily rupture; it is specifically thinned and

Gathering palolo worms in a net at dawn, just before these polychetes shed their eggs or sperms with the rising sun, is an annual event in Samoa. The "night of the big gathering" occurs at a particular phase of the moon between October and December, and in the past it marked the beginning of the Samoan New Year. The collected worms are pooled for a great feast and celebration. The water is so filled with wriggling polychetes that it has been said to look like vermicelli soup. The worms in the net are the detached gamete-filled posterior portions of the Pacific palolo, *Eunice viridis;* they are photopositive (attracted to light) and rise to the surface. The photonegative anterior portions, with head and brain but no gametes, remain on the bottom, retreat to coral crevices, and are active only at night. Each regenerates a new posterior end, which will produce gametes and swarm at the surface the following year. Related species in the Mediterranean and Caribbean have similar habits but their swarming is less markedly synchronized. Species of *Eunice* are the longest of polychetes, with up to 1,000 segments and a length of 3 m. Western Samoa. (K. J. Marschall)

weakened (by the action of hormones) in mature worms that release their gametes by rupture. A number of polychetes develop huge eyes and parapods and flattened, oar-shaped bristles when they become reproductively mature; they abandon their quiet, bottom-dwelling habits and swim vigorously up into the water to spawn and, soon thereafter, to die. In some species, only the rear end of the worm forms gametes and undergoes this metamorphosis, breaking free of the front end, which remains on the bottom, regenerates a new posterior, and survives to reproduce again the following year. Or, many reproductive rear ends are produced by a kind of budding, and each breaks free and swims away.

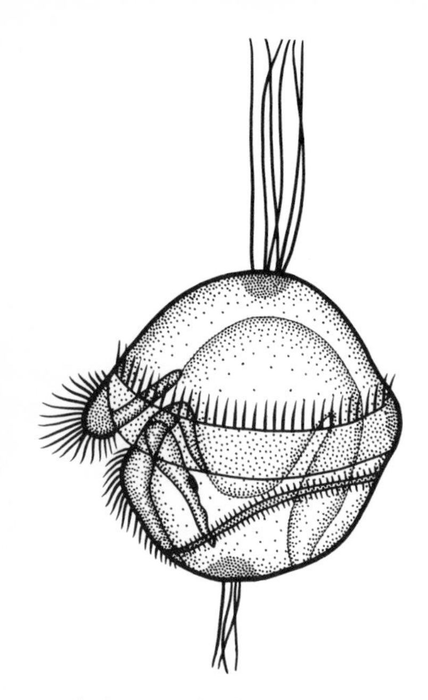

A polychete **trochophore.** (After R. Woltereck.)

The habits of the swimming reproductive forms of polychetes tend to lead these normally solitary and secluded worms to aggregate in large numbers and simultaneously release eggs and sperms, greatly increasing the chances of fertilization. In addition to the specific periodicity of aggregations, the sexes may be brought together by chemical signals. In some nereids the discharge of sperms is set off by a secretion of the spawning female. In the "fire worms" of Bermuda the meeting of the sexes involves the exchange of light signals. The worms come to the surface to spawn each month a few days after the full moon at about an hour after sunset. The female appears first and circles about, emitting a greenish glow bright enough to be readily visible to observers on a shore at some distance. The smaller male then darts rapidly toward the female, emitting flashes of light as it goes. As one or more males swim in tight circles around the female, both sexes shed their gametes into the seawater, where the eggs are fertilized, surrounded by a cloud of luminous secretion. The spent worms, reduced to shreds of tissue, perish.

The fertilized egg of a polychete develops into a ciliated **trochophore larva.** This larva is of considerable theoretical importance because the same type occurs in several phyla. Few animals seem further apart in adult structure than a segmented worm and a snail. Yet their early stages of **development** are almost identical, cell for cell; and the trochophores that result are similar in many respects. Beyond the trochophore stage, however, marked differences begin to appear and the adults are unlike. The close relationship thought to exist between annelids and molluscs would never have been suspected except for the similarities of their development.

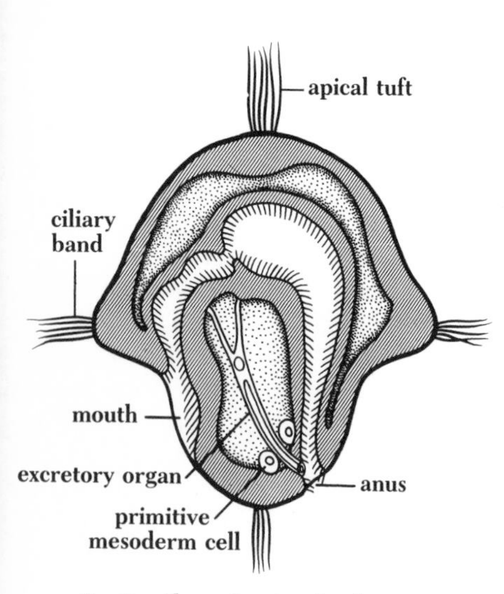

Section through a trochophore. (Modified after C. Shearer.)

The most remarkable of the developmental resemblances that link annelids and molluscs is the origin of the mesoderm in the two groups. As in the roundworms, the early stages of development of certain annelid and mollusc embryos have been followed so closely that each cell has been numbered and mapped. As a result of this extremely painstaking kind of work it is possible to trace the **cell lineage** of any portion of the early embryo. The adult mesoderm comes from a single cell (the *4d* cell), which arises in the same way in both annelids and molluscs. This cell divides into a pair, the **primitive mesoderm cells** (shown lying against the wall of the intestine and near the opening of the larval excretory organ in the diagram of the trochophore). These give rise to two bands of mesoderm in annelids, or clusters of mesoderm cells in molluscs, which finally become hollowed out to form the coelom of the adult.

The development of a polychete trochophore into an adult worm proceeds with the formation of bristles on the elongating lower region of the trochophore, which becomes constricted into segments. The ciliated bands disappear, and the upper region becomes the head. The young worm then settles to the bottom and takes up its adult mode of life, continuing to grow by the addition of new segments in a region just in front of the last segment.

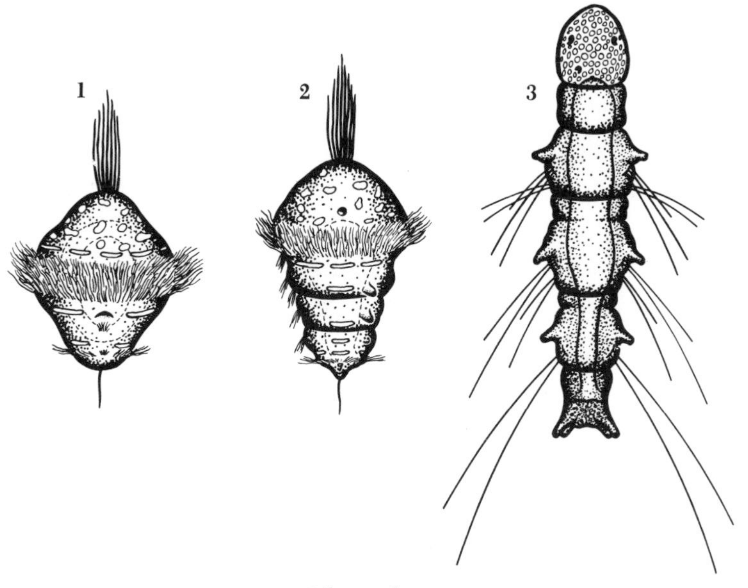

Development of a polychete. 1: Trochophore is elongating at its lower, posterior pole. **2:** Visible indications of the first segments, which in **3** are clearly constricted and already bear the larval bristles. (Modified after D. P. Wilson.)

Oligochetes

The second largest class of annelids is the **Oligochaeta** ("few bristles"), of which the great majority are earthworms. The rest are mostly small or minute worms found in soil or in freshwaters, plus a few marine forms. A photograph of a hydra eating a freshwater oligochete is in chapter 6.

We have seen that oligochetes differ from polychetes in several respects. There are no parapods, and the bristles emerge

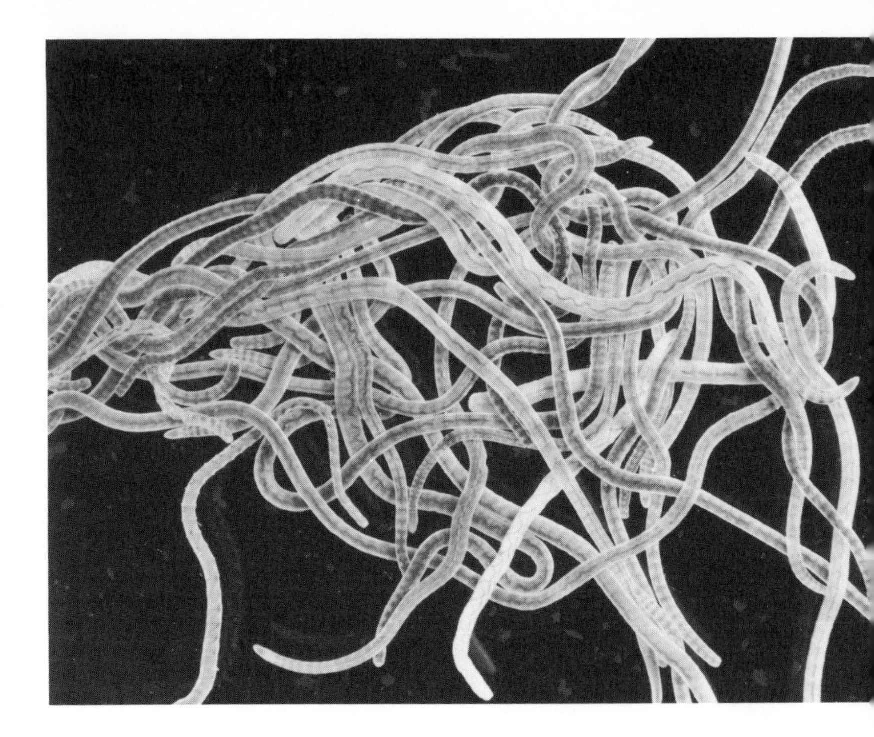

Aquatic oligochetes live on both marine and freshwater bottoms, occupying the same ecological niche as their earthworm relatives; they feed on the organic material in mud and thoroughly mix the surface layers. Many reddish, threadlike aquatic oligochetes have a red respiratory pigment dissolved in the blood and can live in water of low oxygen content. The most familiar of these is *Tubifex tubifex*, found around the world and shown here in laboratory culture. In nature it lives in masses on the bottom of stagnant ponds and deep lakes, each worm waving its posterior end in the water and respiring through the hindgut. The greatest numbers thrive in sewage-laden rivers, and densities of *Tubifex* can be used as a measure of the degree of pollution. *Tubifex* is sold in pet shops as food for aquarium fishes. (R. Buchsbaum)

from pits in the body wall. Whereas most polychetes have separate sexes and the gametes arise from the coelomic lining and develop free in the coelom of many segments, oligochetes are hermaphroditic and the gametes are produced in special organs that occur only in certain segments. Finally, oligochetes have a clitellum that secretes the egg capsule.

Leeches

Among the many features that the leeches, class **Hirudinea,** share with the oligochetes are hermaphroditism and the presence of a clitellum. Leeches, however, have no bristles; and the external ringing of the body does not correspond with the internal segments, of which there is a fixed and smaller number, not separated by partitions. The body is solid, the coelomic spaces being crowded out by the growth of connective tissue. At each end of the body is a sucker, the posterior one being much larger than the anterior, which has the mouth in its center. Despite all these modifications, leeches are closely related to oligochetes and are probably derived from them. There is no character of leeches that is not present in at least some degree in some oligochete, and the two groups are sometimes combined as a single class (Clitellata).

Freshwater leech *Placobdella parasitica* attaches to the skin of snapping turtles and sucks blood with an eversible proboscis; it has no jaws. Removed from the naked skin at the base of the hind leg of a snapping turtle and placed in an aquarium, this one attached itself to the glass by the large posterior sucker and is seen hanging head down. The blood meal is stored in digestive pouches, which show as dark bands (seen through the relatively unpigmented ventral surface). Leeches are hermaphrodites; female reproductive organs can be discerned in the whitish area in the midline, with testes evident as light circles on both sides of the midline. Large specimens are 10 cm (4 in) when fully extended. The dorsal surface is olive green, spotted with orange. Leeches are among the few freshwater invertebrates that may be colorful; at one time their color patterns were used in fabric designs. (S. T. Brooks)

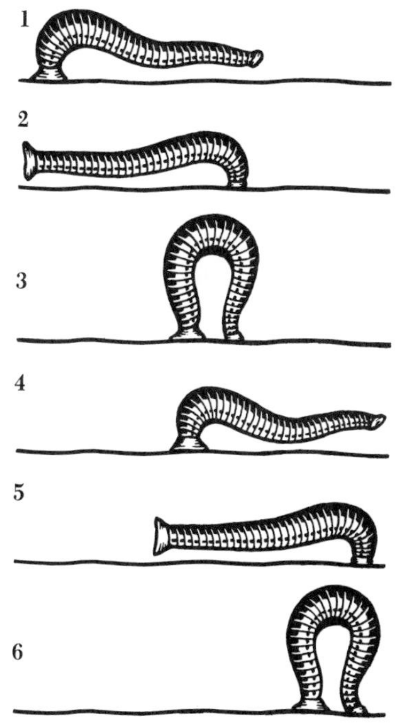

A leech moves by steps, like an inchworm, alternately attaching the anterior and posterior suckers.

Leeches are found mostly in freshwaters, but there are also marine and terrestrial species. Although they mainly suck the blood of other invertebrates or of vertebrates, some of them are predators, swallowing small animals whole. They show some of the adaptations for parasitism that were noted among flatworms, especially the development of clinging suckers. On the other hand, since they do not usually live on a host, but need to move about to locate successive blood donors, they are less modified than flukes or tapeworms. In sucking blood, a leech attaches to some animal by the posterior sucker, applies the anterior sucker to the skin, and makes a wound, often with the aid of little jaws inside the mouth. When it has filled its digestive tract with blood, it drops off and remains torpid while digesting the meal. Large blood meals are few and far between, but the digestive tract has lateral pouches that hold enough blood to last for months. The salivary glands of leeches manufacture a substance

Land leech in a forest in Malaysia is attached to a leaf by its posterior sucker and is making circular searching movements with the narrow anterior end, having sensed the vibrations and odor produced by an approaching mammal (in this case, one with a camera). These small leeches, only 25 mm long before they attach to skin, have sharp jaws and make a wound so painlessly that field workers may be unaware of leeches hanging from their ankles like bunches of grapes. After the leeches drop off, blood continues to trickle from the wounds for as much as 30 minutes. Plantation workers in Southeast Asia and Sri Lanka may suffer serious chronic blood loss. Land leeches are found around the world in wet tropical and subtropical forests. (R. Buchsbaum)

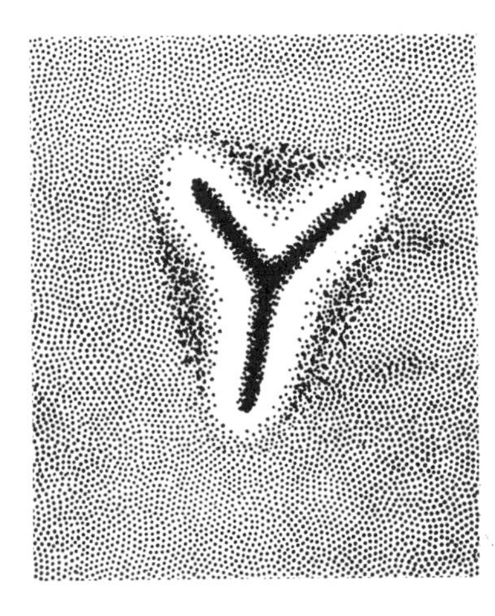

Y-shaped wound made by the three jaws of some leeches. (After Reibstein.)

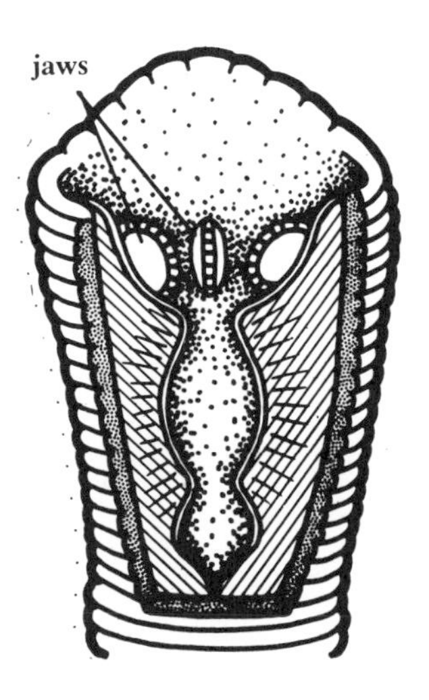

Head of a leech, one side cut away to show three sawlike jaws. (Modified after Pfurtscheller.)

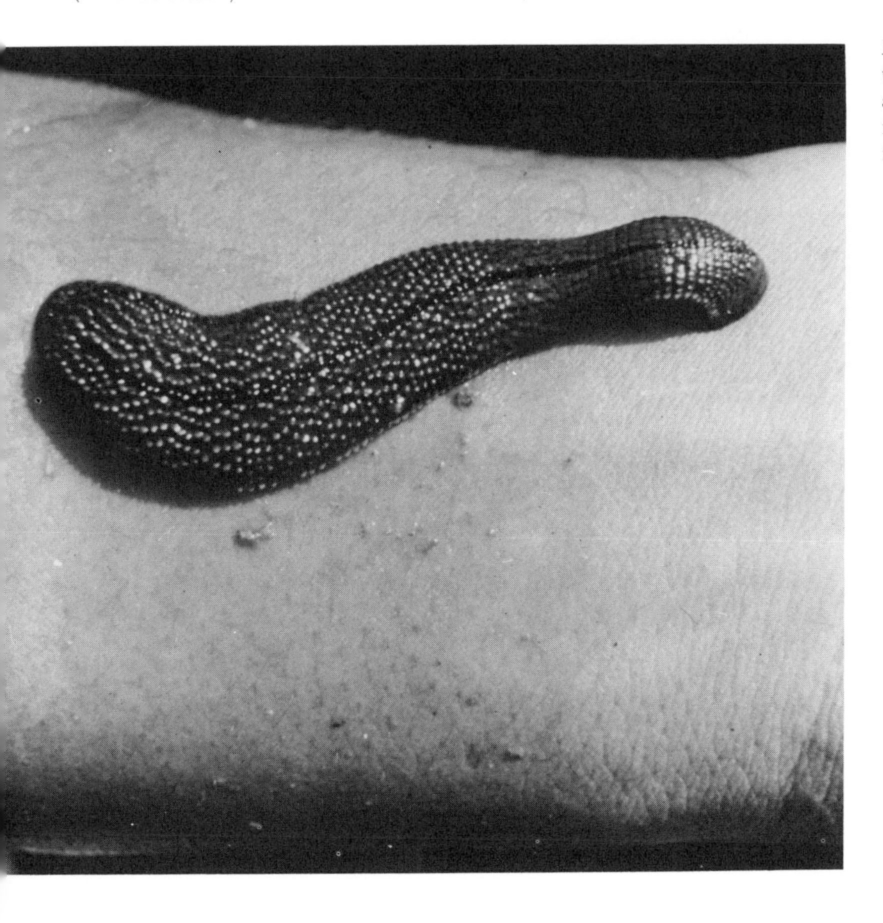

Leech sucking blood, attached to the thin skin on the inner side of the arm of a zoologist who was collecting invertebrates from a freshwater pond in Thailand. (R. Buchsbaum)

called *hirudin*, which prevents the coagulation of the blood while the leech is taking its meal and throughout the following weeks or months during which the blood is stored and slowly digested. This anticoagulant, together with a substance injected by some leeches to dilate the host's blood vessels, causes wounds made by leeches to bleed for a long time after the leech has detached itself.

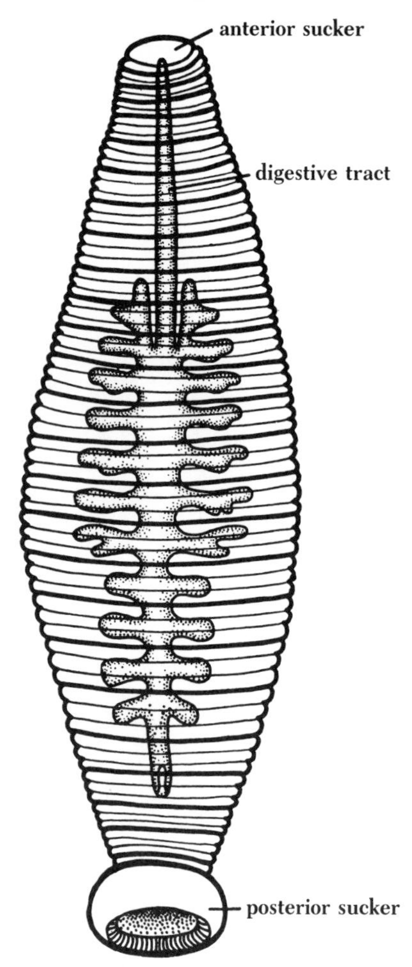

Digestive tract of a leech is provided with lateral pouches in which large quantities of blood, sometimes equivalent to ten times the weight of the leech, can be stored and slowly digested. Symbiotic bacteria aid in digesting the blood. The surface of the leech is thrown into folds, of which only those indicated here by heavy black lines correspond to segments. This animal lives in the gill chambers of a fish. (After E. E. Hemingway.)

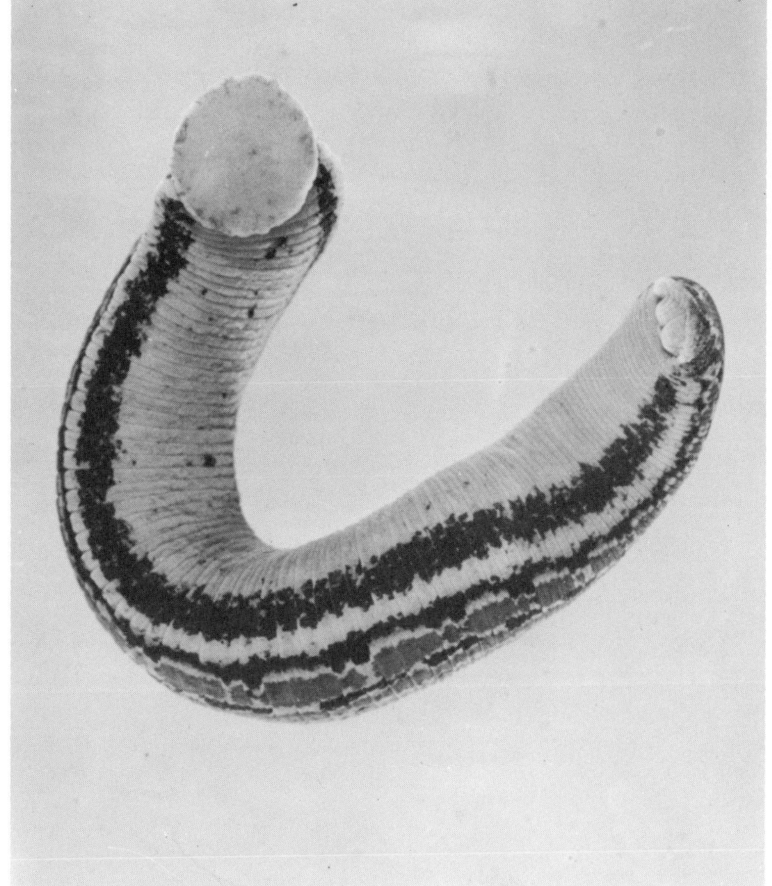

Medicinal leech, *Hirudo medicinalis*, is a European leech that was once used by the millions for bloodletting. Overcollecting and habitat destruction have made it scarce. Though introduced into the northeastern United States, it has not become common. The imported leech shown here was purchased in a Chicago drugstore where leeches are sold for removing blood from black-and-blue bruises around the eyes. *Hirudo* is still in demand in the United States for its hirudin, used as an anticoagulant for some heart patients and in some surgical procedures. The leech hangs by its large posterior sucker; the small anterior sucker (at the right) surrounds the mouth. The ventral surface is pale, the dorsal surface green with brown stripes. This leech was about 12 cm (5 in) long when swimming. (R. Buchsbaum)

The spacious coelom and the segmentation of annelids probably evolved, and are most easily understood, in connection with the habit of active burrowing. Therefore it is not surprising that many of the partitions between segments are reduced or lost in sedentary polychetes. And in leeches, which move in an entirely different way, the partitions and the coelom itself have virtually disappeared. Once established, segmentation offers the same general possibilities as the dividing-up of an animal body into cells, tissues, and organs, namely, the different segments may specialize in different functions. In some annelids the segments are practically all alike, while other annelids have taken advantage of this opportunity to diversify the segments and have developed differentiated body regions. But the opportunists that have really exploited the possibilities of regionalizing a segmented body plan are the arthropods, to be met in the next three chapters.

Lobster guarding its hiding place is a formidable foe. The two large claws of the adult are different, each specialized for its function. The lobster's heavy right claw is a powerful crushing tool; the left claw is used for tearing the flesh of its food. *Homarus gammarus*. Aquarium of the Station Biologique, Roscoff, France. (R. Buchsbaum)

Spiny lobster has no large claws for defense and would seem to be an easy animal to approach. But the body is covered with spines, and the spiny large antennas can deal tearing blows. The flesh is delicious and various species are eaten extensively in southern Europe, Florida, California, and other warm parts of the world. (R. Buchsbaum)

Chapter 17

Lobsters and Other Arthropods: Crustaceans

Asked to name the animals that have been most important to humankind in terms of their effect on the environment, the economy, and the dinner table, most people will name cattle, sheep, horses, and perhaps goats, all mammals. But the animals that occupy first place among all the groups, by the criteria mentioned, are not mammals or any other vertebrates. They are members of the phylum **Arthropoda.**

Of the million or more described species of animals, over three-fourths are arthropods. Arthropods also have the largest numbers of individuals, consume the greatest amounts and the most kinds of food, and occupy the widest stretches of territory and the greatest variety of habitats—and they seem likely to continue to do so. Many of the pestier ones such as houseflies and mosquitoes are all too common and familiar.

One may think of the three main groups of arthropods as having divided among them the three major realms of the planet, or the elements of the ancient Greeks—water, earth, and air. The waters, both fresh and marine, belong to the **crustaceans** such as crayfishes, lobsters, crabs, shrimps, and barnacles. The land has become home to the **arachnids** such as scorpions, spiders, and mites. And the largest group of all, the **insects,** are the only invertebrate group to have conquered the air, though they by no means restrict themselves to this domain.

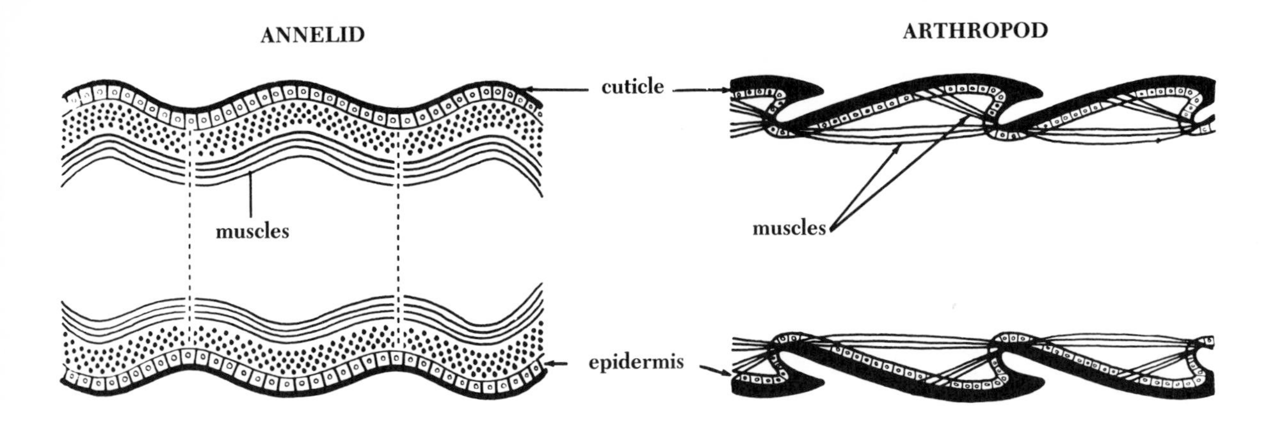

Body wall of annelid and arthropod contrasted. In annelids the cuticle is thin, and the epidermis is underlain by heavy layers of circular and longitudinal muscles. In arthropods the cuticle is heavy, and the muscles occur in separate bundles instead of continuous layers.

The **arthropod body plan** is in some ways an elaboration and specialization of the segmented body plan of annelids. Primitive arthropods are composed of a series of similar segments bearing similar appendages. But in the more modified types, almost every segment of the body has a somewhat different form and function. The outer layer, or **cuticle,** very thin in annelids, is in most arthropods a heavy covering that serves both as a *protective armor* and as an *external skeleton.* The function of the coelomic body cavity as a hydrostatic skeleton has thus disappeared, and with it the coelom, of which only vestiges remain. Horny outer coverings occur in many groups of animals (for example, the covering of a hydroid colony or the cuticle of annelids), but in no case do they appear in so great a variety of structures as in arthropods. Made of the cuticle, in whole or in part, are outer protective coverings, biting jaws, piercing beaks, grinding surfaces, lenses, tactile sense organs, sound-producing organs, walking legs, pincers, swimming paddles, mating organs, wings, and innumerable other structures found among the highly diversified arthropods. This versatile material is to the arthropods what metals and plastics are to industrial peoples, and it is partly to the properties of the cuticle that the arthropods owe their success.

The cuticle is nonliving and is secreted by the underlying epidermis, as are the cuticles of roundworms and annelids. But unlike the soft cuticles of these worms, the hardened cuticles of arthropods do not increase in size as the animal grows. So arthropod growth is accompanied by regular **molting,** during which the inner layers of the cuticle are dissolved and reabsorbed, a new larger cuticle is formed, and the old outer portion is shed. The molt is an occasion not only for growth but also for the repair of damaged parts and for the production of structures appropriate to the next stage as an individual matures. Insects do not molt after the adult stage is reached; but many crustaceans, arachnids, and other arthropods continue to grow and molt throughout life. The molt affects every aspect of an arthropod's life; special *hormones,* released into the blood, initiate molting and coordinate all the changes that take place in physiology and behavior before, during, and after this stressful event.

The cuticle is composed of many different layers and substances, each of which contributes some useful property. A thin surface layer covers the body, and in many arthropods it includes a waxy component that makes the cuticle *waterproof.* Beneath this the bulk of the cuticle is mainly composed of chitin and proteins, and it is both *tough* and *flexible.* Wherever the cuticle is relatively rigid, as it is over most of the surface of an arthropod, the outer portion of the chitin-protein layer is *hardened* by tanning (the formation of chemical bonds between the proteins) and/or by calcification.

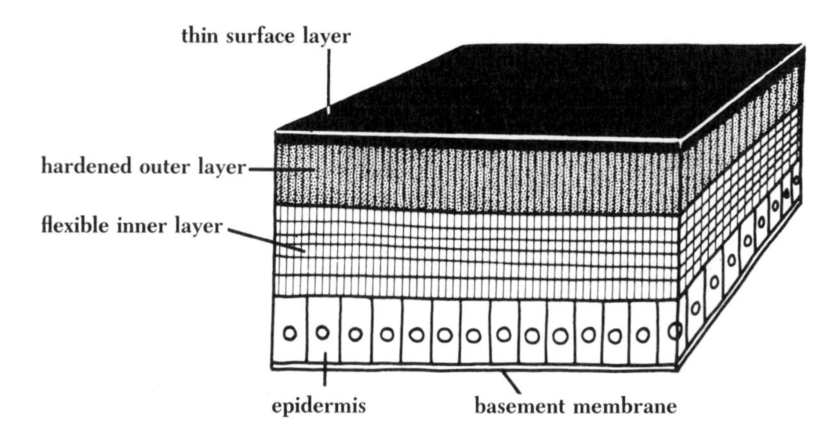

Layered cuticle of an arthropod is secreted by the epidermis.

The hardened cuticle is remarkably strong for its weight and occurs in definitely limited areas, which furnish a *supporting framework* for the tissues within and provide a *surface for the attachment of muscles.* But between the hardened areas, the cuticle remains as flexible membranes, or joints, so that the outer covering of arthropods provides protection and support without sacrificing mobility. This is what makes it so different from the armors of such animals as snails and clams, which have heavy, cumbersome shells that limit movement. The external skeleton, or **exoskeleton,** of the arthropods also differs sharply from the kind of internal framework possessed by vertebrates, an **endoskeleton,** in which the hard supports are surrounded by the soft fleshy parts. We can imagine how it might feel to be an arthropod by mentally putting on a strong, light suit of armor that adheres closely to the skin, and then thinking of our bones being eliminated and our muscles being attached instead to the armor.

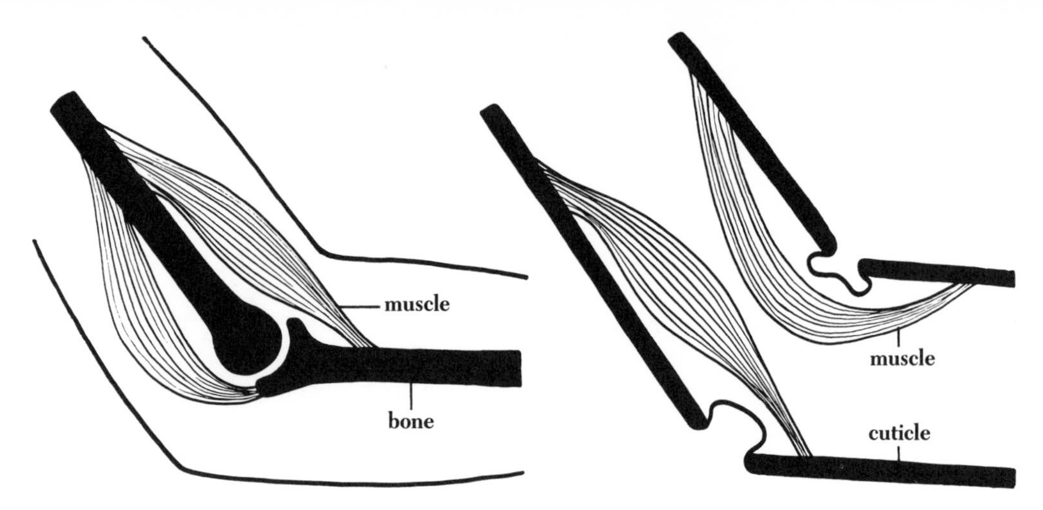

Skeletal and muscular systems of vertebrate and arthropod contrasted. In a vertebrate limb, bones lie internally and have muscles attached to their outer surfaces. In an arthropod limb, the cuticle lies externally and has muscles attached to its inner surfaces.

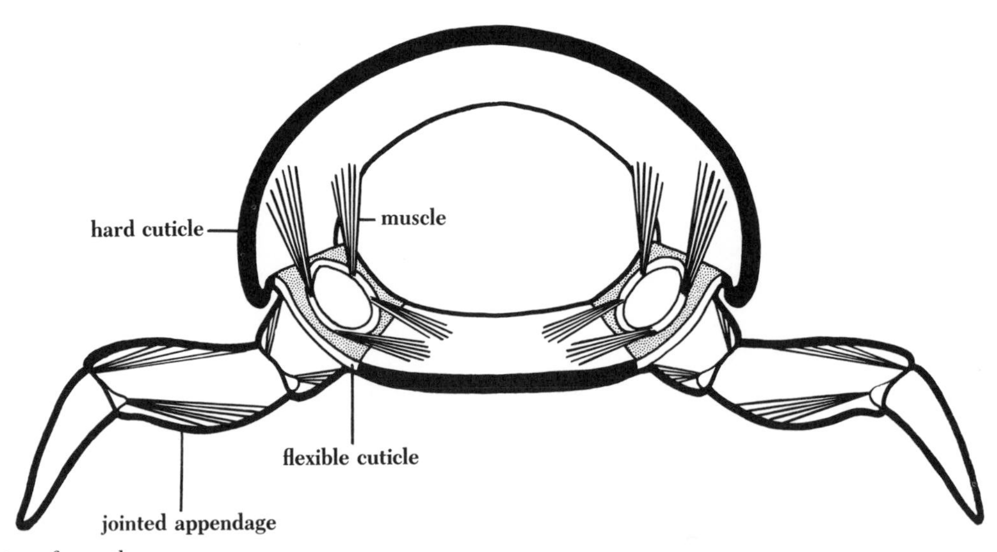

Cuticular exoskeleton of an arthropod consists of thick hardened plates joined by thinner, more flexible membranes. Muscles are attached both to the main framework and to thin cuticular plates that project to the inside. (Modified after R. E. Snodgrass.)

From our suit of armor would project two arms and two legs, for the appendages of vertebrates are four in number. And, though they show a variety of structure in adaptation to different methods of locomotion and to additional services they may perform, such as digging or holding prey, they are primarily locomotory—with the notable exception of the forelimbs of humans. The appendages of arthropods, in contrast, are greater and more variable in number; and some of them have no locomotory function but serve as sense organs, jaws, respiratory structures, or mating organs. The name Arthropods ("jointed feet") refers to the most characteristic structures of arthropods, their chitinous jointed appendages.

The four appendages of all vertebrates—from herrings to hawks to humans—have fundamental similarities in structure. And they are seen to arise in the embryos in the same way from similar structures. Hence the pectoral fins of a fish, the wings of a bird, and the forelegs of a mammal are all said to be **homologous.** The principle of homology is the basis for determining animal relationships. If two animals have clearly homologous parts, the animals are judged to be related. The greater the number of homologous structures, and the more similar the structures and their mode of origin, the closer the relationship. In other words, as part of the definition of homology we assume that the homologous structures of two different animals have come, by a process of modification, from the same or corresponding part of some common ancestor.

Not all structures that resemble each other indicate a common evolutionary origin. Many are alike because they have been adapted to the same functions or environmental conditions; and such structures may arise in entirely different ways in the embryos. They are similar in function but not in basic plan or mode of origin, and are said to be **analogous.** The wing of a bird and the wing of a bee are both used for flying, but they differ fundamentally in structure. One is made of bone and feathers and the other is chitinous, and they do not develop in the same way. They are analogous but not homologous. On the other hand, the wing of a bee is homologous to that of a butterfly. In these insects the wing is essentially the same and arises from a corresponding part of the embryo. In this case the homologous structures also have similar functions. But sometimes homologous structures have different functions: the large pincers of scorpions correspond to appendages that are used by male spiders to transfer sperms to females.

When corresponding structures in different segments of the *same* animal are considered, we say that they are **serially homologous.** One walking leg of a centipede is serially homologous to

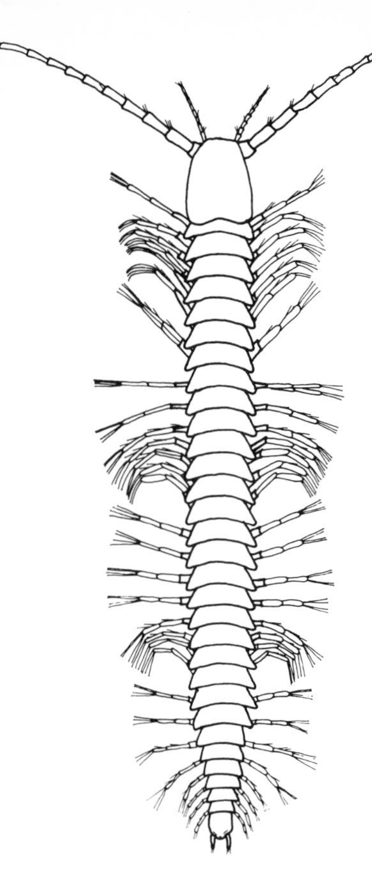

Primitive crustacean with a pair of similar appendages on almost every one of its many segments. This cave-swimmer, *Speleonectes,* belongs to a newly discovered class of crustaceans. (J. Yager)

any other, and all of the walking legs are serially homologous to the centipede's mouthparts as well as to its sickle-shaped appendages that inject poison.

In the most primitive arthropods there is, typically, a pair of similar appendages on each of the many segments, and each appendage serves multiple functions. Among living arthropods, the closest approach to this occurs in certain crustaceans that have, except for a few specialized appendages on the head, a pair of similar flattened appendages on each of the many segments, and each appendage serves for swimming, food collecting, and respiration. In most arthropods, the appendages are fewer in number and have differing and specialized structures that are related to their diverse functions. Certain ones may become disproportionately large or important, while others disappear. Thus to describe the appendages of an arthropod is to tell almost everything about the habits of the animal: where it lives, how it moves, and how it feeds.

We will introduce the arthropods by describing a lobster. In examining the lobster appendages and homologizing the different parts of each, we are better able to understand how a simple flattened swimming appendage can become a chewing mouthpart, a crushing claw, or a stout walking leg.

Lobsters

Aside from minor details, the lobsters are so much like their freshwater kin, the crayfishes, that the description of a lobster applies in general to both animals.

As in other arthropods, and in contrast to annelids, the body of a lobster consists of a *constant number of segments,* grouped into distinct *body regions.* The numbers and regions differ among arthropods, but are fixed within various groups. In lobsters the 5 segments of the **head** and 8 segments of the **thorax** are united into a large **cephalothorax.** Over the dorsal surface and sides of the cephalothorax is a large protective shield, the **carapace,** which consists of a thin layer of tissue covered with exoskeleton. Like much of the rest of the exoskeleton covering the body, that of the carapace is hardened by infiltration with calcium salts. The cuticle also furnishes some internal support to the cephalothorax in the form of thin plates of cuticle secreted by infoldings of the epidermis. These plates protect important organs and increase the area for attachment of muscles. The posterior body region, or **abdomen,** includes 6 segments. The exoskeleton of the abdominal segments is heavily calcified dorsally but more flexible ventrally and between segments, as lobster eaters quickly discover.

The least modified of the serially homologous appendages of a lobster are those of the abdomen. Each consists of a *basal piece* (protopod), which bears at its free end an *outer branch* (exopod) and an *inner branch* (endopod). The number of joints in the three pieces may vary, but the basic plan of this two-branched (biramous) appendage is found throughout most groups of crustaceans, both in highly specialized adult appendages and in those of the simplest larvas. In lobsters the two-branched plan is modified in the appendages of head and thorax by the presence of additional lobes or extensions on the basal piece or by the loss of the outer branch. In lobster embryos, however, almost all the appendages arise as simple two-branched structures.

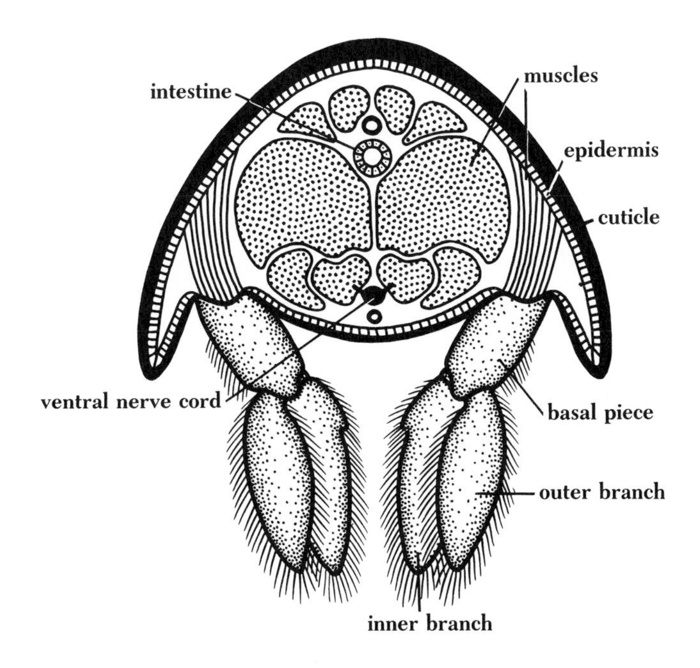

Cross-section of the abdomen of a lobster.

The **head** of a lobster is usually considered to include an anterior nonsegmental portion, the **acron,** and the first five appendage-bearing segments of the cephalothorax. On the acron there is a pair of eyes set on the end of jointed, movable stalks. These are not serially homologous with the appendages, as they arise in a different way. On the first two appendage-bearing segments are **two pairs of antennas,** a distinctive character of crustaceans. The **first antennas** of lobsters are sensory structures with two short whiplike filaments; the **second antennas** have one long sensory filament. The third segment bears the toothed **jaws** (man-

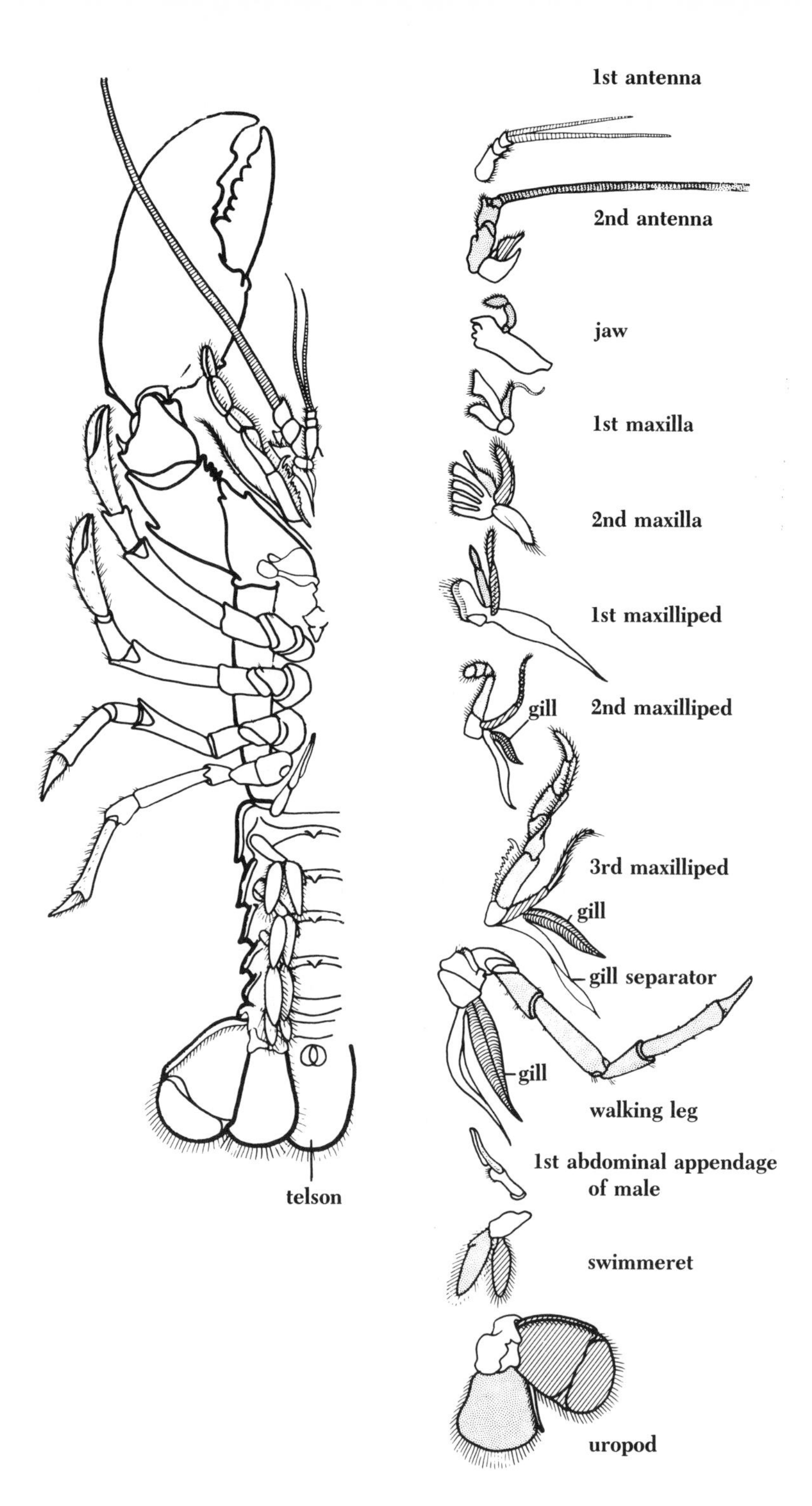

Appendages of a lobster show a marked division of labor. In this drawing, the *inner branch* is stippled, the *outer branch* is marked with diagonal lines, and the *basal piece* and its processes are left unshaded. For example, the long filament of the second antennas is homologous with the inner branches of the other appendages, and the outer branch is represented only by a small scalelike process. The first antennas are considered serially homologous with the rest of the appendages, but their two-branched condition is not. In a lobster embryo the first antennas remain single until long after the other appendages have become two-branched; and even when the larva emerges from the egg, the inner filament is represented only by a small bud from the base of what finally becomes the outer filament.

dibles) for crushing food. On the next two segments are the **first and second maxillas,** which help to shred soft food and pass it to the mouth. Each second maxilla bears a thin, lobed plate, serving as a "bailer" to drive a respiratory current of water over the gills.

The **thorax** has a pair of appendages on every segment. The first three bear the **first, second, and third maxillipeds,** which are sensory and also serve to handle food, shredding it and passing it forward toward the mouth. The fourth thoracic segment bears the big **pinching claws** (chelipeds), used both in offense and defense. The big claws are not symmetrical in lobsters over 3 centimeters long. In the smallest lobsters both claws are slender and have sharp teeth, but as the animal grows, they differentiate further with each molt. One becomes heavier than the

American lobsters, *Homarus americanus,* once a specialty of New England, are now routinely flown to the West Coast, fed regularly, and sold alive in seafood stores (to those not satisfied with the lack of large claws in the spiny lobsters from southern California). These two have the differentiation of the two large claws reversed: one has the crushing claw on the left; in the other it is on the right (the more common condition). The lobsters are dark green when alive, bright red after cooking. (R. Buchsbaum)

other, and its teeth fuse into rounded tubercles; it is used for crushing. The other remains more slender, its teeth become still sharper, and it is used especially for seizing and tearing the food. Each of the next four segments has a pair of **walking legs.** The first two pairs of walking legs end in small pincers that aid in grasping food. The last pair of walking legs have bristly brushes used in cleaning the abdominal appendages. In all of the thoracic appendages, except the first and last, the basal piece bears a thin flap (epipod) to which is attached a gill. The flaps separate and protect the gills, and the walking movements of the legs move the gills and stir up the water in the respiratory chamber under the carapace.

The **abdomen** has a pair of appendages on every segment. Those on the first abdominal segment are different in the two sexes. In the male they are modified to form a troughlike structure used for transferring sperms in mating; in the female they are unspecialized. The next four segments all bear similar two-branched appendages, the **swimmerets,** which aid in forward locomotion and in the female serve as a place of attachment for the eggs. The sixth and last abdominal appendages are the **uropods,** which resemble modified and enlarged swimmerets. Together with the flattened, nonsegmental posterior end, or **telson,** they form a tailfan, used in backward swimming.

Lobsters furnish a striking example of *specialization among appendages* of different segments and, in the case of the big claws, between the right and left sides of the same segment. The flattened, two-branched swimmerets are primitive in form, and they contrast sharply with such specialized appendages as the big claws. In the development of lobster appendages we see how a series of originally similar parts can become differentiated into highly specialized and dissimilar structures that, though no longer functioning in the same way, are still homologous.

The internal parts of a lobster with which some of us are familiar are the large (and very edible) abdominal **muscles.** These are segmentally arranged and include muscles for moving the swimmerets, extensor muscles for straightening the abdomen, and much larger flexor muscles. These last furnish the major source of power for rapid escape movements in which the lobster flexes the abdomen and its tailfan ventrally, with such speed and force that the whole animal shoots backward through the water. In the cephalothorax are numerous muscles for moving the appendages. Like the skeletal muscles of vertebrates, the muscles of lobsters and other arthropods are of the *cross-striated type,* especially suited for rapid contractions. Even the muscles of the digestive tract and other viscera are cross-striated in arthropods.

The **digestive tract** of a lobster or other arthropod consists of three main regions—*foregut, midgut, and hindgut*—of which only the midgut has an endodermal lining. The foregut and hindgut develop as tubular ingrowths of the ectodermal epithelium and so become lined with a layer of cuticle that is continuous with the exoskeleton and is shed when the animal molts. Lobsters are predators and scavengers: they catch live fishes, dig for clams and worms, and have been seen to attack large gastropods, breaking off the heavy shell, piece by piece, to obtain the soft parts. The food is shredded by the maxillipeds and maxillas and then further crushed by the jaws before it enters the mouth. As if this were not enough, part of the stomach (foregut) is specialized as a gizzard, which is lined with hard chitinous teeth and worked by numerous sets of muscles. In the stomach the food is pulverized, strained, and sorted. The smallest particles are sent in a fluid stream to the large digestive glands for digestion and absorption; larger particles go in a steady current to the intestine; and the coarsest particles are returned to the grinding mechanism.

The anterior portion of the stomach is large and bulbous and serves chiefly for storage. The posterior part is mainly for sorting and straining. Between the two lies the grinding region, which reduces the food to fine particles. As these are readily digested in the tubules of the digestive glands, the work of the intestine is less important than in many other animals. This explains how lobsters can get along with such a short uncoiled intestine. Shrimps have a similarly short intestine, familiar to anyone who has removed the dark so-called "vein" (along the dorsal surface) when cleaning shrimps.

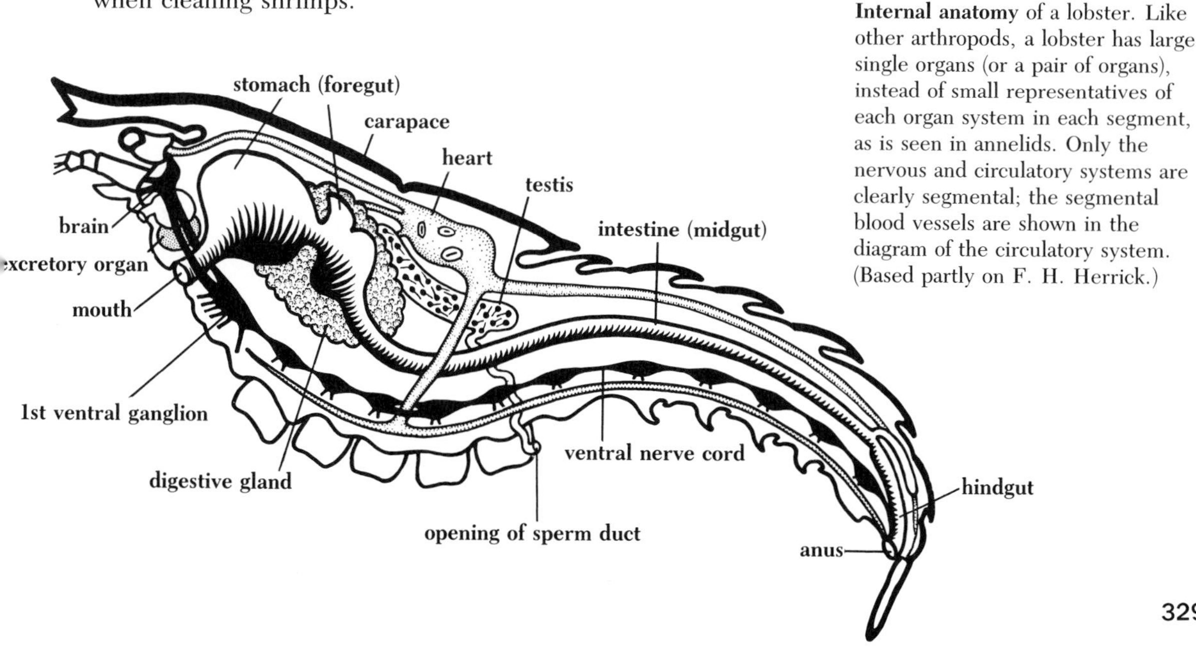

Internal anatomy of a lobster. Like other arthropods, a lobster has large single organs (or a pair of organs), instead of small representatives of each organ system in each segment, as is seen in annelids. Only the nervous and circulatory systems are clearly segmental; the segmental blood vessels are shown in the diagram of the circulatory system. (Based partly on F. H. Herrick.)

The extensive respiratory surface needed to supply the demands of a large and active animal like a lobster is furnished by 20 pairs of **gills,** feathery expansions of the body wall, which are filled with blood channels. The gills are attached to the bases of the legs, to the membranes between the legs and trunk, and to the wall of the thorax. They lie on each side of the body in a chamber enclosed by the curving sides of the carapace. Water enters the chamber under the free edge of the carapace, passes upward and forward over the gills, and is directed out anteriorly in a current maintained by the beating of the flattened plates on the second maxillas.

Cross-section of thorax of lobster shows relations of gill chamber to other organs and the path of the blood through some of the main blood channels. Note the valves of the openings into the heart.

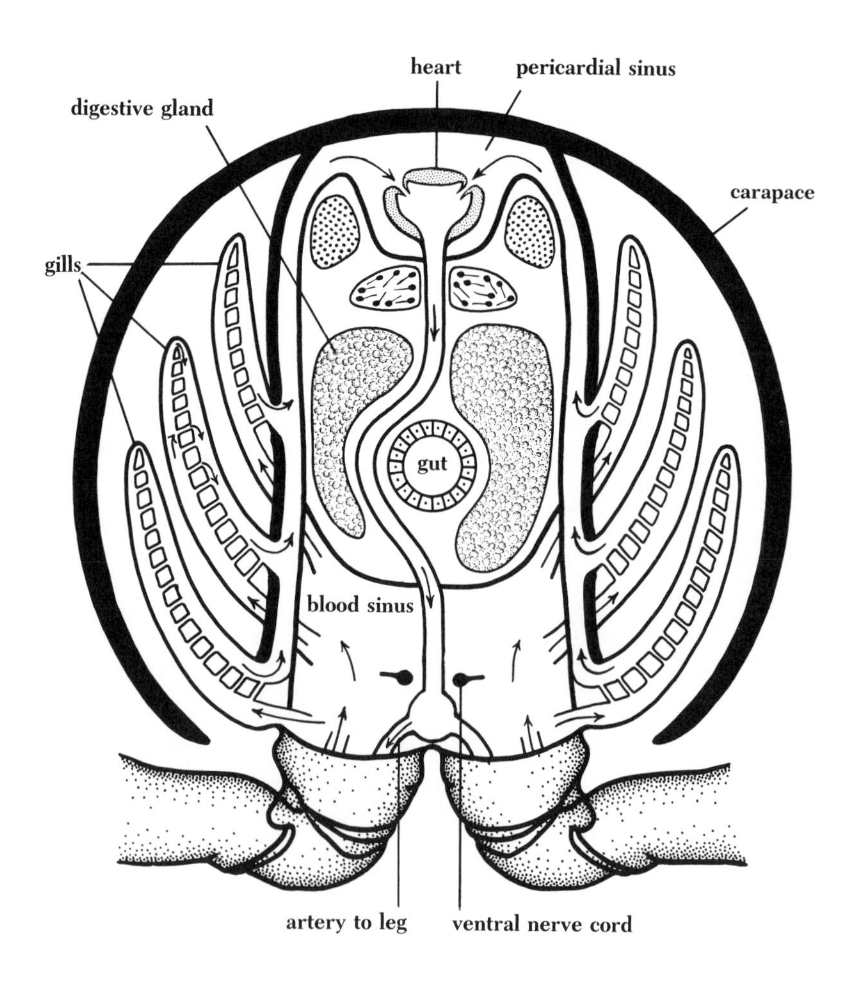

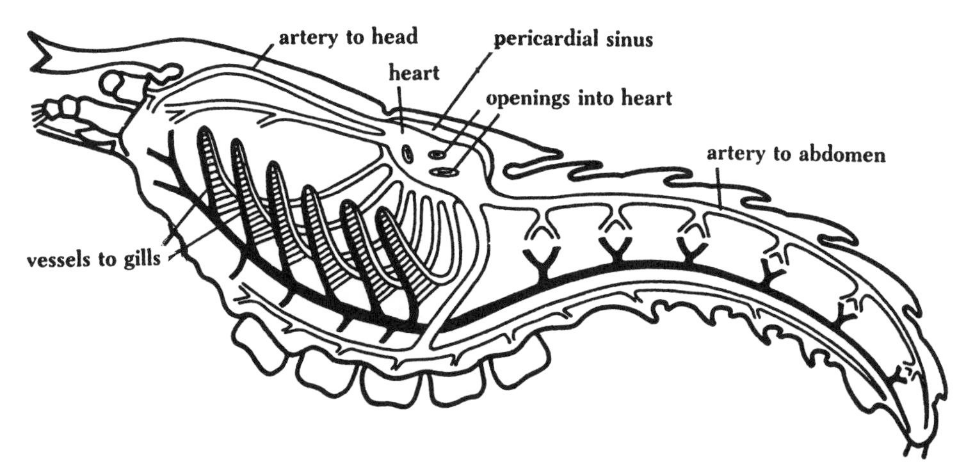

Circulatory system of a lobster, showing the main blood channels. Blood from the tissues flows through the gills before returning to the heart. (Modified after C. Gegenbauer.)

The **circulatory system** of lobsters is an *open* one, as in other arthropods and in molluscs. A muscular heart lies dorsally in a sinus filled with blood. In the sides of the heart are three pairs of openings through which blood from the pericardial sinus enters the relaxed heart. When the heart contracts, valves prevent the blood from going out these openings; instead, it is driven into arteries that go to the tissues of the body. The smallest branches of the arteries open into blood sinuses. Together, the blood sinuses constitute the main body cavity (hemocoel) of an arthropod. Blood returning from the tissues collects in a large ventral sinus and from there enters the gills, where it gives up carbon dioxide and takes on oxygen. Then it is returned, through a number of channels, to the large pericardial sinus that surrounds the heart.

There is a single pair of **excretory organs,** each consisting of a hollow sac, a glandular mass, and a coiled tube leading to a bladder, which opens through a pore at the base of the second antenna. The excretory glands control the volume and salt content of the blood and excrete some wastes, but most nitrogenous waste is lost as ammonia through the extensive surfaces of the gills.

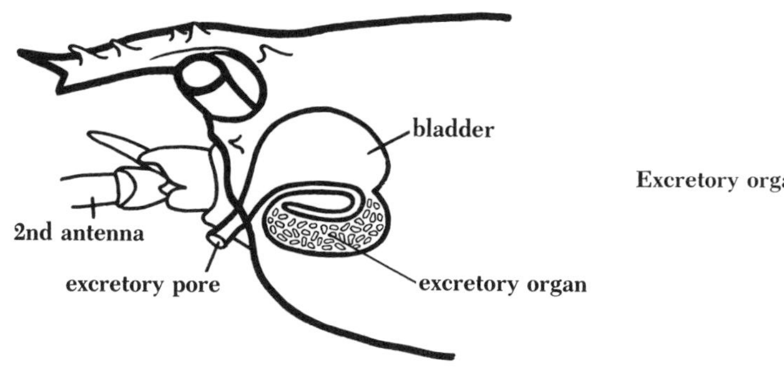

Excretory organ of a lobster.

Nervous system of a lobster is much like that of an annelid, with a dorsal brain, a ring of tissue around the esophagus, and a ventral, double, ganglionated cord.

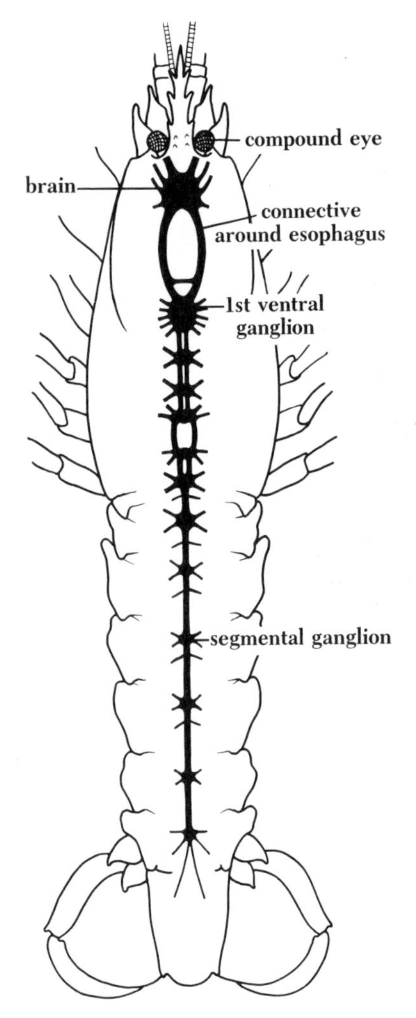

The general pattern of the **nervous system** is like that of annelids. The large brain lies dorsally in the head near the eyes. From it a pair of connectives pass ventrally, one on either side of the esophagus, and unite below the digestive tract at the first ventral ganglion, a compound ganglion from which a double nerve cord extends backward, enlarging into paired ganglia in every segment.

The most conspicuous sense organs are the **compound eyes.** These are composed of hundreds or thousands of closely packed visual units (ommatidia), each with its own lens, light-sensitive

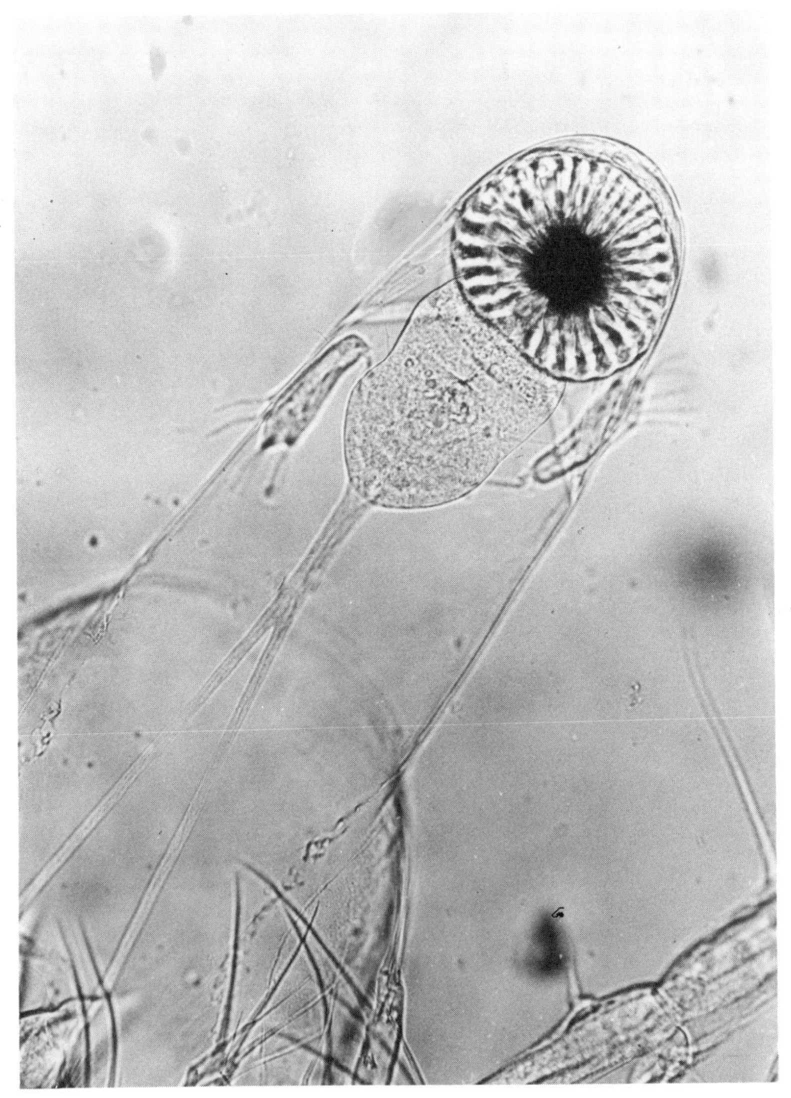

Compound eye of a small crustacean (*Leptodora*, a cladoceran). The transparency of this animal also reveals the brain, lying behind the eye, and the muscles that move the eye. (R. Buchsbaum)

cells, shielding pigment, and nerve fibers that go to the brain. Not all arthropods have compound eyes, but these eyes are unique to arthropods and are (along with the camera eyes of squids) among the mostly highly developed eyes of invertebrates. The compound eye is especially well suited to detecting *motion;* and it has been suggested that for an extremely small eye, the compound type is more acute than the camera design.

For lobsters, which are most active at night and live at depths where there is seldom enough light for clear vision even in daytime, the eyes are probably secondary in importance to the **sensory bristles** that are distributed all over the surface of the antennas and other appendages—from 50,000 to 100,000 of these bristles occurring on the big claws and walking legs alone. The bristles and other sense organs signal touch, water currents, pressure waves, and diverse chemicals. A lobster keeps track of its own orientation and movements by sense organs located in the soft membranes (at the joints and between segments) and by a pair of **statocysts** in the bases of the first antennas. Each statocyst is a water-filled sac that opens to the outside by a fine pore. On the floor and walls of the sac are rows of fine sensory bristles in contact with a mass of tiny sand grains cemented together. The position of the mass on the bristles gives the lobster information about its position with respect to gravity, and any movement of the lobster causes the mass to shift over the bristles.

> To study the function of the statocysts, one investigator performed a very ingenious experiment. He obtained a shrimp that had just molted and therefore had no sand grains in its statocysts. He put the animal in filtered water and supplied it with iron filings. The shrimp picked up the iron filings and placed them in the statocysts. Then, when the investigator turned on a powerful electromagnet *above* the animal, it turned over on its back.

The **reproductive system** consists of a pair of ovaries or testes, which lie in the dorsal part of the body and from which a pair of ducts leads to the external openings at the bases of the second walking legs in female lobsters, fourth walking legs in males. During mating, the male deposits sperms into a special receptacle near the female pores. As in many arthropods, the sperms are nonflagellated and nonmotile. (The flagellar tails of the motile sperms of some arthropods are the *only* motile flagella or cilia present in any arthropod cells.)

The fertilized eggs are fastened by a sticky secretion to the swimmerets of the female and are kept well aerated by movements of the swimmerets during the months that she carries them. A young lobster hatches from the egg as a free-swimming

Larval stages of a lobster. *Left:* The first larval stage is about 8 mm long. The appendages are all two-branched similar structures. The swimmerets at this stage are only small buds beneath the cuticle, and the larva swims about at the surface by the rowing action of the flattened, fringed outer branches of the thoracic appendages. *Right:* The fourth larval stage is about 15 mm long and resembles an adult lobster. Like the first stage, it swims at the surface feeding on small organisms; but forward swimming is now by means of the swimmerets. The outer branches of the legs are reduced and no longer visible; the inner branches are differentiated, though right and left big claws are still similar. (After F. H. Herrick.)

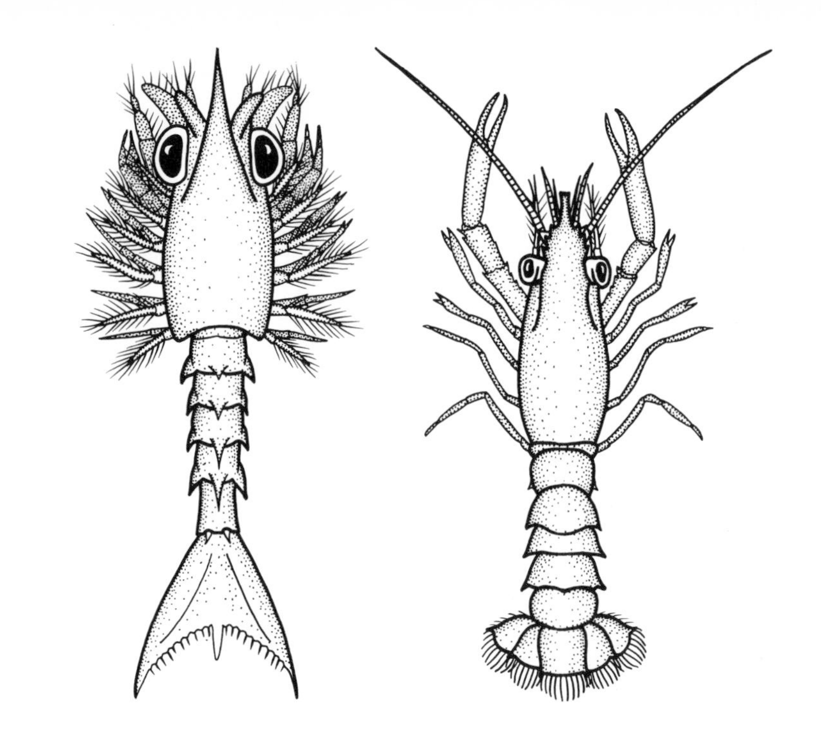

Mariculture of lobsters is promising at the government-supported experimental laboratory at Bodega Bay, California. Raised in individual plastic cubicles to prevent conflict (and cannibalism), these young lobsters will grow to salable size. It is still uncertain whether they can be reared on a commercial scale at a cost that is competitive with that of wild-caught lobsters. But mariculture becomes increasingly competitive as wild populations dwindle. (R. Buchsbaum)

larva and goes through a series of changes before it comes to resemble the adult. Lobsters continue to grow and molt throughout their lives.

Crustaceans

The name Crustacea was originally used to designate those animals having a hard but flexible "crust," as contrasted with those having a rigid shell like that of oysters or clams. Since nearly all arthropods have a hard, flexible exoskeleton, we now use more distinctive criteria for assigning an animal to the subphylum **Crustacea,** of which lobsters are members. Crustaceans may be roughly distinguished as mostly aquatic arthropods with gills, jaws, and two pairs of antennas. Lobsters are giants among crustaceans; the great majority are measured in millimeters and have thin noncalcified exoskeletons.

There are so many different types of crustaceans, not to mention individual species (which number about 35,000), that no one could think of common names for them all, and most are referred to as some kind of shrimp, crab, or lobster. These loose tags are assigned according to the general shape of the animal and are often shared by types that are only distantly related. The lobsters we have described are of the genus *Homarus*, distributed around the North Atlantic, and are more closely related to freshwater crayfishes than to some other animals called lobsters (spiny lobsters, slipper lobsters). "Lobster" usually means a good-sized crustacean with a stout trunk and abdomen but, because it sounds elegant, it is used rather more broadly by menu writers than by most people. "Crab" is commonly used for crustaceans with prominent pincers and compact bodies (but is also applied to some arthropods that are not crustaceans). And a "shrimp" can be almost any crustacean but especially a long, slender one with many similar appendages. We imagine primitive crustaceans to have been shrimplike.

The earliest crustaceans are believed to have lived in seawater, and the group is still predominantly marine. The oceans are the easiest place to live in and require the fewest adjustments on the part of inhabitants. The salt concentration of animal tissues and body fluids is much closer to that of seawater than freshwater. And because of their tremendous volume, the seas provide relatively constant salt content, oxygen content, and temperature throughout the year, although the absolute amounts vary with depth and from place to place. Crustaceans are so abundant in the oceans that they have been called the "insects of the sea," and there is hardly any way of life in the sea not followed by some member of this diversified subphylum.

Left: **Mantis shrimp,** or stomatopod, is seen here in an aquarium in Florida. In nature stomatopods live mostly in shallow tropical and subtropical waters. Some make burrows in sand or mud and await their prey, spearing worms and fishes with a large appendage (seen at the front) that unfolds like a jackknife, and with lightning speed. Another kind of stomatopod lives in holes in rocks, and clubs its prey rather than spearing it. Stomatopods are themselves good eating for humans and are savored in the islands of the Pacific. (R. Buchsbaum)

Right: **Isopods** are named from their many similar legs but usually recognized at a glance by their dorsoventrally flattened bodies. Most are marine, living on the bottom or on vegetation, but some live in freshwaters and others are terrestrial. The wood-boring isopods shown here, called gribbles, are tiny marine crustaceans that occur in immense numbers in old submerged wood. They chew and digest their way along, doing enormous damage to pier pilings around the world. *Limnoria quadripunctata*, only 3 mm long, is a common gribble of the U.S. West Coast. (C. A. Kofoid)

Below, right: **Amphipod** is flattened from side to side, like this beachhopper, *Orchestoidea californiana*, 25 mm (1 in) long. It is common on exposed sandy beaches in California and may be seen scavenging in picnic leavings or cast-up seaweeds. If a mass of damp, rotting seaweed is disturbed, amphipods hop out in all directions and then head back in. During sunny hours beachhoppers take refuge in burrows in the sand, which they defend vigorously against usurpers. Most amphipods are marine, living on the bottom, on seaweeds, or as pelagic forms. But there are many freshwater amphipods and a few in moist soils. (R. Buchsbaum)

Below: **Commercial shrimp,** *Penaeus*, is large (up to 20 cm). Members of this genus are collected by the ton for human consumption in Florida, South America, the Mediterranean, Australia, and other warm parts of the world. Seen here in an aquarium at the Gulf Specimen Co., Panacea, N.W. Florida. (R. Buchsbaum)

Above: **Krill** is a term for enormous swarms of little shrimplike crustaceans, the euphausids. Such swarms in the Arctic and Antarctic oceans may extend over many square kilometers of surface water, providing an abundance of food for many invertebrates, fishes, and seabirds, and most notably for the great baleen whales. The fisheries of Norway, Japan, and Russia are preparing to exploit krill for human consumption and for fertilizer, but the natural ups and downs of these and many other marine organisms make them an unreliable source for burgeoning human populations to depend on. (In 1984, for example, the Antarctic swarms of krill virtually disappeared.) Seen here is the most abundant and most studied of the euphausids, *Euphausia superba*, about 50 mm (2 in) long. (Preserved specimen, Scripps collection, lent by W. A. Newman; photo, R. Buchsbaum.)

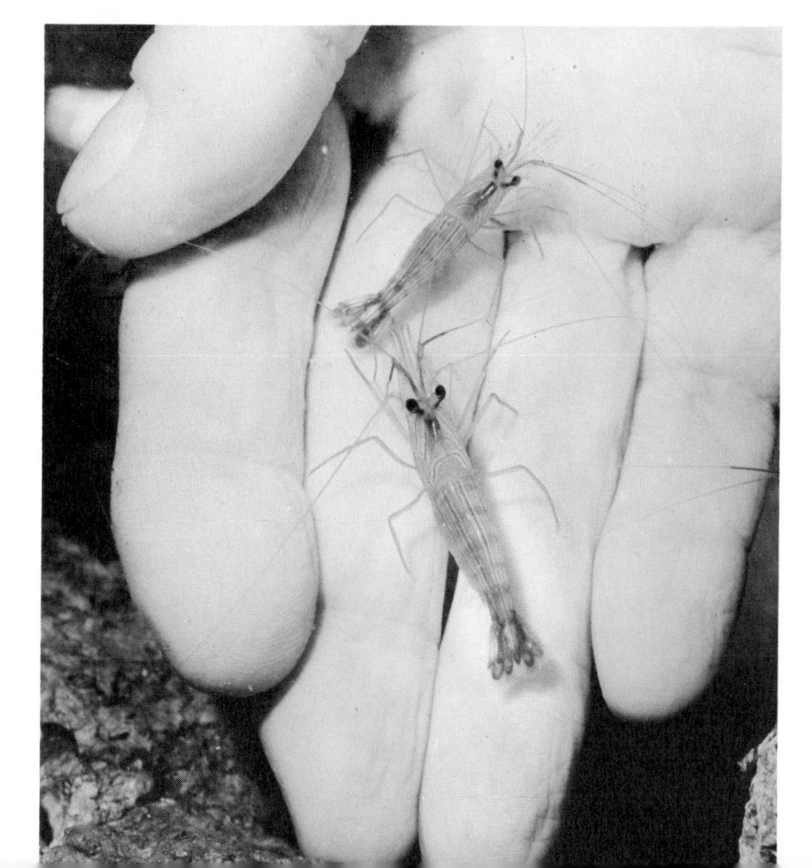

Right: **Red-striped cleaning shrimps** pick parasites from fishes. Attracted to a hand dipped into a small aquarium, these two are picking away at the skin. *Hippolysmata wurdemanni*. N.W. Florida. (R. Buchsbaum)

Crab has a small flat abdomen that is folded under and fits into a shallow depression in the underside of the cephalothorax. Crabs run sideways on four pairs of walking legs. The Dungeness crab, *Cancer magister,* chief commercial crab of the U.S. Pacific coast, is being overcollected. It is found offshore on sandy bottoms from the Aleutians to Baja California. This one came out onto an Oregon beach to mate. (R. Buchsbaum)

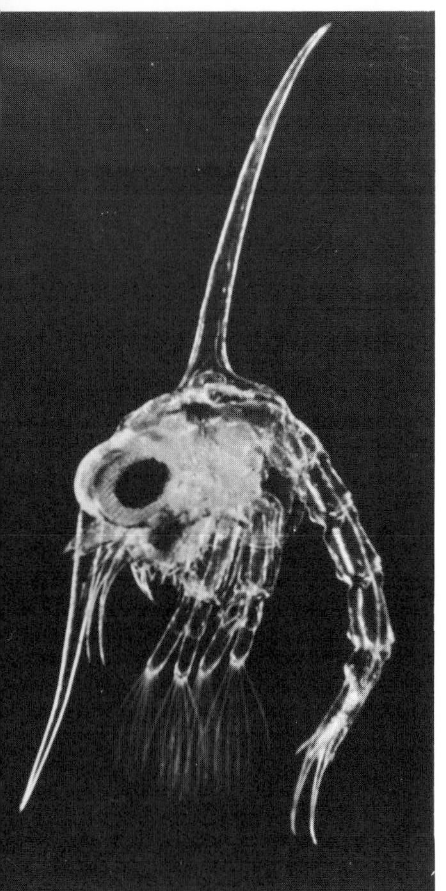

Left: **Zoea of a crab** is the free-swimming larval stage that hatches from the egg and molts several times, each time growing a little larger and more like the adult. Seen here is the second zoeal stage of the velvet swimming crab, *Macropipus puber.* The carapace bears long spines, typical of crab zoeas; presumably they slow sinking and discourage some predators. England. (D. P. Wilson)

Above: **Fiddler crab,** so named because males have one larger claw, half the weight of the crab or more and often brightly colored. Standing beside the opening to his burrow, a male waves his large claw conspicuously, signaling other males to stay out of his territory or attracting receptive females. The waving display is species-specific and probably prevents interbreeding of different species. In females both claws are of equal size. Fiddler crabs occur in vast numbers on most tropical sandy or muddy shores. Australia. (J. S. Pearse)

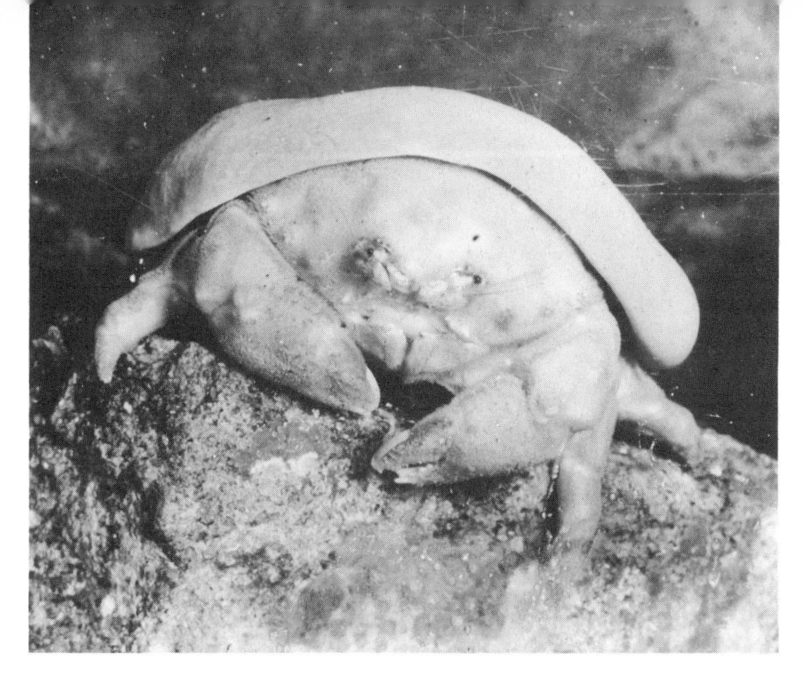

Left: **Sponge crab,** *Dromia vulgaris,* disguises itself from predators with a covering of sponge. It hollows a piece of sponge to fit its back and holds it on with the last two pairs of legs. The sponge *Suberites* offers chemical protection as well as camouflage. In an aquarium *Octopus* will not attack *Dromia* unless *Suberites* has been removed. Roscoff, France. (R. Buchsbaum)

Right: **Ghost crab** on a sandy beach in Fiji has left its burrow at night to scavenge and to hunt small prey. Named for the hornlike extensions on its eyes, *Ocypode ceratophthalmus* has 30,000 facets in each eye. The speed of these fleet crabs has been clocked at 2.1 m/sec (7 ft/sec). This and other species of *Ocypode* are found on warm sandy beaches around the world. (R. Buchsbaum)

Burrowing ghost crab on a Panama shore was throwing sand from its tunnel when interrupted by human footsteps. It retreated deep into the tunnel for at least half an hour before reappearing at the opening. Most ghost crabs inhabit tropical shores, above high-water mark, and dig deep burrows down to the level of moist sand. (M. Buchsbaum)

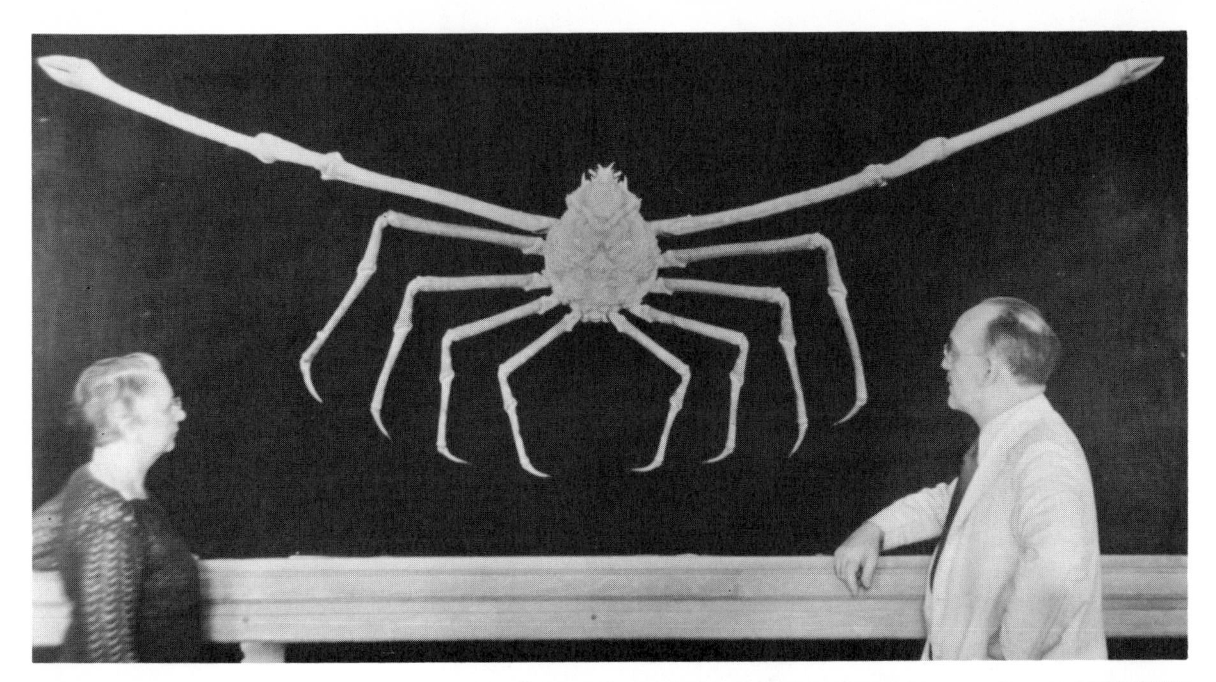

Giant spider crab from waters off Japan is the largest of crabs. This dried specimen of *Macrocheira kaempferi* shows the long clawed appendages at the front, spanning up to 3.65 m (12 ft), and the four pairs of walking legs. The abdomen is turned under, and this dorsal view shows only the cephalothorax, up to about 45 cm (18 in) across. (Buffalo Museum of Science)

Hermit-crab has a long, soft abdomen that it inserts into the empty shell of a marine snail. As the hermit-crab grows, it must adopt larger and larger shells, sometimes by dispossessing another hermit-crab. Only the hard, calcified cephalothorax with its two pairs of walking legs is protruded from the shell—or quickly withdrawn when danger threatens. The last two pairs of legs are tiny. *Dardanus megistos*, seen here on a tidal flat at Heron Island in the Great Barrier Reef of Australia, may be more than 30 cm (12 in) long. Smaller but similar hermit-crabs on all temperate shores are important scavengers. (R. Buchsbaum)

Mole-crab, or sand-crab, *Emerita talpoida,* inhabits sandy beaches from Cape Cod to Yucatan. It is one of the few beach residents that can brave the rigors and exploit the food-bearing currents of the zone of breaking waves. With its almost cylindrical body dug in at a 45° angle, and facing seaward, it holds out its feathery second antennas, which filter small organisms from the returning current as the waves wash back to the sea. A related species lives on the U.S. West Coast. Most sand-crabs are tropical or subtropical. (R. Buchsbaum)

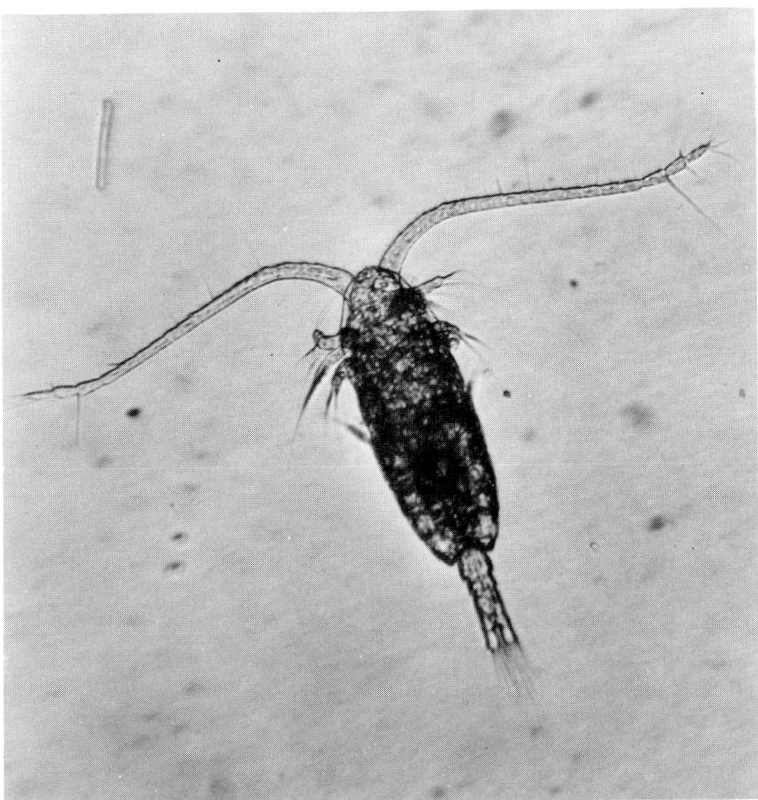

Marine copepod is a microcrustacean 2 mm long. Often called the "insects of the sea," copepods are the most numerous of all animals. As primary consumers, they feed on the microscopic producers (photosynthetic protists) that form the broad base of the marine food pyramid. Copepods are themselves eaten by secondary consumers, from small or larval invertebrates and young fishes in the plankton to great baleen whales. Large copepods are also secondary consumers, devouring vast quantities of the eggs and larvas of various pelagic animals (often copepod predators). Many species of copepods are bottom-living types or are parasitic on invertebrates or vertebrates, especially fishes. (R. Buchsbaum)

Among the most highly modified crustaceans are the **barnacles,** sessile marine animals that live attached to rocks, wooden pilings, ships, and the bodies of other animals. Most people are surprised to find barnacles among the arthropods, assuming them to be molluscs because of their thick calcareous shells. Early zoologists, too, classified them with the molluscs until their relationships were demonstrated beyond argument by a close study of their structure and especially their development. A young barnacle larva just hatched from the egg is free-swimming. Its three pairs of appendages and other characters identify it clearly as a **nauplius,** a larval type characteristic of crustaceans. After swimming about for a time and passing through several molts and stages, it settles on some solid object and attaches by its head end. In spite of their extreme modifications, adult barnacles can be recognized as arthropods by their chitinous jointed appendages, which are two-branched as in most other crusta-

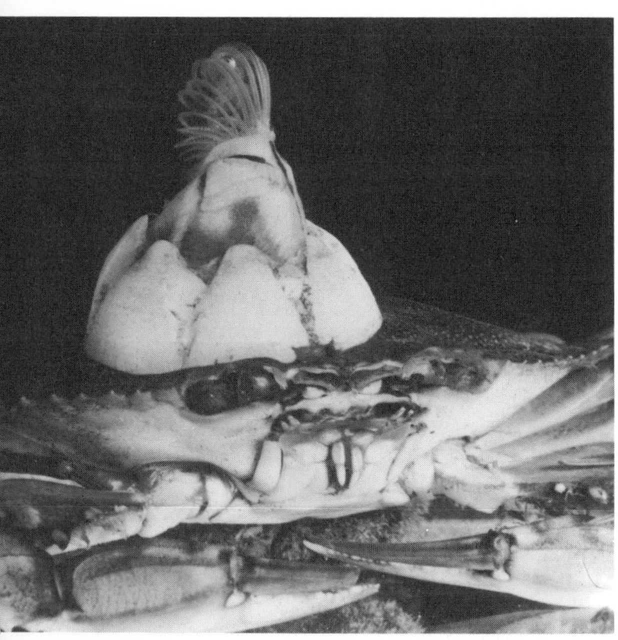

A barnacle has been described as an animal that sits on its head and kicks food into its mouth with its feet. This acorn barnacle has the broad base firmly cemented to hard substrate (in this case, the back of a blue crab). Overlapping calcareous plates form a sturdy wall around the sides of the barnacle, and two pairs of plates at the top open and close the opening through which the feeding appendages are extended (as here) and then withdrawn in a steady rhythm. Florida. (R. Buchsbaum)

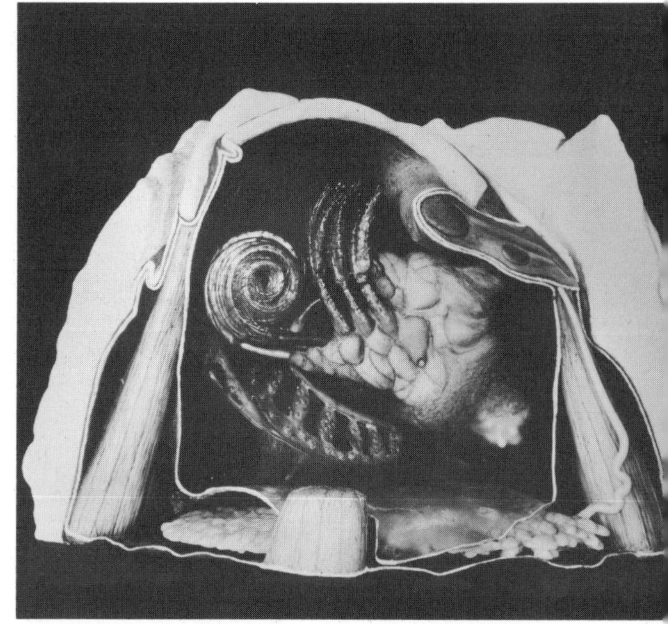

Model of an acorn barnacle, a side view with half the wall of plates removed. This barnacle is closed, as if exposed to air during low tide. When a submerged barnacle is open and feeding, the long, coiled appendages are slowly extended by being pumped full of blood and quickly withdrawn by muscles. They transfer food to the shorter appendages, which in turn pass it to the mouth. (American Museum of Natural History)

Acorn barnacles cover a lobster that has been confined in an aquarium for a long time and apparently has not grown or molted normally. Barnacles may live on almost any hard surface on which the last larval stage chooses to settle, often close to other barnacles of its species. Helgoland. (F. Schensky)

Stalked barnacles, or goose barnacles, are similar in structure and life history to acorn barnacles. Stout, jointed appendages show the animals to be arthropods. This cluster of Pacific goose barnacles, *Pollicipes polymerus*, tore loose from its attachment to a surf-beaten rock in Monterey Bay, California, and was photographed in an aquarium at Hopkins Marine Station. (R. Buchsbaum)

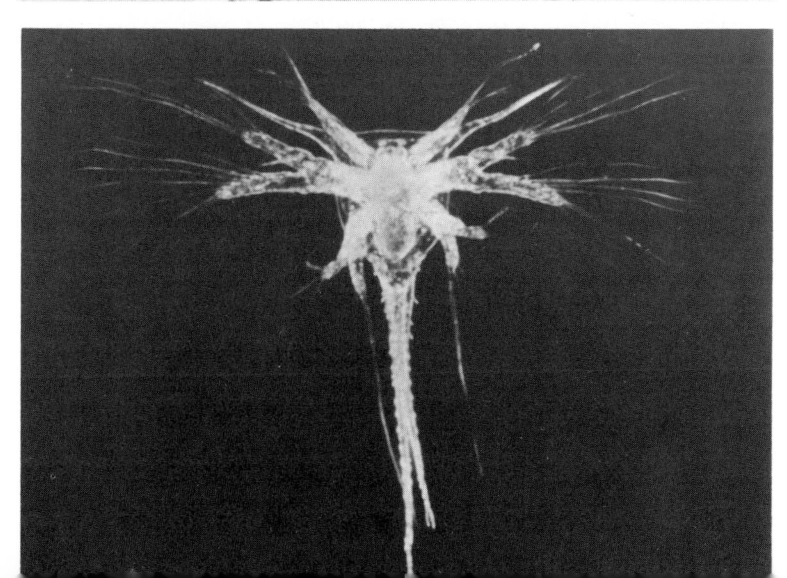

Nauplius larva of a stalked barnacle, *Lepas*, confirms the crustacean affinity of the sessile adult barnacle. Like the nauplius of other crustaceans, it has three pairs of appendages. (D. P. Wilson)

ceans and are heavily fringed with bristles. The appendages are thrust out of the shell and sweep through the water like a casting net, entrapping small animals and organic fragments. With few exceptions, barnacles are hermaphroditic; as in many other groups of animals, this is thought to be associated with their sessile life, which limits contacts between individuals. Another group of curious marine crustaceans closely related to barnacles are the **rhizocephalans,** which have a free-swimming nauplius but soon fasten onto a crab and, in the adult stage, send rootlike processes into every part of the host's body. The parasitized crab can neither molt nor reproduce, but remains alive, nourishing its unwelcome guest until the parasite has completed its life cycle.

Parasitic crustaceans related to barnacles are rhizocephalans, *Sacculina carcini,* seen here as two bulbous sacs protruding through the cuticle on the underside of a European shore crab *(Carcinides maenas)* at Roscoff, France. Each sac is filled with the eggs of the parasite, which grows inside the crab as a branching rootlike system of tubules pervading all the blood spaces of the host, even to the tips of the legs. Similar rhizocephalans are found in crabs of both American coasts, and various kinds parasitize other crustaceans, including barnacles. (J. Vasserot)

Freshwater crustaceans have done quite well when we consider that this medium is a more difficult one for all groups of animals. To invade the rivers that connect directly with the ocean, a crustacean not only must become adjusted to the lowered salt content but must be able to maintain itself against the downstream current. This is possible only for crustaceans that are relatively large, such as freshwater crayfishes and crabs, and that do not release their young as fragile, free-swimming larvas that are easily swept downstream and back to the ocean. Most freshwater crustaceans are small and fragile even as adults, and they do not live where there are strong currents but in quiet ponds or lakes, to which they must be carried by winds or in mud on the feet of water birds. Small ponds may dry completely in summer and freeze solid in winter, and even sizable lakes are subject to violent fluctuations in temperature. Crustaceans have adapted to these rigorous conditions partly by the development of thick-shelled eggs that resist drying and freezing.

Fairy-shrimps are primitive crustaceans, most of which live in temporary ponds, roadside ditches, or marshes that dry up in summer. They lay thick-walled eggs that withstand summer drying and winter freezing, lying dormant in the mud until they hatch out as naupliuses when wetted by the first heavy spring rain. The leaflike legs with feathery bristles serve both for locomotion and for straining microscopic food from the water. Fairy-shrimps swim along slowly on their backs and feed as they go. All three seen here are females with brood pouches filled with eggs. *Eubranchipus vernalis*, about 25 mm (1 in) long. Pennsylvania. (R. Buchsbaum)

Brine-shrimp, *Artemia salina*, about a centimeter long, is a fairy-shrimp that lives in salty inland lakes and in evaporating basins dammed off from the ocean for commercial production of salt. Brine-shrimps are abundant in the Great Salt Lake in Utah, but never occur in the ocean. The female seen here has a brood pouch filled with eggs. The dried eggs of brine-shrimps are sold widely in pet shops; when they hatch (as advanced naupliuses) and develop into adults, they provide live food for aquarium fishes. (R. Buchsbaum)

Waterflea, *Daphnia,* 1 to 3 mm long, gets its common name from the jerky way in which it "hops" through the water, by vigorous strokes of the branched second antennas. *Daphnia* and other freshwater cladocerans strain microscopic food from the water by means of bristly legs. They are themselves prey to many small invertebrates and fishes, and are the most important primary consumers in the economy of lakes, exceeding even the copepods. Visible here are the prominent eye and the translucent carapace that covers body and appendages but is open ventrally. Above the long, food-filled intestine is the oval muscular heart, and behind the heart is the brood pouch containing a developing offspring. (P. S. Tice)

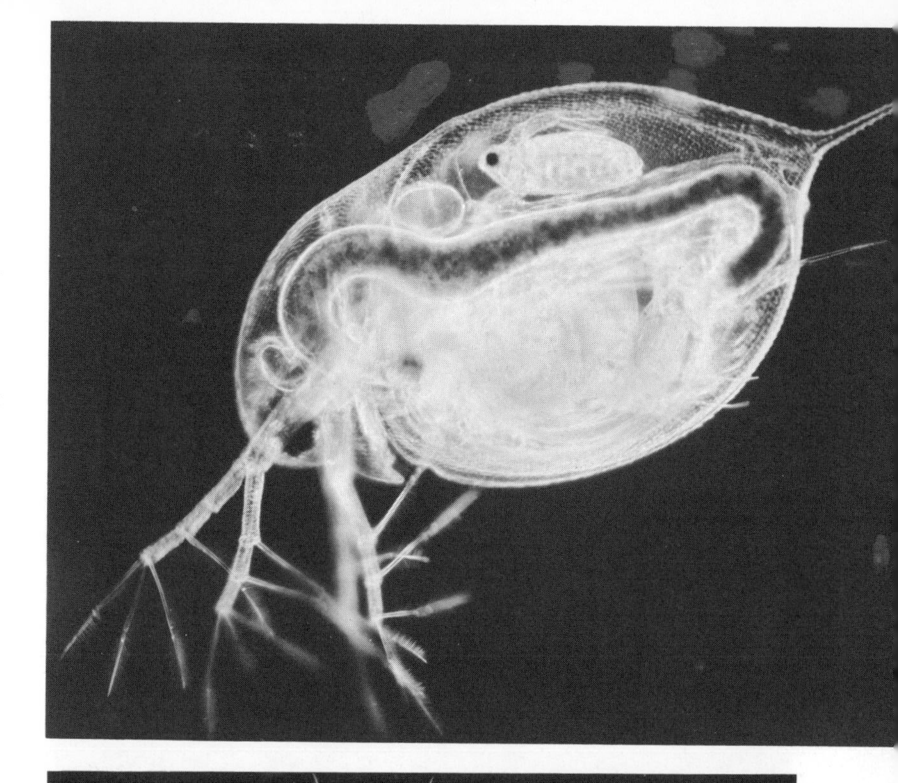

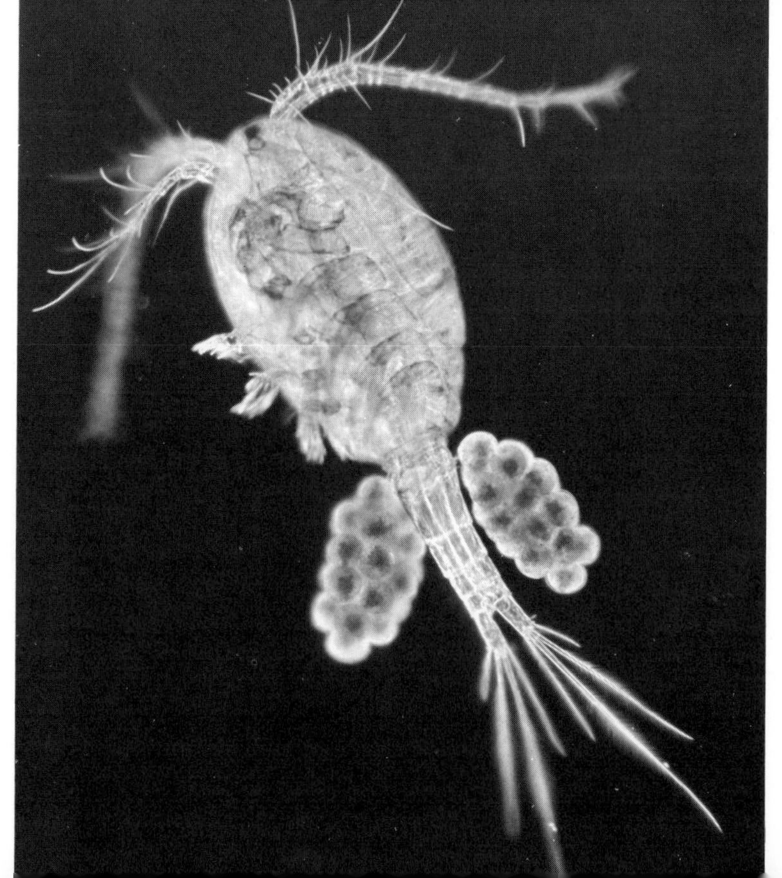

Freshwater copepod, *Cyclops,* is named for its median eye. This mature female, almost 2 mm long, has two egg sacs attached to the front of the abdomen. *Cyclops* swims with the thoracic legs and uses the long sensory first antennas as rudders to direct movement. Copepods feed on a variety of protists and even minute animals, and in ponds and lakes are second only to cladocerans as links in the food chain between the microscopic organisms and the small invertebrates and fishes. (P. S. Tice)

Ostracod has a bivalved, hinged, almost opaque carapace that resembles a clam shell only about 1 mm long. At the left protrude the two pairs of antennas, both sensory and active in swimming. At the ventral gape are seen the clawed tips of two pairs of thoracic legs with which the ostracod crawls along surfaces. Ostracods are common in both seawater and freshwater. (R. Buchsbaum)

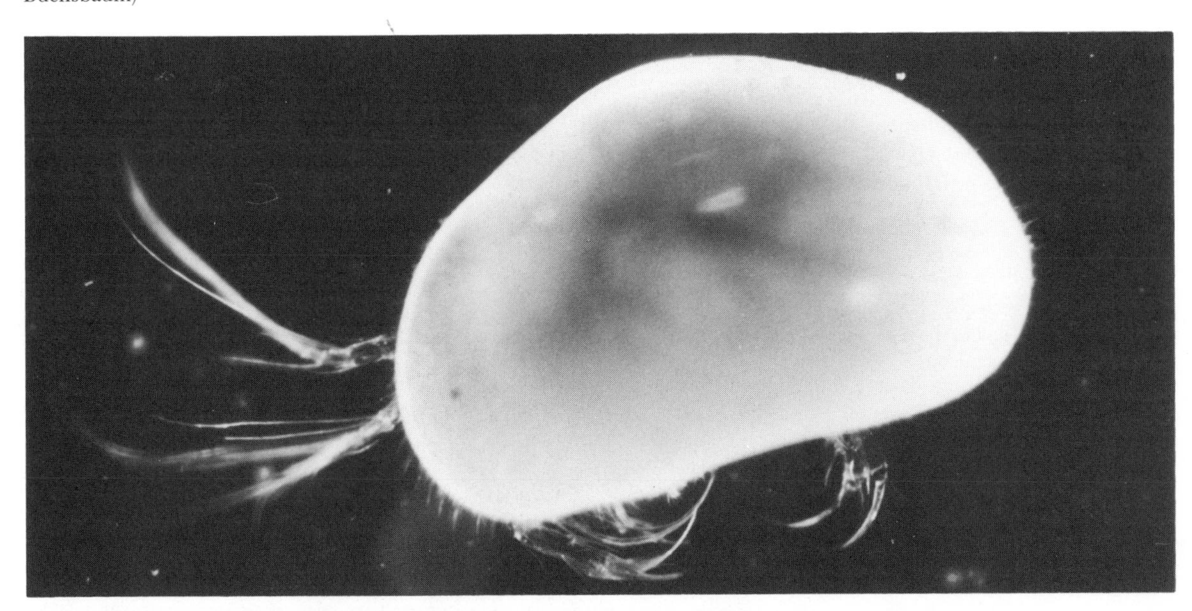

Crayfish, like a lobster, is a scavenger and predator, feeding on decaying organic remains but also catching small animals. Crayfishes walk slowly headfirst along the bottoms of streams and ponds, but can shoot backward to escape a threat by suddenly contracting the powerful abdominal muscles, which are (as in lobsters) the part that people eat. This female is brooding young beneath the abdomen.

Right: **Swamp crayfish** beside the chimney that surrounds its burrow. The burrows are 30 to 100 cm deep and have a water-filled cavity at the bottom. They are built in swamps and meadows, often far from a surface stream. We have seen such chimneys, with bright blue young crayfishes sheltered inside, in the center of Pittsburgh, Pennsylvania, a city in which underground streams come near the surface in some gardens. (C. Clarke)

Below: **Head of a crayfish,** showing the stalked compound eyes and jointed antennas. The first pair are short and two-branched; the second pair are long and single. Two pairs of antennas are a distinguishing character of crustaceans. Insects, centipedes, and millipedes have one pair; arachnids and other chelicerates have none. (P. S. Tice)

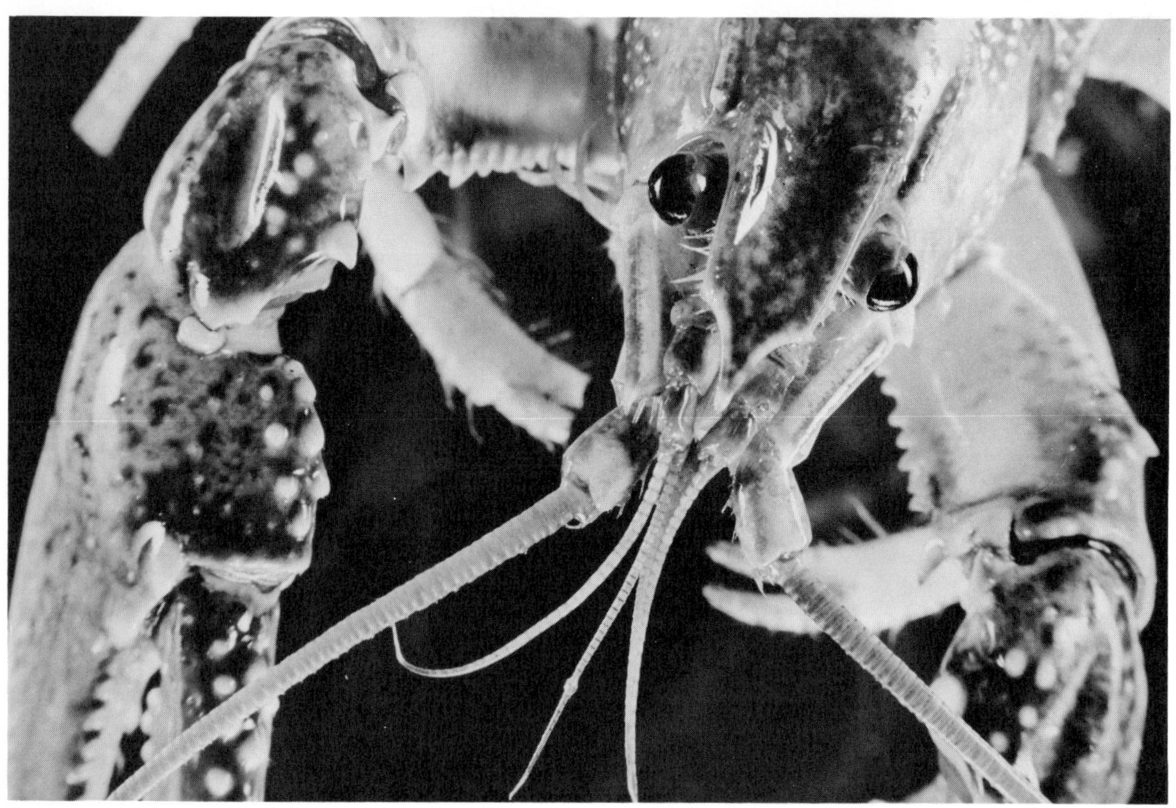

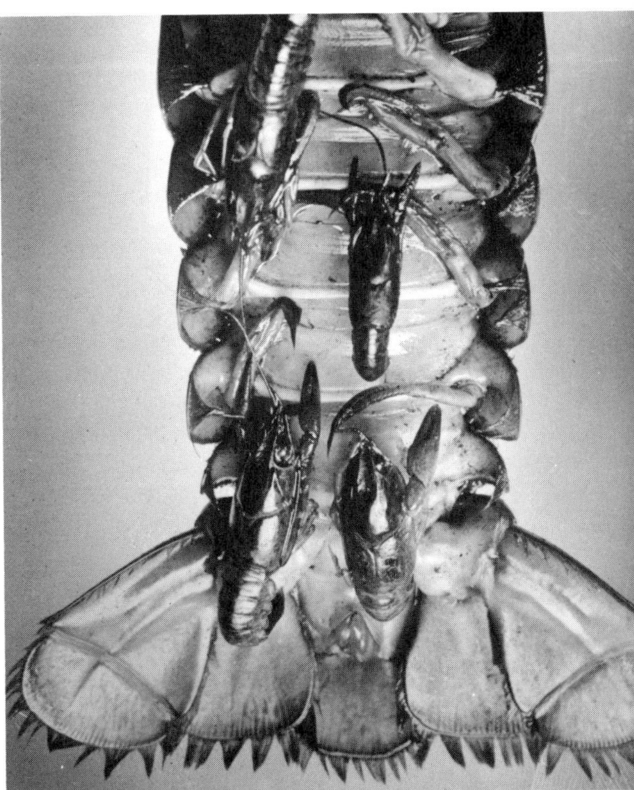

Eggs of a crayfish (*left*), develop attached to the abdominal appendages. They hatch into **young crayfishes** (*right*), that look like miniature adults and cling for a time to the abdominal appendages of the mother. (C. Clarke)

Adapting to **life on land** is a still more difficult step. Temperature ranges are even more extreme, drying is a constant threat, and respiratory mechanisms must be modified to use air. The few crustaceans that do fairly well on land manage to a large extent by avoiding these problems. Land crabs live in warm, moist places and emerge from their burrows to feed mainly at night. They must return to the sea to release their swimming larvas. The isopods called woodlice or pillbugs are more widespread. They still depend on moist hiding places but do not have any aquatic stage. In addition to gills, these crustaceans have developed a system of invaginated air-filled respiratory tubes that are remarkably like those evolved by arachnids and insects, the two main arthropod groups that have specialized in land life.

Rocky-shore isopod, *Ligia occidentalis,* which reaches 35 mm in length, is common on steep rock faces and in rock fissures or caves on the California coast and even along the Sacramento River. Popularly known as rocklice or sea slaters, members of the worldwide genus *Ligia* have left the sea and would drown if submerged for long. They scurry about on damp rock, but keep just above the spray line when surf is strong. At low tide they venture into the intertidal zone to scavenge. (R. Buchsbaum)

Opposite: **Land crab** has not made a complete transition to terrestrial life. It must pass through its earliest stages in the sea; it still has gills for much of its respiratory exchange; and like other crustaceans it lacks the outer waxy layer of the cuticle that slows evaporation in insects and arachnids. This species, *Cardisoma guanhumi*, can be found hiding under logs in the moist forests of Panama, roaming even several kilometers inland. It feeds at night and during rains. (R. Buchsbaum)

The most land-adapted crustaceans are the isopods called by such names as sowbugs, woodlice, or pillbugs. They are abundant in the ground litter of gardens and forests, where they reproduce, far from any body of water, by brooding their young in a fluid-filled pouch. Land isopods still depend for most of their respiratory exchange on abdominal appendages that serve as gills, and must channel dew or rain drops to these respiratory surfaces; but they also have internal air tubes that resemble those found in insects and arachnids. Vulnerable to drying, they hide by day in moist refuges and come out at night to feed on plant matter. When abundant, they play an important role in the formation of humus. (R. Buchsbaum)

Chapter 18

Arthropods on Land: Arachnids

Adapting from aquatic to land life is not easy. The handful of terrestrial invertebrates so far described—land planarians, nematodes, earthworms, land snails, and a few crustaceans—are all restricted to moist habitats. Land crabs, as already mentioned, must seek out water to reproduce, for their young stages are aquatic. Even among those groups of terrestrial arthropods and vertebrates that have succeeded in occupying relatively dry habitats, the great majority (including humans) are still utterly dependent for their survival on reliable sources of water to drink. These animals reign over the dry habitats on earth only because they have evolved specific behaviors, structures, and physiological measures that enable them to carry out a policy of strict water conservation. Prominent among this landed gentry are the arthropods called arachnids.

Opposite: **The arachnid body,** as exemplified by a spider, can readily be distinguished from that of an insect by the division of the body into only two regions, the first bearing four pairs of jointed walking legs; insects have three body regions and three pairs of walking legs. Spiders lack the antennas of insects, but the pair of short leglike pedipalps, extended in front, are sensory, as are the many large bristles on the legs. Seen here is a female shamrock spider, *Araneus trifolium,* common in North American meadows. She builds a symmetrical orb web of silk and waits quietly for small prey to be ensnared. Many spiders build no webs but roam about actively seeking prey, as do other arachnids. Of the nearly 70,000 species of arachnids, about half are spiders. The rest are divided into ten other orders, some of which will be illustrated. (R. Buchsbaum)

Arachnids

The class **Arachnida** includes the spiders, the scorpions, the ticks and other mites, the harvestmen ("daddy longlegs"), and a few minor groups. No other class of animals is less loved by most people. There is some basis for this dislike, in that scorpions and spiders can inject a poison that produces painful, though seldom serious, results in humans; some ticks suck human blood and spread disease, and certain other mites are parasites in human skin. But relatively few people in large cities have ever had a single unpleasant experience with an arachnid. The sinister reputation of a group like the spiders, which do little harm and much good (by killing insect pests), is based on nothing more than a vague fear of animals that run rapidly and live in dark places.

Hardly any description will fit all the orders; but, in general, arachnids are terrestrial arthropods with a body divided into two main regions: a **cephalothorax** bearing six pairs of appendages, of which four of the pairs are walking legs, and an **abdomen** that has no locomotory appendages, though it may have some other kind. The *four pairs of walking legs* usually serve as a convenient, if superficial, way of distinguishing arachnids from insects, which have only three pairs. But the difference between the groups is much more deep-seated. Arachnids differ from crustaceans and insects in having *no compound eyes,* though they have simple ones, and *no antennas,* though the long first legs or other appendages may be primarily sensory in function and the whole body is covered with delicate sensory bristles. Also, arachnids have *no jaws* homologous with those of crustaceans or other arthropods. None of the arachnid appendages is completely specialized for chewing, but on the bases of one or more of them are sharp chewing processes. In front of the mouth arachnids have a pair of **cheliceras** (kuh-LISS-er-uhs), appendages that end in pincers or (in spiders) sharp fangs. These distinctive appendages give their name to the arthropod subphylum **Chelicerata,** to which arachnids belong, together with two marine groups. Behind the mouth is a pair of **pedipalps,** appendages that serve a sensory function, as in spiders, or are used for seizing prey, as in scorpions. Almost all arachnids are predators; and among the various orders either the cheliceras or the pedipalps are the important weapons of offense, but never both in the same animal.

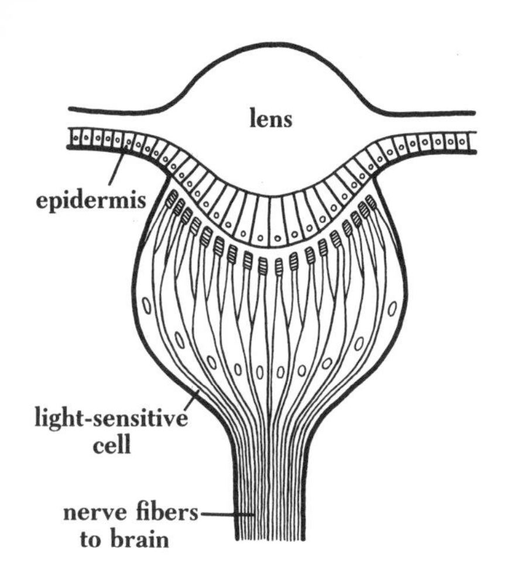

Simple eye of a spider is so called because the light-sensitive cells share a single lens, in contrast to the compound eyes of crustaceans and insects. The lens is made of cuticle and secreted by the underlying epidermis. (After Hentschel.)

Whipscorpion comes out of hiding at night from under stones, loose bark, or litter. There are simple eyes on the front of the carapace, but these nocturnal arachnids depend mostly on tactile information from the long whiplike first pair of walking legs, which explore the ground like feelers. The large, stout pedipalps seize small arthropod prey—and can also deliver a strong pinch to human fingers. Whipscorpions also defend themselves by spraying a stream of acid liquid (mostly acetic acid) from the rear end—hence the common name "vinegaroon." The secretion is irritating to human skin and eyes. The segmented abdomen ends in a long filament (here directed to one side). *Mastigoproctus giganteus,* photographed at the Desert Museum in Tucson, Arizona, occurs across the southern United States and is the largest of whipscorpions, reaching 8 cm (3 in). Most members of this order live in tropical America and Asia. (R. Buchsbaum)

Harvestman is so named because in northern temperate climates these arachnids mature in July and August and are most often seen at harvest time. Another common name is "daddy longlegs," an apt name for the kind of harvestmen with extremely long legs that make it easy to climb among leaves. Others have sturdy legs no longer than the body. In this order of arachnids the oval body has a cephalothorax broadly joined to a clearly segmented abdomen. (R. Buchsbaum)

Sunspider is the common name applied to solifugids that are active in the daytime. This one, from a California garden, was climbing about in low shrubs seeking insect prey. Solifugids have two large eyes on the head but get most of their information from the long bristles that cover the body and appendages. They run with the large pedipalps and first pair of legs held out as feelers. Solifugids run very fast, "like the wind," and a common name applied to the whole order is "windscorpions." Unlike scorpions, windscorpions have no rear stinger, but the large pinching cheliceras (visible in this photo) can deliver a bite that is painful and sometimes becomes seriously infected. Day-active solifugids may come into city streets, and the nocturnal majority are attracted to lights at night, sometimes coming to campfires or entering tents to the dismay of campers. Most solifugids are desert dwellers in Eurasia and Africa, but other species occur in Europe, Central and South America, the West Indies, Florida, and the southwestern United States. Large solifugids (up to 5 cm long) may add to their usual insect diet an occasional toad, lizard, small bird, or mouse. (R. Buchsbaum)

Spiders are by far the largest and most widely distributed order of arachnids. A generalized description of a spider, though applying in many respects only to this one group, will give some further idea of arachnid structure and habit. In a spider the cephalothorax is covered by a shield, the carapace, on which are set the simple eyes, usually eight in number. The cheliceras are sharp and pointed and are used for capturing and then paralyzing the prey by injecting a poison. Ducts from a pair of poison glands lead through the cheliceras and open near their tips. The pedipalps look like small legs but are sensory, and their expanded bases help to hold and compress the prey. In a mature male the pedipalps are modified for transferring sperms to the female. The *four pairs* of walking legs end in curved claws.

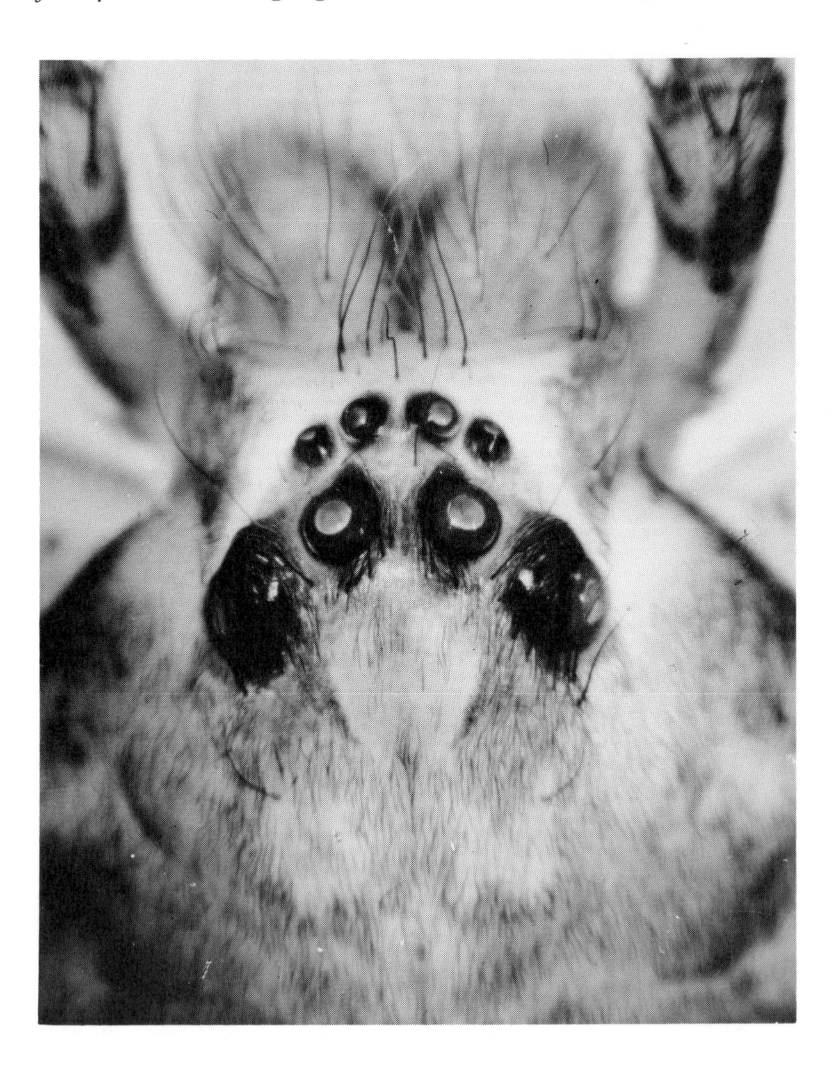

Eight simple eyes of a spider are usually arranged in two rows. The lens is surrounded by a ring of dark pigment. Some spiders have only six eyes. (P. S. Tice)

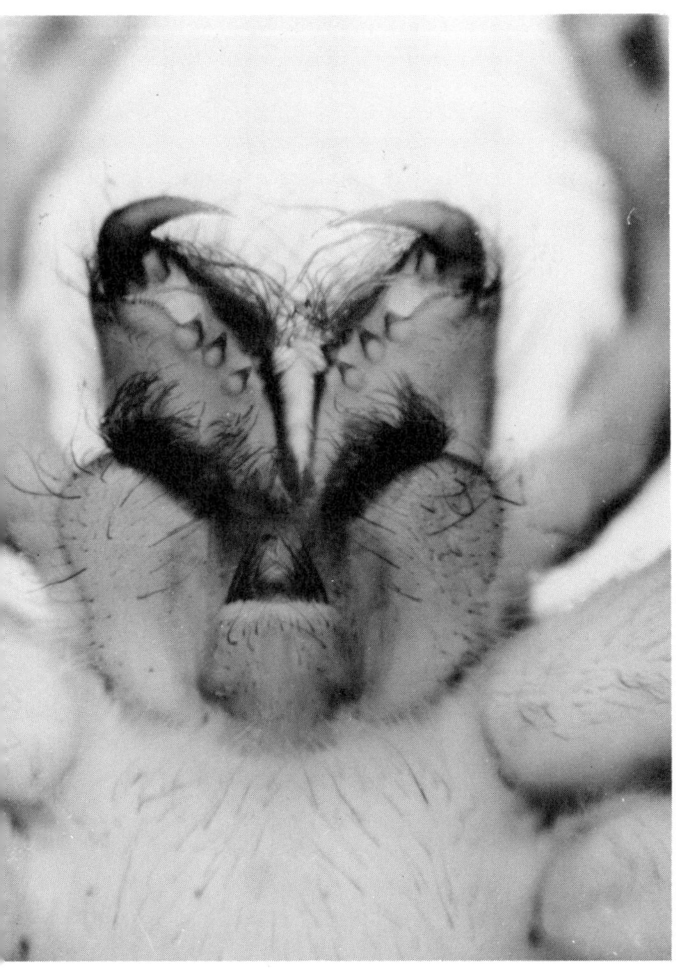

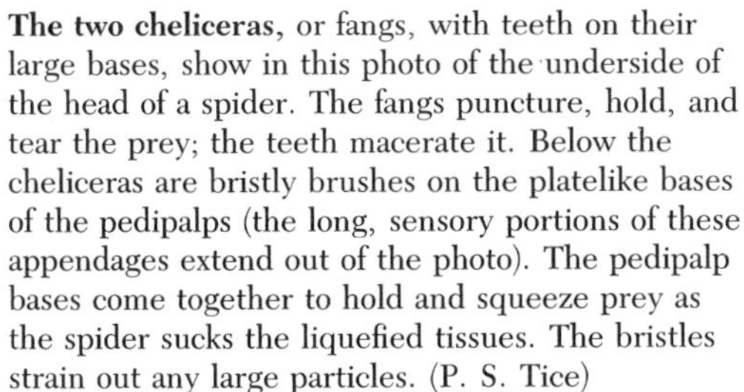

The two cheliceras, or fangs, with teeth on their large bases, show in this photo of the underside of the head of a spider. The fangs puncture, hold, and tear the prey; the teeth macerate it. Below the cheliceras are bristly brushes on the platelike bases of the pedipalps (the long, sensory portions of these appendages extend out of the photo). The pedipalp bases come together to hold and squeeze prey as the spider sucks the liquefied tissues. The bristles strain out any large particles. (P. S. Tice)

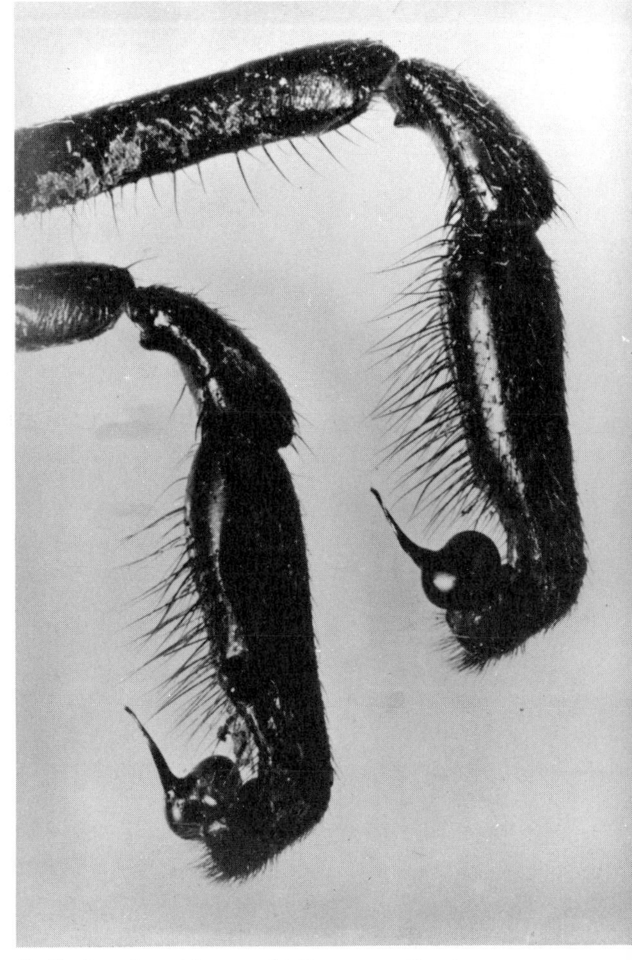

Pedipalps of a spider are chiefly sensory, but in maturing male spiders the tips enlarge and become modified for copulation, as seen here. The male spins a tiny silk web and deposits on it a fluid droplet filled with sperms. He then dips the palpal tips in the droplet until each is filled. Later, at the time of mating, the male inserts his palpal tips into two openings in the female and fills her sperm receptacles, where the sperms are stored until the eggs are mature and ready to be fertilized and laid. (L. Passmore)

The abdomen seldom shows any evidence of external segmentation and has no appendages except the **spinnerets,** of which there are usually three pairs. The spinnerets are fingerlike organs that have at their tips a battery of minute spinning tubes (sometimes a hundred or more on each spinneret), from which fluid silk issues. Silk is a protein that polymerizes under tension; thus the silk hardens as it is stretched into a thread. The spinning tubes connect with several kinds of silk glands that produce different kinds of silk for spinning various parts of the web, making a protective cocoon for the eggs, binding the prey, and so forth. Some of the tubes produce not silk but a sticky fluid that makes certain threads of the web adhesive.

A spider, when an insect becomes entangled in its web, apparently not only feels the tugging and can locate the prey in the web but can even distinguish the movements of different kinds of prey, which it approaches in different ways. If the prey is judged likely to escape, the spider hurries to the scene, seizes the struggling animal at once, and promptly injects poison. If the prey is large and formidable, the spider may use a more indirect method, first throwing swathes of silk over the prey so that the paralyzing bite can be delivered with less risk.

Black and yellow argiope, *Argiope aurantia,* is an orb-weaver common in U.S. and Canadian gardens. It hangs head down in the web. The zigzag pattern in the center of the web is a signal conspicuous to birds and warns them against flying through and getting their feathers soiled with sticky threads. It saves the spider from frequent rebuilding of the web. The body of a full-grown female measures 25 mm (1 in); the male, as in many species, is much smaller, only 6 mm long. (C. Clarke)

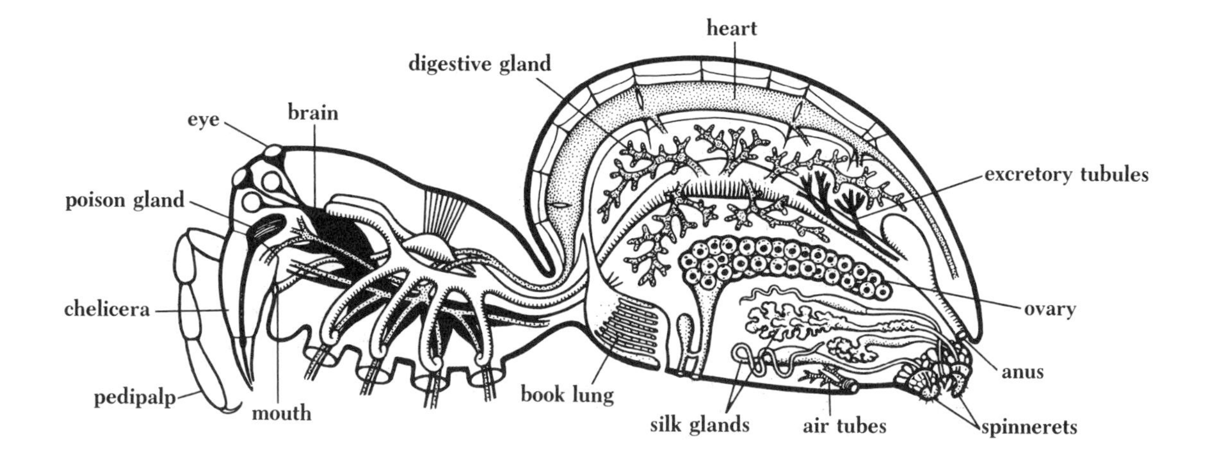

Diagram of a spider. As in other arthropods, the circulatory system is open; the long heart lies dorsally in a large sinus and receives blood through openings in its sides. (Adapted mostly from C. Warburton.)

A spider has a minute mouth and does not swallow solid food. Instead, the animal introduces digestive secretions through the wounds made by the cheliceras. The liquefied tissues of the prey are then sucked up by means of a sucking stomach (aided by the squeezing action of the bases of the pedipalps and sucking action of the pharynx). Beyond the stomach the digestive tract gives off several pairs of pouches, which increase the digestive and absorptive surface, and a large digestive gland, which branches extensively and occupies most of the spider's abdomen. This gland is the main digestive organ and is capable of taking up large quantities of food at one time, storing it, and then gradually releasing it for use. This enables spiders to go for long periods without taking food, though they must have water quite often.

The chitinous **exoskeleton,** which serves lobsters so well in the water, also seems made to order for spiders on land. Land life requires, among other things, a relatively *impermeable outer*

Just after molting this trapdoor spider *(left)* has a pale delicate cuticle and is helpless; in a day or so the cuticle darkens and hardens. The old cuticle *(right)* was first loosened by a molting fluid, then split along the sides and shed. Trapdoor spiders are among the primitive spiders that continue to molt as adults; most spiders do not. (L. Passmore)

covering to reduce drying of the moist tissues within and a fairly rigid *supporting framework.* In vertebrates the covering is furnished by scales or heavy skin, and the framework is an internal bony skeleton. In spiders and insects the cuticle fills both jobs, its outer layer becoming relatively waterproof.

Nevertheless, much water is inevitably lost by spiders (and other land animals) in two functions: respiratory exchange and excretion. **Respiratory exchange** requires an extensive surface, and no biological surface permeable to respiratory gases is impermeable to water. However, water loss is minimized in spiders because their two kinds of respiratory structures are open to the outside only through narrow slits or pores. The **book lungs** are air-filled sacs into which hang (like the pages of a book) a series of leaflike folds of the body wall. The folds are held apart by supports so that air can circulate freely between them. Spaces within the folds are filled with blood and communicate with blood sinuses of the abdomen. The **air tubes** (tracheas) are branching air-filled channels that open through small pores on the abdomen and bring air to the tissues. Most spiders have both kinds of devices for exposing a large respiratory surface to air, but some have only book lungs and some have only air tubes.

The **excretory system** consists of tubules that open into the intestine and excrete nitrogen mainly in the form of guanine. Because this substance is insoluble in water, its excretion involves little water loss. In many spiders there are also excretory organs that open near the bases of the legs.

Tarantulas are large hairy spiders that hide during the day in the cracks of trees and under logs, stones, or debris, and at night come out to stalk their prey, mostly large beetles. The tarantulas common in the southern and southwestern states of the United States reach a length of 5 cm (2 in); their bite is said to have the feel and effect of a sharp pinprick and is not dangerous. Their shed hairs are irritating to human skin. Most people insist they are revolted by the long legs and hairiness, but no one on record has ever objected to these same characteristics in a Russian wolfhound. Tarantulas are sold as pets, and females have lived 20 years in captivity, but it is better to leave them in the wild. (L. Passmore)

Tropical tarantula, of Panama, seen here about life size, is a giant among spiders, but its poisonous bite usually produces only local effects in humans. Some South American tarantulas of the Amazon Basin have a legspan of 18 cm (7 in) and can catch nestling birds, lizards, or small snakes. (R. Buchsbaum)

Below, left: **Trapdoor spider burrows** open by a door constructed of silk and soil and hinged with silk. The size of each door indicates the diameter of the burrow and the size of its spider occupant. The doors have all been opened by the photographer. When closed they match the soil surface. Vibrations caused by approaching prey alert the spider. (L. Passmore)

Above: **Trapdoor spider,** *Bothriocyrtum californicum,* has just pounced on a land isopod. (L. Passmore)

Cellar spider hangs head down in an irregular web of almost invisible threads, positioned to seize tiny crawling insects that pass along the edge of a washbasin. The delicate body and long slender legs of *Pholcus phalangioides*, one of the "daddy longlegs" spiders, make it vulnerable to drying, and it frequents bathrooms and cellars, leaving the drier, heated rooms to the sturdier, shorter-legged, common house spiders, which also hang upside-down in an irregular web. Deserted webs, covered with dust, become visible as cobwebs and are a minor nuisance. But all the common house-inhabiting spiders are harmless to humans and should be welcomed and protected for their help in keeping down such household pests as mosquitoes, house flies, clothes moths, and silverfishes. (R. Buchsbaum)

Hunting spider roams freely to catch prey. Here it has pounced on a katydid and sunk its poison fangs into the insect, immobilizing it while the spider liquefies the tissues and sucks up the fluid food. Panama. (R. Buchsbaum)

Black widow spider, *Latrodectus mactans,* occurs around the world in warm climates. In the female the abdomen is bulbous and shiny black. And in the northern Mediterranean race it is marked with red spots on the upper surface. In the United States the abdomen is all black above but has on the underside a red hourglass-shaped mark. In some parts of the western states this red mark is missing. The black widow causes almost all the fatal spider bites in the United States—about 5 a year, though these are only 5% of those inflicted. Three related "widow" species in the United States are not likely to come into contact with humans. The black widow female is about 15 mm long. The male is much smaller, has longer legs, and does not feed or bite; the two round black structures, which look in this photo like stalked eyes, are the sperm-filled tips of the pedipalps of the male, approaching the large female to mate. (L. Passmore)

The other dangerous spiders in the United States are the "brown spiders" of the genus *Loxosceles.* Only the brown recluse spider, *Loxosceles reclusa,* with a violin-shaped mark on the carapace, is likely to enter houses and come into contact with humans, usually when it hides in a towel or among clothes in a closet. The bite does not heal readily, and it may ulcerate and require skin grafting. Rare fatalities of children have been reported. This southern spider has spread northward into the central and eastern states of the United States.

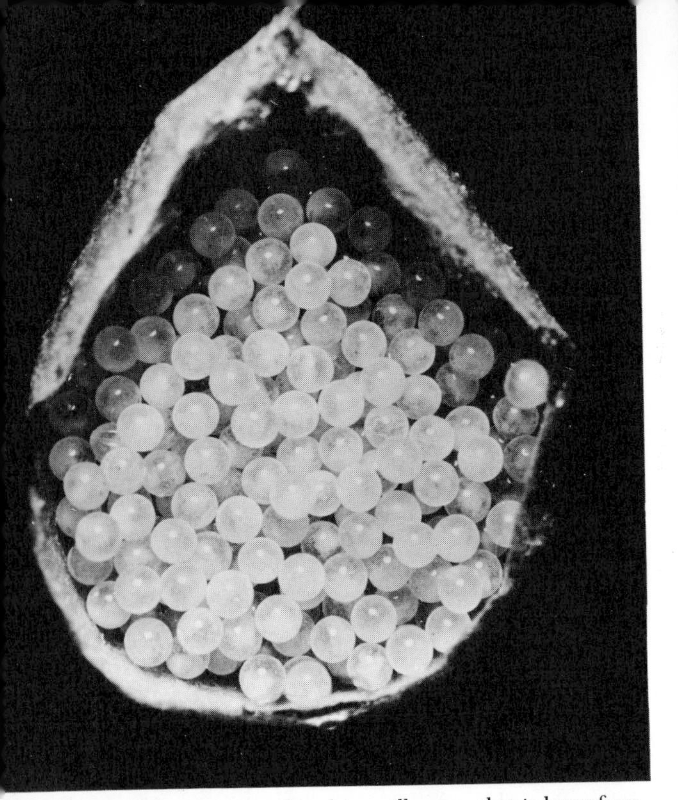

Spider eggs are enclosed in a silken sac that is hung from the web or from some solid object. Or it may be carried about by female hunting spiders, which have no fixed home. This is the opened egg sac of a black widow spider. (L. Passmore)

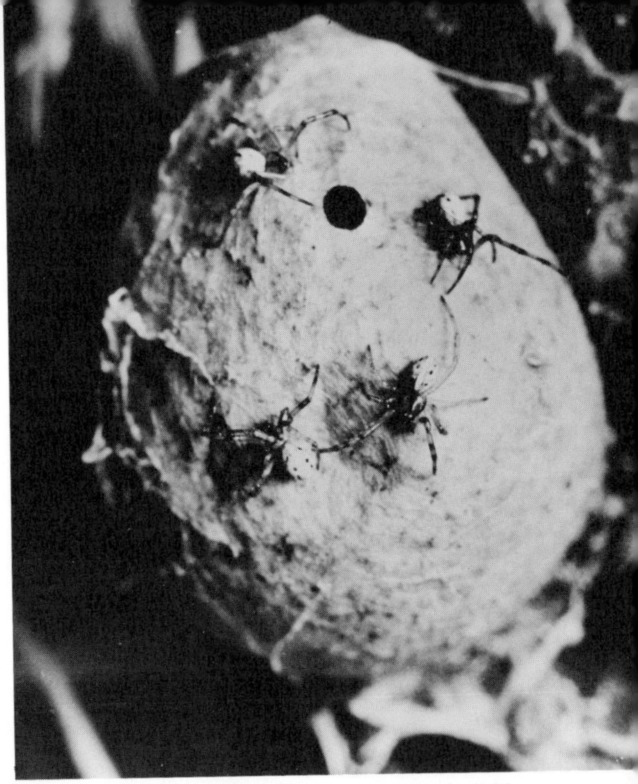

Spiderlings, just emerging from the egg sac, look like miniature adults, though they may not yet have the distinctive adult shapes and colorings. These are young black widow spiders. (L. Passmore)

Opposite: **Scorpion.** Held by an expert, this scorpion has no chance of defending itself by wielding its stinger. but no untrained person should pick one up. In Arizona, where the most dangerous U.S. species occur, the increasing encroachment of people on scorpion habitats has created problems. Houses invite scorpions to enter and share the shade in summer, warmth in winter, and a steady supply of water. Scorpions come out to feed at night on small arthropods and hide by day. In scorpion country one should not walk barefoot in a darkened house or put on shoes in the morning without first shaking them out. (R. Buchsbaum)

Scorpions are easily recognized: at the front end they have large armlike pedipalps with strong pincers for seizing prey, and at the tip of the rear end is a distinctive stinger that injects poison. The cephalothorax bears the small cheliceras (with pincers that tear food), the large pedipalps, and four pairs of walking legs that end in tiny claws. In the middle of the carapace are two simple eyes, but scorpions depend more on the sense of touch provided by sensory bristles on the pincers. The abdomen is long and conspicuously segmented. It is divided into a wide anterior portion and a posterior, slender portion that looks like a tail; at its tip are the anus and the terminal stinger. Scorpions hide by day under stones or bark or make burrows in the ground. At night they prey on other arachnids and on insects, and large ones may even attack small rodents. Most scorpions live in the hot dry areas of the world, but some are common in the moist tropics. Scorpions are found even in the southern Alps and occur in three-quarters of the continental area of the United States, being absent only from the northeast.

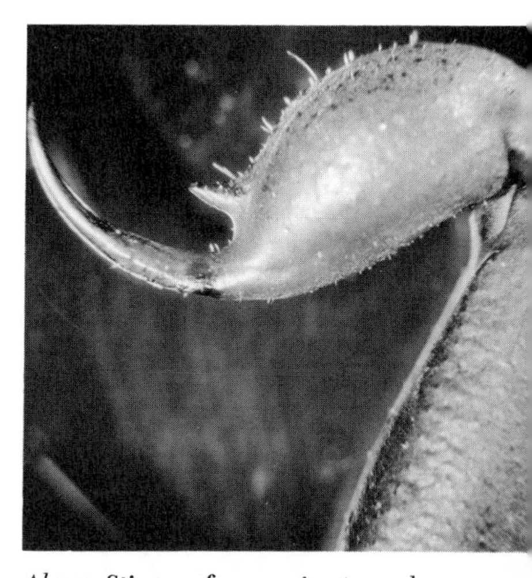

Above: **Stinger of a scorpion** is used in defense and also when necessary for quieting struggling prey. The muscles that move the stinger are located in the last abdominal segment, which also bears the anus. The curved spine of the stinger injects poison from two glands housed in the bulbous portion. In Arizona, Mexico, and North Africa there are species in which the venom has a potent neurotoxin, and human fatalities occur, almost all in children. A sting that causes only a local redness and swelling is likely to subside soon. A sting that shows no redness or local swelling, but marked hypersensitivity at the sting site, followed by numbness in that area, signals the possibility of generalized symptoms and serious consequences to follow. Medical attention should be sought; antivenin may be necessary. But in most scorpion species the poison is no more toxic than that injected by a bee or wasp sting, and in the United States and Canada fatalities from these stinging insects far outnumber deaths caused by spiders or scorpions. (P.S. Tice)

Mites, as their name implies, are small and easily overlooked. Yet this group includes the most dangerous of arachnids, the blood-sucking **ticks.** For the bite of a tick to cause death by itself is rare, though not unknown; the main danger to humans from ticks and other parasitic mites is that they spread diseases that may be fatal. The disease organisms transmitted by these arachnids include viruses (encephalitis, Colorado tick fever), rickettsias (Rocky Mountain spotted fever, scrub typhus, Q fever), bacteria (tularemia), spirochetes (relapsing fever), and sporozoan protozoans (babesiasis). Less serious but intensely irritating are the "chiggers" of the southern United States and other warm, moist areas of the world. These attack human skin, causing severe itching, as do "scabies" mites that burrow into the skin and live there. Similar species are responsible for "mange" in dogs, cats, horses, and other domestic mammals and for a variety of skin disorders in poultry. Domestic stock also suffer from tick-borne diseases and may become infected with tapeworms and parasitic nematodes through bites or by accidentally swallowing certain mites. Minute mites inhaled in household dust are a major cause of human dust allergies.

Central nervous system of a mite, in which the ganglia are fused in a single mass. Within the mite, it appears so large as to make one fear that these tough little animals are also brainy. But its great relative size results only from the fact that the number of nerve cells can be reduced only so far in an active animal, no matter how small. The small size and relatively large surface area of mites also accounts for the absence or scant development of respiratory and circulatory systems.

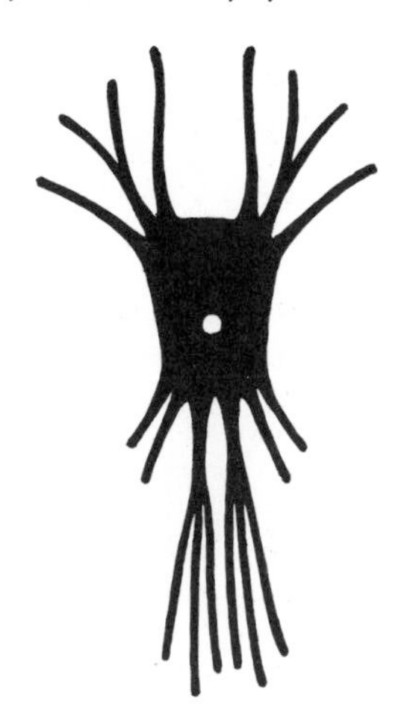

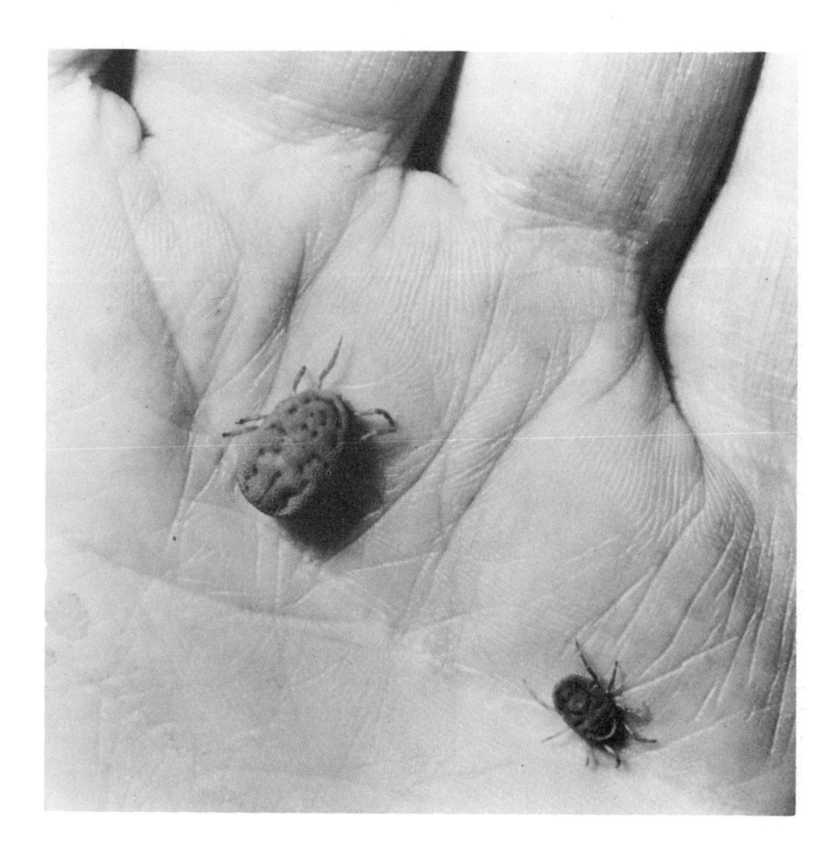

Soft ticks are large, blood-sucking mites, some of which parasitize humans and their domestic animals, thus transmitting serious diseases. Every year they cause millions of dollars of damage to cattle alone. The two seen here were kept for five years without food in the U.S. Public Health Service Laboratory. During this period of starvation they maintained within their bodies the organisms that cause relapsing fever when transmitted to humans by the bite of the tick. (Science Service)

Mites, like other arachnids, are fluid feeders, sucking up tissue fluids or solid food that they have liquefied by means of digestive secretions. The cheliceras, pedipalps, and other parts together form the "beak," which is barbed and must be carefully disengaged when you are removing a tick. Mites are unusual among the mainly predatory arachnids in that many feed exclusively on plants, piercing the cells and sucking out the contents. In doing so, they spread viral and fungal infections from plant to plant, and some mites are significant pests of agricultural crops, gardens, and house plants. On the other hand, the majority of mites are harmless to human interests and the many free-living predatory mites are important agents of pest control.

Hard tick has a hard shield covering the anterior part of the dorsal surface in females and the whole surface in males. The wood tick seen here, from a forest in Panama, has its beak inserted through human skin and is sucking blood. Hard ticks are widespread in the world. and many transmit the microorganisms that cause diseases such as encephalitis, Rocky Mountain spotted fever, scrub typhus, tularemia, and others. They are also vectors of disease in domestic stock. (R. Buchsbaum)

Wood mite, from rotting wood, is only 0.8 mm long. Such mites help to recycle fallen trees. The enormous mite populations in leaf litter and soil play an important part in breaking down and recycling plant and animal organic matter. On a forest floor there may be more mites than insects and all other arthropods combined. Mites also play a role helpful to humans by feeding on grain-eating insects in stored grain. (R. Buchsbaum)

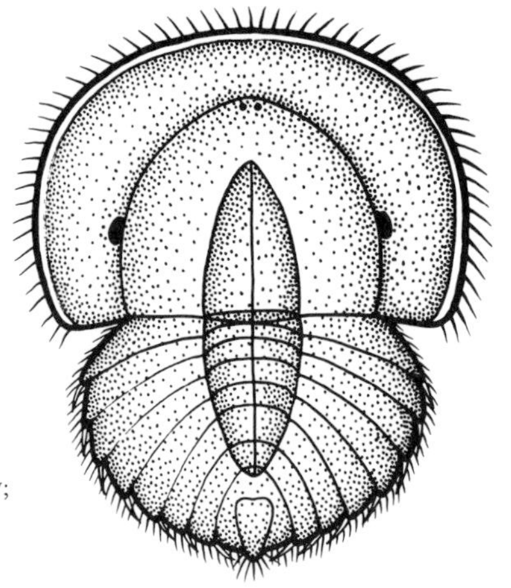
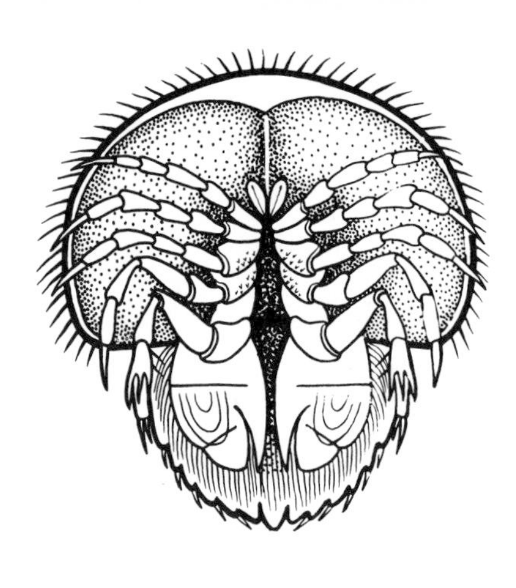

Free-swimming **larva of a horseshoe-crab** lacks the long tail spine of the adult. *Left:* dorsal view; *right:* ventral view. (After J. S. Kingsley.)

Mating pair of horseshoe-crabs are moving up a beach in Delaware Bay, where every spring thousands come out of deeper water into the shallows and beaches to mate. The smaller male clings to the female, and she tows him everywhere she goes. When ready to lay, she makes a "nest," a hollow in the sand near high-tide mark, and deposits her eggs. The male releases sperms into the water over the eggs and then at last lets go his grip on his mate. Later the hatching young work their way to the sand surface at high tide. *Limulus polyphemus,* seen here, occurs along the Atlantic coast of the United States and in the Gulf of Mexico to Yucatan. It has become important as the source of a reagent, made from *Limulus* blood, that is the most accurate means of detecting contamination by bacterial endotoxins in treated drinking water and in pharmaceuticals used in medical treatment. Horseshoe-crabs, one of several Asian species, are sold in markets of Bangkok, Thailand, but only the eggs are eaten and only in seasons when they are not toxic. (R. Buchsbaum)

Horseshoe-Crabs

Horseshoe-crabs are not crustaceans but marine chelicerates, the only living representatives of the class **Merostomata** and the only chelicerates with compound eyes. These animals are often referred to as "living fossils" because they have changed so little from types that lived over 400 million years ago. No one can say with any certainty how they have been able, with "no modern improvements," to survive in competition with the diversity of

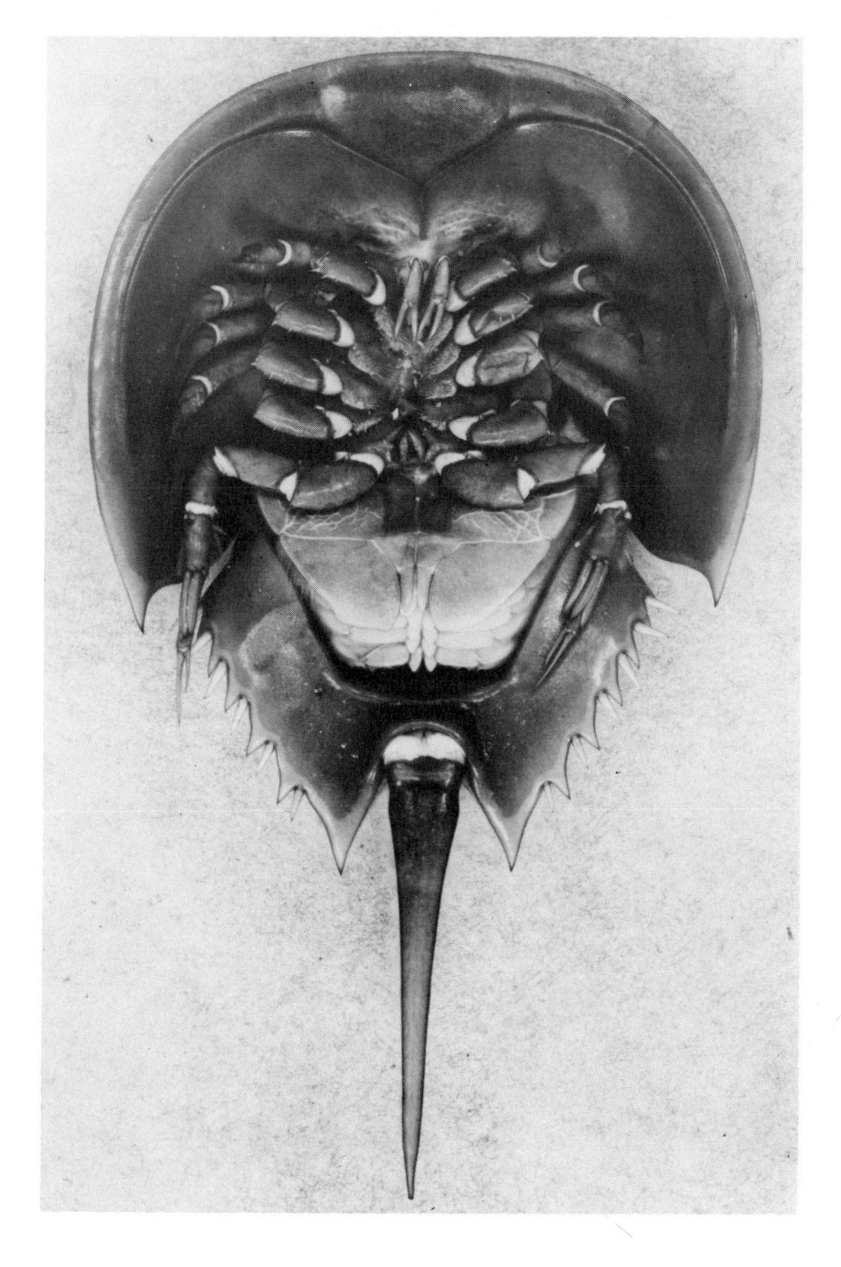

Underside of a horseshoe-crab shows the tiny pincers of the cheliceras and the five pairs of stout walking legs, all but the last pair with pincers used in seizing prey. The flat abdominal plates bear the leaflike book gills. The tail spine is used by an overturned horseshoe-crab to right itself. *Limulus polyphemus* reaches a meter (39 in) from the front margin to the tip of the tail spine. (R. Buchsbaum)

Extensive collecting of horseshoe-crabs during the nineteenth and twentieth centuries greatly reduced populations from former levels. Dried and ground, the horseshoe-crabs were used as fertilizer and as pig and chicken feed. In 1865 some 1,200,000 were taken from 1.5 km of the shoreline of Delaware Bay, and records show 4 to 5 million large horseshoe-crabs collected annually around 1930. This photo was taken in 1924 near Bowers Beach, Delaware. (Delaware State Division of Historical and Cultural Affairs)

crustaceans and other animals that are commonly described as more "highly developed," "advanced," and so on. Such phrases often refer to specialized characters, and perhaps the stability of horseshoe-crabs as a type results from their very lack of specialization. Their shallow-water habitats along sandy or muddy shores are readily available. They are able both to walk along the bottom and to swim by flapping their abdominal appendages. And they feed on a variety of worms, molluscs, and any other animals they find by burrowing in the sand.

First walking leg of a horseshoe-crab, showing the **chewing process** on its base.

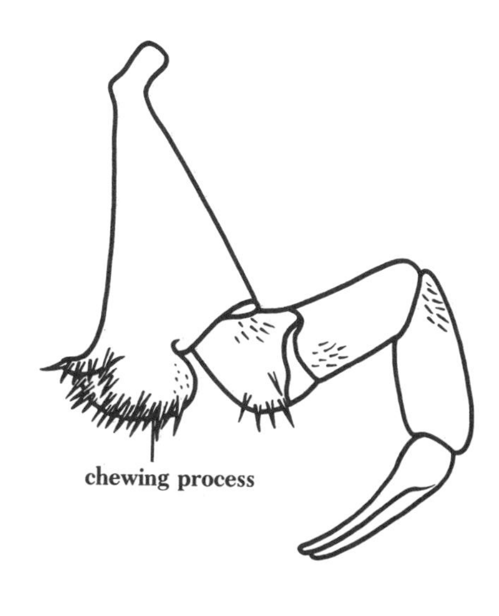

Aside from their interest as archaic forms, horseshoe-crabs give us some idea of what the aquatic ancestors of modern arachnids may have been like. As in arachnids the body is divided into *cephalothorax* and *abdomen;* and the thorax has six pairs of appendages, of which the first is a pair of pinching *cheliceras* and the other five pairs are walking legs. All except the last pair also end in pincers, used to pick food from the soft bottom, and have spiny processes on their bases, used to shred the food and pass it forward to the mouth. Attached to the flattened abdominal appendages are *book gills,* groups of thin plates in which blood circulates. These are so similar in plan to the book lungs of terrestrial arachnids as to suggest strongly that book lungs evolved from book gills.

Sea-spider is not a spider or even an arachnid but a member of the wholly marine chelicerate class **Pycnogonida.** Pycnogonids feed with a sucking proboscis on soft invertebrates such as cnidarians. The large proboscis in *Pycnogonum stearnsi* (which extends just in front of the four dark eyes in this face view) is larger than the tiny abdomen at the other end of the segmented trunk bearing four pairs of clumsy legs. This individual was crawling about on a hydroid, feeding on the polyps, but members of this central California species are often found at the base of large green sea anemones, with the proboscis inserted, and are conspicuous because of their pink color. Pycnogonids occur mostly in cold seas, and there are giant species on Antarctic bottoms, with long legs that span 30 cm (12 in). But most are measured in millimeters and are all but impossible to see as they creep about slowly, on long slender legs, feeding among hydroid branches. (R. Buchsbaum)

Chapter 19

Airborne Arthropods: Insects

Most of the impressive statements that are made about arthropods—their dominance over all other animals in number of species, the tremendous economic and medical consequences they have for humans—are really about insects. Insects outnumber other arthropods about 8 to 1, and the study of insects, **entomology,** is often the subject of a separate course for college biology students and the field of a separate department in a university or museum.

Insects *are* marvelously diverse in their diets, habits, forms, and especially their colors. On the other hand, although insect diversity appears overwhelming (even to the point of being discouraging), getting acquainted with the insects is not so hopeless as it may seem at first. Over 85% of insect species belong to only 4 of the 30 or so orders, and these dominant orders are familiar ones: the beetles; the flies; the moths and butterflies; and the wasps, bees, and ants. Indeed, insects as a group are in many respects less diverse than crustaceans. The name "insect" comes from a Latin word meaning "incised" and refers to the fact that insects generally have a sharp division between the head and thorax and between the thorax and abdomen. All insects have these same body divisions, and almost all have the same number of segments and the same appendages on each. Grasshoppers are fairly typical representatives.

Opposite: **Grasshopper** is typical of insects in having a pair of prominent antennas, two large compound eyes, several pairs of complex mouth parts, two pairs of wings, three pairs of walking legs, and a conspicuously segmented body covered with a nearly water-proof cuticle important for land life. (R. Buchsbaum)

Grasshoppers

The **head** of a grasshopper has two compound eyes and three simple ones. The compound eyes are similar in structure to those

Diagram of a **generalized insect,** with antennas that are simple filaments, unspecialized walking legs, and two similar pairs of wings. The differences among the various insect groups show up most conspicuously in modifications of these structures. Also, most insects have fewer abdominal segments than are shown here, owing to loss or fusion at the posterior end. (After R. E. Snodgrass.)

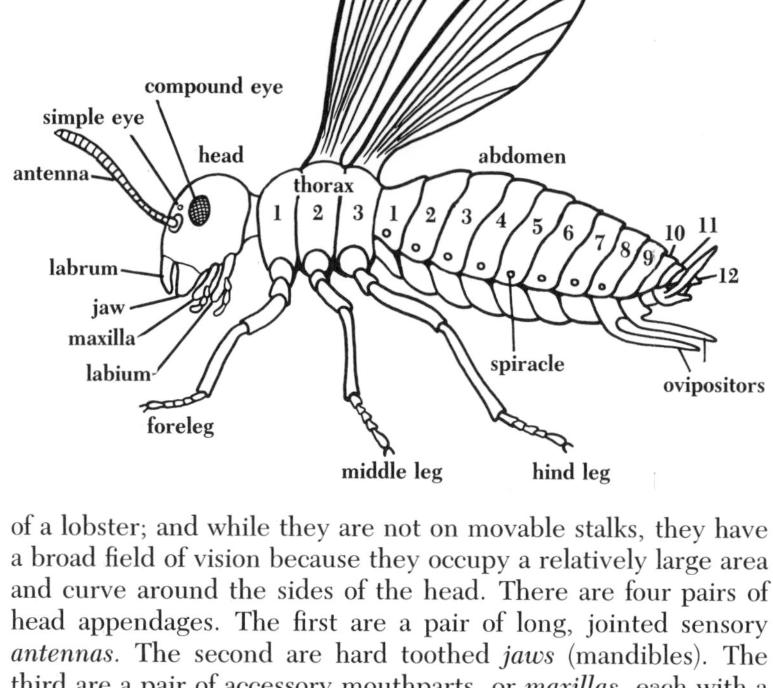

Middle leg of a grasshopper shows the parts of a typical insect leg. (After R. E. Snodgrass.)

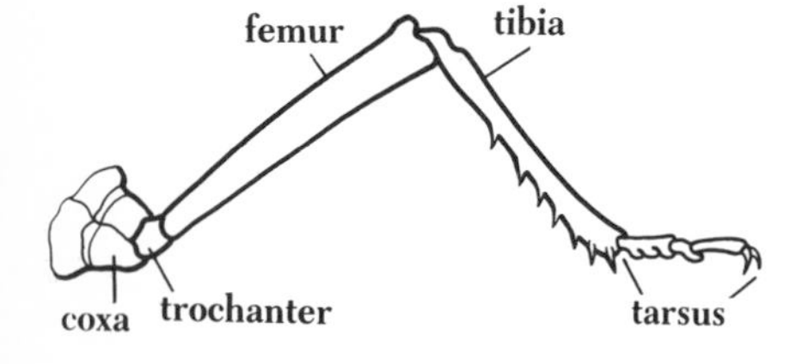

of a lobster; and while they are not on movable stalks, they have a broad field of vision because they occupy a relatively large area and curve around the sides of the head. There are four pairs of head appendages. The first are a pair of long, jointed sensory *antennas.* The second are hard toothed *jaws* (mandibles). The third are a pair of accessory mouthparts, or *maxillas,* each with a jointed sensory palp. The fourth pair are fused into a single plate, the *labium,* which has on each side a sensory palp and which forms the lower lip of the mouth. The upper lip, or *labrum,* in front of the mouth, is a simple hardened outgrowth and does not form from paired appendages. The head is mostly covered by a hard skull-like capsule, and the segmentation is obscure, but the four pairs of appendages show that at least four segments are present.

The **thorax** bears *three pairs of legs,* a character of all insects. The thoracic segmentation is partly concealed dorsally by a chitinous shield and by the wings, but the three thoracic segments are clearly visible ventrally and at the sides. The legs are composed of a characteristic series of articulated parts with joints between. Each leg ends in two curved claws flanking a fleshy pad that aids in clinging to smooth surfaces. The first two pairs of legs are typical walking legs, like those of most insects. The third pair is specialized for jumping; one part (the femur) contains muscles for jumping and is enlarged out of proportion to the rest of the leg. The *two pairs of wings,* located on the second and

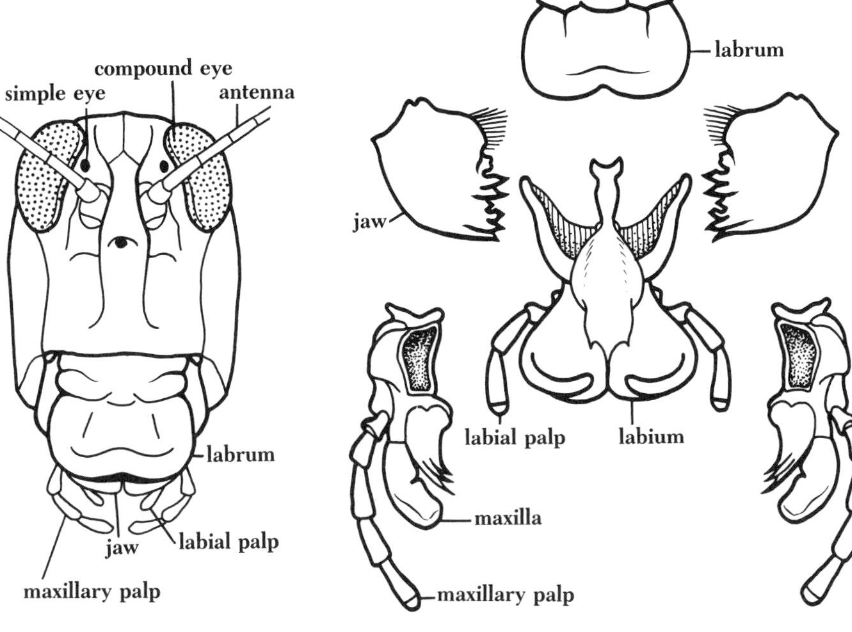

Left: Front view of the **head of a grasshopper.** *Right:* The mouth parts removed from their attachments. (After R. E. Snodgrass.)

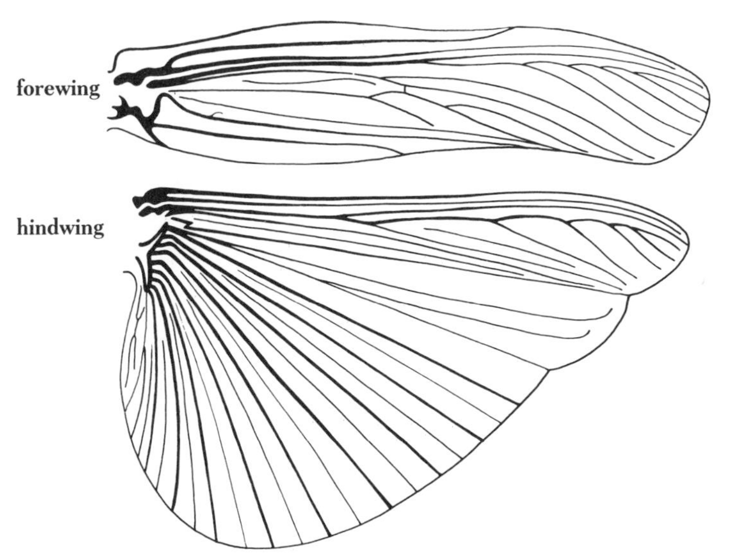

Wings of a grasshopper. Only the larger veins are shown. (After R. E. Snodgrass.)

third segments of the thorax, differ from each other. The forewings are narrow and hardened, and they serve as a cover for the hindwings. Both pairs are used in flying. The hindwings are broad and delicate, spread wide in flight, but folded like a fan to fit under the first pair when the grasshopper comes in for a landing. The wings are made of cuticle and are stiffened by thickenings called *veins.*

The **abdomen** has no appendages except those at the posterior end, which are associated with mating and egg laying. The abdomen contains most of the soft internal organs, since the thorax is nearly filled with the muscles that move the legs and wings.

As in other arthropods, the **digestive tract** consists of foregut, midgut, and hindgut. The foregut and hindgut are lined with cuticle, which is shed and renewed with the exoskeleton at each molt. The foregut of a grasshopper contains special regions that cut and grind the quantities of vegetation that make up the diet. The midgut, which lies mainly in the abdomen, has no cuticular lining and serves as the main organ of digestion and absorption. The junction of the midgut with the hindgut is marked by the attachment of long excretory tubules. Thus the hindgut receives the waste materials of digestion and excretion.

The foregut starts at the mouth, receives a secretion from the *salivary glands,* and continues as a narrow esophagus, which leads to the *crop,* a large thin-walled sac in the thorax. On the inner walls of the crop are transverse ridges armed with rows of spines that probably serve to cut the food into shreds. The crop is mainly a storage sac that enables a grasshopper to eat a large quantity at one time and afterward digest it leisurely. From the crop the food passes into a muscular *gizzard* lined with chitinous teeth. At the posterior end of the gizzard is a valve that prevents the food from passing into the midgut before it is thoroughly ground and also prevents food in the midgut from being regurgitated. Digestion probably begins in the crop, for the food entering that organ is already mixed with salivary secretion, and it

Internal anatomy of a grasshopper, a female individual. (Based on R. E. Snodgrass.)

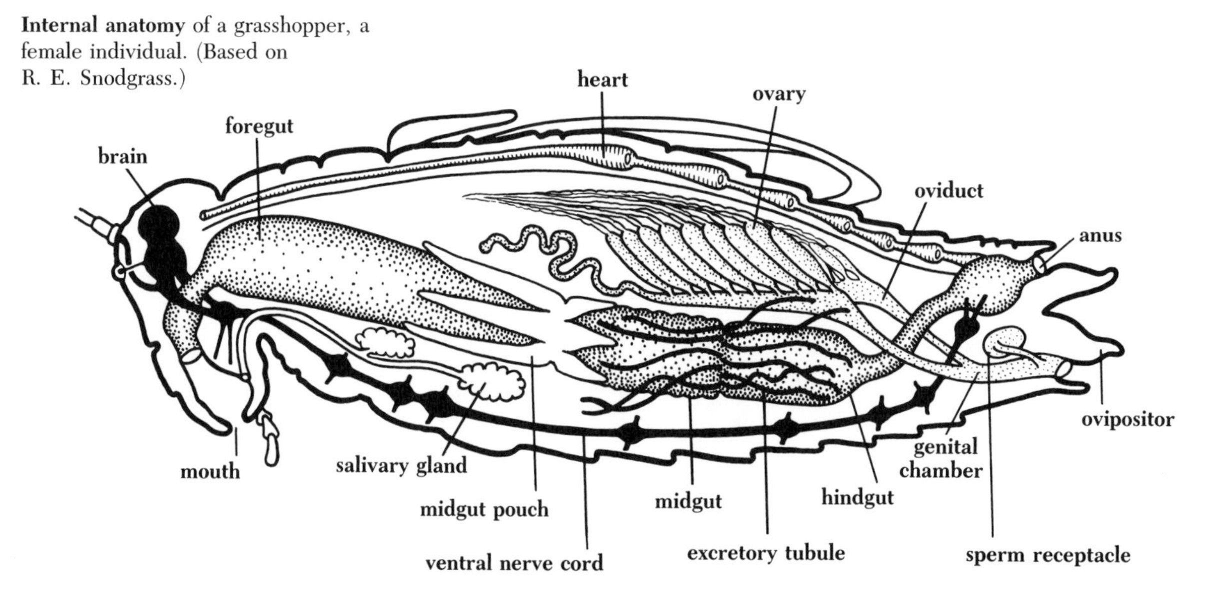

also receives some digestive juices that pass anteriorly from the midgut. Opening into the anterior end of the midgut of a grasshopper are six pairs of pouches that secrete digestive juices and also aid in absorption.

The **excretory system** consists of a number of tubules (called Malpighian tubules, from the name of their discoverer) that lie in the blood sinuses and extract nitrogenous wastes from the blood. Fluid from the tubules is poured into the hindgut, and here much of the water and some of the salts are reabsorbed. The material remaining, mostly crystals of uric acid, is expelled from the anus as dry waste. This system is much like that in spiders and serves the same important function in conserving water.

Also similar to a system we saw in spiders, but independently evolved and more highly developed in insects, is the system of **air tubes** (tracheas). The air tubes lead from paired openings that lie at the sides of the thorax and abdomen. The ten pairs of these

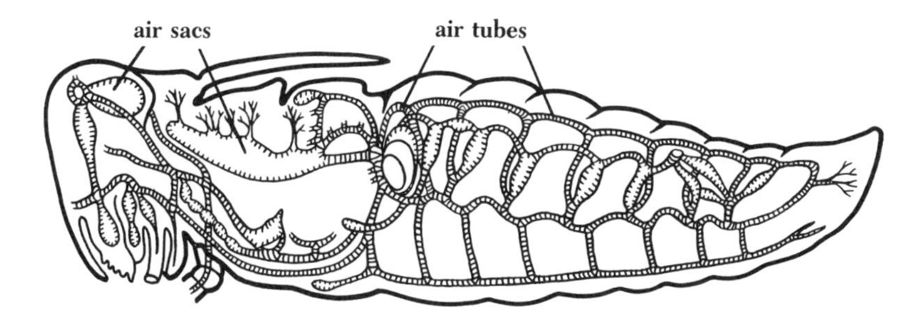

Respiratory system of a grasshopper, showing only the main air tubes and air sacs. (After Vinal.)

openings, or *spiracles*, are guarded by fine hairlike bristles that keep out dirt and by valves that are opened or closed to regulate the flow of air. Closing the valves aids in decreasing the evaporation of water. The air tubes are prevented from collapsing by spiral thickenings in the cuticle that lines their walls. From a system of longitudinal and transverse main trunks, smaller branches ramify to all parts of the body, eventually becoming so small (less than one micrometer, or 1/25,000 of an inch in diameter) that groups of the finest ones (tracheoles) are made by a single cell. Here and there the larger tubes widen into air sacs. The air moves chiefly by diffusion, but muscular breathing movements, which alternately compress the air sacs and then allow them to expand, aid in exchanging the air. The greater the muscular activity, the greater the pumping action on the air sacs and the better the circulation of air. In a grasshopper the first four pairs of spiracles open only at inspiration, and the remaining six pairs open only at expiration; this facilitates the flow of air. In the deepest branches oxygen moves by diffusion alone, first along the tubes, and then into the surrounding blood spaces and tissues. Carbon dioxide leaves by the reverse route. Small amounts of oxygen and carbon dioxide also may be exchanged through thin parts of the body surface.

The **circulatory system,** relieved of the responsibility for carrying respiratory gases, is much less extensive than in a lobster or a spider. In fact, there is only one vessel, the long contractile *dorsal vessel,* composed of the tubular *heart,* which pumps the blood forward, and its anterior extension, the *aorta.* In each segment through which it passes, the heart is dilated into a chamber perforated on each side by a slitlike opening through which blood enters. Blood leaves the heart through a series of excurrent openings and through the aorta, which carries blood into the head and there ends abruptly. The blood flows out into sinuses among the tissues, bathes all the muscles and soft organs, and finally returns to the heart. Although the system is an open one, the flow is given some direction by a series of partitions, and the slow rate of blood flow suffices for the redistribution of food and collection of wastes. The blood also serves as a reservoir for food and water; it contains cells that destroy bacteria and other parasites; and blood pressure plays a part in hatching from the egg, in molting, and in expansion of the wings by filling the veins with blood.

The **nervous system** is a ventral, double, ganglionated cord. The brain lies above the esophagus and between the eyes. It is joined to the first ventral ganglion by a pair of connectives that encircle the gut. The *brain* has no centers for coordinating muscular activity; after removal of the brain the animal can walk, jump, or fly. As in other invertebrates, the brain serves as a sensory relay that receives stimuli from the sense organs and, in response to these stimuli, directs the movements of the body. It also exerts an inhibiting influence, for a grasshopper without a brain responds to the slightest stimulus by jumping or flying—a very unadaptive kind of behavior. And even in the absence of any external stimulation, the animal without a brain displays incessant activity of the palps and legs. In addition the brain is responsible for certain complex behavior patterns and for modifying them by learning. The *first ventral ganglion* controls the movements of the mouth parts and exerts a general excitatory influence. The *segmental ganglia* are connected and coordinated by nerves that run in the cords, but each is an almost completely independent center in control of the movements of its respective segment (or segments) and appendages. In some insects these movements have been shown to continue in segments that have been severed from the rest of the body. An isolated thorax is

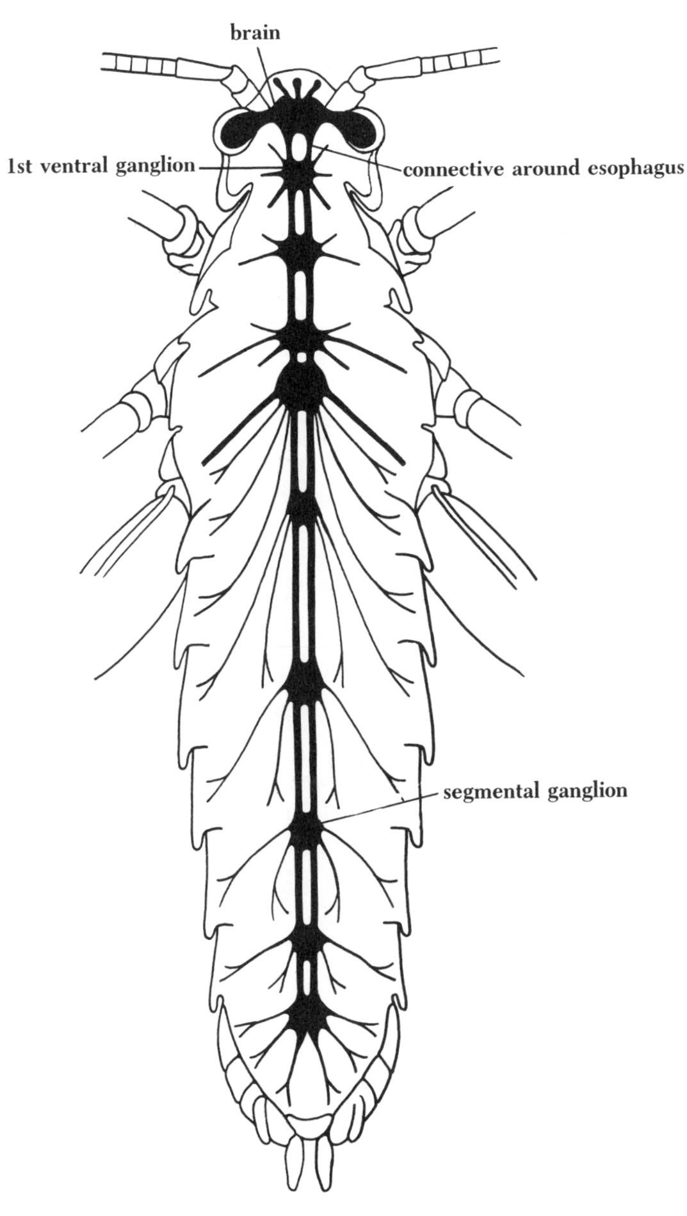

Nervous system of a grasshopper. (Modified after R. E. Snodgrass.)

capable of walking by itself, and an isolated abdominal segment performs breathing movements.

The **reproductive system** of an insect includes a pair of gonads and their associated ducts. A male grasshopper has a pair of testes, which discharge sperms into a sperm duct; glands secrete seminal fluid into the duct, which opens near the posterior end of the body. A female grasshopper has a pair of ovaries, with oviducts that join a common genital chamber. During copulation

Mating grasshoppers. The smaller male is astride the female. Among the various kinds of insects, the sexes often locate or identify each other by means of signals particular to each species: odors, sounds, body colors and shapes, luminescence, special behaviors. (P. Leibman)

the male introduces sperms into the female's seminal receptacle, a sac in which they are stored until the time for egg laying. When mature, the eggs pass down the oviduct; yolk and shell are secreted around the still unfertilized eggs. A small pore is left, however, through which a sperm may enter. As the eggs pass into the genital chamber, they are fertilized by sperms expelled from the sperm receptacle. A conspicuous set of stout appendages, the ovipositors, near the posterior end of the abdomen, are used for digging a hole in the ground in which to lay the eggs. Most insects do not care for their eggs, except for selecting a favorable laying site—often a specific food plant or the body of an animal on which the young may feed when they hatch.

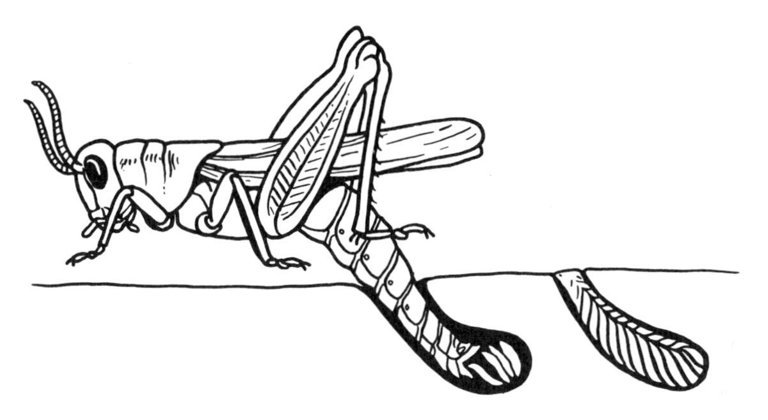

Grasshopper laying eggs in the ground. On the right is a completed batch of eggs. (After Walton.)

The **development** of insects is rather different from that of most other animals and is strongly influenced by the large amount of yolk in the egg. Instead of dividing into two cells and then into four, and so on to form a blastula, the zygote nucleus divides many times without the division of the cytoplasm. The nuclei then move to the periphery, and cell membranes appear between them, thus forming a layer of cells. Subsequently, some of the cells pass inward and give rise to endoderm and mesoderm; the remaining outer layer is the ectoderm. Besides the nervous system, the ectoderm gives rise to the epidermis, tracheal system, foregut, and hindgut, and secretes the cuticle associated with each. The midgut develops from endoderm. The mesoderm becomes divided into segmental blocks; and within most of these, paired coelomic sacs appear. The coelomic sacs later break down and do not form the adult body cavity, which is not a coelom but a hemocoel, filled with colorless blood.

Development of an insect.
1: Fertilization. **2:** The nuclei migrate to the periphery. **3:** Cell membranes appear between the nuclei, resulting in a single-layered embryo that corresponds to the blastula of other animals. **4:** When segmentation first appears, the segments are not differentiated into body regions. **5:** Later, the anterior-most segments are incorporated into the head; the segments of the thorax enlarge and grow legs; and the rudimentary appendages of the abdomen mostly disappear. (After R. E. Snodgrass.)

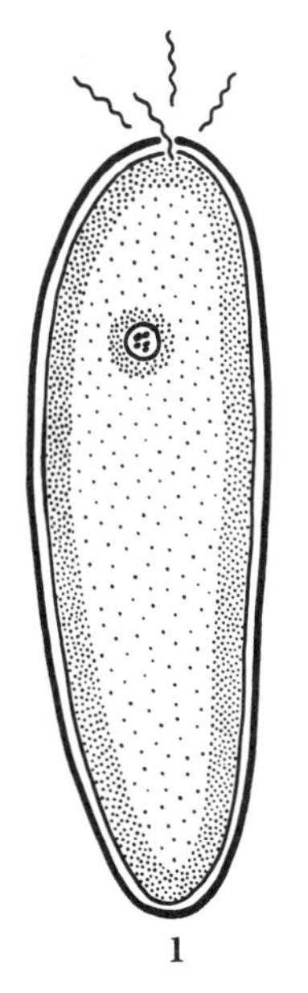

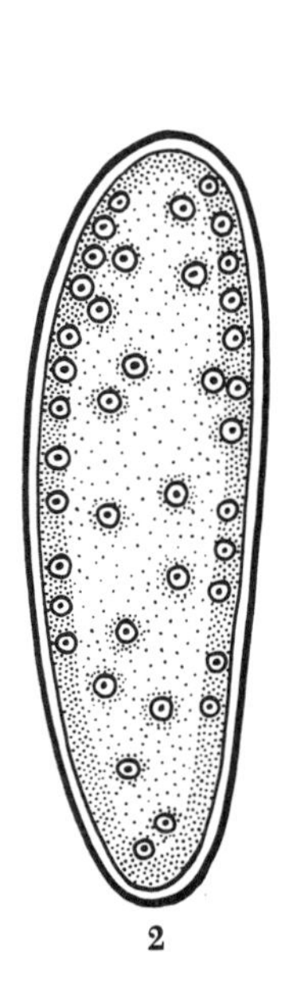

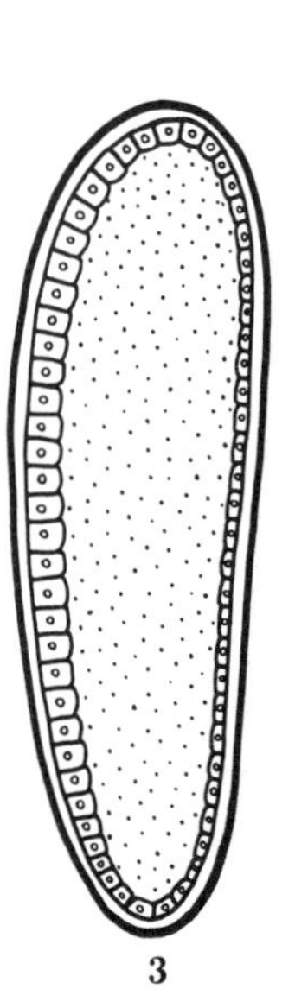

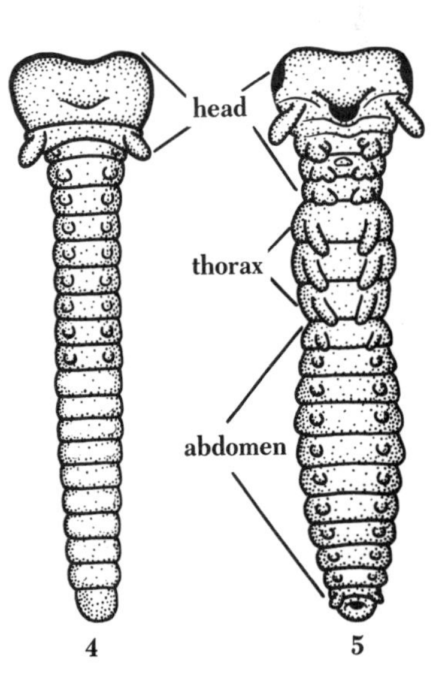

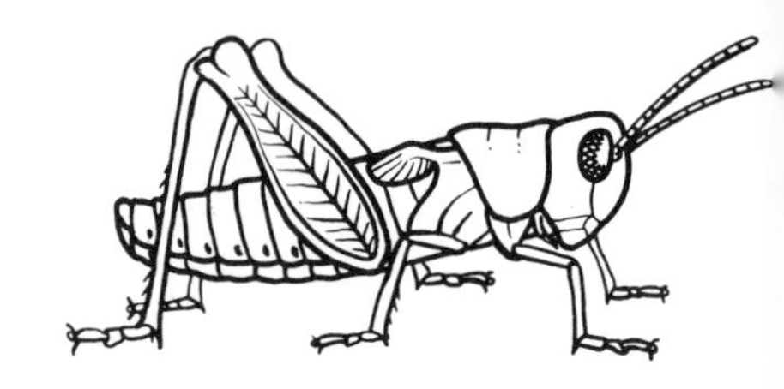

Gradual metamorphosis of a grasshopper from first nymphal phase to adult. In the adult the head is smaller in proportion to the rest of the body, and the wings are fully developed. (Combined from several sources.)

The young grasshopper, known as a **nymph,** hatches from the egg in a form that resembles the adult, differing mostly in the disproportionately large size of the head and in the lack of wings and reproductive organs. It feeds upon vegetation, grows rapidly, and undergoes a succession of **molts,** five in most grasshoppers. At each molt the old exoskeleton, thinned by dissolution and reabsorption, is discarded. It is ruptured and shed, not by gradual growth (which merely stretches the membranes between the segments), but by a combination of muscular contractions and the uptake of quantities of water and air. The newly emerged insect is soft and white; and since it is in a precarious condition, it usually retires to a safe place until the soft cuticle hardens and darkens. With each successive molt, differentiation continues, until the final molt results in the adult form.

Variations in Insect Structure

To have observed insects at all is to have noticed that they vary tremendously—from flattened, crawling cockroaches that feed on scraps of food, to flying butterflies that suck nectar from flowers, to swimming beetles that chase animal prey. These differences are mostly in external structures; internally, insects are more alike.

Variations in the **digestive tract** are related mostly to what the animals eat. Grasshoppers and cockroaches, which feed on solid food, have a well-developed gizzard equipped with hard plates and spines. Insects that suck juices have no gizzard. Honeybees suck up nectar into a honey stomach, which corresponds to the crop in grasshoppers. The nectar, destined for storage in the hive, is prevented from leaving the honey stomach by a valve, developed from a region that corresponds to the gizzard in grasshoppers.

The great majority of insects obtain air for **respiration** through a system of air tubes; but a few, such as tiny collembolans, have no air tubes and exchange respiratory gases through

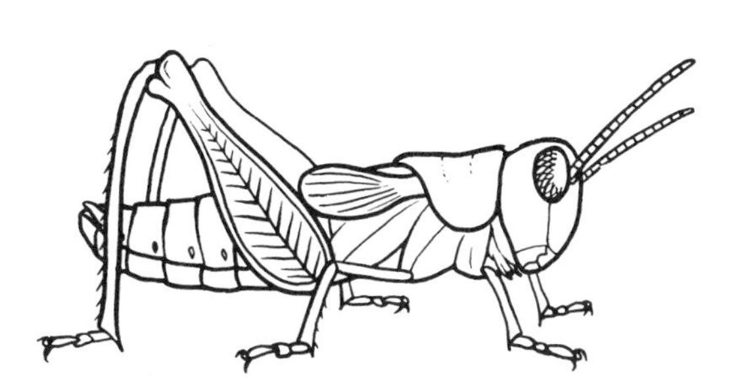

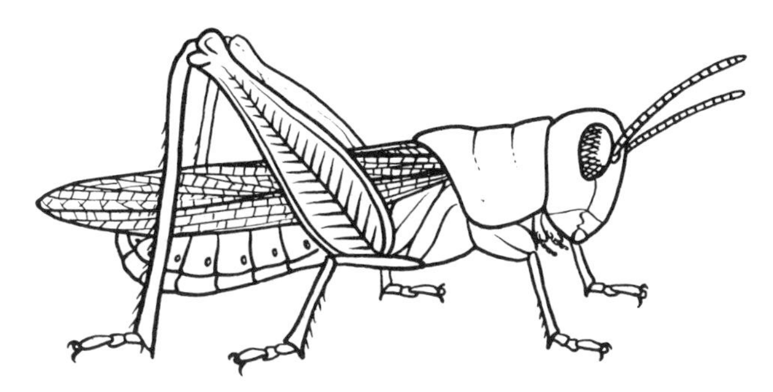

the body surface. Aquatic insects have special respiratory adaptations. Diving bugs and beetles may carry an air bubble down with them. Many aquatic insect larvas have air tubes but no open spiracles, and they obtain oxygen by diffusion through the cuticle of thin-walled gills. Parasitic insect larvas that live in the fluids

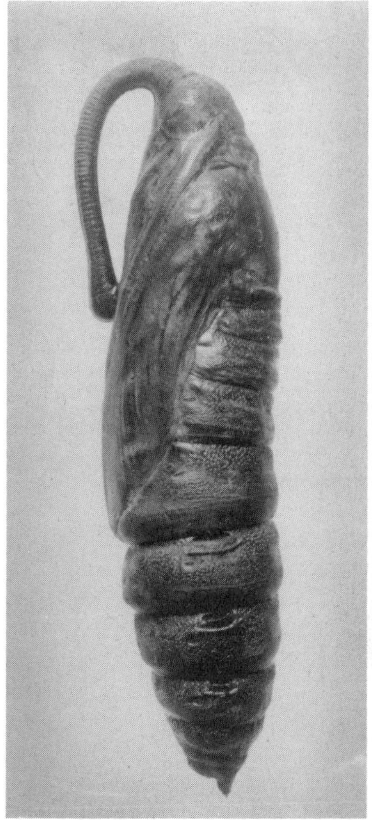

Respiratory exchange in most insects is by means of a system of air tubes that communicate with the outside through open spiracles along the sides of the body, not only in adults but also in immature stages such as the caterpillar larva *(left)* and the pupa *(above)* of a sphinx moth. (Larva by R. Buchsbaum, pupa by C. Clarke.)

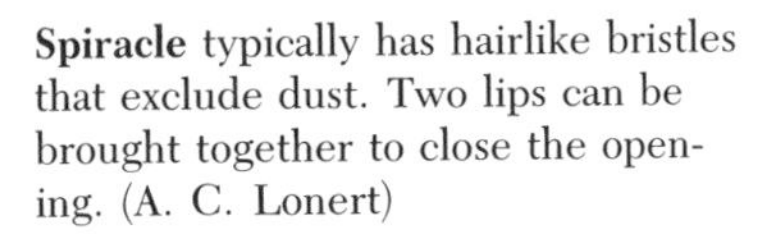

Spiracle typically has hairlike bristles that exclude dust. Two lips can be brought together to close the opening. (A. C. Lonert)

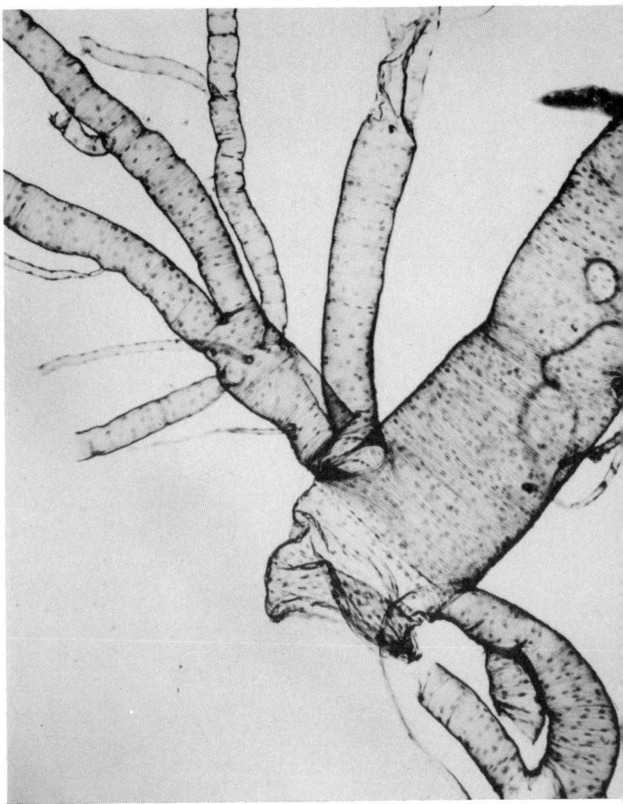

Air tubes, highly magnified, show thickenings of the cuticular lining that keep the walls of the tubes from collapsing. (A. C. Lonert)

and tissues of their hosts may also operate with closed spiracles, but some maintain access to air from the outside or from the respiratory system of an air-breathing host.

The essential parts of the **reproductive system** of most insects are as described for grasshoppers, but some show modifications of both structure and life history. Male and female aphids mate in the fall and produce fertilized eggs in the standard way, but the young that hatch in the spring are all female, and their all-female offspring arise by *parthenogenesis* from unfertilized eggs. They are nourished and develop within the mother's body and are born alive. Aphids multiply rapidly in this way all during the summer, and in the fall sexual forms are again produced. *Hermaphroditism* is extremely uncommon in insects, but occurs in certain scale insects. The hermaphrodites are able to fertilize their own eggs, but may also mate with and be fertilized by the

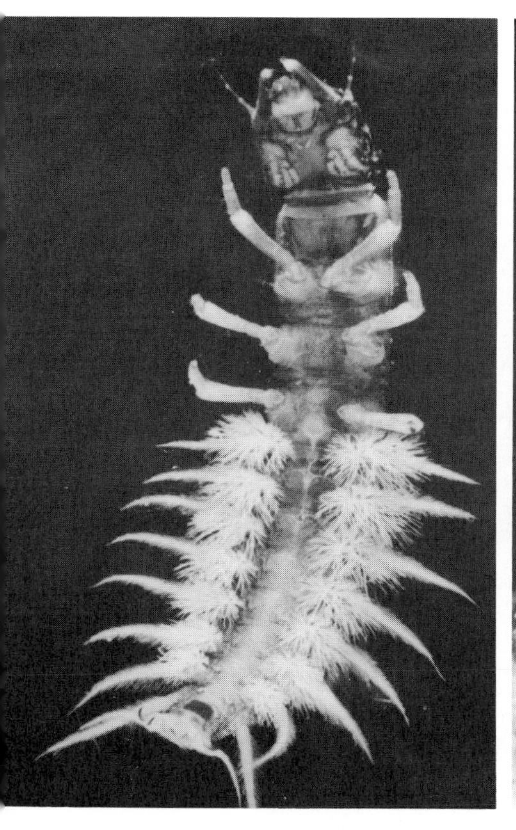

Gills. The aquatic larva of a dobsonfly, *Corydalus*, exchanges respiratory gases by diffusion through air tubes contained within thin-walled tracheal gills. (P. S. Tice)

Scale insects on a leaf. These are sessile females with a dark brown hemispherical cuticular shield that covers the soft wingless body. They remain fixed throughout their adult life, with mouth parts permanently inserted into the host plant, from which they suck juices. One female here has just released a brood of tiny white nymphs, and these can be seen crawling about, among and on the sessile females. California garden. (R. Buchsbaum)

rare males, which arise from unfertilized eggs. Scale insects also serve as an extreme example of external differences between the sexes, or *sexual dimorphism.* Mature females (and hermaphrodites) are wingless and sessile, so modified that they can hardly be recognized as insects but look like flat scales, bumps, or bits of cottony fluff on the plants from which they suck fluids. The tiny males do not feed, but are mobile, with or without wings.

Although conservative in their internal anatomy, insects present a dazzling show of external variation that attracts collectors and fascinates biologists. There are, of course, the easily observed differences in body shape, color, and size. But among the variations that are most important in adapting the animals to their different ways of life are those of the sense organs and appendages.

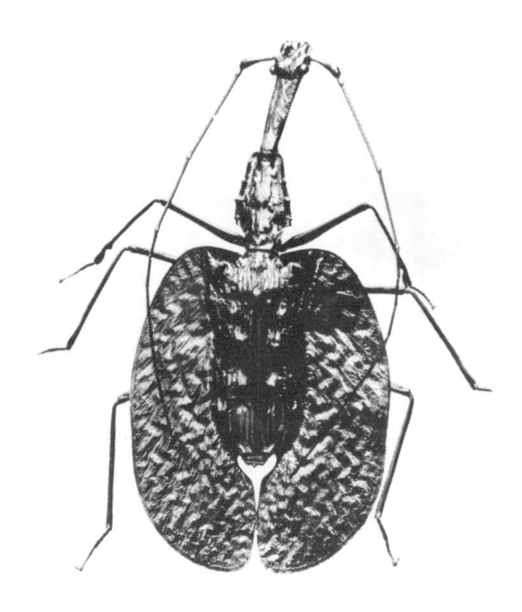

Insects vary in shape and color more than any other comparable group of animals. Many insects resemble their surroundings. Some look strikingly like the green leaves, dried leaves, twigs, bark, or flowers on which they live. The katydid, *above*, which lives in dense foliage in Brazil, is compressed from side to side and is green and leaflike. The beetle, *opposite, above right*, which lives under bark in Malaysia, is compressed dorsoventrally and is brown in color. The Kallima butterfly of India, *opposite, bottom right*, when at rest looks very much like a dead leaf. The walkingstick, *opposite, left*, is greatly elongated and well disguised among twigs, whereas the beetle, *bottom right*, has a compact body and is one of many insects with a striking surface pattern that contrasts with its habitat. In many cases such patterns and bright colors appear to be a warning of distasteful or toxic tissues.

Most insects have as **sense organs** a pair of compound eyes, three simple eyes, and a pair of antennas. In addition, the mouth parts may bear jointed sensory projections, called palps, and the body is clothed with a variety of sensory bristles, scales, pits, and so on. There may also be special organs of smell or hearing.

The **compound eyes** of insects vary from the huge ones of dragonflies with thousands of facets to the tiny ones of some ants with fewer than a dozen. Although the eyes can adapt to light and dark, insects that are habitually active only by day or by night have differing eye structure that suits their particular schedule. And groups of facets in different parts of the eye may vary for different kinds of vision; for example, separate groups of facets may have specialized sensitivities to color, ultraviolet, or polarized light.

Sense organs on the head of this insect (a praying mantis) include a pair of large compound eyes, three simple eyes (ocelli) in the center, and antennas that are long and segmented. (R. Buchsbaum)

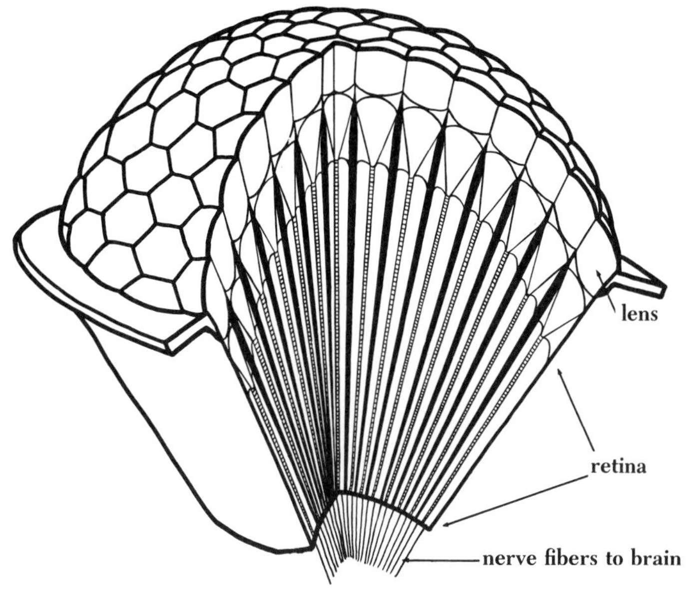

Compound eye of an insect is composed of many visual units, each with a small bundle of light-sensitive cells and a lens. In this diagram, a sector has been cut away. (Partly after R. Hesse.)

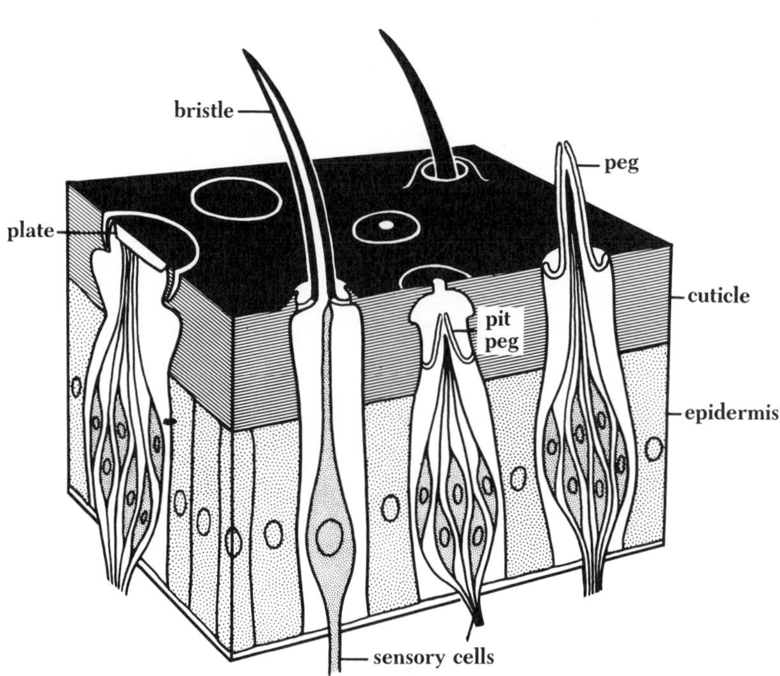

The surface of an insect antenna is studded with several kinds of **sense organs.** The *bristle* responds to touch and also registers other types of mechanical stimulation, such as wind during flight. The *peg, pit peg,* and *plate* are probably sensitive to various chemicals and serve the senses of smell and taste. (The epidermal cells that shape the cuticle of the sense organs and surround the sensory cells have been omitted for simplicity; white spaces mark their location.)

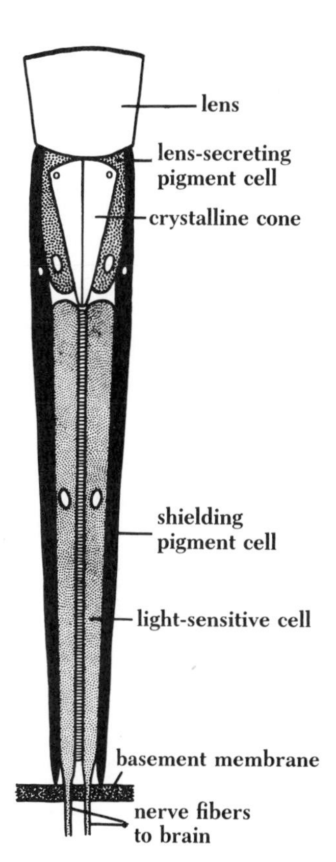

Single unit of a compound eye is called an *ommatidium.* The lens and crystalline cone are refractive bodies. The light-sensitive cells are surrounded by a shield of pigment cells, which exclude oblique rays of light. This ommatidium represents the type present in day-flying insects, active in bright light.

The **antennas** are the chief site of the sense of *touch*. They may be very long, as in cockroaches, crickets, and katydids, or they may be tiny, as in dragonflies, which depend mostly on sight. The touch receptors of the antennas are the fine hairlike bristles with which they are clothed. The bristles are stiff and are joined by a delicate cuticle at their bases to the surface of the antenna. The antennas also bear receptors for the senses of smell and taste and others sensitive to humidity and to temperature.

Not only on the antennas, but scattered over the bodies of most insects are numerous tactile bristles. Other sense organs may occupy rather unusual places. For example, the **auditory organs** of a grasshopper are on the sides of the abdomen just above the base of the third legs, while those of a katydid are in the "knees" (near the upper end of the tibia) of the first pair of legs. **Taste organs,** which enable insects to detect water and to

Compound eyes of a dragonfly occupy most of the head and each is composed of nearly 28,000 separate units. The antennas, which look like mere bristles, one beneath each eye, apparently play a minor role. In a housefly the eyes have only 4,000 units and in some ants there is only one. Some nocturnal and cave insects have no compound eyes at all and rely chiefly on their well-developed antennas. (P. S. Tice)

Antennas of a male cecropia moth. (P. S. Tice)

Auditory organ in the foreleg of a katydid. (C. Clarke)

distinguish sweet, salty, sour (acid), and bitter substances, occur not only in the mouth but also on the antennas, palps, feet, and egg-laying appendages.

Grasshoppers represent a fairly generalized group of insects in having **mouthparts** of the *biting and chewing* type. This is the most primitive and widely distributed kind, present also in beetles and in many other orders of insects. Two other main types of mouthparts are common: *sucking* mouthparts, as in butterflies, and *piercing and sucking* mouthparts, as in cicadas. In most butterflies the jaws are rudimentary, and the two maxillas are greatly elongated, each forming a half-tube, so that when they are held together they form the long sucking proboscis of the adult, through which nectar and other liquids are sucked up by muscular pumping. The proboscis is extended only when the insect is feeding; when not in use, it is coiled under the head. The piercing beak of a cicada consists of three pairs of appendages that correspond to the jaws (mandibles), maxillas, and labium of a grasshopper. The long tubular labium is not inserted into the food; it serves only as a sheath for the other mouth parts, being grooved on its dorsal surface to form a channel in which lie the mandibles and maxillas, which do the piercing. The mandibles are long, fine stylets with minute teeth at the end. The maxillas

The mouthparts of different kinds of insects consist of the same basic parts but are modified for various methods of feeding. The large and formidable jaws of a female stag beetle deliver a bite that can draw blood from human fingers and provide her with an effective defense. The much larger jaws of the males, branched like the antlers of a stag, are weaker as defensive weapons; they are used mostly against other males of the same species, in fights during which each tries to lift and overturn the other. Stag beetles feed by licking up sap and other plant juices with a hairy labium. (P. S. Tice)

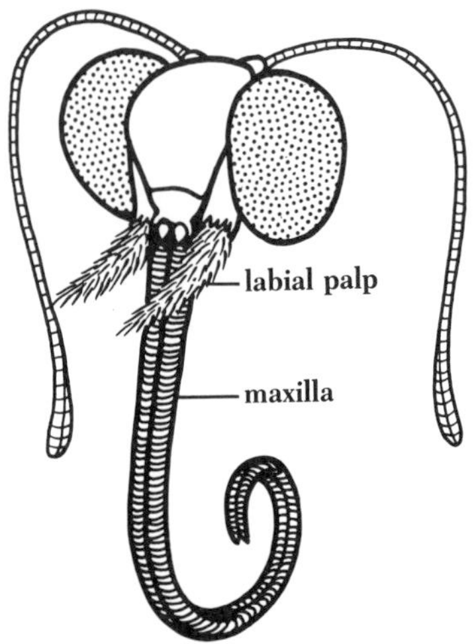

Head of a butterfly, showing **sucking mouthparts.**

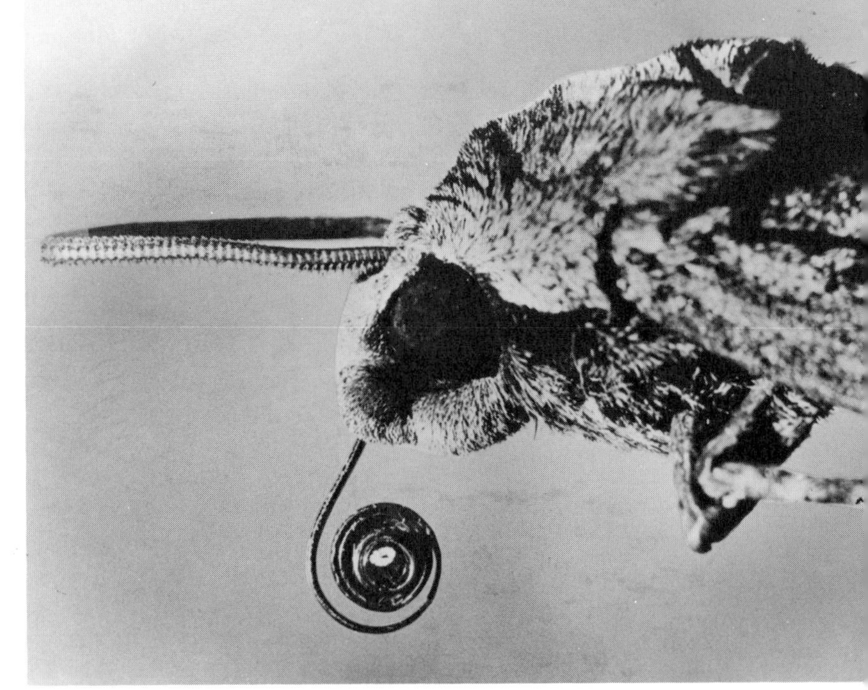

Sucking tube of moths and butterflies consists of two elongated maxillas, each forming half the tube. (A. C. Lonert)

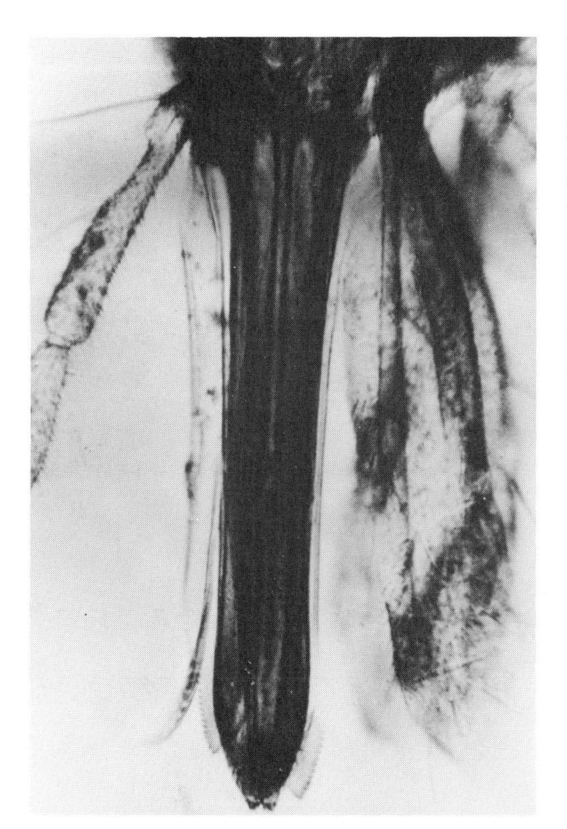

Piercing and sucking mouthparts occur in insects that feed by drawing blood or plant juices. The elongated parts are stiff and sharp and used to make the wound, or are modified as a tube through which liquid food is drawn up. Piercing mouthparts are usually fine stylets or flat blades, which may have saw edges, as in the beak of a "punkie" or "sand fly" seen here. As in moths, butterflies, and houseflies, the sucking action is provided by muscles in the head.

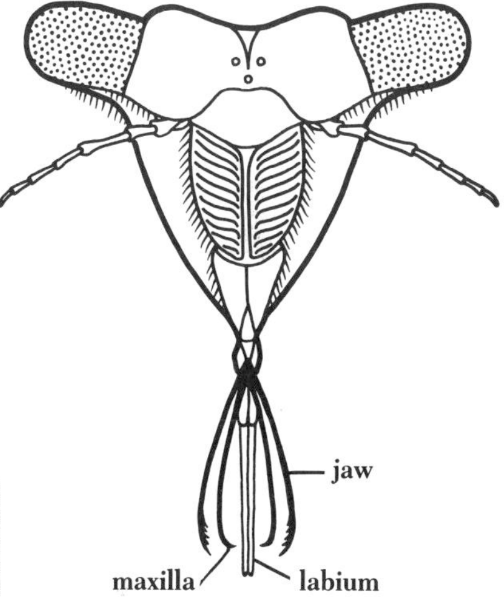

Head of a cicada, showing **piercing and sucking mouthparts** spread apart.

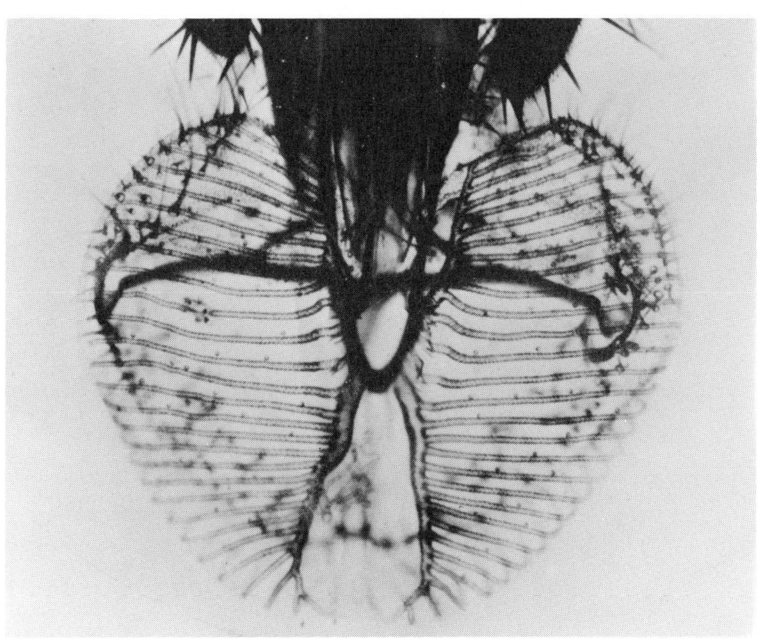

Sponging tongue of a house fly is the expanded two-lobed tip of the labium. The food is sucked up through an opening at the end of the cleft between the two lobes.

are similar but hooked at the tips; each is crescent-shaped in cross-section, and the two are fastened together by interlocking grooves and ridges to form a channel through which the food is sucked up. Mosquitoes and other biting flies, as well as fleas and bed bugs, have the mouthparts adapted in various ways for feeding on blood. Other insects have still other modifications of the mouthparts, such as the sponging tongue of a house fly. The larval stages of some insects have mouthparts and feeding habits entirely different from those of the adult.

The thoracic **legs** of insects are modified in a variety of ways, but are usually composed of the same basic parts (shown and named in the drawing of a grasshopper leg). Grasshoppers have walking legs with terminal pads and claws for clinging to vegetation or other objects. House flies have adhesive pads at the tips that enable them to walk up smooth vertical surfaces, such as glass. Water beetles have flattened legs, fringed with bristles, for swimming. But the legs may serve other functions besides locomotion. The walking legs of a honey bee are modified for collecting pollen and for grooming. Each is quite different from the others, so that together they constitute a complete set of tools for collecting and manipulating the pollen upon which the bees feed.

Legs of insects are six in number and are borne on the thorax, the middle region of the body. All are composed of the same five jointed parts and most end in two curved claws. Some have pads that enable the insect to walk on smooth or vertical surfaces. The least specialized legs are simple walking legs, all three pairs much alike, as in this darkling beetle, *Eleodes*. California. (R. Buchsbaum)

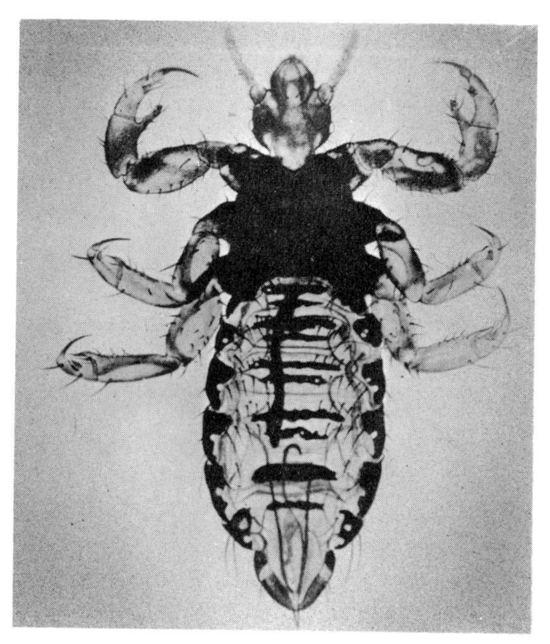

Grasping legs of the human body louse, *Pediculus humanus*, enable this blood-sucking ectoparasite to cling to body hairs or to clothing. (U.S. Army Medical Museum)

Swimming legs of a water-scavenger beetle are flattened and fringed with bristles. These aquatic beetles have no gills but breathe by coming to the surface at intervals to obtain fresh air, which they carry down with them as a film under the wingcovers and among the fine bristles on the undersurface of the body. (C. Clarke)

Food-collecting legs of a honey bee are also used for walking. Pollen clings to fine bristles on legs and body and is transferred to "baskets" on the hindlegs, here seen well loaded. (C. Clarke)

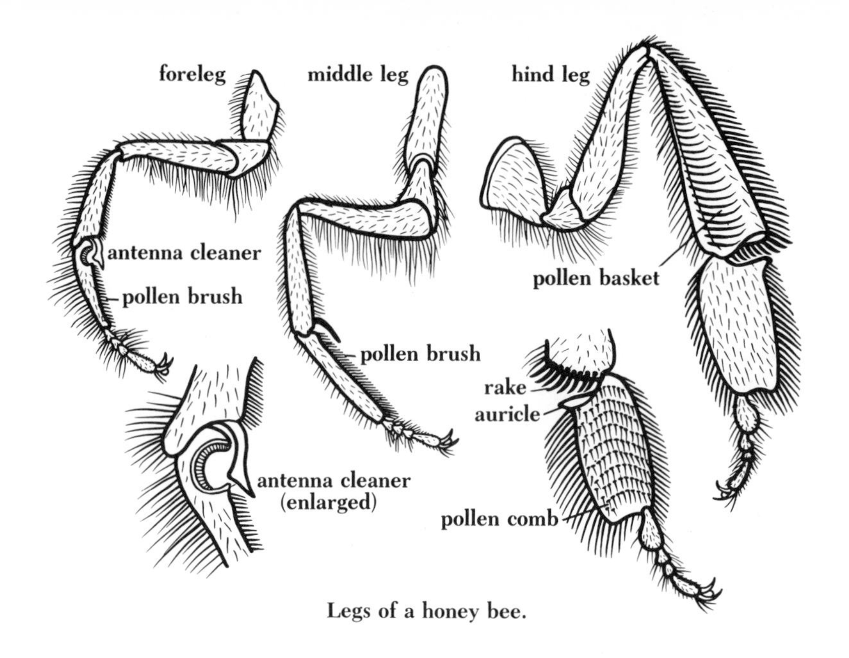

Legs of a honey bee.

On the **foreleg,** along one edge of the inner surface of the tibia and the large first part of the tarsus, is a fringe of short, stiff, unbranched hairs that form a *pollen brush* used for collecting the pollen grains that become caught among the feathery hairs of the body when the bee visits flowers. The pollen brushes of the forelegs clean pollen from the front part of the body, including the compound eyes. The first part of the tarsus also has a semicircular notch that is lined with a comblike row of bristles and is closed by a clasplike projection on the end of the tibia. The antenna, held firmly in place by the clasp, is cleaned by drawing it through the notch. Comb and clasp together are called the *antenna cleaner.* The **middle leg** is the least specialized of the three. The large first tarsal part is wide and flat and covered with stiff hairs that form a brush for removing pollen from the forelegs and thorax. The **hind leg** is most elaborately equipped; it carries the load of pollen. Rows of *pollen combs* on the inner surface of the very large and flattened first part of the tarsus scrape the pollen from the middle leg and posterior part of the abdomen. The *rake,* a series of stout spines on the lower end of the tibia, removes the pollen from the combs of the opposite leg, and it falls on the *auricle,* a flattened plate on the upper end of the first part of the tarsus. The leg is then flexed slightly, so that the auricle is pressed against the end surface of the tibia, compressing the pollen and pushing it onto the outer surface of the tibia and into the *pollen basket.* The pollen basket is formed by a concavity in the tibia, which has, along both edges, long hairs that curve outward. Pollen clings together and to the basket hairs because it is moistened with sticky nectar regurgitated from the mouth. When the baskets are loaded, the bee returns to the hive and deposits the pollen in special wax cells. Combined with sugars and other substances, the pollen mixture becomes "beebread," which provides a source of protein for both adults and larvas.

The **wings** of insects are flattened, two-layered expansions of the body wall and at first consist of the same parts: cuticle and epidermis. Later the two opposing layers meet, except along the channels in which lie nerves, air tubes, and blood spaces. Eventually these channels form the "veins" of the wing; the epidermal cells degenerate; and the adult wing is almost completely made of cuticle, though it may have a circulation of blood and also nerves connected to sense organs on the surface.

Not all insects have wings. Several primitive groups (proturans, collembolans, diplurans, thysanurans) never developed them. And in some orders one or both pairs have been lost secondarily. For example, flies (dipterans) have only the first pair, the second being reduced to a pair of knoblike structures; and fleas and lice have lost both pairs. In addition, among the orders that typically have two pairs of wings, there are wingless members. In the social insects certain castes lack wings; and in certain groups the males have wings while the females are wingless, as was mentioned for scale insects, or the opposite.

Wings of insects are typically two pairs of membranous extensions on the thorax. Wings in which the numerous veins form a fine network, as in this dragonfly, are considered to be relatively unspecialized. (R. Buchsbaum)

One pair of wings is found in the crane fly and other flies (order Diptera). The wing veins are large and few in number, a relatively specialized pattern. The second pair of wings is represented by a pair of stalked knobs, the halteres, which assist in maintaining balance during flight. (R. Buchsbaum)

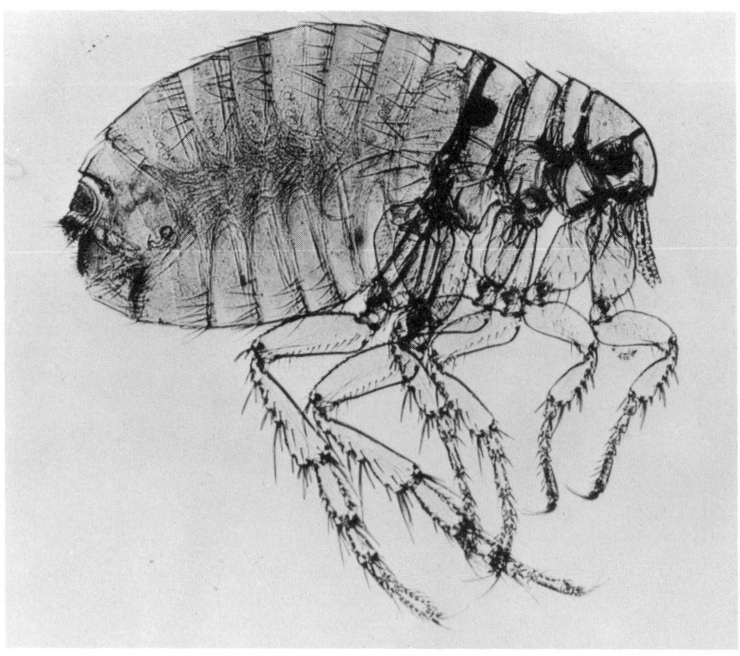

Absence of wings may be primitive, or the result of secondary loss as in fleas, such as this dog flea. Fleas (order Siphonaptera) are small, wingless insects with piercing and sucking mouth parts and complete metamorphosis. (U.S. Army Medical Museum)

Thickened forewings of beetles, usually called "wing covers," form a protective armor over the membranous hindwings and most of the body. This elaterid beetle was photographed with forewings and hindwings spread out as if in flight. (P. S. Tice)

The two pairs of wings may beat independently, as in dragonflies, or they may be mechanically hooked together, as in bees. In beetles, the forewings are thick, stiff wing covers, and they move very little during flight but contribute stability and important surface area for lift, as do the rigid wings of an airplane. In general, small insects beat their wings with higher frequency than do large ones: the wings of some tiny flies beat 1,000 times per second. Such fantastic rates require highly specialized mechanisms of control for the nerves and muscles involved, and also high rates of metabolism, which in turn depend on relatively warm temperatures. Small insects cannot fly in cold air, but larger insects are more independent of the external cold. They can generate enough heat (by doing preflight warm-up exercises) and retain it by insulation (the "furriness" of bees and moths) to take off on cold mornings.

The **abdomen** of adult insects bears appendages only at the posterior end; these are modified for mating or egg laying, or as sensory projections, or in other ways. In a female grasshopper they are used for digging a hole in the ground in which to lay eggs. In ichneumons (wasplike hymenopterans) the egg-laying apparatus is long and sharp; and when the ichneumon senses the presence of a beetle larva within a tree, the apparatus is used to drill a hole in the wood and deposit eggs in the body of the larva. When the eggs hatch, the young ichneumon larvas feed on the beetle larva. In a honeybee worker, the egg-laying apparatus is modified as a sting connected with poison glands, as some of us know from painful encounters with these insects.

Abdominal appendages at the rear of an earwig are large and hardened. They form curved pincers that serve in capturing prey, in defense, and in mating. (R. Buchsbaum)

Metamorphosis

Not all insects undergo a development that involves changes radical enough to be termed **metamorphosis.** The primitively wingless insects hatch from their eggs in practically the adult form. They grow larger and later add parts to the antennas and other appendages, but such changes are no greater than those undergone by most animals in their development. Some of the secondarily wingless insects, such as lice, have abandoned their metamorphosis along with their wings. The development of grasshoppers is an example of **gradual metamorphosis.** The *nymphs* lack wings but generally resemble the adults in having compound eyes, in using similar mouthparts to eat the same type of food, and in sharing adult habitats. The developing wing buds are external and visible at an early stage; and the most conspicuous external change at the final molt is the full development of the wings. Dragonfly nymphs also have external wing buds and resemble adults in being carnivores with large compound eyes, but their metamorphosis involves greater changes because they live in the water and have gills and prey on aquatic animals, whereas the adults are terrestrial and air-breathing and catch other flying insects.

Gradual metamorphosis in *Magicicada septendecim,* a periodical cicada, takes 17 years from egg to adult. **1.** This **adult female** has made a slit in a tree branch with her sharp, sawlike ovipositor and is inserting her eggs. In years of heavy egg laying, whole forests or orchards may turn brown, but later all except the smallest trees recover.
(J. C. Tobias)

2. The eggs hatch in about 6 weeks into minute nymphs only 1 mm long, which fall to the ground and bury themselves. The nymphs suck juices from the fine roots of trees. (C. Clarke)

3. The nymphs emerge from the ground after 17 years. Protected during all this time from extremes of weather and from predators, they have passed through 5 nymphal stages, molting between stages and growing from 1 mm in length to 25 mm. As many as 40,000 nymphs may emerge at about the same time from the ground under a large-sized forest tree. They come out at dusk, when insect-eating birds are turning in for the night, and crawl up the nearest vertical object. (C. Clarke)

4. Adult emerging *(left)*, from a slit down the back of the nymphal cuticle, is soft and white with bright red eyes. Its wings are still folded accordian-style. After the adults emerge, the empty golden nymphal coverings can be seen everywhere clinging to trees. (P. Knight)

5. Newly emerged adult *(right)* with wings partly expanded. By the next day the cuticle of wings and body has hardened and darkened, and the cicada can fly. On the second or third day the males begin their resounding concert, females are attracted, and pairs mate. Adults live three or four weeks. (P. Knight)

1. Gradual metamorphosis in water is seen in the dragonfly nymph, which lives at the bottom of ponds, lakes, and streams, preying on small animals. Note the large compound eyes and the developing wings. (L. W. Brownell)

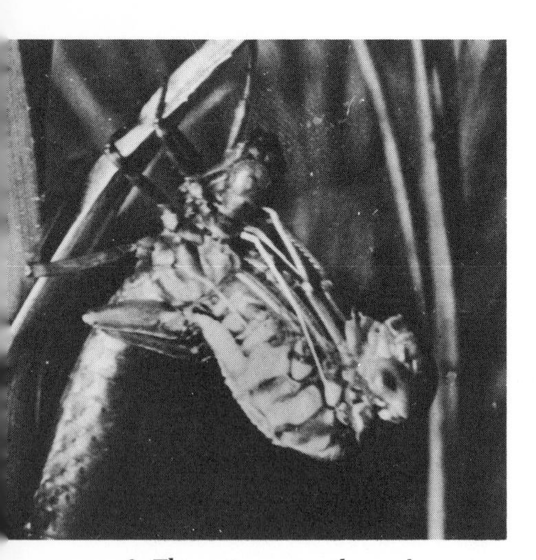

2. The mature nymph crawls up a plant that extends above the water. The cuticle splits down the head and thorax, and the adult emerges.

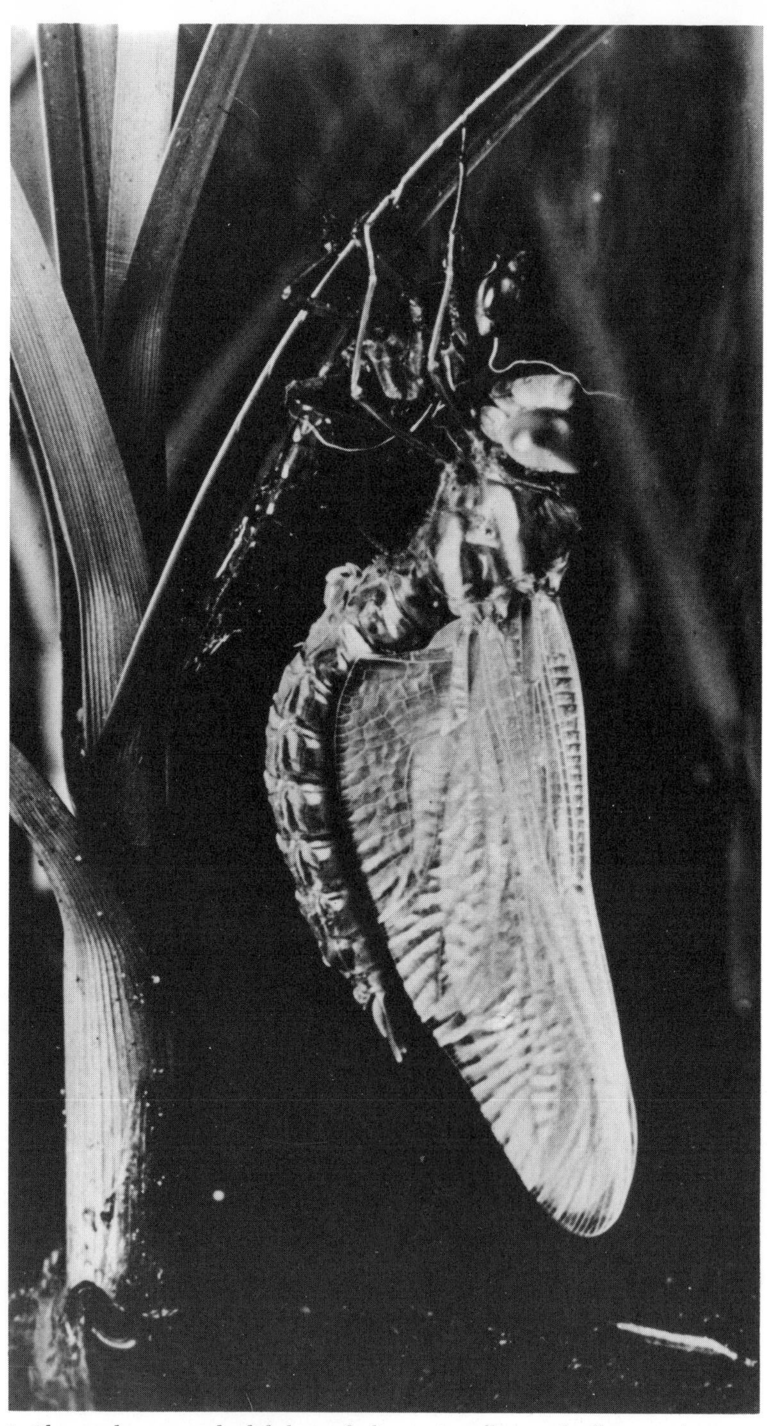

3. The newly emerged adult has soft, limp wings that gradually expand as blood is pumped into the many wing veins.

4. The wings harden as the dragonfly hangs by its legs, which can be used only for grasping vegetation or for seizing prey, not for walking.

Complete metamorphosis is illustrated here by the development of the monarch butterfly, *Danaus plexippus.* (All photos by C. Clarke.) **1. The egg,** a little more than a millimeter high, is cemented to the underside of a milkweed leaf. The larva chews its way out, devours the eggshell, and starts feeding on milkweed plants.

The most radical changes of all are those undergone by insects with **complete metamorphosis,** such as butterflies and their larvas called *caterpillars.* The larvas of these insects have small internal wing buds, not visible from the outside, and are so different from the adult, not only in habit but also in gross structure, that the radical metamorphosis to the adult must be preceded by an externally quiescent stage, or *pupa.* During the pupal stage, the immature tissues are modified or, in some insects, almost completely broken down and reorganized into the adult form. Other insects with wormlike larvas include beetles, flies, wasps, bees, and ants. But not all larvas are wormlike. An antlion larva has a broad, flattened body; it digs a pit in the sand and lies buried just below the center of the pit, with only a pair of large pincerlike jaws protruding. When an ant or other crawling insect stumbles into the pit, the sand slides under its feet, carrying it down into the waiting jaws of the larva. The adult is a slender, winged form related to lacewings. One possible advantage in this type of life history, which includes a nymph or larval stage with structure and habits very different from those of the adult, is that it enables the developing individual to exploit two different food sources and to use the resources of two separate habitats.

2. The larva (called a caterpillar in moths and butterflies) increases its weight 3,000-fold in 15 days. A large caterpillar can chew up a milkweed leaf in 4 minutes.

3, 4, 5. The larva changes into a pupa. Attached by silk to the underside of a leaf, the larva hangs head down and sheds its cuticle, revealing the pupa already formed within.

6. The pupa (called a chrysalis in some butterflies) appears inactive, but inside the whole structure of the insect is being reorganized.

7. The newly emerged adult clings to the pupal covering while the wings expand and harden enough to permit flight.

1. Antlion uses its jaws to pierce small prey (mostly ants) and suck their juices. This antlion is one of several that were collected in the Indiana Sand Dunes at the south end of Lake Michigan; they lived for many months in a pan of sand on the desk of one of the authors, building pits and catching ants dropped into the pan. Finally they pupated. *Myrmeleon.*

2. Pupa, exposed by cutting open its sand-encrusted, pea-sized cocoon.

3. Adult antlionfly is a delicate nocturnal flier that hides in shrubs by day. It is a member of the order Neuroptera and has four wings with a dense network of veins, as is characteristic of this group. (All photos by R. Buchsbaum.)

Since there is some tissue reorganization during the molting of grasshoppers and other insects with gradual metamorphosis, the changes in the pupa may be interpreted as a more extreme type of molting. Indeed, both molting and metamorphosis are controlled by some of the same **hormones.** Like other arthropods, insects begin each molt with the secretion of *molting hormone* (ecdysone), which is distributed in the blood and coordinates the many changes that occur during the molt. The growing larva also secretes *juvenile hormone,* which prevents metamorphosis and ensures the production of larval structures at each molt. When the larva has reached a certain size, however, it stops secreting juvenile hormone, and metamorphosis proceeds, first with the formation of a pupa and finally with the last molt to the adult stage. Adult insects do not molt or grow.

Newly molted cockroach nymph lies helpless beside the empty cuticle it has just cast off. The new cuticle is soft and elastic. Soon it will become hardened and darkened. Panama. (R. Buchsbaum)

That normal insect development depends on the proper timing and combination of hormones can be demonstrated by experimentally altering their balance. If the glands that secrete juvenile hormone are surgically removed from a larva, it undergoes metamorphosis at the next molt, long before it would normally do so and regardless of its small size. Conversely, if juvenile hormone is supplied to a larva that is ready to mature and has stopped secreting the hormone itself, metamorphosis is inhibited. The larva continues to grow and, when eventually permitted to metamorphose in the absence of juvenile hormone, it produces a giant but otherwise normal adult.

These experiments suggest that it should be possible to devise *hormonal insecticides*. Indeed, certain plants appear to have done so. Substances isolated from these plants interfere with the action of juvenile hormone and cause insect larvas to metamorphose prematurely when very small, forming miniature (and abnormal) pupal and adult stages. Other substances isolated from certain plants mimic juvenile hormone and prevent metamorphosis. Various substances that act like juvenile hormone can be synthesized in the laboratory and appear to have possibilities for commercial use in insect control. Compared to conventional toxic insecticides, synthetic hormones can be made to act more specifically on certain insect types, such as one family of bugs; they seem to be nontoxic to other animals (noninsects) and to plants; they are biodegradable but stable enough for use; and insects are less likely to become resistant to products that closely resemble their own hormones than to conventional toxins. Problems in the development of hormonal insecticides probably lie mainly in the application of just the right quantities at critical times. For example, too small an amount of juvenile hormone might simply delay metamorphosis, permit further growth, and result only in larger and more voracious pests!

Behavior

Just as hormones secreted within the body carry information among the tissues, coordinate their activities, and produce specific responses, certain chemicals secreted by insects outside the body carry information to other members of the species and produce specific behaviors. These chemicals are called **pheromones.** The most common pheromones are sex attractants, sex-specific "perfumes" usually secreted by a female to announce her location to males. The results can be spectacular, bringing males to females over distances of several kilometers. One female pine sawfly, held in a cage in a pine tree, attracted over 7,000 males in 5 hours. Sex attractants occur in many insects, but are especially widespread and well studied in butterflies and moths. The

Antennas of male moth are large and branched, providing enormous surface for the olfactory sensory receptors that respond to female scent and guide the male toward it. (R. Buchsbaum)

large feathery antennas of male moths are covered with sense organs that respond only to the sex attractant of females of the same species. Male moths are so rigid in their response to this irresistible substance that they will crowd around a drop of it absorbed onto a piece of paper and completely ignore a real live female in an airtight glass container nearby. Other kinds of reproductive pheromones passed between individuals stimulate sexual maturation or copulation; and still others serve, among social insects, to identify different types of individuals, to signal danger, or to mark trails and food sources.

The possible influence of trail-marking pheromones must be carefully excluded from experiments on **learning** in which ants, for example, are tested in a maze. If individuals leave chemical marks that they or others later follow, they will falsely appear to have learned the maze pattern. However, experiments in which this pitfall has been strictly avoided show that ants are capable of learning to find their way through a complex maze involving many turns and choice points. Despite their industrious reputation, the ants learned the maze faster if it was an obstacle on their way home to the nest than if they were made to pass through it on their way out to forage from a food box. Compared to rats that were tested in the same maze pattern (built to a

Army ant workers, and a few large-headed soldiers, accompany their large queen around and around the periphery of a box, following the pheromone trail she continues to lay down. *Eciton.* Barro Colorado Island, Panama. (R. Buchsbaum)

larger scale), the ants needed 3 to 4 times more trials to master the maze. There were, however, individual differences in both groups: the most successful ants took only twice as many trials as the least successful rats. Learning by honey bees has also been convincingly demonstrated. Bee workers learn not only the location of certain patches of flowers, to which they return repeatedly, but also the times of day at which the flowers are open. Although limited kinds of learning have been demonstrated in some other insects as well, most of the behavior of insects is instinctive and stereotyped.

The behavior of social insects varies with caste (worker, queen, etc.), and caste is determined by a combination of heredity, hormones, pheromones, and nutrition. In honey bees, a few female larvas that are fed a special substance develop into queens while the rest become sterile workers that perform the various tasks of the colony (cleaning, brood care, defense of the hive, foraging, etc.) successively as they age. The few males do no work in the colony. In termites, on the other hand, there are usually equal numbers of males and females in up to 7 castes of workers, soldiers, and reproductives, and the caste into which each young nymph develops is determined primarily by the levels of hormones and pheromones to which it is exposed. In some species the workers are not a permanent caste but nymphs that later develop into soldiers or reproductives. Kinds of behavior important in maintaining termite societies include the mutual licking and exchange of feces that goes on constantly between individuals, serving to distribute pheromones throughout the colony and also, in wood-eating termites, to provide newly molted workers with a supply of the flagellated protozoans (see chapter 4) that digest the cellulose in their diet.

The insects are divided into 5 classes: 4 small classes of primitively wingless insects and 1 huge class containing the roughly 30 orders of winged insects. The members of an order have in common the same general structure, including similar mouthparts and wings, and they almost always show the same type of metamorphosis. In spite of differences in shape, size, and color, they have enough features in common so that typical members of an order can be classified at a glance as dragonflies, termites, beetles, and so on. In the photographs that follow, the major orders of insects are presented, with illustrations of some of their features, their common representatives, and a few of the ways in which they affect the lives of other animals, including humans.

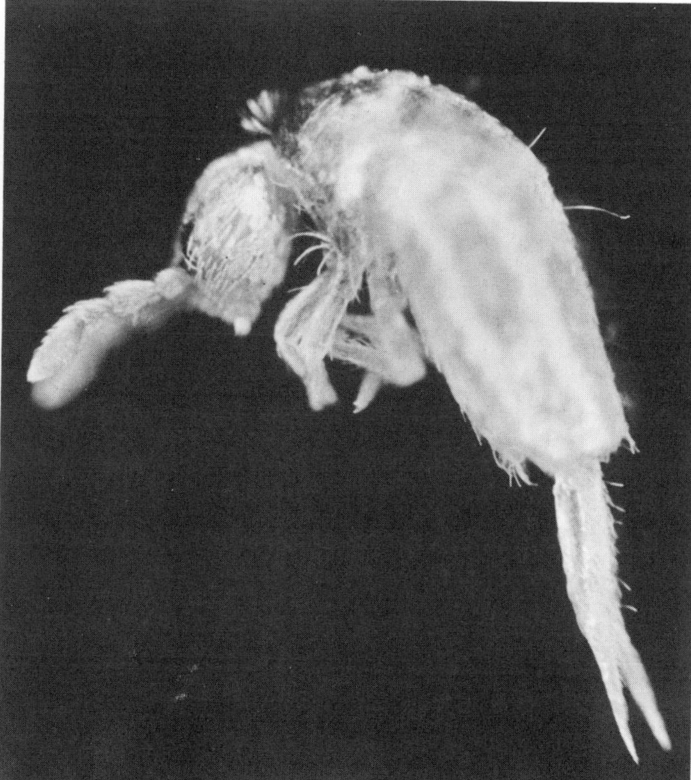

Springtails (class **Collembola**) perform high jumps by means of a pair of abdominal appendages that form a springing mechanism. The spring is held in place under the abdomen by a catch formed from another pair of abdominal appendages, *left*. When the catch is released, the insect springs into the air, *right*. Collembolans are minute insects, primitively wingless, and develop without metamorphosis. They have simple chewing mouthparts and a group of simple eyes on each side of the head. Most feed on decaying matter and live under stones, rotting wood, and leaves. (P. S. Tice)

Bristletail is a member of the class **Thysanura.** Both common and class names come from the two or three many-jointed filaments that most thysanurans bear at the rear end. All have simple chewing mouthparts, are primitively wingless, and develop without metamorphosis. Most bristletails are found under stones or in ground litter. The firebrat, *Thermobia domestica*, 13 mm long, is shown here eating holes in filter paper. It frequents warm spots such as fireplaces and steam pipes. Silverfishes are similar thysanurans found more often in damp places such as basements, bathtubs, and sinks. They eat paper and book bindings. (R. Buchsbaum)

Mayfly adults are seen in late spring and during the summer around bodies of freshwater in which they lay their eggs and where the aquatic nymphs feed and gradually metamorphose. In some species the adults emerge, mate, lay their eggs, and die all in a single evening; in others this takes a few days. The ephemeral nature of these delicate adults suggested the order name **Ephemeroptera.** The adult does not feed. It derives its energy, and the female provisions her eggs, from the food reserves stored by the long-lived nymph. Mayfly species that emerge by the millions in some areas attract much attention as they swarm about street lamps or lighted house windows.

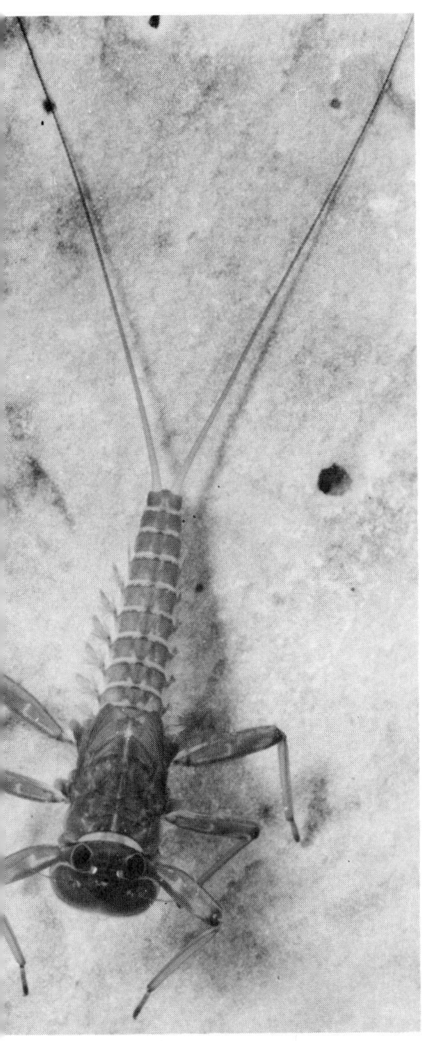

Mayfly nymph resembles the adult in having an elongate body, compound eyes, and long tail filaments. But the nymph is aquatic, with feathery tracheal gills along the abdomen. It feeds mostly on plant materials, and it may take as long as three years to metamorphose. Mayfly nymphs form an important part of the diet of freshwater fishes, as do also the bodies of millions of adults that fall into the water when they die after the eggs are laid. Many fishing "flies" are modeled after mayflies. (R. Buchsbaum)

Damselflies and their close relatives, the dragonflies (see earlier photos of metamorphosis), comprise the order **Odonata.** Both groups have aquatic, predaceous nymphs that metamorphose gradually and finally emerge as adults with four membranous wings having many crossveins. Dragonflies hold the wings out at the sides when perching, while damselflies bring the wings together, parallel with and a little above the long slender abdomen, as seen here. The compound eyes are larger in dragonflies, occupying much of the head in these strong fliers with good vision. In the less active damselflies the eyes are smaller and project at the sides of the head. Dragonfly nymphs respire through tracheal gills within the rectum; damselfly nymphs have three platelike tracheal gills projecting from the rear of the abdomen. (C. Clarke)

Katydid is nocturnal and lives in trees, in contrast to the grasshoppers, which feed by day and stay close to the ground. Katydids have long, hairlike antennas, auditory organs on the forelegs (see earlier photo in section on sense organs), and a long, flattened ovipositor. They are noted for the songs of the males. (R. Buchsbaum)

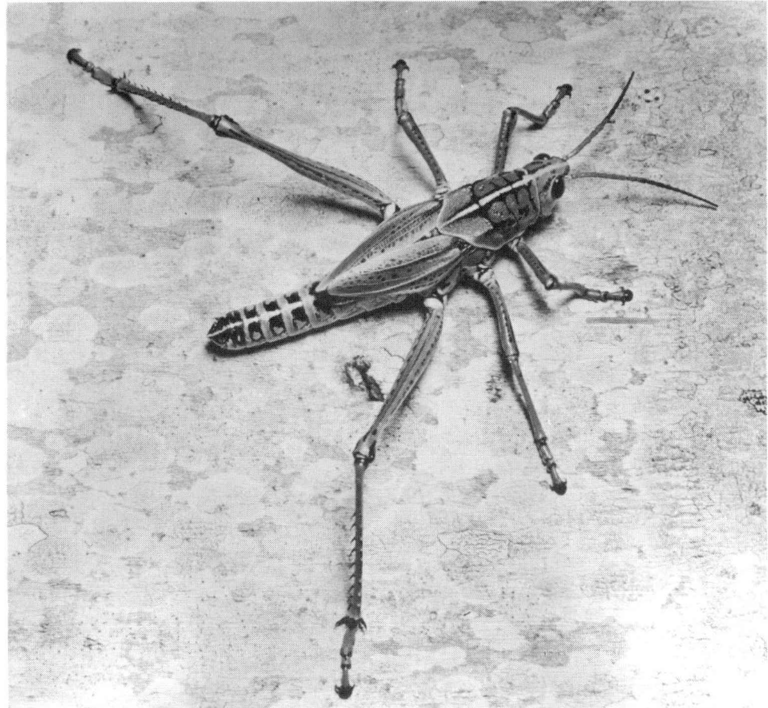

Grasshopper belongs to the largest grouping in the order **Orthoptera,** which also includes katydids and crickets. All have chewing mouthparts and gradual metamorphosis. Some have small wings, or no wings, but when wings are present the forewings are stiff and serve to cover the many-veined membranous hind wings, which are folded like a fan when at rest. This Florida lubber grasshopper is typical in having short, stout antennas, auditory organs on the first abdominal segment, and greatly enlarged femurs in the hind legs for jumping to escape predators. (R. Buchsbaum)

Field cricket is more like a katydid than like a grasshopper—both in anatomy and in feeding mostly at night. It lives in open fields or about houses, where it reveals its presence by chirping. (L. W. Brownell)

Tree crickets. Their high-pitched trill is among the most conspicuous insect songs of summer nights. (C. Clarke)

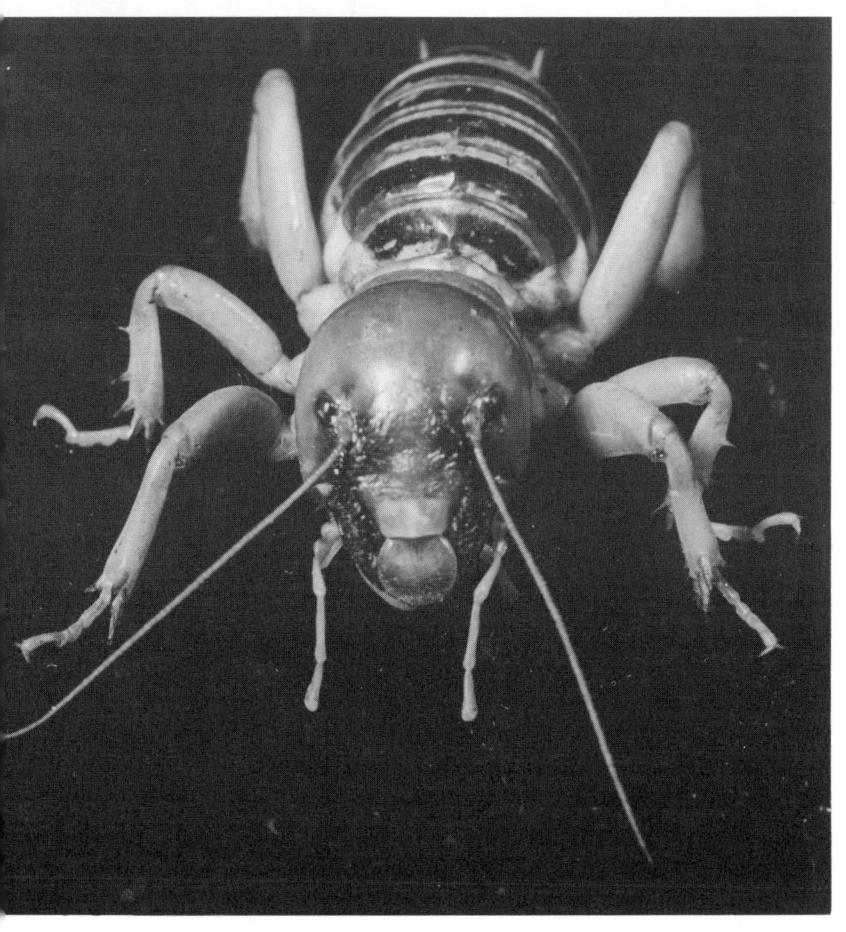

Jerusalem cricket, *Stenopalmatus*, is a wingless orthopteran, common in California. It uses its stout legs to burrow in soil, feeding on roots and tubers. Despite its large size (50 mm) and formidable appearance, it is harmless to humans. (R. Buchsbaum)

Walkingstick was hard to discern among the vines on a tree trunk in Panama until it moved and was picked up. Most members of the order **Phasmida** look like green or brown twigs; some tropical ones are flattened, winged, and look like green leaves. Both types move slowly or remain quiescent in postures that make them inconspicuous. The largest insect in the United States is a walkingstick that reaches 18 cm (7 in) and is found in the southern states; tropical walkingsticks may be almost twice as long. (R. Buchsbaum)

Praying mantis is named for the upraised position in which it holds the large forelegs, modified for seizing prey. Most mantids, order **Mantodea,** are tropical, and many of these look like green or brown leaves. Temperate forms are elongate, like this Chinese mantid, 10 cm (4 in) long. Introduced into the United States, it is sold widely for biological control of garden pests, but it is just as likely to eat useful insects. (R. Buchsbaum)

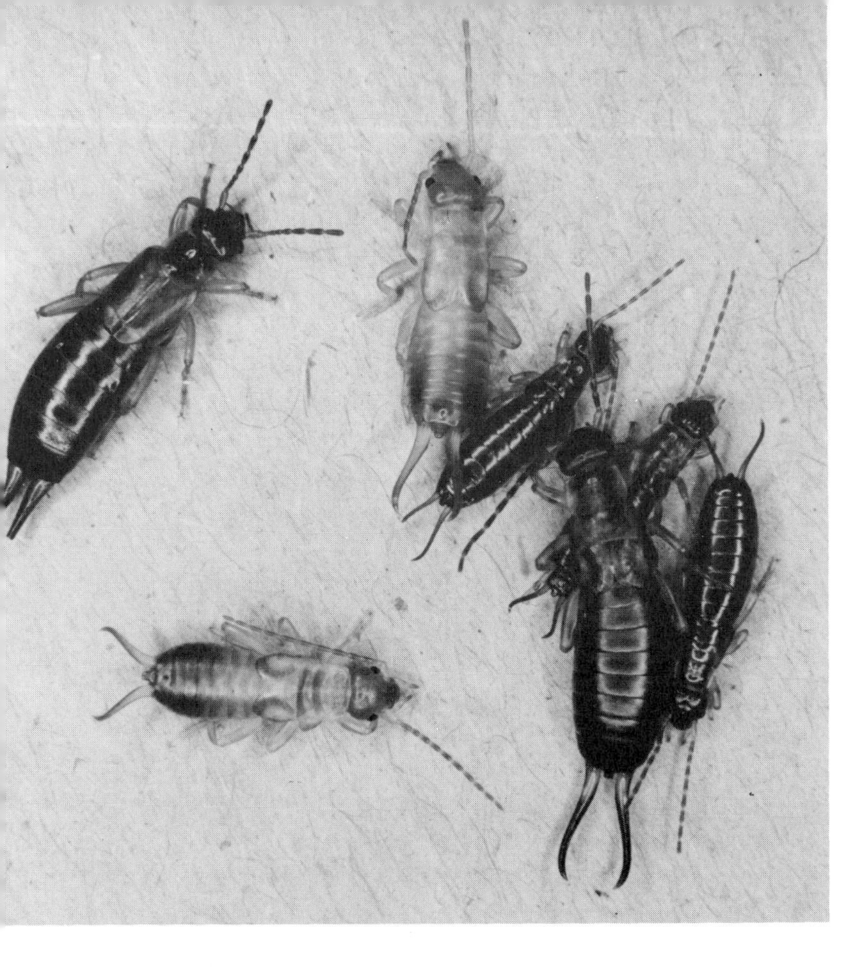

Earwigs (order **Dermaptera**). The European earwig, introduced into the United States, flourishes in California, where it does serious damage to vegetable crops, fruit trees, and ornamental garden plants. These flattened insects have a tough, shiny cuticle and a movable forceps that is used in defense, in capturing prey, and in mating. They live in little colonies, and the females guard their nymphs. By day they hide under litter and in plant crevices; at night they come out to feed. Two of those seen here have just molted. (R. Buchsbaum)

The walkingsticks, mantids and earwigs are sometimes submerged in the large order Orthoptera. Unlike orthopterans, these insects do not have jumping legs, but their wings and chewing mouthparts are similar to those of orthopterans. The nymphs closely resemble the adults and metamorphosis is gradual. Also included in the orthopteroid orders are the cockroaches and termites.

Large tropical cockroach, *Blaberus giganteus*, was removed from its hiding place in a large wooden clock in the forest laboratory of the Smithsonian Institution in Panama. It seemed unruffled, waving its long antennas and standing its ground as only an insect 75 mm (3 in) long can afford to do. Members of the order **Blattodea** are primarily tropical and forest-dwelling, but some species have penetrated forests and houses in all but the coldest parts of the world. In forests they hide under bark, among fallen leaves, in rotting wood, and in every kind of crevice. Their flattened bodies, and the chitinous shield covering the underslung head, make it easy for them to enter the narrowest openings. They emerge mostly at night to scavenge with chewing mouthparts. (R. Buchsbaum)

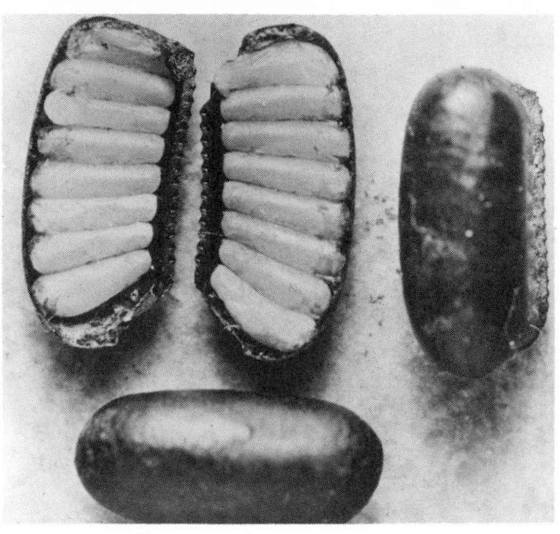

Oriental cockroach, *Blatta orientalis,* originally from Asia, has been spread around the world with shipments of food. The female seen here *(left)* has an egg case, with hardened covering, protruding from the genital opening; she carries it about until the eggs hatch. *Right,* an opened egg case reveals two rows of eggs. (L. Passmore)

Newly hatched nymphs of the oriental cockroach resemble the adults except for size, wing development, and maturity of reproductive organs. Like other orthopteroid insects, they will undergo gradual metamorphosis. Cockroaches may carry disease organisms on their bodies and in the fecal pellets they drop into human food, so they are always under suspicion as a health hazard. But they are sturdy insects, easy to raise, and they serve well in many kinds of laboratory research into learning, nervous conduction, effects of chemicals, and so on. (L. Passmore)

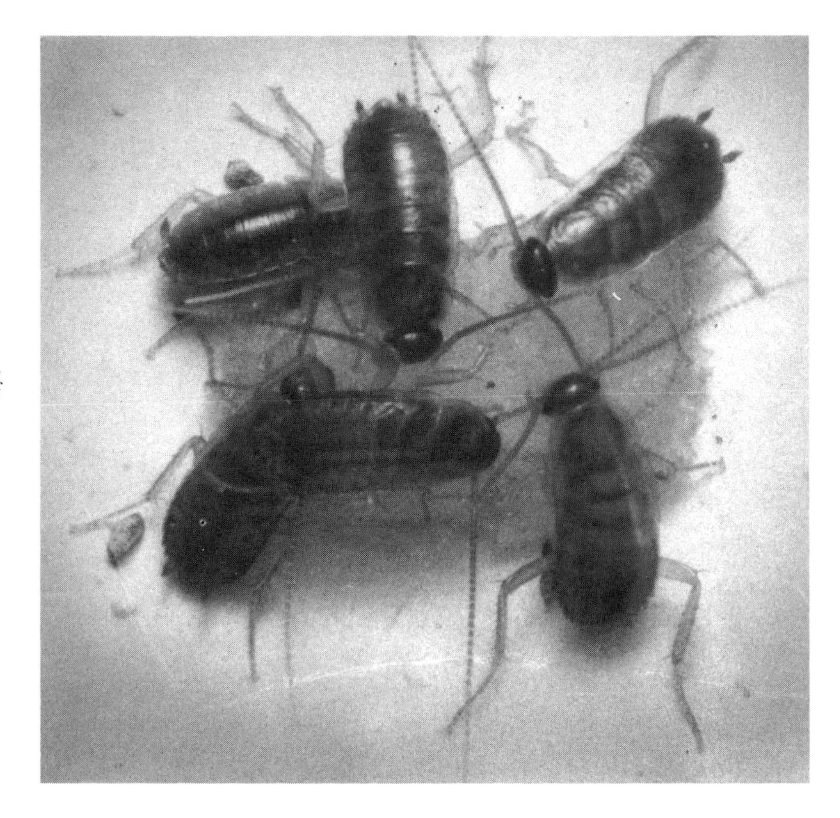

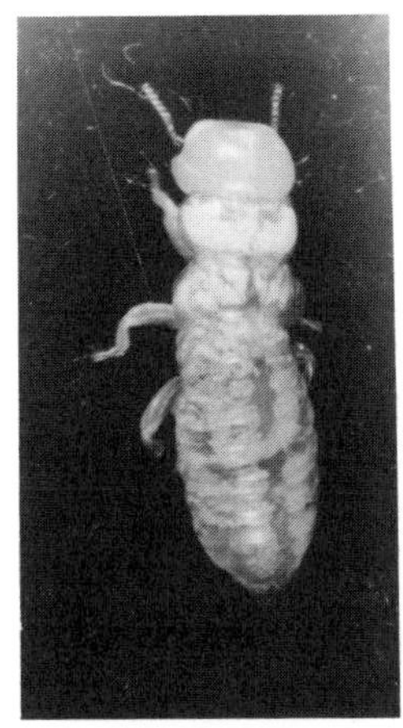

nymph worker

soldier

reproductive

Termites, order **Isoptera,** all live in social colonies composed of differentiated castes, regulated mostly by pheromones (see section on behavior earlier in this chapter, p. 410). *Workers* build and maintain the nest, collect the food, and feed and care for the other members. *Soldiers* defend the colony. Certain *reproductives* provide the fertilized eggs. Others, with newly developed wings, fly from the nest to found new colonies. Having shed their wings, each founding pair digs a nest and mates. These photos are of a species of *Kalotermes,* in which the workers are not a permanent caste but nymphs that later develop into soldiers or reproductives. (See working nymphs from this same Florida colony in the photo that heads chapter 2.) Termites are best known for the enormous damage that some species do to wooden buildings. The cost of controlling termites and repairing their damage comes to hundreds of millions of dollars a year in the United States. (R. Buchsbaum)

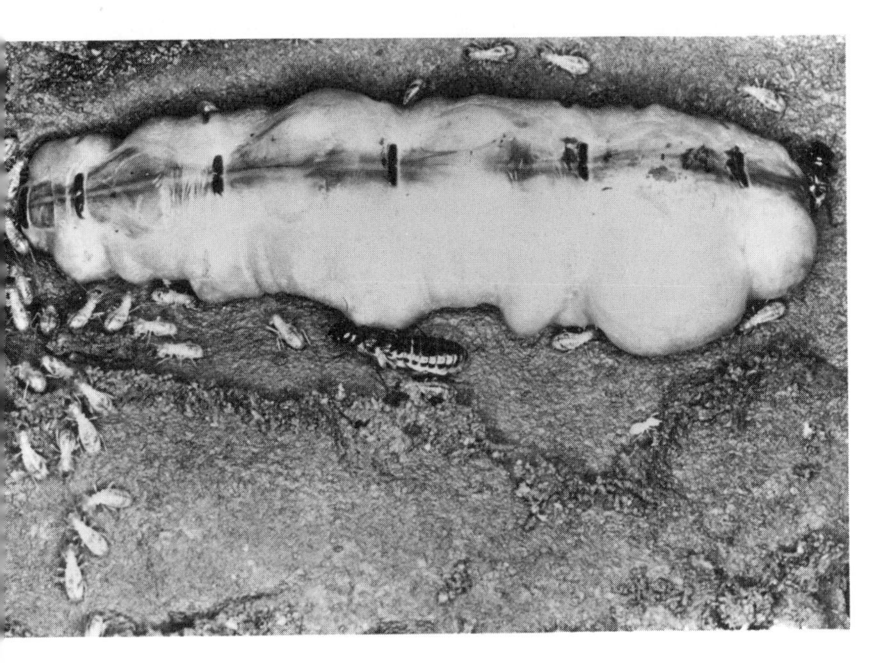

Termite queen in an opened reproductive chamber near the center of a large termite mound in Ghana. Her head and thorax, at the right, appear tiny next to her enormous abdomen, swollen by the enlargement of her ovaries. Some African termite queens reach 10 cm (4 in) in length and can lay 36,000 eggs a day or 13,000,000 in a year, the greatest fecundity in any terrestrial animal. About midway of the length of her abdomen, the male reproductive, or king, lies close at her side. Both are fed and licked by the many permanently juvenile workers that here are clustered mostly about the queen's rear end. The workers receive the eggs and carry them to the brood chambers of the nest. (L. Pittman)

Nasute soldiers, instead of the usual large-jawed type, are found in certain termite species. The jaws are vestigial and the head is extended as a pointed snout from the tip of which a fluid can be squirted. Exposed to air, the fluid becomes viscous and sticky and gums up small enemies, usually ants. A few seconds before this photo was taken, the nest had been poked, making an opening about 1 cm across and promptly bringing forth this array of snouted nasutes ready to do battle. A few nonsnouted workers can be seen at the top of the opening about to repair the damage. The entangling fluid acts also as an alarm pheromone, summoning nearby soldiers to join in. This nest of *Nasutitermes ephratae* was attached to a tree trunk in Panama. In grassland this same species builds mound nests. (R. Buchsbaum)

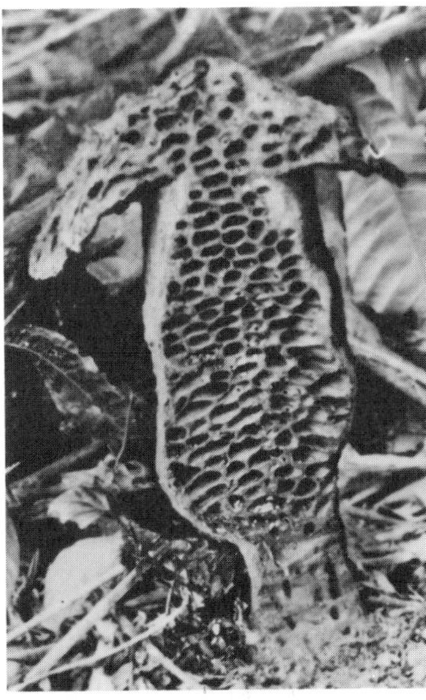

Termite nest, *Cubitermes loubetsiensis*, intact and cut open to expose its elaborate system of chambers. The caplike roof is neatly adapted to shed rain in the tropical rain forest. (H. O. Lang)

Termite skyscraper of an Australian species, *Amitermes meridionalis*, can be 4 m high. The north-south axis is also about 4 m, the east-west axis only 1 m at its widest part, the base. This north-south orientation of the long axis provides a minimum of exposure to midday sun, long supposed to help keep the nest temperature lower. But recent studies show that nests built in shady areas are similarly oriented. The explanation for the construction of these nests requires further study. (G. F. Hill)

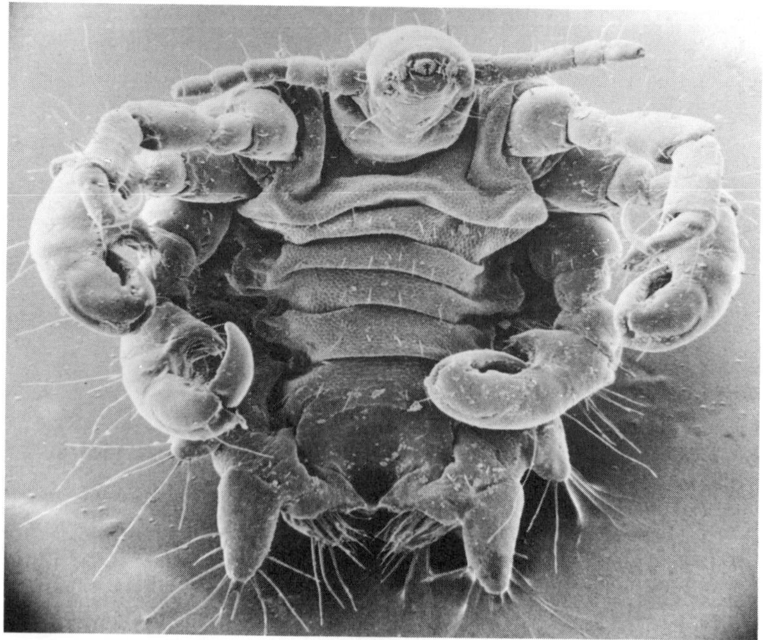

Sucking lice, order **Anoplura,** are ectoparasites that use piercing mouthparts to suck the blood of terrestrial and marine mammals. The stylets are withdrawn into the head when not in use. Anoplurans are secondarily wingless and have clawed legs that grasp the hairs of the host. They hatch as nymphs and metamorphose gradually. Those shown here are adults of the only two species that suck human blood. *Above,* a **common human louse,** *Pediculus humanus,* about 3.5 mm long. It comes in two varieties that behave differently. The head louse favors the fine hairs of the head, to which it cements its eggs. The body louse often clings to and lays its eggs in the host's clothing, and it flourishes in cold climates, where both bathing and change of clothing are likely to be less frequent. It is spread on clothing and bedding; and during extreme crowding, as in wartime, infected body lice become vectors of the microorganisms that cause epidemic typhus and other fevers. *Below,* a **crab louse,** *Pthirus pubis,* lives chiefly among the coarse pubic hairs and lays its eggs on them. This louse can cause intense irritation and itching but is not known to transmit microbial disease. In massive infections it may reach the armpits or even the beard and eyelashes. Scanning electron micrograph. (P. B. Armstrong)

Above: **Aphids** are tiny soft-bodied insects that suck plant juices, then exude the excess from the anus as honeydew, a sweet fluid. In this photo ants are attending wingless female aphids and feeding on their honeydew. Some species of aphid are tended by ants as we herd cows. The eggs are collected in the fall and overwinter in the ant nest. The newly hatched nymphs are carried by the ants to a suitable plant, and later nymphs are transferred to other plants. This fondness for aphids is not shared by farmers or gardeners, to whom aphids are among the worst of all plant pests, doing enormous damage and transmitting plant viruses. Aphids are members of the order **Homoptera** ("homogeneous wings"). Homopterans vary greatly in size and shape, ranging from small scale insects, leafhoppers, mealybugs, and spittlebugs, to large lanternflies and cicadas. All have piercing and sucking mouthparts and feed on plants. (R. Buchsbaum)

Below: **Treehoppers** jump when threatened, as do their relatives, the leafhoppers, which include some of the most important little pests of farms and gardens. In these treehoppers, the hard shield over the front part of the thorax is greatly enlarged and extends over the abdomen and wings, almost obscuring them. The resemblance of these homopterans to the thorns of the twig on which they rest presumably helps to disguise them from predators.

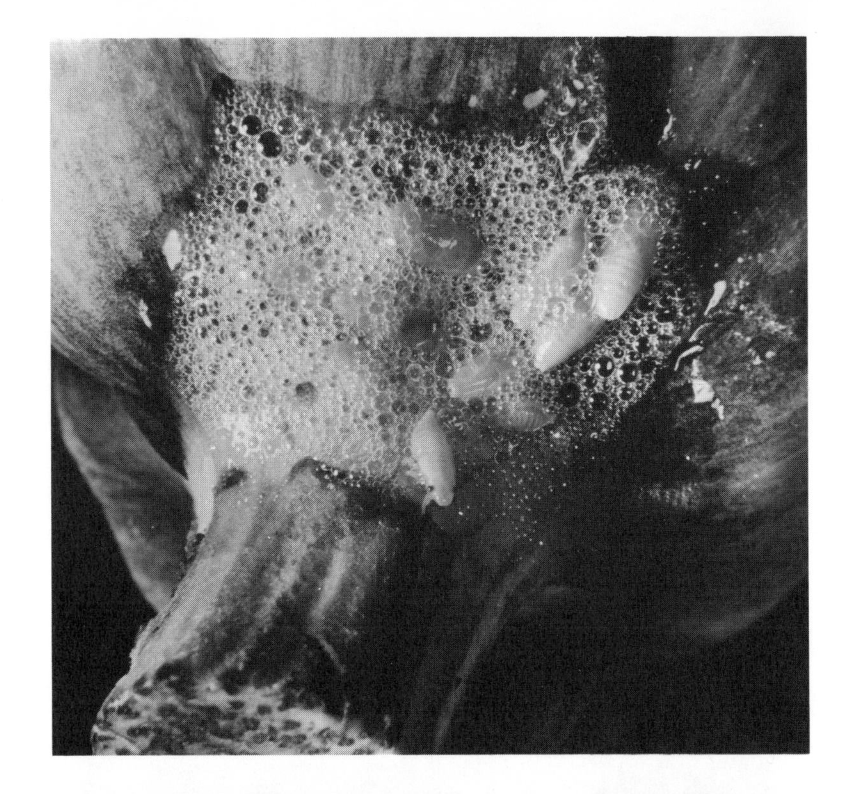

Spittlebug nymphs produce conspicuous masses of spittle, a mucilaginous froth that completely covers them, protecting them from predators or the drying of hot sun. So sheltered, they suck plant juices and gradually metamorphose. Those seen here were exposed by wiping away the spittle at the base of an artichoke. The flying adults resemble leafhoppers and occur on many plants in fields and gardens. (R. Buchsbaum)

Mealybugs are wingless female homopterans covered with a white powdery secretion. Unlike their close relatives, the sessile scale insects, mealybug adults retain their legs and can move about. In sufficient numbers, they can do great damage as they suck plant juices. The delicate males are gnatlike with only one pair of wings. (C. Clarke)

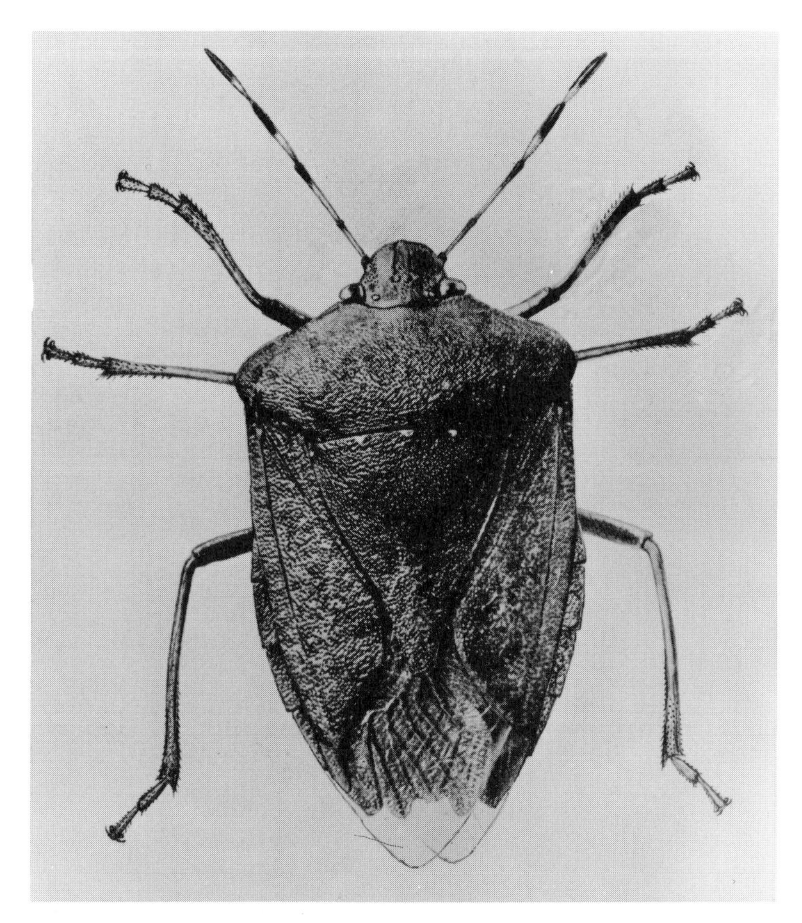

Stink bug sucks the juices of farm and garden plants; it often leaves behind a fetid odor, and this explains the disagreeable taste of some berries in a box of otherwise good ones. To most people a bug is almost any kind of insect, but the term is usually applied by entomologists only to members of the order **Hemiptera,** in which the two pairs of wings are not alike, the forewings being thickened and leathery in their front half, the hind wings entirely membranous. Hemipterans have piercing and sucking mouthparts and gradual metamorphosis. Most are plant suckers like this stink bug, and many are agricultural pests; but others suck insect blood, and these often benefit farmers by preying on pest species. Some members of this group attack other animals, including humans, and may act as vectors of disease. (U.S. Bureau of Entomology)

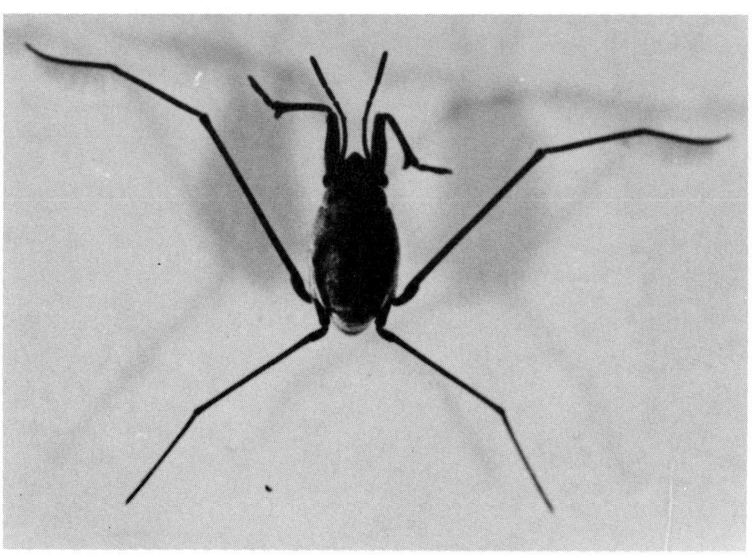

Marine water strider is the only kind of insect that lives in the open ocean, often far from land. It is wingless and strides about on the surface film with two pairs of long middle and hind legs, grasping prey at the surface with the short forelegs. The oceanic species are of the genus *Halobates;* all other genera are found in freshwaters, where the familiar slender water striders dimple the surface film with the tips of their feet and feed on insects that have fallen into the water. (R. Buchsbaum)

Right: **Bed bug** is flattened and wingless and about 6 mm long. These bugs suck blood from birds and mammals. They are pests in houses, hotels, and military barracks. They come out at night to suck blood from sleeping victims and hide during the day in cracks in walls, under wallpaper, or beneath mattresses. The bite is irritating but bed bugs are not important vectors of disease. Preserved specimen. (U.S. Army Medical Museum)

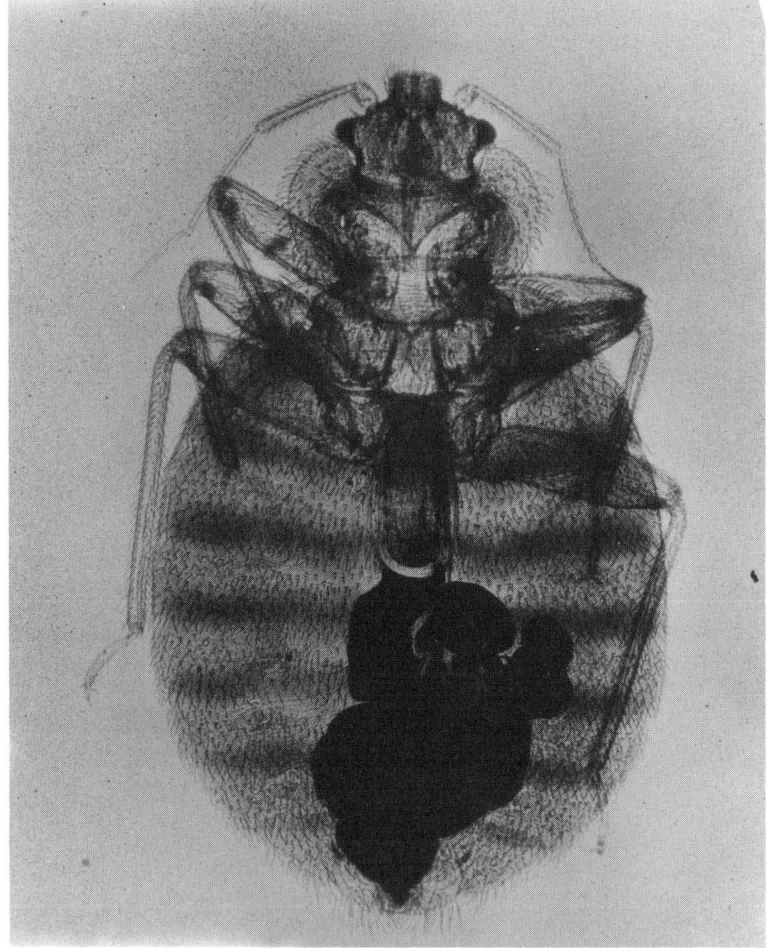

Giant water bugs include the largest of the various kinds of aquatic hemipterans. Some in the United States reach 5 cm (2 in), and a South American species is twice as big. The forelegs grasp insects and snails, or even tadpoles and small fishes. The middle and hind legs are flattened for swimming. The smaller bug here is a male, with eggs cemented to his back by a female. He will carry the eggs until the nymphs hatch out. These bugs fly from one pond or lake to another, and are often attracted to electric lights. Preserved specimens. (P. S. Tice)

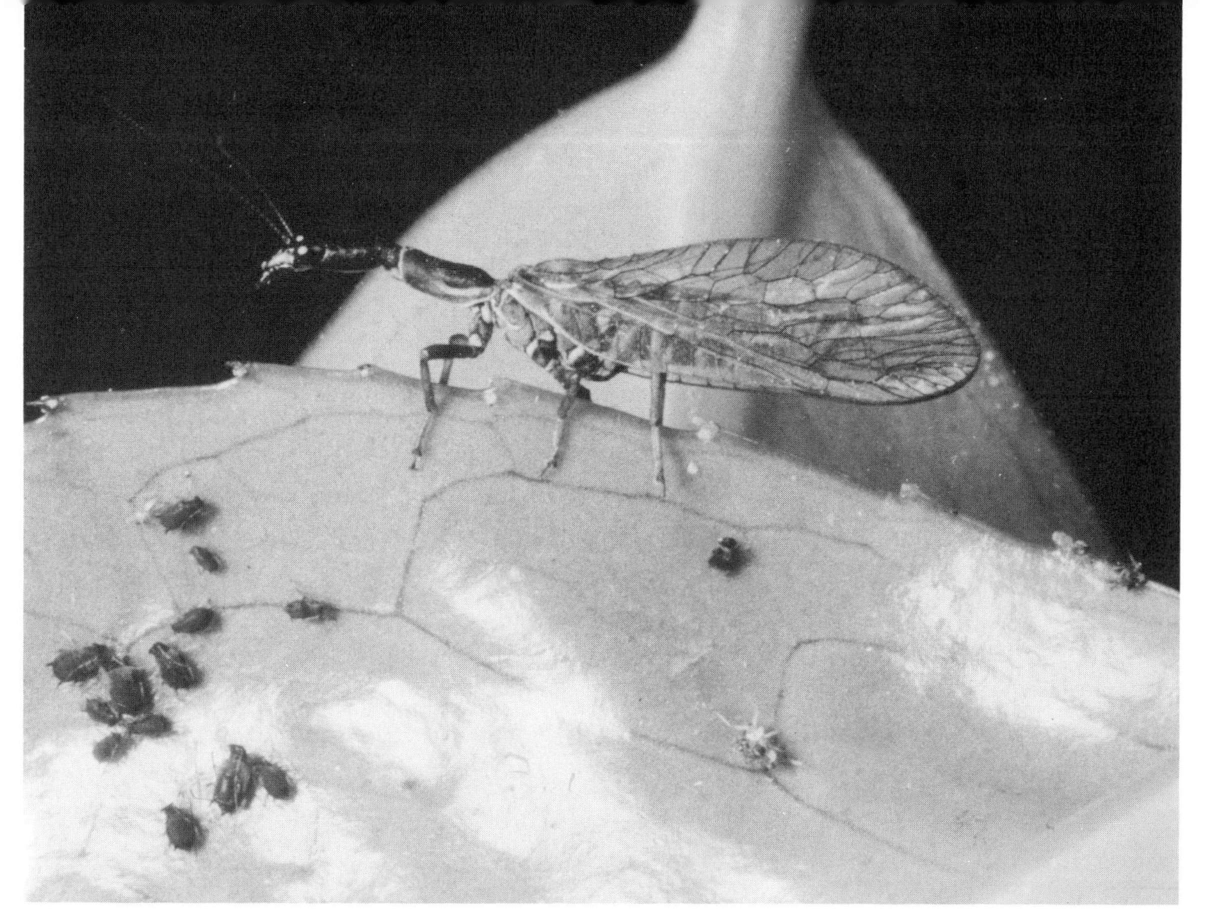

Snakefly, named for the serpentlike appearance of the front end is a member of the order **Neuroptera,** characterized by four membranous net-veined wings, chewing mouthparts, and complete metamorphosis. In the United States snakeflies live only in the far western states. They are common on trees and shrubs and especially welcome in orchards and gardens for their help in controlling small plant pests. This snakefly was 25 mm from head to wingtips and had been devouring aphids. (R. Buchsbaum)

Dobsonfly, a magnificent neuropteran with a wingspread of 10 cm (4 in) or more, is best known for its aquatic larva, called a dobson or hellgrammite by those who use it as fishing bait. The large-jawed larva lives under stones in rapid streams and catches small animals, especially the aquatic nymphs of other insects. It breathes by tracheal gills (see earlier photos of respiratory variations). After almost three years it goes on land to pupate, emerging as a winged adult. (R. Buchsbaum)

Golden-eyed lacewing is a neuropteran easily recognized by the metallic color of the eyes and by the lacy greenish wings, about 1 cm long. A garliclike odor repels birds and insect predators. The spindle-shaped **larva,** *lower left,* called an aphislion, is a welcome sight in gardens and farms as it feeds voraciously on aphids and scale insects, sucking their juices with piercing mouthparts like those of antlions (see earlier photo of metamorphosis). The **pupa,** *upper left,* has just used its chewing mouthparts to cut a lid out of its spherical silken cocoon. When it emerges it will crawl about briefly before changing into a flying adult. (C. Clarke)

Eggs of a lacewing are laid in groups, attached to leaves by long stalks. Presumably this discourages the first-hatched aphislions from devouring the unhatched eggs of their siblings. (K. B. Sandved)

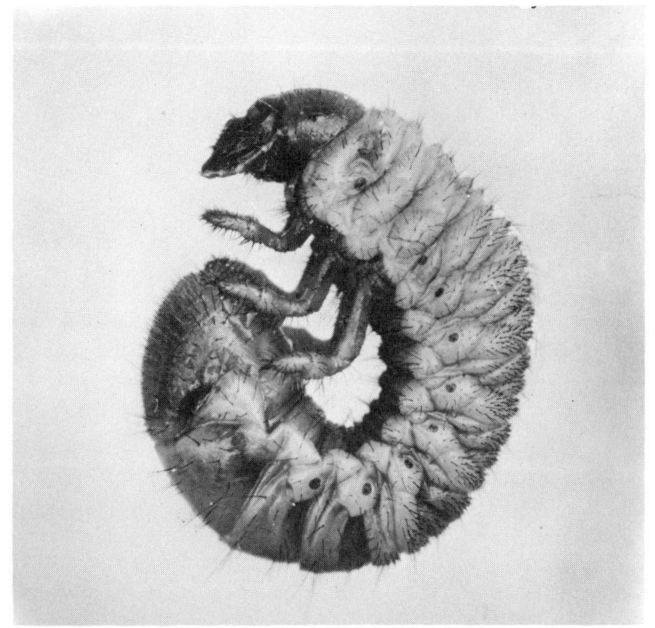

Japanese beetle, *Popillia japonica,* is typical of the order **Coleoptera,** in which hardened forewings protect the membranous hindwings folded underneath. Beetles, with 330,000 species, are the largest order of insects. Like neuropterans (and other insects in the rest of this photographic series), beetles have complete metamorphosis. The Japanese beetle, introduced into the United States, has become a serious pest. *Above, left,* the **adult** chews on foliage, flowers, and fruits, severely damaging fruit trees and many other wild and cultivated plants. *Above,* the **larva,** called a grub, feeds on the roots of grasses and overwinters in the soil; it harms lawns and golf courses. *Left,* the **pupa.** (U.S. Bureau of Entomology)

Cotton-boll weevil with long snout inserted into a cotton boll (seedpod). It lays its eggs in the hole made by feeding, and the larvas that hatch destroy the boll. This beetle is the most economically devastating of all plant pests in the United States. Yet in the aptly named town of Enterprise, Alabama, there stands a monument "in profound appreciation of the boll weevil and what it has done as the herald of prosperity." What it did was to force cotton-based southern agriculture to diversify into more profitable crops such as peanuts. The long-snouted beetles called weevils are the largest family (60,000 species) in the animal kingdom. Many are pests of both growing plants and stored starchy foods. (U.S. Bureau of Entomology)

Ladybird beetles are favorites of farmers and gardeners. Even urban children have been taught to treat them kindly and send them on their way. These small beetles feed chiefly on tiny homopteran pests—aphids, scale insects and mealy bugs. Wintering aggregations (as seen here) are collected by the bushel from high California hills and sold to citrus growers. In early spring the beetles are liberated in orchards; others make their own way down into the valleys. Later, during the dry season, the new generation returns to high, cool refuges. The bright coloring of these beetles, usually yellow, orange, or red with black spots, is a warning signal of their distastefulness. (N. Burnett)

Hercules beetles display the striking differences that may occur between the sexes of the same species. The male has a long horn on the thorax, a shorter one on the head; together they are used in conflict with other males during the mating period. The female is smaller and has no horns. These tropical beetles are seen here about life size.

Flour beetle, *Tribolium.* This 3-mm long brown beetle can be found feeding and rapidly multiplying in flour, cereals, nuts, dried fruits, and other dry foods. Related, but much larger (15 mm) are beetles of the genus *Tenebrio,* which also grow in cereals. The larvas of *Tenebrio molitor* are yellow and 25 mm long; called mealworms, they are sold as fish bait and as live food for pets. (R. Buchsbaum)

Whirligig beetles are usually seen in small aggregations on the surface of ponds and quiet streams. They dart around each other in graceful curves. The eyes are divided, the upper half for seeing in air, the lower half for seeing in water. (C. Clarke)

Fireflies are not flies but beetles with soft wing covers held loosely over the body. On summer nights the intermittent flashing of fireflies serves to bring the sexes together. That the attractant is light and not odor is easy to demonstrate. A female confined in an opaque but perforated box attracts no males; one placed in a corked glass bottle does. *Above, right:* The light organs are in the rear segments of the abdomen. *Right:* Photo made with the light of a firefly whose abdomen can be discerned toward the right. (R. Buchsbaum)

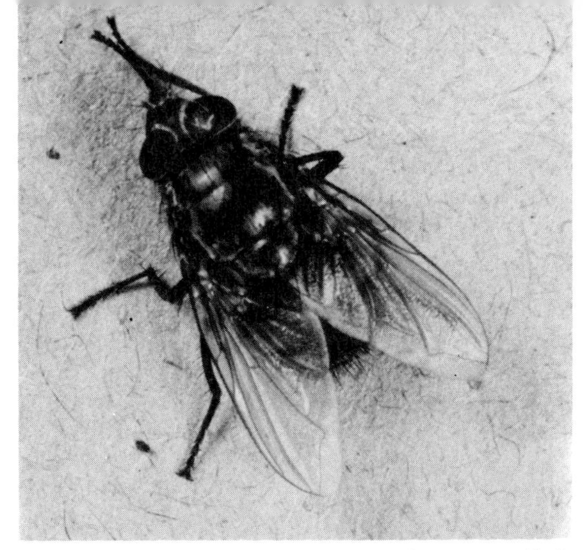

Blow fly. The many kinds of flies—including some called mosquitoes, gnats or midges—comprise the order **Diptera.** Flies (unlike other insects with "fly" in their common compound names, such as mayflies, butterflies, etc.) have only one pair of membranous wings; the second pair is represented by stalked knobs (see photo of crane fly in section on insect wings). Most adult flies lap or sponge up liquid food.

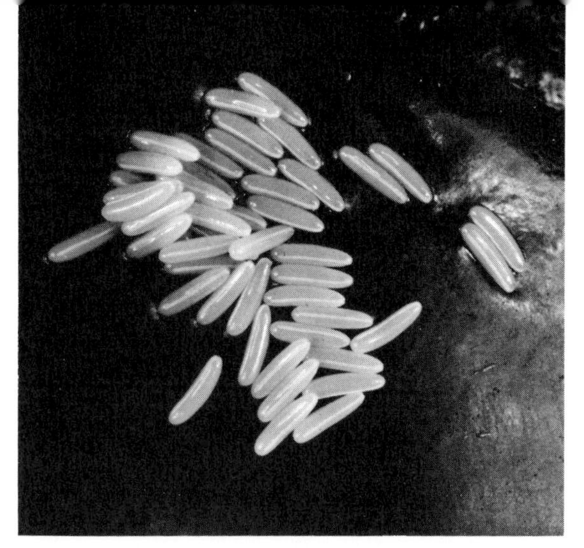

Eggs of a blow fly, laid on the surface of a chicken liver left in an open dish. Blow flies live outdoors but often enter houses and lay their eggs on meat, cheese, or other foods.

Larvas of a blow fly. Wormlike fly larvas, called maggots, burrow into and feed on decaying matter. These maggots made a soupy mess of the chicken liver on which they hatched, and they had to be removed to a clean surface for their portrait. The carcass of a vertebrate, filled with hundreds or thousands of maggots, will in a few days be reduced to bone, tough skin and tendons, and hair.

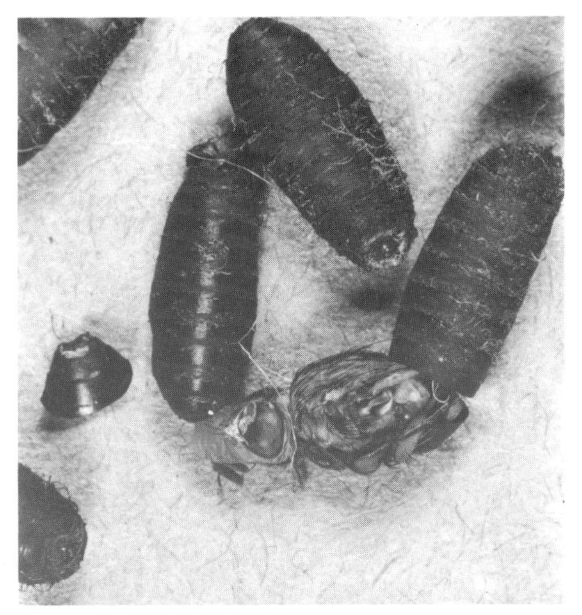

Pupas of a blow fly cannot move, and appear quiescent, but inside the brown, thickened, last larval covering, the larval structures are undergoing a thorough remodeling into the structures of the adult. (All photos on this page, R. Buchsbaum.)

Larvas of a bot fly, attached to the stomach lining of a horse by their hooklike mouthparts, suck blood. If thousands are present, the horse may die. The adult flies do not feed, but even their presence makes horses nervous and irritable. The big, fuzzy flies lay their eggs on skin or hairs, usually on a horse's forelegs. And when the eggs are licked, the larvas hatch and are swallowed. They attach, feed, and finally pass out with the feces and pupate in the soil. (U.S. Bureau of Entomology)

Mediterranean fruit fly is a medium-sized fly with mottled wings that are held outspread. It occurs in many warm countries around the world. During brief invasions of Florida and California it has inflicted many millions of dollars of damage on citrus and other orchards. (S. Whitely)

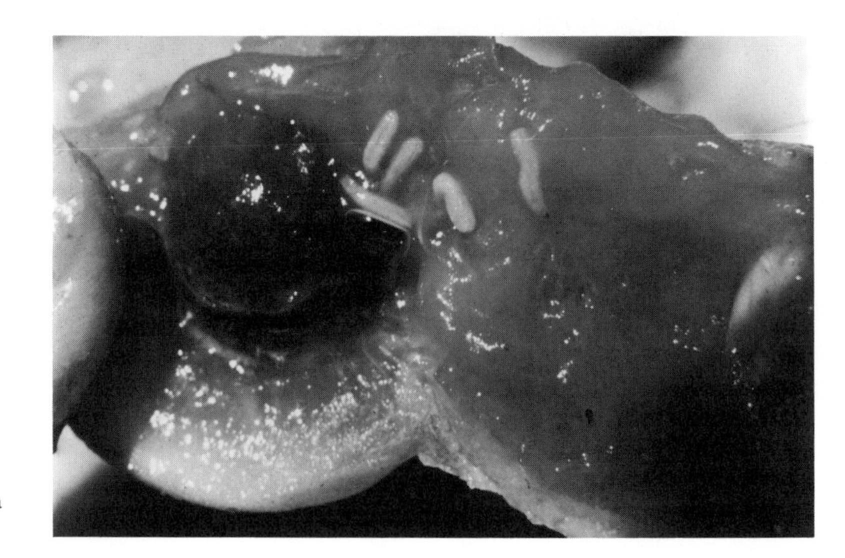

Larvas of the Mediterranean fruit fly are the culprits in the extensive damage done to orchards by the fly. This photo shows larvas found in an opened apricot in California during a brief invasion. (S. Whitely)

Drosophilid flies, although often called fruit flies, do not feed on fruit. They are attracted to the odors of damaged and fermenting fruits, in which they lay eggs. The larvas feed on yeasts that grow on exposed fruit tissue. These flies can be a nuisance in households and at picnics, but they are not a threat to fruit growers. The tiny (only 2 to 4 mm long) flies are the source of much of our knowledge of genetics, the science of heredity. *Drosophila melanogaster* has been raised by the millions in laboratories, each breeding pair in a separate bottle that contains sterile food inoculated with yeasts. Such a breeding pair is shown here, the female at the left. (R. Buchsbaum)

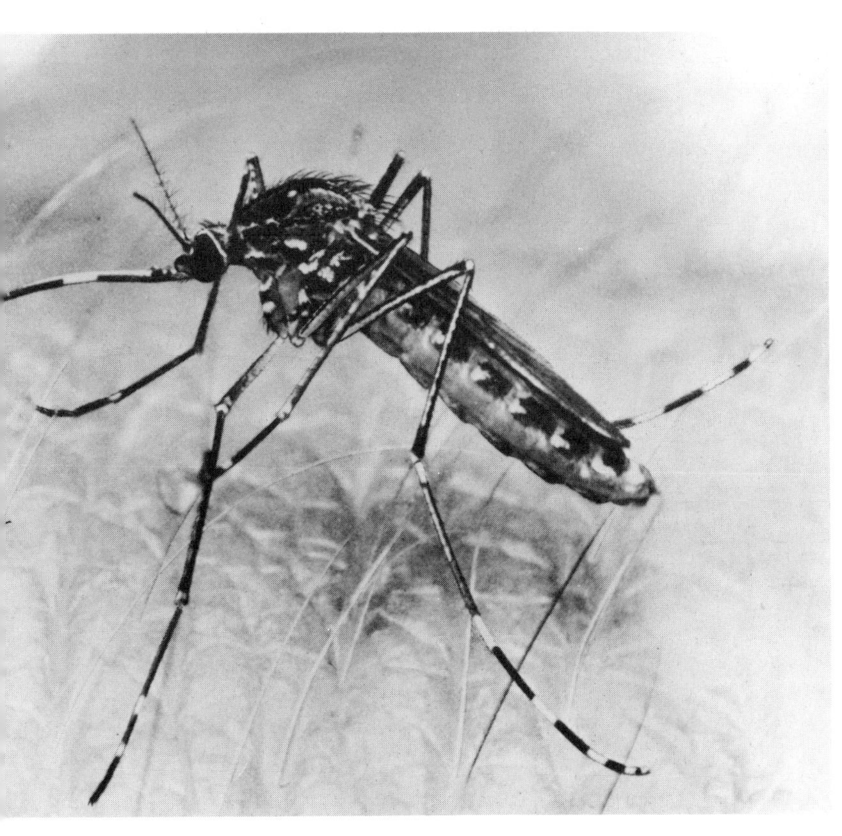

Mosquito poised to insert its long piercing and sucking mouthparts in human skin. Mosquitoes are the invertebrates with the greatest effect on human health and history. The most studied of all arthropods, they are said to have caused the fall of Greece, the outcome of many wars, and the long delay in the building of the Panama canal. Half of all deaths in human history have been attributed to the mosquito-borne diseases malaria and yellow fever. Other important mosquito-borne diseases in the tropics are elephantiasis, caused by filarial nematodes, and dengue, caused by a virus. Only females suck blood; male mosquitoes feed on nectar and other sweet plant juices. The female of *Culex tarsalis,* seen here, feeds on sparrows and other small birds that harbor the virus of western equine encephalitis in the western United States; in sucking the blood of horses and humans it transmits the virus. The common house mosquitoes in the United States are also of the genus *Culex* but their bite is not hazardous. (E. S. Ross)

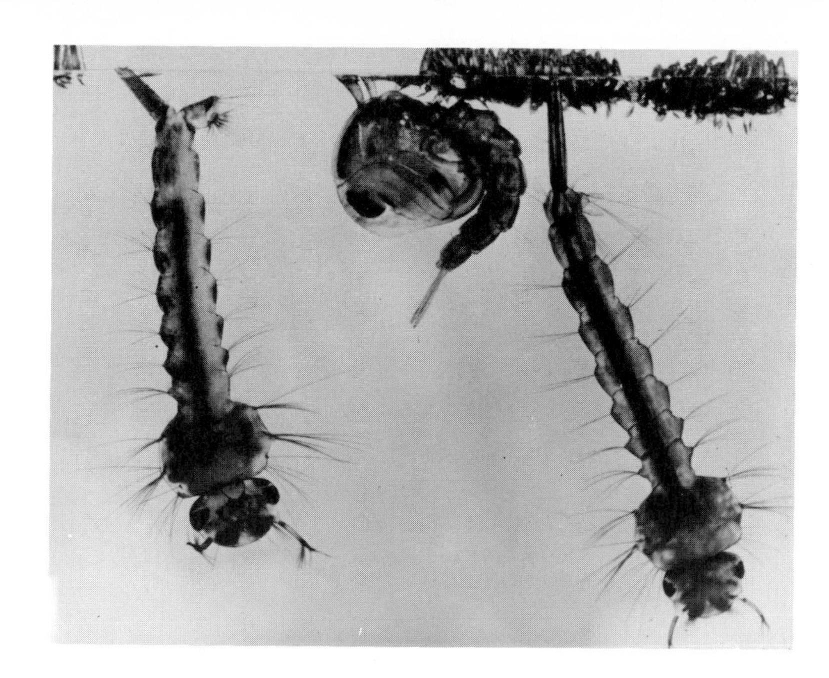

Development of a mosquito is always aquatic. Here, three clusters of **eggs** float at the surface. Two long slender **larvas,** at rest, hang head down with the breathing tube at the posterior end protruding through the surface film. At the fourth molt the larva transforms into a **pupa** with large head and thorax and a slender, flexible abdomen; two breathing tubes project from the anterior thorax. The pupa can swim about, using two leaflike appendages at the tip of the abdomen, but does not feed. In a few days the winged adult will work its way out of the pupal covering, using it as a raft until the wings dry and it can fly away to lead a terrestrial life. *Culex tarsalis.* (E. S. Ross)

Tiger moth is fairly typical of the large order **Lepidoptera.** Most of the 100,000 species are night-flying moths, but the order also includes day-flying butterflies and skippers. Almost all are quite similar in basic structure and habit, having four membranous wings covered with tiny flattened scales and a long proboscis with which they suck nectar (see earlier drawing of butterfly mouthparts). The caterpillar larvas, though of diverse colors and many bizarre forms, are similar in having chewing jaws with which to feed on the various parts of terrestrial plants. Many lepidopteran larvas are serious pests of trees and of field crops, as well as of stored foods. (L. W. Brownell)

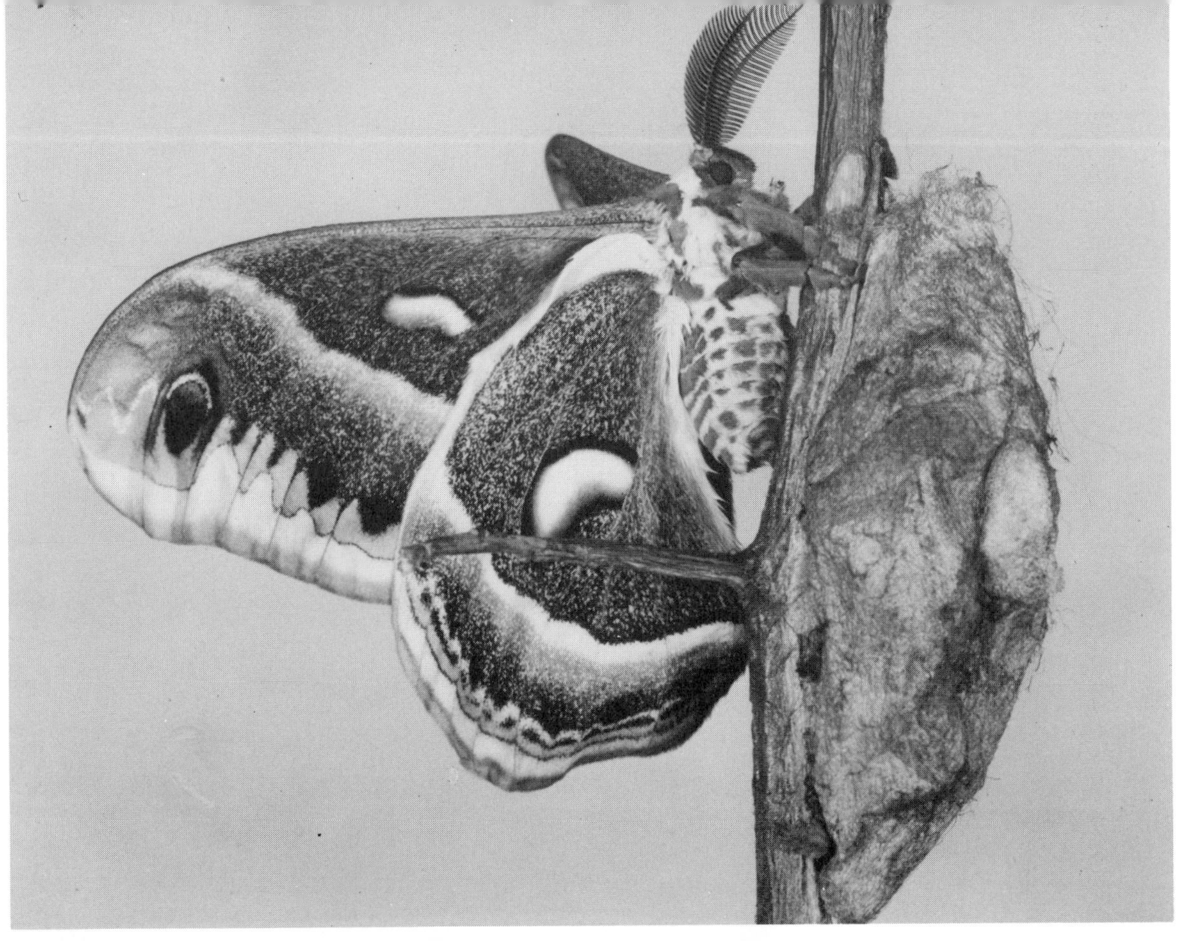

Cecropia moth has a wing span that reaches 15 cm (6 in). This one, its wings fully expanded, rests on a twig that bears the silken cocoon from which the moth emerged and waits for the wings to harden enough to support flight. The caterpillar is greenish and reaches 10 cm in length. Some large Asiatic members of the same family produce commercially valuable silk. (R. Buchsbaum)

Pupa of a cecropia moth in an opened cocoon. The outlines of the feathery antennas and folded wings can be seen through the pupal covering. (L. Keinigsberg)

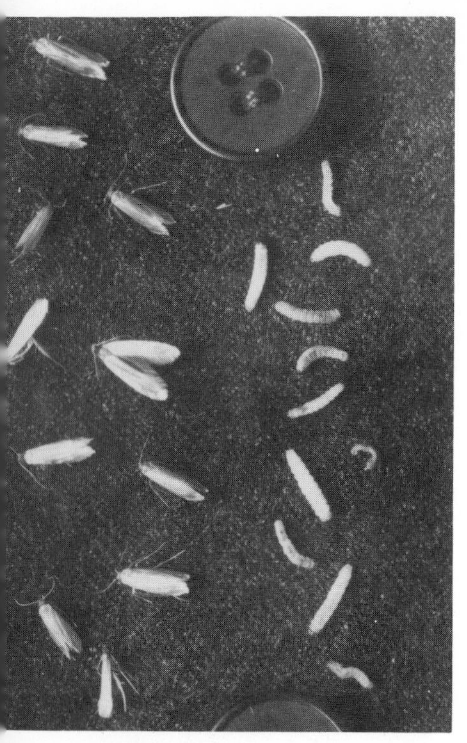

Clothes moths have a wing span of only about 15 mm. Nevertheless, these drab, harmless little adults strike terror in householders who know that their fluttering presence is a signal that, somewhere in the house, moth larvas soon will be—or are already—eating holes in woolen clothes or blankets. (U.S. Bureau of Entomology)

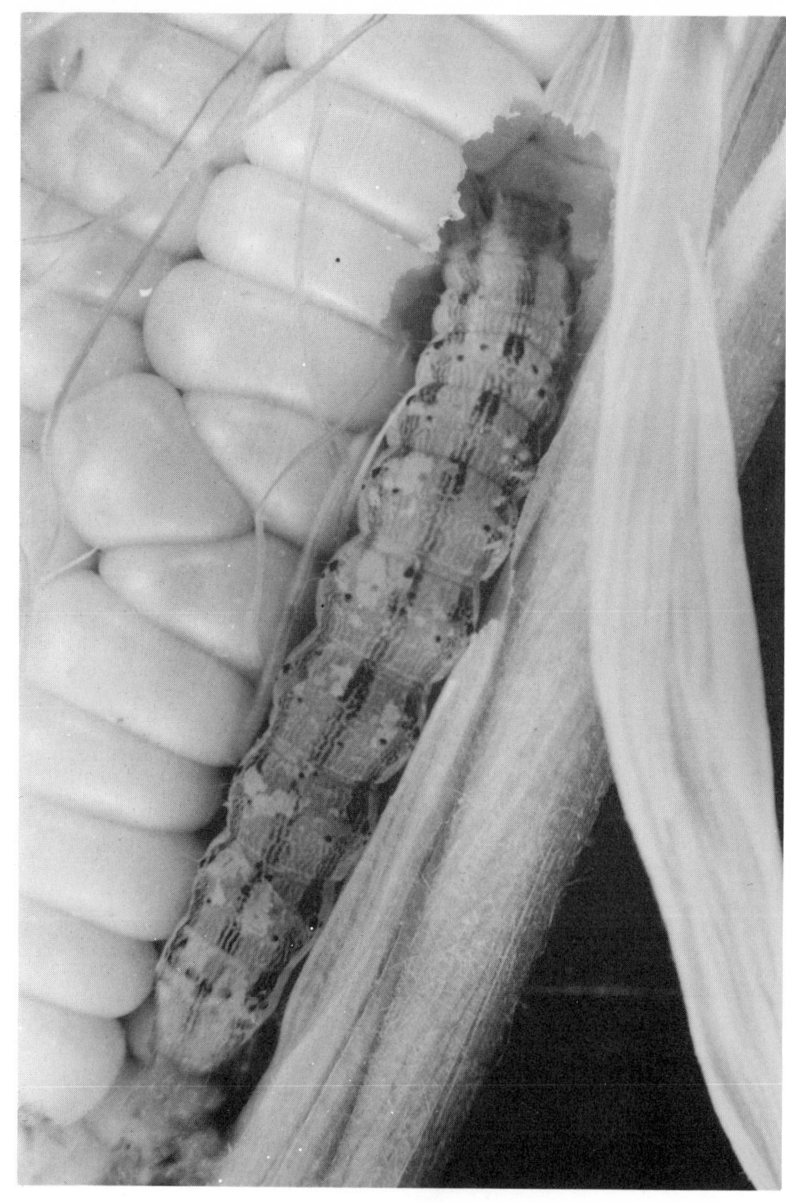

Corn earworm damages corn, tomato, and cotton crops. The moth lays its eggs in the corn silks, and the caterpillars that hatch feed downward into the tips of the ears, eating the young kernels and exposing surfaces on which molds then grow. The full-grown larva leaves the ear and pupates in the ground. (P. S. Tice)

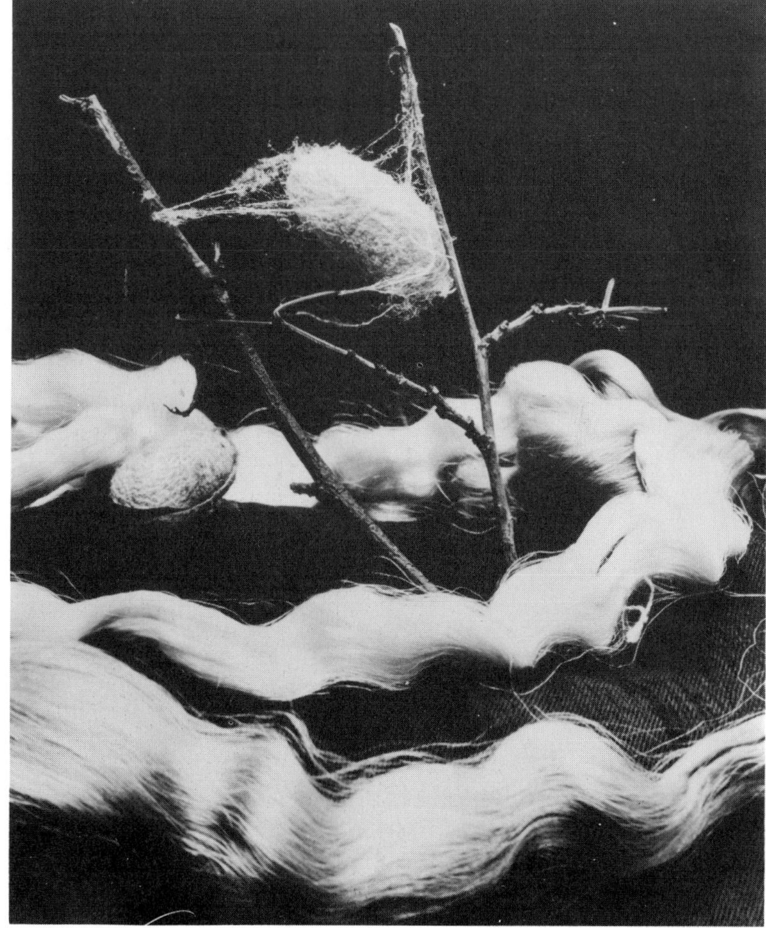

Above left, **Silkworm moths,** *Bombyx mori,* have become thoroughly domesticated and are no longer known in the wild. Originally native to China, where attempts to export silkworm eggs from the country were punishable by death, they were finally smuggled out and now are cultivated in various warm countries, mainly China, Japan, India, and the Mediterranean area. The caterpillars, *above right,* are fed on mulberry leaves until ready to pupate in silk-wound cocoons, two of which are seen at *right* beside hanks of raw silk. The silk filament forming a cocoon may reach 1,200 m in length. (C. Clarke)

Butterfly feeding, its proboscis fully extended as it sucks plant fluid. Butterflies are readily distinguished from most moths by their day-flying habit, knobbed antennas, more slender body, and resting posture with the two pairs of wings held together and vertical. The distinctive colors and patterns of butterfly wings help to bring the sexes together and, in toxic or distasteful butterflies, ward off experienced predators. (K. B. Sandved)

Scales of a butterfly wing overlap like shingles on a roof and are loosely attached, so that they come off like powder when the insect is handled. A lepidopteran that flies into the sticky threads of a spider web often escapes by the loss of a few scales left adhering to the web. The scales of the wings and the "hairs" of the body and appendages are modified bristles. (P. S. Tice)

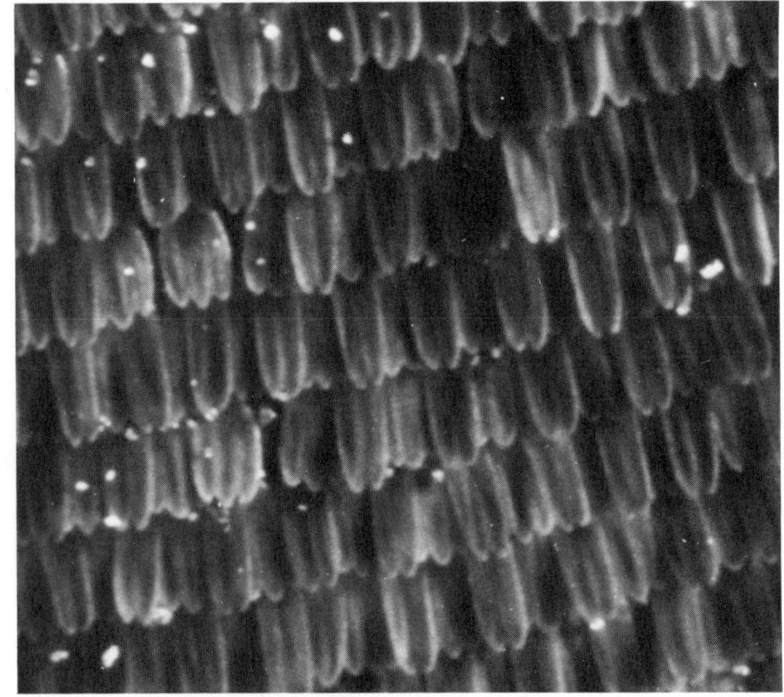

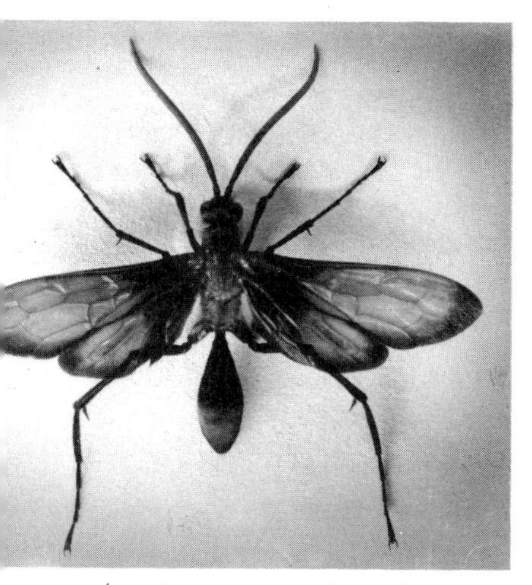

Wasp has two pairs of membranous wings with few veins. Each small hind wing has along its front edge a row of minute hooks that fit into a groove in the rear margin of the larger forewings. This coupling of the wings is a character of all members of the order **Hymenoptera,** a name derived from Hymen, the ancient Greek god of marriage. In addition, the wasps, ants, bees, and most other hymenopterans (except sawflies) are "wasp-waisted," having a marked constriction in the front portion of the abdomen. *Pepsis.* Southwestern United States. (L. Passmore)

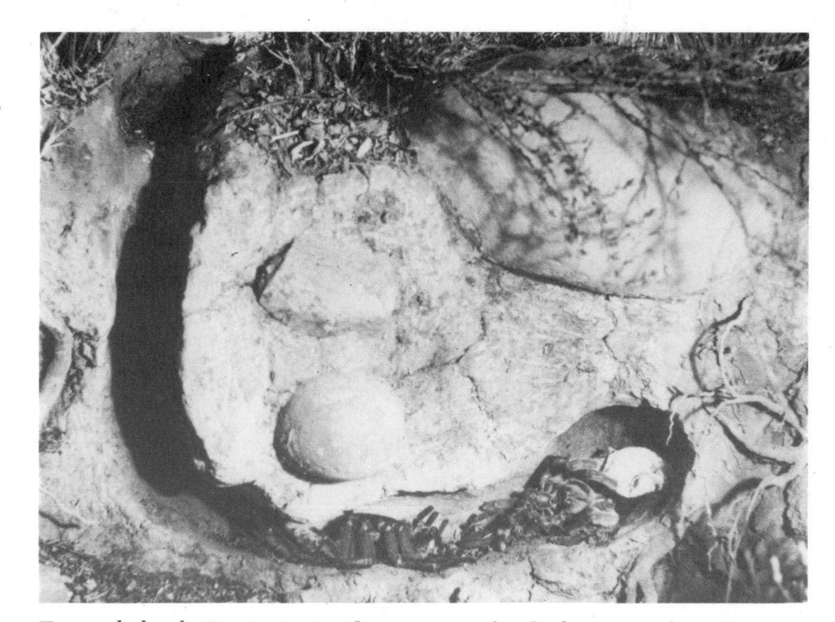

Tarantula-hawk, *Pepsis,* is one of many wasps that feed on nectar but provision their nests with spiders. And being a giant wasp (up to 5 cm long), it can tackle large spiders. *Above,* a female is subduing a tarantula, which is paralyzed by the powerful venom of her sting. Sometimes the wasp loses the struggle and is eaten by the tarantula. *Below,* a burrow containing two paralyzed but still living tarantulas. The female wasp will lay a single egg in each spider, then fill in the burrow. Hunting at night, she may lay 20 eggs in a season. When the larvas hatch, they feed on the spider, grow rapidly, and pupate in the burrow. The newly emerged adults must dig their way out. (L. Passmore)

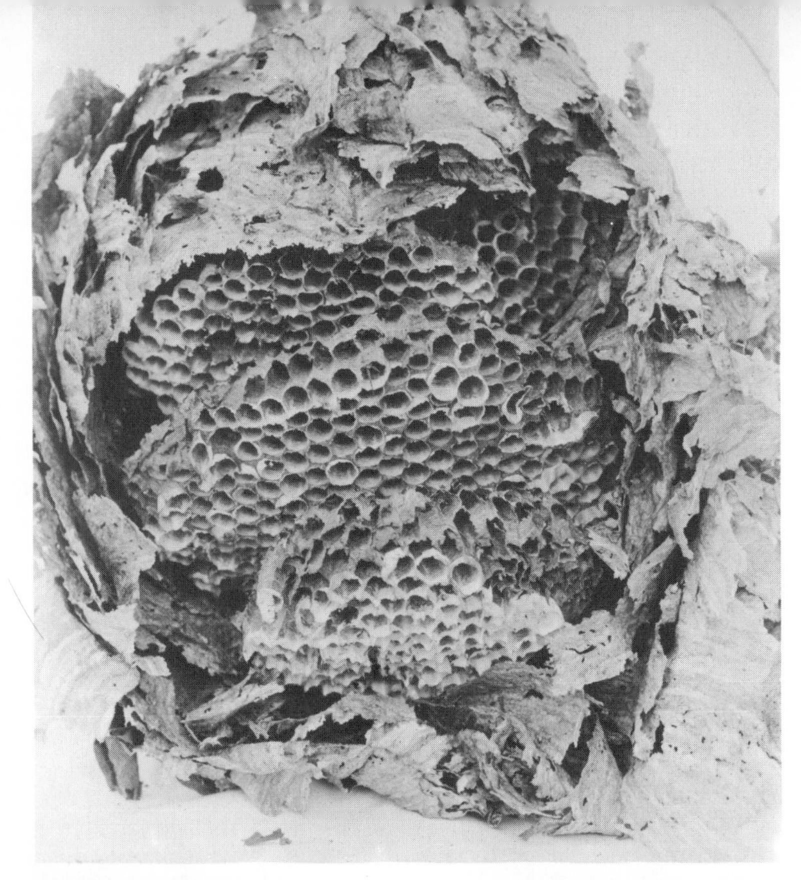

Hornet nest, made from bits of weathered or decayed wood mixed with saliva, is hung from tree branches or from the eaves of buildings. This nest was built by bald-faced hornets, large social wasps that are black with white markings; it has been cut away to expose many tiers of hexagonal cells in which the young are reared by sterile female workers. The mated queen lays all the eggs. Late in the season new queens and males are produced; they mate, and in temperate climates all members of the colony die except the queens, which hibernate and found new colonies in the spring. The closely related yellowjackets are also social but usually build their papery nests in holes in the ground or in hollow logs. Hornets and yellowjackets drink the juices of flowers or fruits, but feed their larvas on chewed-up insects and spiders. Hornets usually do not sting unless the nest is threatened; yellowjackets are more easily provoked. (C. Clarke)

Oak galls, induced on oak trees by minute brown wasps that are seldom seen. In the spring the wasp inserts its eggs into oak tissue, and the plant walls off the parasite within the gall, an overgrowth of tissue on which the developing larva feeds and by which it is well sheltered. By fall the fleshy green and red gall ("oak apple") has become tan and papery, as seen here, and the larva pupates inside without building a cocoon. The adult makes a hole in the gall as it exits. (L. W. Brownell)

Cocoons of a braconid wasp project from the body of a sphinx moth caterpillar. The eggs of this small wasp were laid, by means of a long ovipositor, beneath the skin of the caterpillar. The larvas fed on the host tissues and then gnawed their way through the body wall, constructing external silken cocoons. The adult wasps will emerge and fly away as the exhausted caterpillar slowly dies. (R. Buchsbaum)

Portrait of an ant reveals no special clues as to why the ant family should be more widespread and more numerous in individuals than other insects. It may be that ant behavior and social organization hold the answer, as all but a few parasitic ants live in colonies with castes that show marked differences in structure and behavior. This ponerine ant, from a forest floor in Panama, is near the large end of the size spectrum for worker ants; it is about 25 mm long, has large sturdy jaws, and is predatory, especially on termites. Most ants are smaller and eat vegetation or scavenge decaying plant and animal matter. (R. Buchsbaum)

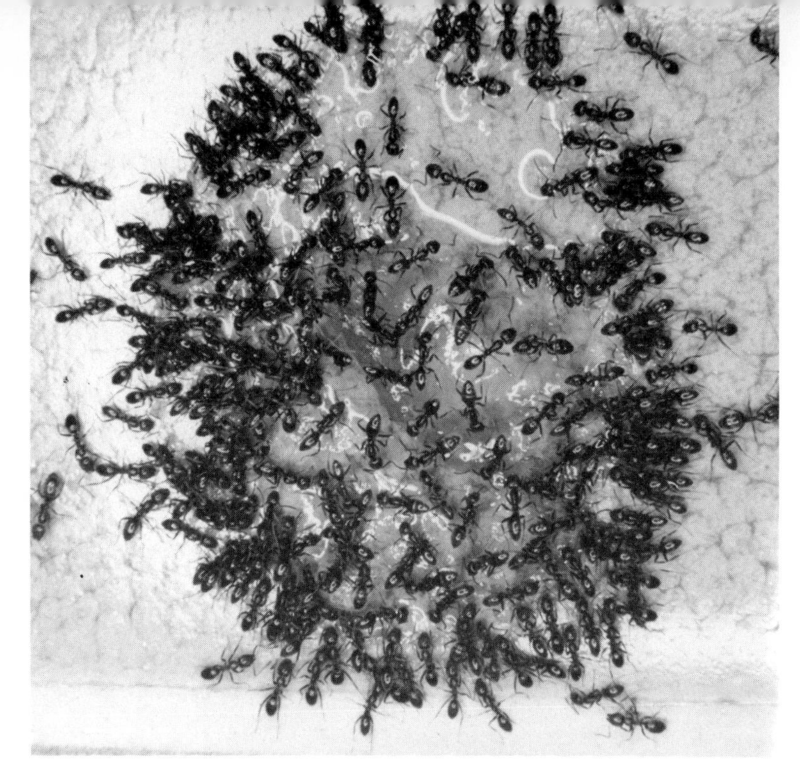

Argentine ants, introduced about a century ago from South America, have become a common household nuisance in the warm areas of the United States. These workers, feeding on a drop of honey in a California kitchen, are less than 2 mm long and dark brown. The long antennas, elbowed at a sharp angle, are a characteristic of ants. (R. Buchsbaum)

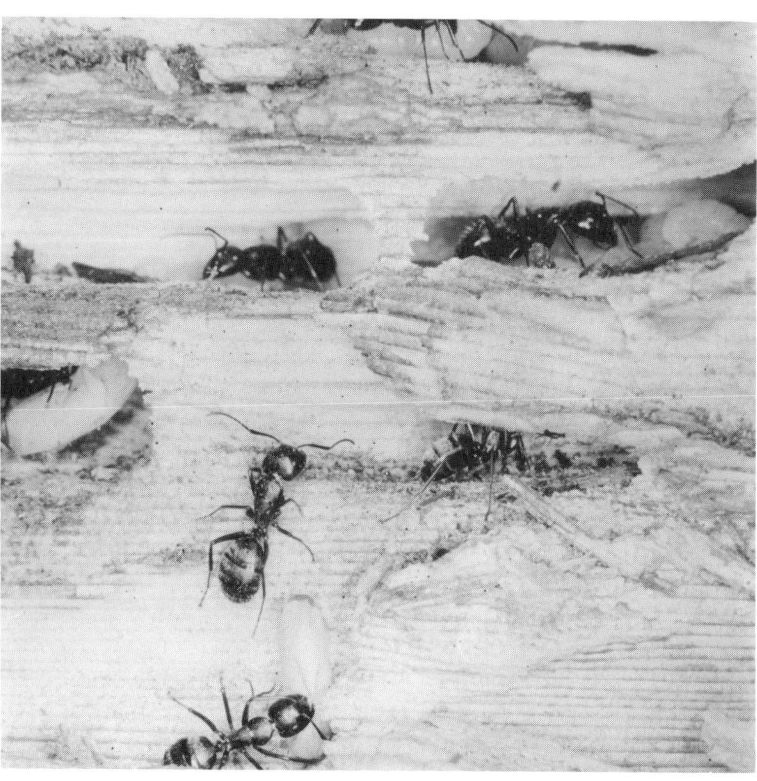

Carpenter ants nest in exposed decaying wood, including weathered wooden posts, utility poles, and the outer windowsills and porches of old houses. They excavate chambers in which to rear their young (some of which are seen here), but they do not eat the wood. Carpenter ants forage for food, favoring sweet plant juices and animal remains. Pennsylvania. (R. Buchsbaum)

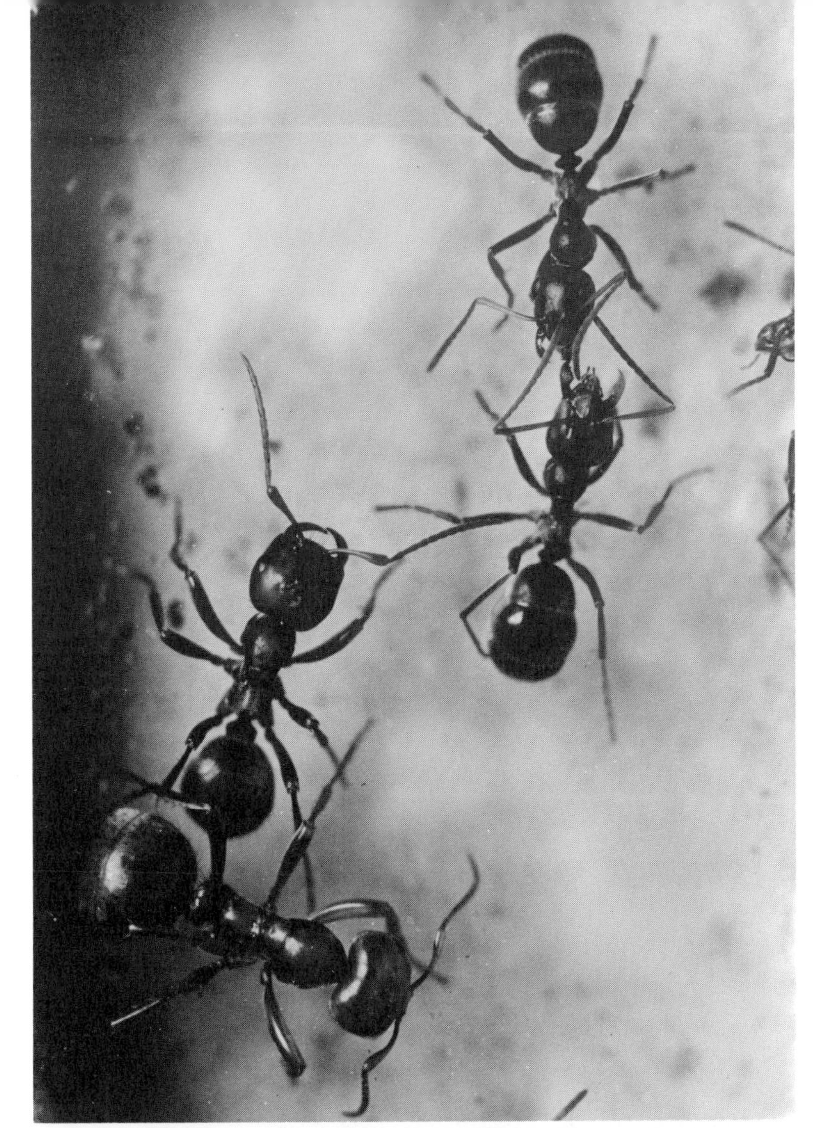

Slave-making ants (two large ants at left with sickle-shaped jaws useful only for fighting) cannot feed themselves. They raid the nests of another ant species and carry off the pupas. On emergence, these adopted slaves (two smaller ants at right with biting jaws) serve as workers in the colony. Illinois. (R. Buchsbaum)

Leaf-cutting ants cut pieces of leaves from plants and carry them to their large underground nest. There the pieces are chewed up and formed into moist balls on which is grown the fungus that the ants feed on. They can defoliate a 5-m tree in a day. From the nest, trails lead out in all directions. Seen here is a portion of a trail with workers carrying pieces of leaves that average 5 times their own weight. Two large-headed soldiers stand guard. Leaf-cutting ants occur in warmer parts of the United States, but are best known in the American tropics. Panama. (R. Buchsbaum)

Army ants occur in the southern United States, but only in the tropics do they form colonies numbering millions of individuals. They search for prey, especially other insects, on the forest floor or may ascend into a tree to raid a wasp nest while the wasps wait helplessly near by. When a vast army overruns a house that happens to be in their path, the human residents simply exit until the ants have moved on, then return to a home that has been thoroughly cleaned of animal pests. Army ants are nomadic, moving from place to place at intervals, and each time rebuilding their temporary but remarkable nest out of the living bodies of the ants themselves. The nest, or bivouac, usually built under an overhanging log, is complete with runways leading to chambers where they care for and feed their young. When the colony packs up and moves on, workers transport the young stages. Panama. (R. Buchsbaum)

Honey bees are the most appreciated of all hymenopterans, not only for the honey and wax they produce, but even more for their essential role in pollinating the flowers of fruit trees and of other crops. Workers of *Apis mellifera* are seen here, some depositing pollen in the hexagonal wax cells of a vertical comb in a hive. This European species has been introduced into the United States and many other countries.
(R. Buchsbaum)

Collecting honey and wax from a hive requires skill. Most beekeepers, when opening a hive, calm the bees by introducing smoke into the hive and wear protective clothing, especially a hat and veil to cover the head. The stinger at the rear of the worker bee is a modified ovipositor. Bumble bees and stingless bees also store honey, but only certain species of the honeybee genus *Apis* store appreciable quantities and will nest in hives provided by humans.
(D. Stone)

Honeybee worker, a sterile female. In hymenopterans, unlike termites, the workers are all females, which hatch from fertilized eggs; males hatch from unfertilized eggs and do no work in the colonies. This recently dead specimen is still covered with pollen grains caught on the bristly body and legs. When collecting nectar and pollen, from flower to flower, bees inadvertently pollinate the flowers. The elongated mouthparts are here extended and widely separated, the long tongue drooping. (C. Clarke)

Bumble bee has a longer tongue than a honey bee and can suck nectar from, and so incidentally pollinate, important crops that cannot be served by honey bees. The dense covering of long bristles on bumble bees conserves body heat, an important factor in helping these large and heavy bees to get airborne in the cool of morning. (R. Buchsbaum)

Centipedes and Millipedes

Related to the six-legged insects, or hexapods, and probably not unlike their ancestors, are several groups of many-legged, mainly terrestrial arthropods called **myriapods.** Except for their more abundant legs, myriapods are much like wingless insects. Both have unbranched appendages and similar internal anatomy, including air tubes for respiratory exchange. But unlike the flamboyant and ubiquitous insects, myriapods are quiet ground dwellers, mostly keeping to moist shelters under logs and rocks or burrowing in soil, and seldom intruding on human affairs; only a few pose any medical or economic threat. The body consists of a distinct head, followed by a long trunk consisting of numerous similar segments and bearing the many legs.

Centipedes ("hundred-legged") form the class Chilopoda. All are predators. They feed on any small animals they can catch, such as snails and slugs, earthworms, and soft insects. Large tropical centipedes can even subdue small lizards, snakes, birds, and mice. The appendages of the first trunk segment are modified as a pair of **poison claws,** which seize the prey and inject venom through their perforated tips. Centipedes have a pair of sensitive antennas and hunt mostly by touch, in the dark. The simple eyes play little part and are not even necessary for the negative reaction of centipedes to bright light, which still occurs when the eyes are completely covered with opaque paint. The appendages of the head include a pair of jaws and two pairs of maxillas.

Centipede with a pair of short, stout legs on each of the many similar segments that make up the long trunk. *Scolopendra,* seen here, has 21 pairs of legs. Australia. (O. Webb)

Poison claws of a centipede (seen here in an enlarged view of the underside of the anterior end) are not jaws but modified appendages of the first trunk segment. This centipede from Bermuda was 15 cm long; some tropical ones are over 30 cm. Even small centipedes should be handled carefully, for they can deliver a painful bite, and the bite of a large one may hospitalize a human adult and be dangerous to children. (P. S. Tice)

Each of the trunk segments, except for the first and the last two, has a pair of walking legs, better termed running legs, for the animals move very fast. A house centipede was clocked at 42 centimeters per second. The number of legs ranges from 15 to 177 pairs and the smooth coordination of so many rapidly moving legs is no mean accomplishment. Moreover, a centipede can compensate for lost legs without any noticeable limp and will regenerate them at the next molt.

Though terrestrial, centipedes have never come to terms with the problem of water loss. A centipede such as *Lithobius* can survive many hours completely immersed in water but will die in a few hours in an uncovered dish of dry earth. The cuticle is quite permeable, the air tubes have no closing mechanism, and the excretory products are not suited to water conservation. It is the behavior of centipedes that enables them to get along

House centipede differs from other centipedes in having long delicate legs and compound eyes. House centipedes live in damp places in houses, commonly basements and bathrooms. Their bite is not dangerous and they should be welcomed, as they feed on cockroaches, silverfishes, flies, and other household pests. (R. Buchsbaum)

on land. Even stronger than their negative reaction to light is their positive response to contact, which tends to keep them in moist, confined situations. A specimen of *Lithobius* placed in a glass dish will run about ceaselessly; but if some narrow, transparent glass tubing is added to the dish, the centipede will soon come to rest inside the tubing, which affords a maximum of contact with the animal's surface.

Centipedes lay their eggs in the soil, and some kinds stay with their eggs to guard them. The eggs are rich in stored yolk that nourishes the young through several molts until they are able to hunt for themselves.

Millipedes do not have the "thousand legs" that their name implies but they do have very large numbers. The maximum recorded was 752 legs, but numbers around a hundred are more common, and some millipedes get by with a few dozen. The many legs of millipedes do not make them particularly speedy (in fact, they walk rather slowly) but provide great power for burrowing through the soil. Each pair of legs represents a segment of the trunk; but during development the segments fuse in pairs, so that each narrow body ring in the trunk of an adult bears *two* pairs of legs. This is the meaning of the name of the group, class Diplopoda. The fusion of the segments and the narrowness of the rings shorten the body and make it less likely to buckle under the force of burrowing.

Woodland millipedes of the eastern United States, *Pseudopolydesmus*, have a body flattened in cross-section as is typical of millipedes that move among crevices. (R. Buchsbaum)

Head of a millipede *(opposite)* has a pair of antennas and sometimes, as seen here, a cluster of simple eyes near the base of each antenna. The head is tucked under during burrowing, and the millipede pushes through the soil with a hardened cuticular plate visible behind the head. **Rear end** of the animal *(left)* shows the terminal anus and a number of typical trunk rings, each with two pairs of legs. (P. S. Tice)

Eggs of a millipede were laid in a nest made of earth and were guarded by the mother. (R. Buchsbaum)

Giant millipede from East Africa is 20 cm long, but otherwise it resembles small types common in temperate woodlands and gardens, with the smooth body rounded in cross-section as is typical of millipedes that burrow through soil. (R. Buchsbaum)

In contrast to the predatory centipedes, most millipedes are herbivores; they feed mainly on decaying vegetation and contribute significantly to soil formation. Only when they eat the living roots of young plants do they sometimes become garden and agricultural pests. The mouthparts include a pair of jaws and a pair of maxillas, fused to form a lower lip; the second maxillas are lacking. Millipedes do not bite and are mostly quite safe to handle. When picked up, their first response is usually to curl up into a tight spiral or ball, although many can secrete toxic or unpleasant substances, including hydrogen cyanide, from pores along the body.

Chapter 20

Annelid-Arthropod Allies

Among the nations of the world, it is the superpowers that most often make the headlines and get most of the attention, while smaller countries lying in the shadows of these great neighbors may be every bit as interesting and often more charming places to visit, but are often overlooked. Among the phyla of animals, likewise, some of the minor ones have been omitted from this book. However, this chapter will touch briefly on four small phyla that appear to be allies of the annelids and arthropods and that may shed some light on the evolutionary relationship of these two superpowers.

Opposite: **Peripatuses** are animals much talked about but seldom seen, even by those who carefully search under every likely rotten log or mound of leafmold in the tropical rain forests where peripatuses are known to live. Of extremely retiring habits, they come out only at night to capture small animal prey. When poked with a finger, both mother and offspring shot out a sticky defensive secretion. The large female shown (about 12 cm long) gave birth to two young, but one ate the other. *Macroperipatus geayi.* Panama. (R. Buchsbaum)

Peripatuses

Animals that come closer than any others to being the "missing link" between any two phyla are the members of the small phylum **ONYCHOPHORA** (on-i-KAH-for-a). The phylum name means "claw-bearers," but the distinctive combination of characters that onychophorans share with the wormlike annelids and the leggy arthropods might be better conveyed by the common name "walking worms." In practice, however, onychophorans are usually referred to as peripatuses, a name derived from *Peripatus*, the first genus described for the group. These rare animals are found in moist places under logs and leaf litter in the tropical or south temperate forests of the Malay Peninsula, New Guinea, Australia, New Zealand, the Caribbean, Central and South America, and Africa. Their occurrence only in local regions of such widely separated parts of the world suggests that they were

Leg of a peripatus is soft and plump and quite unlike those of arthropods, but the claws resemble insect claws.

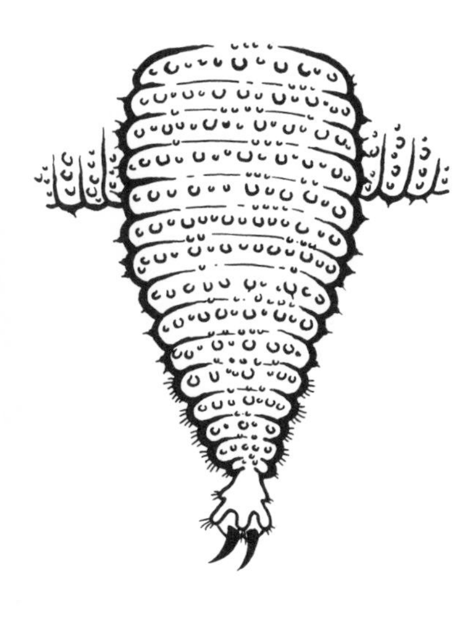

Air tubes of a peripatus are lined with thin chitinous cuticle, like those of arthropods, but their distinctive arrangement indicates that they evolved independently. (Modified after K. C. Schneider.)

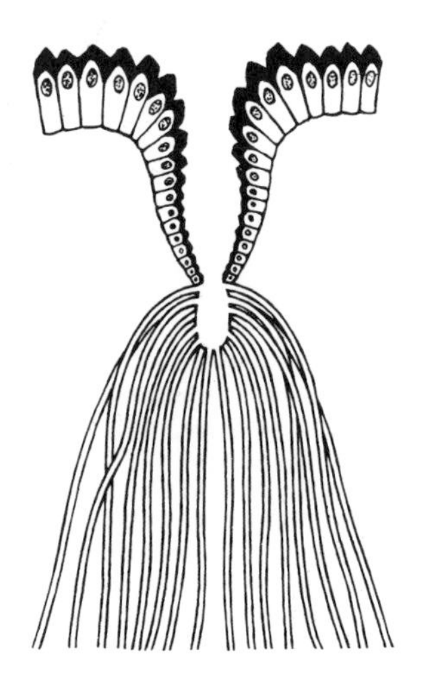

probably more widespread in the past. Neither the peripatuses as we find them now, nor any other living animals, could be ancestral to any group as old as a phylum. But there is little doubt that they are descendants from a line that branched off close to the primitive annelid-arthropod stock.

A peripatus looks much like a caterpillar, usually about 5 to 10 centimeters long, with soft velvety skin and many pairs of legs. Unlike typical annelids and arthropods, peripatuses show no external segmentation, though there is a pair of **legs** for each internal segment of the body. The legs end in claws that superficially resemble the claws of insects, but differ from arthropod legs in that they are soft and plump and not jointed.

The outer covering is a thin flexible **cuticle,** which is chitinous like that of arthropods. It is regularly molted and replaced by growing individuals. The cuticle is ridged and covered with microscopic projections that give it a velvety texture and that prevent it from being readily wetted by water. Beneath the epidermis that secretes the cuticle are continuous **layers of muscles,** as in annelids, a sharp contrast to the typically discrete muscle bundles of arthropods.

Peripatuses usually come out at night and feel their way around by means of two sensory **antennas** on the head. There is also a pair of small eyes, which show similarities to those of both annelids and arthropods. When attacked, a peripatus will squirt a sticky defensive secretion from a pair of glands that open on two head appendages, the **slime papillas.** Peripatuses feed on small animals such as insects, cutting into the prey by means of a pair of chitinous **jaws.** Digestive juices secreted onto the prey liquefy the tissues, which are then sucked up. These three pairs of appendages (antennas, slime papillas, and jaws) define the head, which is not otherwise set off from the long trunk.

The **internal anatomy** is a mixture of annelid-like and arthropod-like structures. The digestive tract is simple and not particularly distinctive; both ends are lined with cuticle, as in arthropods. The circulatory system is open and like that of arthropods. A long contractile dorsal vessel, the heart, extends the length of the body. From the heart, the blood flows into large sinuses in the tissues and finally collects in the space surrounding the heart, which it enters through openings in the heart wall, a pair for almost every body segment. The most arthropod-like character of all is the respiratory system, consisting of air tubes that open from the external surface and extend throughout the body, piping air directly to the tissues. Although such structures occur nowhere else in the animal kingdom except in land arthropods, their varied structure indicates that they have arisen independently in the peripatuses and in several groups of land arthropods, and they cannot be taken as evidence that peripatuses are related to any of these groups.

The air tubes of peripatuses differ from those of most arthropods in several important respects. In arthropods there are relatively few openings in the body wall, and they usually have closing mechanisms. The openings lead into large air tubes that branch repeatedly, the branches decreasing in size and ramifying throughout the body. In peripatuses a large bundle of unbranched air tubes arises directly from each external opening and penetrates deep into the tissues of the body. The external openings are necessarily numerous and scattered over the body, and each is a small pit that lacks any kind of closing mechanism.

The loss of water through this exposed system of air tubes is considerable. Experiments designed to test water loss under comparable conditions showed that a peripatus loses water about as fast as an earthworm, 20 times as fast as a millipede, 40 times as fast as a smooth-skinned caterpillar, and 80 times as fast as a cockroach. Although the thin cuticle of peripatuses is relatively permeable to water, water is lost mostly through the system of air tubes. Water is replaced by drinking, by the fluids in prey, and by uptake through the surface of special thin-walled sacs that can be everted and pressed against moist substrates.

The nervous system is more diffuse than in annelids and arthropods. From the brain run two widely separated ventral nerve cords, which show only small thickenings in each segment and are connected by many fine cross-strands.

The most annelid-like character of peripatuses is the **excretory system.** This consists of segmentally arranged pairs of coiled tubes that open by external pores at the bases of the legs. The inner end of each tube opens into a small coelomic sac from which fluid is swept into and along the tube by beating cilia, as in annelids. As has been mentioned, motile cilia occur nowhere in arthropod systems (except in some arthropod sperms). The major mode of nitrogenous excretion, however, resembles that of most land arthropods; the midgut lining excretes uric acid, which is eliminated with the feces.

The reproductive system is also coelomic and ciliated. The sexes are separate, and the eggs are fertilized within the body of the female. In some species eggs are laid, but in most forms the embryos develop within the female, nourished either by food stored in the egg or by nutrients the mother supplies through a placentalike connection. Well-developed young are born. Since internal fertilization and development of the eggs are adjustments to land life and have evolved independently in terrestrial animals of many groups, they have no special significance for the evolutionary relationships of peripatuses.

At every level of classification there are some animals to be found with structures resembling those of two otherwise discrete groups. This is a situation that follows naturally from what we know of the continuous nature of the process of evolution. But it always creates difficulties in classification. Peripatuses have been particularly controversial. They have sometimes been placed with the annelids, but today are more likely to be included in the arthropods. Either way, the definition of the phylum becomes undesirably diluted, and it seems more satisfactory to place these unique animals in a phylum of their own.

Notice that in each of these representatives of three phyla, the many appendages are all much alike and move in a series of waves. (R. Buchsbaum)

Annelid (polychete)

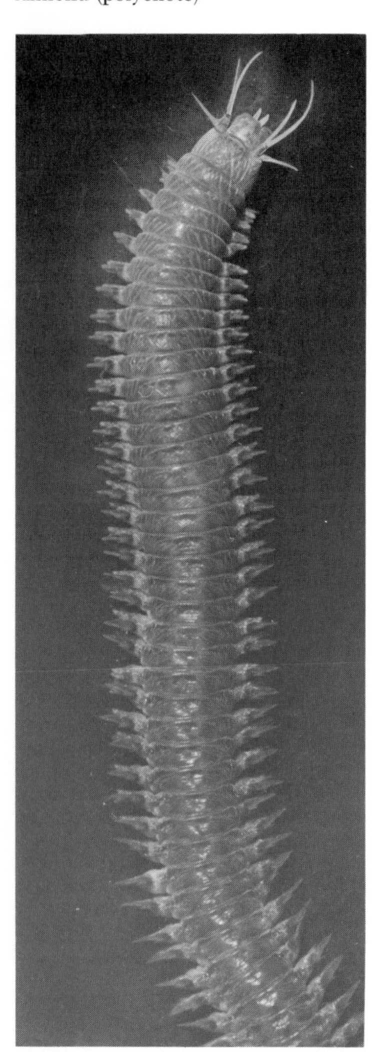

Onychophoran

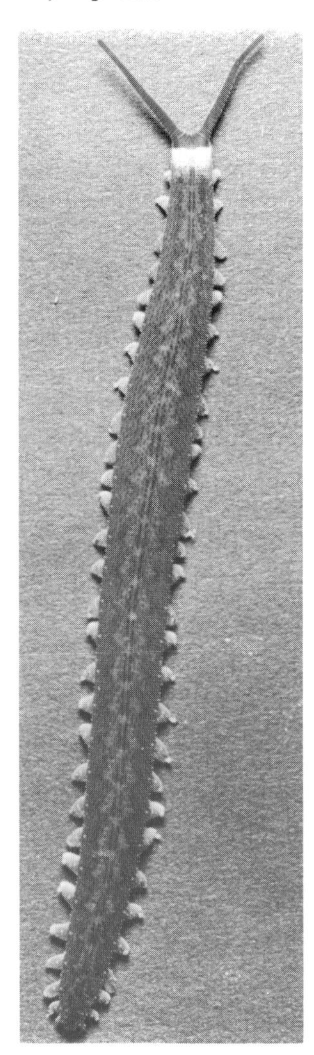

Arthropod (millipede)

Tardigrades

To appreciate both common and scientific names of the phylum **TARDIGRADA,** one need only watch the short, plump little "water bears" as they slowly crawl about on plant debris. Their four pairs of stumpy clawed legs can best be compared to those of peripatuses, which they resemble in some other ways also, though on a different size scale. Tardigrades are almost all microscopic, the largest barely exceeding a millimeter in length. These tiny animals are found in sea sediments and in lakes and ponds, but mostly inhabit temporarily wetted places on land such as in mosses, lichens, roof gutters, and cemetery urns. They are well adapted to withstand drying, even for years, and become active again within a few hours of being wetted. Tardigrades feed mostly on plant juices, sucking them from plant cells pierced by two sharp stylets protruded from the mouth, much as in nematodes.

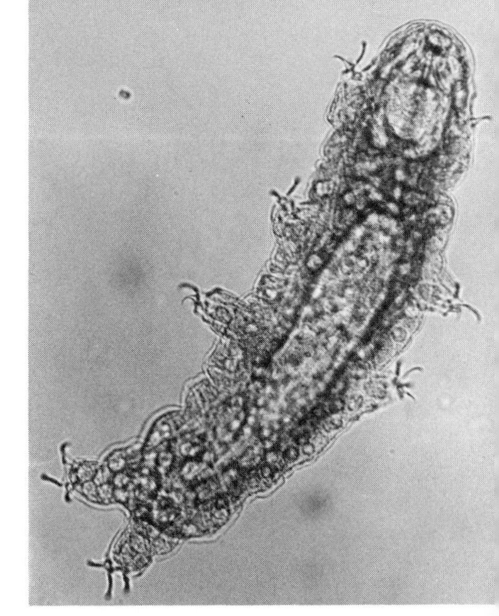

Tardigrade, *Hypsibius*, washed from moss close to a freshwater pond. About 0.7 mm long. (R. Buchsbaum)

Echiurans

The plump marine worms of the small phylum **ECHIURA** are seldom seen outside of the burrows that they make in sand or mud or the other crevices in which these soft, defenseless animals take shelter. They feed by collecting organic particles and tiny organisms, either by catching suspended food particles in a mucus net or by sweeping particles from the sediment into a ciliated, scoop-shaped proboscis that has given them the common name "spoonworms." The mouth opens at the base of the pro-

Echiuran with a ribbonlike proboscis about as long as its plump body. In some echiurans, the proboscis is many times longer than the rest of the body; in others, it is very short. N.W. Florida. (R. Buchsbaum)

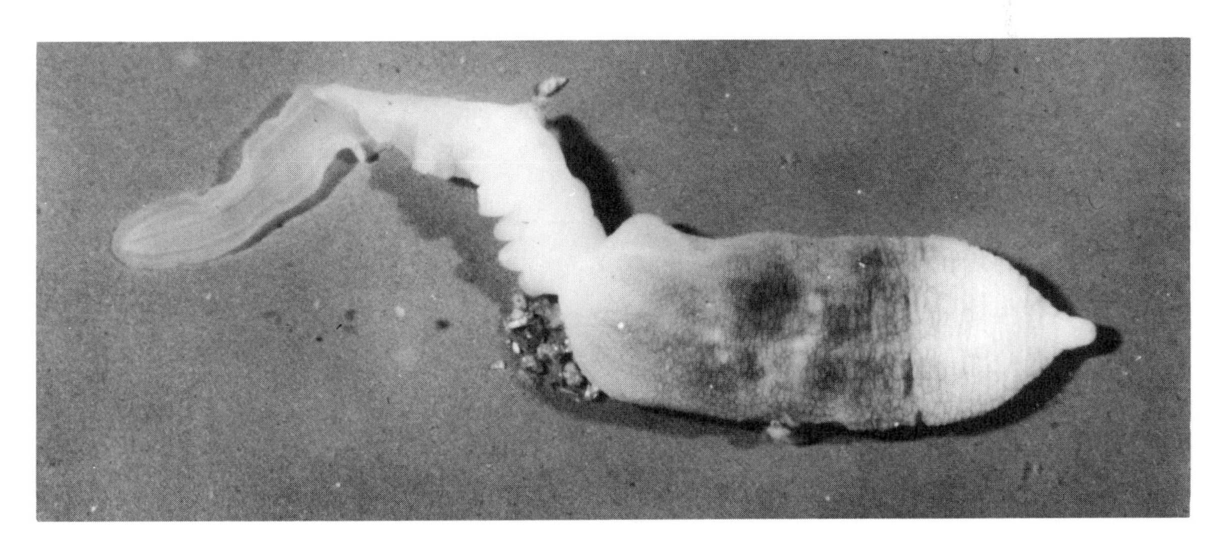

boscis, and the anus at the rear end of the animal. Echiurans have a muscular body wall surrounding a large, fluid-filled coelom, a ventral nerve cord, a simple closed circulatory system, and excretory organs that open into the coelom through ciliated funnels. Thus they are organized much like annelids—with the notable exception that they are *not* segmented. Eggs and sperms are easy to collect from certain echiurans, and many laboratory studies have been made of their development. It resembles that of annelids, and their swimming larvas are trochophores.

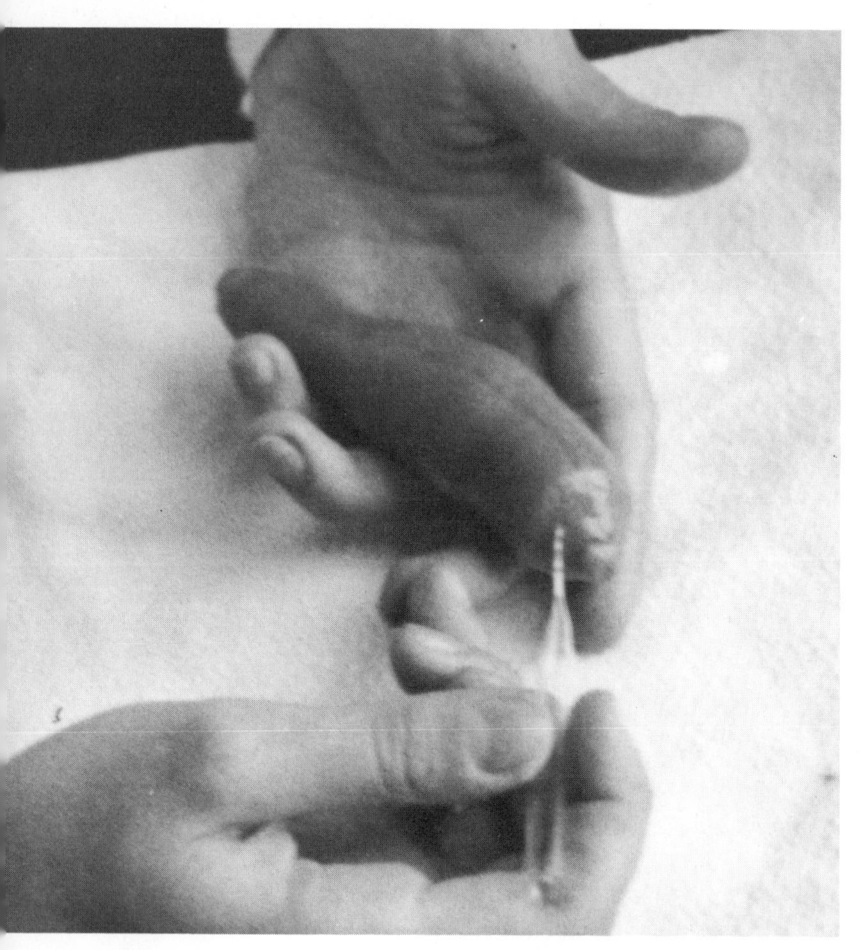

Collecting gametes from *Urechis caupo*, a large echiuran of sandy or muddy bottoms along the U.S. West Coast. Eggs or sperms are easily obtained by inserting a glass pipet into one of two genital openings near the anterior end. Many of the early events of development, which occur in some form in all animals, have been studied in this worm. Hopkins Marine Station, Pacific Grove, California. (R. Buchsbaum)

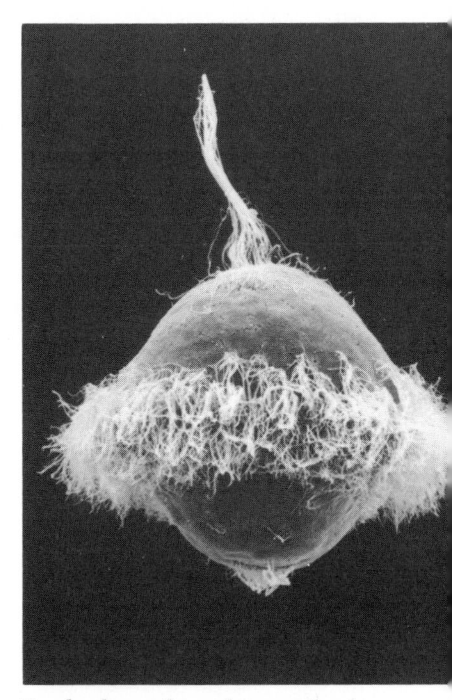

Trochophore of an echiuran, *Urechis caupo*, resembles those of polychete annelids. The cilia of the apical tuft and the ciliary band look like white fur in this scanning electron micrograph. Before it settles, the larva will elongate to a wormlike form. Diameter about 0.1 mm. (C. B. Calloway and R. M. Woollacott)

Sipunculans, *Themiste pyroides.* In the sipunculan at the left, the tentacles are extended; in the one at the right, they are withdrawn. These sipunculans were living in holes in a rock made by rock-boring clams. Oregon. (R. Buchsbaum)

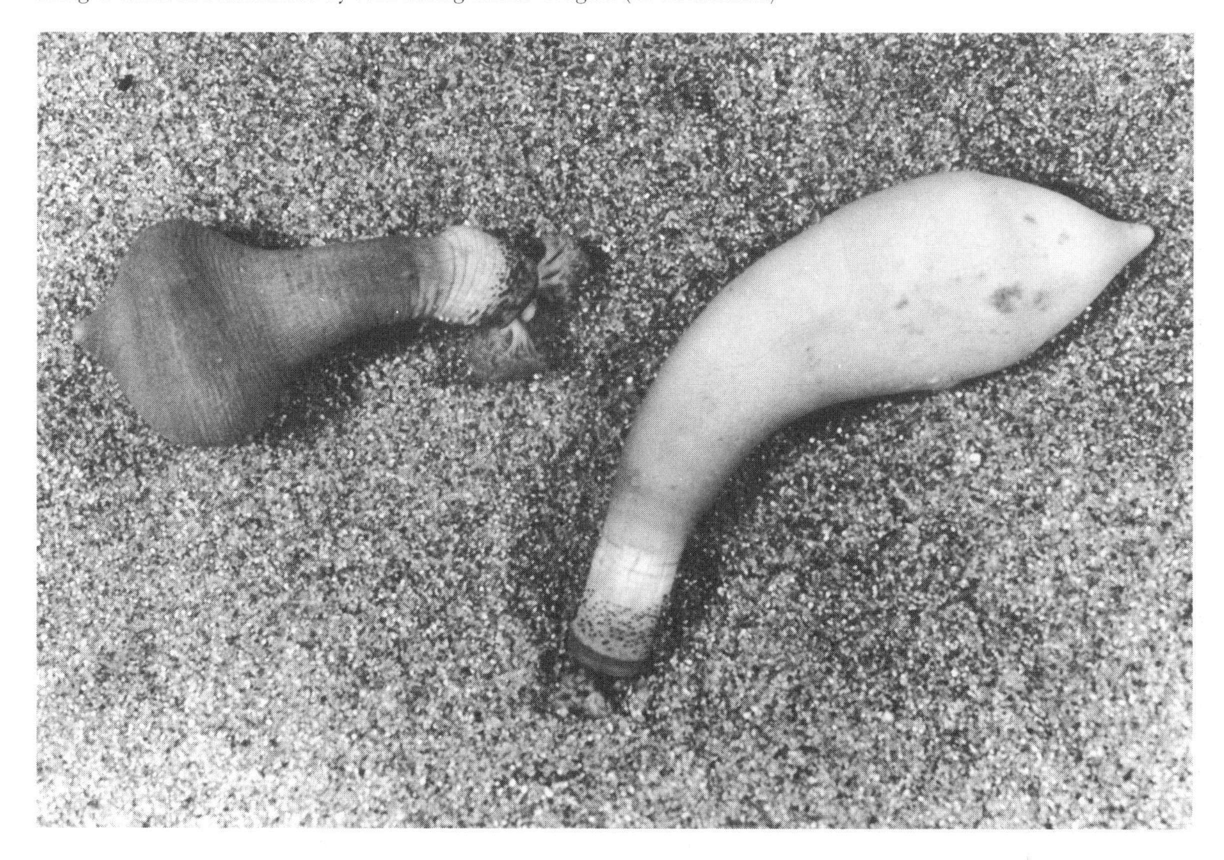

Sipunculans

The common name "peanut worms" is often applied to worms of the phylum **Sipuncula.** Sipunculans can withdraw the narrow anterior end of the body into the fatter, more posterior portion, and in this position the bulgy body resembles a peanut; or the contracted body may assume a rounded shape much like that of a single peanut seed. The anterior end is used in burrowing, and it bears tentacles that gather organic food particles. Some sipunculans ingest large quantities of sediment and digest the organic content, as do earthworms. The mouth opens at the anterior end, and the gut spirals back on itself such that the anus also opens near the anterior end, on the dorsal surface. Otherwise, the unsegmented, muscular sipunculans resemble echiurans (and annelids) in many respects.

Chapter 21

Spiny-Skinned Animals

With the arthropods we reached a peak of invertebrate evolution and might have closed our story. But organic evolution has not followed the rules of good dramatic style, presenting us with a strong main theme that proceeds to a neat climax. Instead it provides us with enough materials to construct many different stories, each ending in a grouping of animals with a distinctive design for living. The differences among the phyla make it easier to divide our book into discrete chapters, and to arrange them in a sequence that suggests levels of complexity, but not necessarily close relationship. The phylum to be described here seems to have gone off in another direction from that followed by the molluscs, annelids, and arthropods covered in preceding chapters. Indeed, the body plan of the adult members of the phylum **Echinodermata** ("spiny-skinned") is utterly different from that of any other phylum. The best clue to the relationship of echinoderms with other groups is in their larvas, which resemble the larvas of acorn worms. This is surprising, inasmuch as acorn worms are allied to the phylum in which humans and other vertebrates have been included. Of this, more will be said in the next chapter.

The phylum Echinodermata is exclusively marine and is divided into six classes: the well-known sea stars, brittle stars, feather stars, sea urchins, and sea cucumbers, and the newly discovered sea daisies. The echinoderms most familiar to everyone, even to those who have never been near the seashore, are **sea stars** (also called starfishes).

There are nearly two thousand species of sea stars, but most look quite alike and live similar kinds of lives. Almost all are predators and scavengers that feed on a variety of animals that they find as they slowly move along on seashores or ocean bottoms.

Opposite: **Ten-armed sea star** is less common than the typical five-armed kinds. In some species, the normal number is 6, in others it is 25 or more. In the species shown here the usual number of arms is 9 or 10, but the range is 7 to 13. *Solaster endeca.* Mount Desert Island, Maine. (R. Buchsbaum)

Most sea stars live along rocky coasts, where the hard substrate furnishes a place of attachment for the animals on which they prey, and where sea stars can effectively use their tube feet for hanging on. When the tide goes out, sea stars may be seen stranded temporarily in shallow tide pools, or clinging to rock walls. *Asterias.* Maine. (R. Buchsbaum)

The body of a sea star consists of a central **disk** from which radiate a number of **arms.** There are usually *five* of these, but some species have a larger number of arms arranged radially around the disk. The **mouth** is in the center of the disk on the under surface. There is no head, and the animals can move about with any one of the arms in the lead. The radial symmetry of adult sea stars is superficially like that of cnidarians, which have tentacles radiating out from a central disk. However, as will be seen later, larval sea stars are bilaterally symmetrical, and the adult radial symmetry almost certainly is derived secondarily from a basic bilateral symmetry.

Sea stars are remarkable in that the body may be unyieldingly *rigid* under some circumstances, as when a curious collector is trying to pull one off a rock; but amazingly *flexible* at other times, as when an overturned sea star rights itself with a display

A sea star rights itself by bending its arms and pulling with the tube feet. *Pisaster ochraceus.* Monterey Bay, California. (R. Buchsbaum)

of agile bending and twisting. This dual nature is provided by a skeletal system of **calcareous plates,** bound firmly together but remaining as separate elements, embedded in fibrous connective tissue that can become either stiff or yielding. The plates bear many calcareous knobs or spines, some of them movable. The skeletal plates and spines form an **endoskeleton** embedded in the body, as is our own, that differs from the exoskeletons of arthropods and the shells of molluscs, which usually lie outside the body.

Locomotion in sea stars is by means of a hydraulic-pressure mechanism, known as the **water vascular system** and unique to echinoderms. This system of canals is open to the surrounding water through minute holes in the calcareous **sieve plate** (madreporite) on the upper surface. Descending from the sieve plate is the **stone canal** (so named because its wall is stiffened by calcareous rings), which connects to the **ring canal** encircling the mouth. From the ring canal arise the **radial canals,** one in each arm. Along the length of each arm a radial canal connects, by short side branches, with numerous **tube feet**—hollow, thin-

Water vascular system of a sea star showing essential features. The ring canal may have one or more sets of vesicles of unknown function; they are omitted here.

sieve plate
radial canal
ampulla
tube foot
stone canal
ring canal

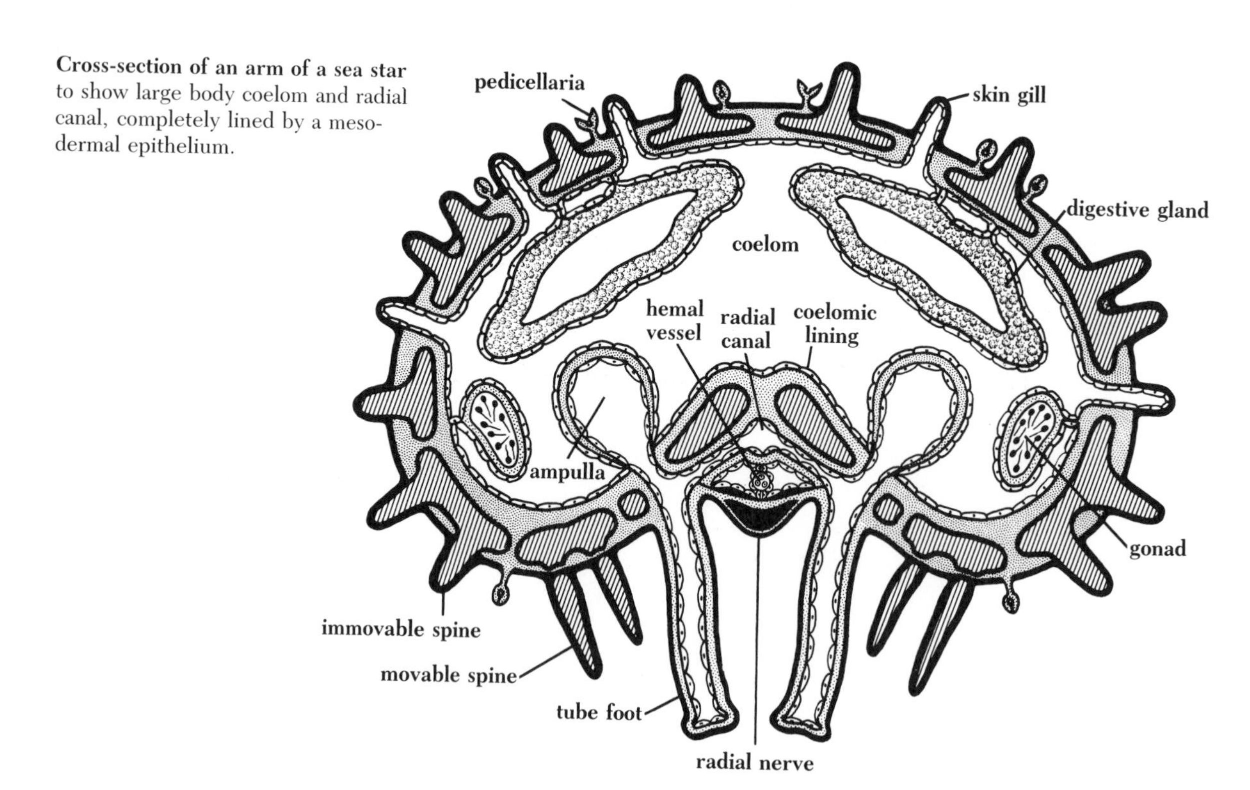

Cross-section of an arm of a sea star to show large body coelom and radial canal, completely lined by a mesodermal epithelium.

Rows of tube feet radiate from the central mouth of this sea star, as in other echinoderms. The tube feet are extended by hydraulic pressure and retracted by muscular contractions. Under surface of *Asterias*. Maine. (R. Buchsbaum)

walled, cylindrical tentacles that extend out from the body, each usually ending in a sucker. Each tube foot is provided with a rounded muscular sac, the **ampulla.** When an ampulla contracts, the water it encloses is prevented by a valve from flowing back into the radial canal and is forced into the tube foot. This extends the tube foot, which attaches to the substrate with mucus and with its sucker. Next, the longitudinal muscles of the tube foot contract, shortening it, forcing the water back into the ampulla, and pulling the animal forward. Of course, one tube foot is a very weak structure; but there are hundreds of them, and their combined effort is capable of moving the sea star along, sometimes quite briskly, with the many tube feet acting as a train of little legs.

Sand stars live on sandy sea bottoms where they hunt animal prey and hide by burrowing just below the surface of the sand. Their pointed tube feet are thrust forcefully into the sand during burrowing and lack the suckers used by most sea stars to cling to rock. *Above:* Sand star on the surface. *Below:* Moments later after the animal has nearly buried itself. *Luidia.* Panacea, N.W. Florida. (R. Buchsbaum)

The radial canals and their tube feet extend along the lower surface of each arm, outside two rows of calcareous plates that form a V-shaped groove (usually called the "ambulacral groove," from a Latin word meaning a "walk," because the groove with rows of tube feet on each side reminded someone of a flower-lined garden walkway). The ampullas extend between the plates into the central body cavity, while the tube feet project out of the groove. On either side of the groove are rows of movable spines, which can be brought together to close over the groove. These spines protect the tube feet and the nerves that coordinate the activity of the water vascular system.

The **nervous system** is simple, as in all echinoderms. There is no brain, or even ganglia that might coordinate activities. A **nerve ring** encircles the mouth and connects with five **radial nerves,** which extend the length of the arms, below the radial canals of the water vascular system. The nerve ring relays impulses between the radial nerves so that one arm can lead and the others follow. Over the whole surface of the animal there is a rich **nerve net** that is sensory and also coordinates local movements of the surface and spines. At the tip of each arm are deli-

Regeneration of lost arms is common in many species of sea star, and animals with one or more regenerating arms can often be seen feeding, righting themselves, and otherwise behaving like normal animals. *Asterias.* Maine. (R. Buchsbaum)

cate sensory tentacles (thought to be sensitive to food and other chemical stimuli) and a pigmented eyespot (sensitive to light and shadows).

Sea stars move rather slowly, but they have no difficulty running down their prey, for most **feed** on slow-moving or sessile animals, especially snails, clams, mussels, and oysters. Anyone who has ever tried, barehanded, to open a live clam or oyster will wonder how it can be done by a sea star not much bigger than the clam it attacks. The sea star mounts the clam in a humped-up position, attaches its tube feet to the two valves of the shell, and begins to pull. The clam responds by closing its shell tightly. But the sea star can use its numerous tube feet in relays and is able to outlast the clam. When the shell opens even slightly, the sea star turns the lower part of its **stomach** inside out, extending it out the mouth and then into the clam through the small crack between the two halves of the shell. Once the stomach is inside the clam, digestive enzymes are poured in so the clam is digested without even being swallowed. The digested material is transported along ciliated tracts into the sea star's stomach and then into the five pairs of **digestive glands,** one pair in each arm, connected to the upper part of the stomach. Very little indigestible material is taken in by this method of feeding, and the digestive glands assimilate and store most of the ingested nutrients; there is practically no intestine; and the anus, opening on the upper surface, passes very little material. Small clams or snails may be taken whole into the stomach, and the shells or other small indigestible remains are usually ejected through the mouth. Some species of sea star have no anus at all.

Sea star eating a clam. *Left:* Humped up over a clam in a tide pool, a sea star applies its tube feet to the two valves of the shell and pulls them open just enough to allow its stomach to enter between them. *Right:* Lifted out of the water, the sea star is held to show the thin stomach everted into the clam shell, where it is digesting the soft tissues inside. *Asterias.* Maine. (R. Buchsbaum)

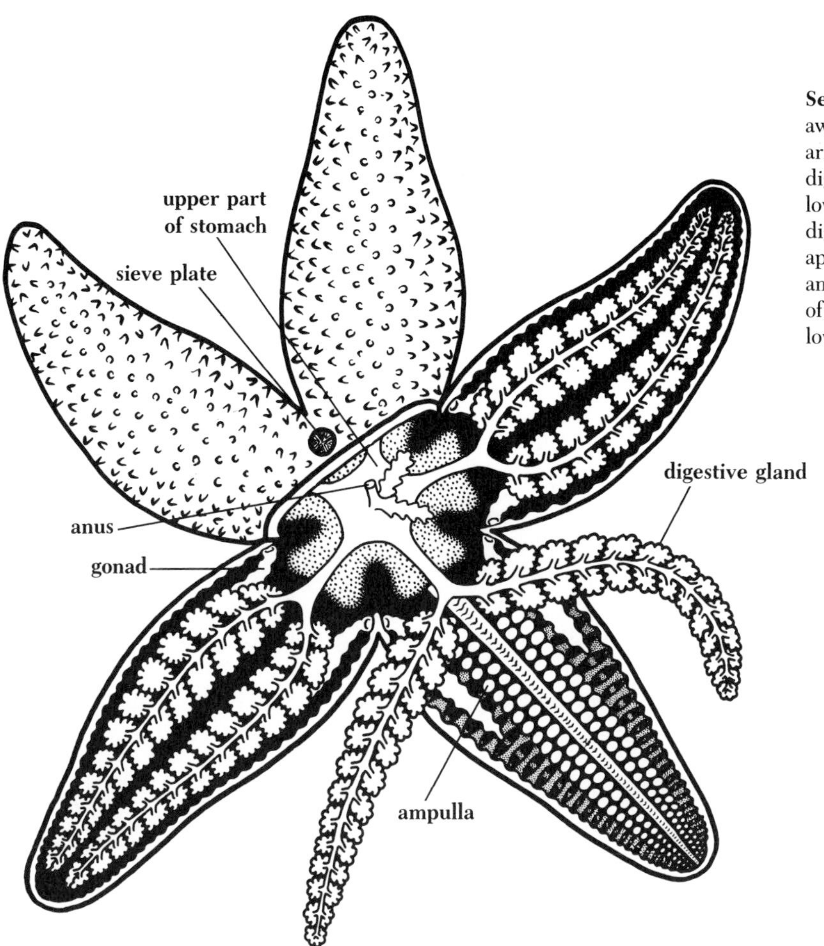

Sea star with body wall dissected away from upper surface of three arms and most of the disk, to show digestive system. In the arm on the lower right, the two branches of the digestive gland have been spread apart to show the rows of ampullas and the two gonads. The upper part of the stomach is unshaded; the lower part is stippled.

Crown-of-thorns sea star, *Acanthaster*, feeds on corals in the tropical Pacific and Indian oceans. During periods when these sea stars become especially numerous, they sometimes destroy whole reefs. Fiji. (R. Buchsbaum)

Running between the water vascular system and the radial nerves in the arms are **hemal vessels,** and the fluid in these vessels is pushed along by a pulsating, heartlike sac located under the sieve plate. But materials are distributed mainly in the large **body coelom,** which is filled with fluid nearly identical to seawater and that bathes most of the internal organs. The fluid is kept in constant motion by cilia on the coelomic epithelium. Nutrients from the digestive tract that diffuse into the coelom are transported quickly to other organs such as the gonads. In addition, oxygen diffuses through the tube feet and ampullas into the coelomic fluid for rapid consumption by internal organs and tissues. Many small fingerlike projections, the delicate thin-walled **skin gills,** pass between the calcareous plates and extend from the surface; these also allow oxygen to enter the coelom and carbon dioxide to diffuse out.

The surface of many sea stars contains numerous **pedicellarias.** These small pincerlike structures, occurring singly or in clumps around the bases of spines, apparently protect the sea star from small animals that might settle and grow on its surface. In addition, by catching and crushing small animals creeping along the surface, they may provide nutrients to the epithelial cells.

There is *no specialized excretory system.* Ammonia, carbon dioxide, and other soluble metabolic wastes are flushed from the body through the tube feet and skin gills. Insoluble waste materials, including disintegrating cellular debris, are engulfed by ameboid cells in the coelomic fluid; these cells then leave the body through the walls of the skin gills.

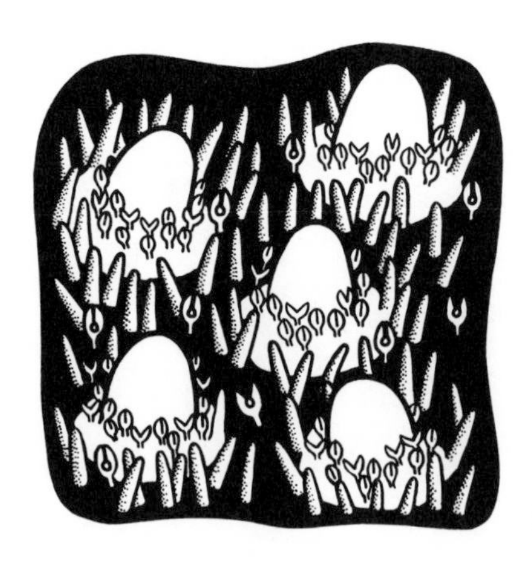

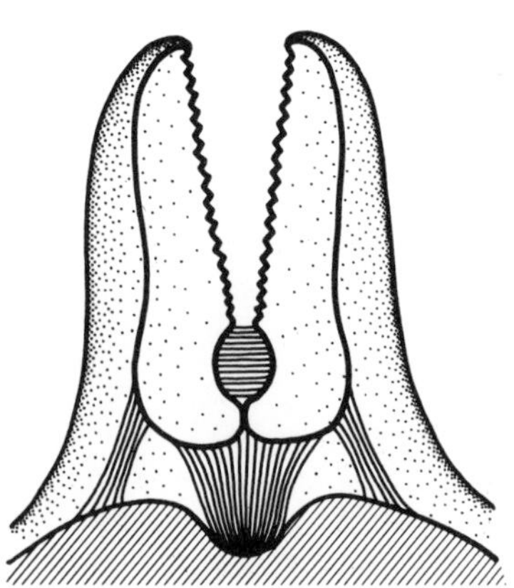

Left: Small portion of the **surface of a sea star** to show the delicate *skin gills* occupying the spaces between the large rounded *spines.* Tiny pinching *pedicellarias* occur around the bases of the spines and among the skin gills. *Right:* **A pedicellaria** is opened and closed by muscles. (After L. Cuenot.)

Most sea stars have separate sexes; individuals are either males or females. Two **ovaries** or **testes** lie in the corners of each arm and open directly to the exterior. As in many other marine animals, the eggs and sperms of sea stars are shed directly into the seawater, where fertilization occurs.

Animals that shed, or *spawn*, their eggs and sperms directly into the seawater are faced with a problem: eggs and sperms are single cells that cannot survive long in seawater; unless they meet within a short time, they perish. It is not surprising, therefore, that there are a variety of mechanisms that such animals use to **synchronize their reproductive activities.** Many animals, for example, are sensitive to the fluids released during spawning by other members of their species, and they spawn when they detect materials associated with eggs and sperms in the water. Thus, spawning by one animal (in response to a slight change in water temperature or some other stimulus) can lead to a chain reaction with "epidemic spawning" of others in the area.

However, for this synchronizing mechanism to work, all the animals must be full of eggs or sperms and ready to spawn at the same time. Many sea stars possess an *internal calendar* that controls the time when they produce eggs and sperms. Experiments have shown that they set the calendar by detecting seasonal changes in daylength. They can be "tricked" into producing eggs and sperms at any time a researcher wishes simply by placing them on the appropriate schedule of daylengths.

Since eggs and sperms are more easily obtained from mature sea stars and sea urchins than from most animals, they have been much used by biologists to study the mechanisms of fertilization and early development. Sperms surround the egg, and after the first one enters (which it does rapidly, leaving its flagellum behind), a fertilization membrane rises over the surface of the egg and prevents other sperms from following. The sperm and egg nuclei fuse, completing fertilization. Each nucleus brings the hereditary contributions of one parent to the new individual.

The fertilized egg divides into two equal cells, and these divide into four, eight, sixteen, and so on until a hollow **blastula** is formed that is ciliated and hatches out of the fertilization membrane. By the end of the first day, the free-swimming blastula has been transformed, by an infolding of the cells at one pole, into a two-layered **gastrula** with an outer ectoderm and an inner endoderm. This infolding crowds out most of the old **blastula cavity** and produces the **primary gut cavity,** which will later become the cavity of the digestive tract. The opening into this new cavity is called the **blastopore.** Through the gastrula stage the development of most animals is essentially alike, although it differs in rate of development, amount of yolk present, equality of division of the cells, and the particular way in which the gastrula becomes two-layered. As development proceeds, differences between the many kinds of animals increase.

In echinoderms two small out-pocketings pinch off from the endoderm, near the top of the primitive gut cavity. The resulting sacs eventually give rise to the **coelom** and its derivatives, including the water vascular system. The remainder of the primitive gut cavity forms an esophagus, stomach, and intestine, with the blastopore serving as the larval anus. The esophagus bends ventrally, meets an ingrowth from the ectoderm, and breaks through to form the larval mouth. The cilia on the surface concentrate along a definite band around the mouth and a free-swimming, **bilaterally symmetrical larva** is formed. In sea stars, the ciliary band elongates and separates into two bands, and the resultant larva, called a **bipinnaria,** swims and feeds in the plankton.

Early development of a sea star. (After Delage and Hérouard.)

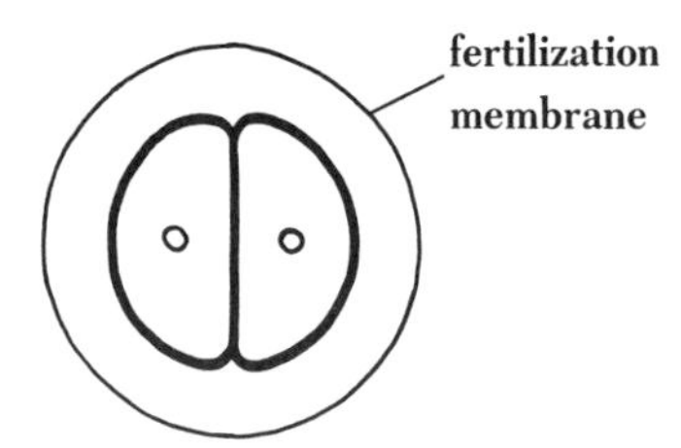

1. **Two-cell stage.**

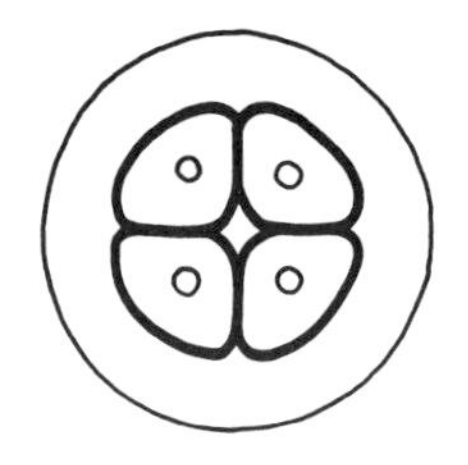

2. **Four-cell stage.**

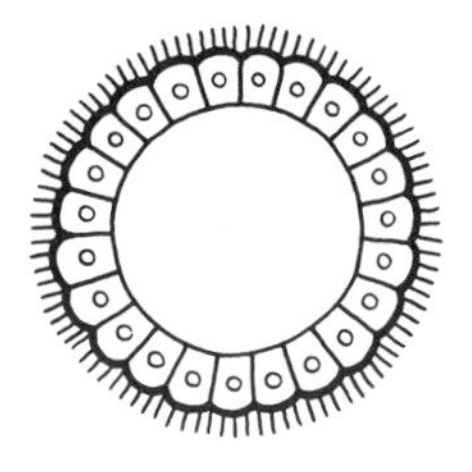

3. **Ciliated blastula.**

4. **Early gastrula.**

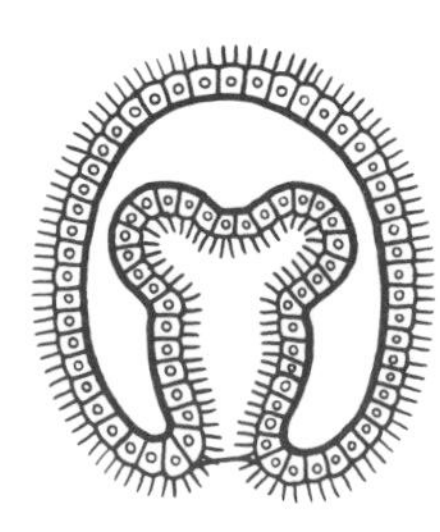

5. **Beginning of outpocketing of coelomic sacs from primitive endoderm.**

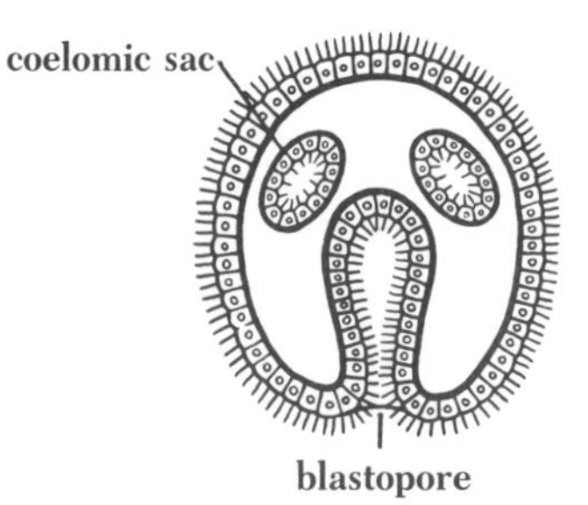

6. **Coelomic sacs are pinched off.**

While drifting and feeding in the currents for weeks to months, the larva slowly changes in preparation for **metamorphosis.** The posterior parts of the coelomic sacs surround the stomach and will form the general body cavity of the adult. The left anterior coelomic sac grows around the esophagus, buds off five lobes that form the radial canal rudiments of the water vascular system, and sends a tube to the surface that will be the stone canal. When ready to metamorphose, most sea star larvas attach temporarily to some solid object and change rather quickly. The larval mouth and anus close, and a new mouth breaks through on the larva's left side, while a new anus opens on the right side, thus producing an adult axis at right angles to the larval axis. Tube feet grow out from the radial canal rudiments, attach to the substrate, and the tiny sea star takes its first steps and crawls away.

Later development of a sea star. (Mostly after S. Hörstadius.)

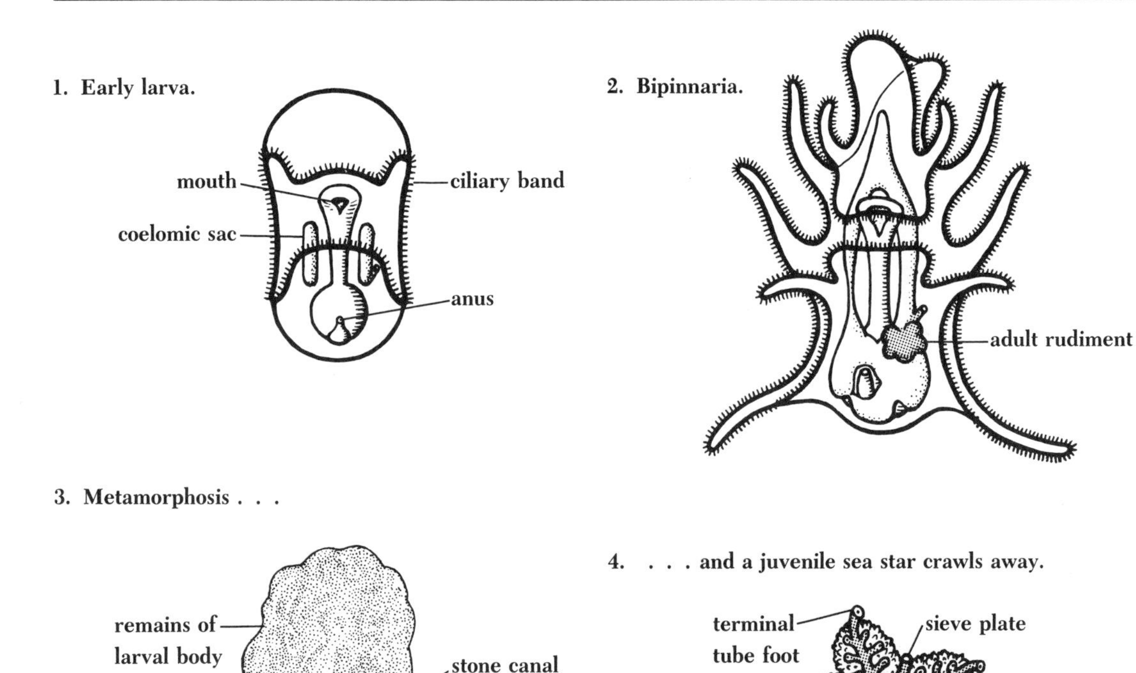

4. . . . and a juvenile sea star crawls away.

The symmetry of an adult sea star is called **secondary radial symmetry** because it develops secondarily from a bilaterally symmetrical larva. The earliest known fossil echinoderms also were not radially symmetrical, but they were not bilaterally symmetrical either. Rather, they possessed peculiar asymmetrical shapes—some seemed to be bilateral animals that were lying on their right side and feeding with tentacles from their left side. Many later fossil forms were completely attached and radial. As we have seen before, bilateral symmetry best suits the needs of fast-moving animals and radial symmetry seems best in sedentary or sessile animals, which must meet the environment on all sides. It is considered likely by many biologists that ancestral echinoderms were bilateral animals that became radial as they took up a sedentary, suspension-feeding habit, and that modern free-living echinoderms such as sea stars were derived from such sedentary ancestors, whose adult radial symmetry they still retain.

Sea urchins seem very unlike sea stars, yet they have the same fundamental structure. Sea urchins look like large animated burs with long, sharp, movable **spines** that provide protection and aid

Sea urchin. These echinoderms live mostly on rocky shores where they feed on attached and drifting seaweeds. Their suckered tube feet are long and slender, extend well beyond the tips of the spines, and are used both for locomotion and for snaring pieces of food. Five teeth in the mouth bite the food into small pieces. *Strongylocentrotus.* Maine. (R. Buchsbaum)

Spines of sea urchins are used for locomotion as the animals "walk" on their spine tips; spines also offer protection from predators. The spines may be short, stout, and effective for burrowing into solid rock, as in the short-spined sea urchin *(Echinometra)* sitting in its hole, *left;* or they may be long, brittle, and very sharp, as in the needle-spined sea urchins *(Diadema)* that move about in the open, *right*. Great Barrier Reef, Australia. (O. Webb)

Slate-pencil sea urchin has very thick spines that are used to wedge the animal in holes within coral rock and make it very difficult to dislodge. *Heterocentrotus*. Great Barrier Reef, Australia. (O. Webb)

The test, or skeleton, of a sea urchin is globular and made up of closely fitted interlocking plates. There are five double rows of holes in the test through which the external tube feet connect to the internal ampullas and radial canals. The round protuberances articulate with spines, here removed, in a ball-and-socket arrangement. In the living animal, the spines and test are covered by living tissues. *Strongylocentrotus.* Monterey Bay, California. (R. Buchsbaum)

the tube feet in locomotion and feeding. Instead of numerous small calcareous plates embedded in connective tissue, sea urchins have larger plates that fit closely together and form a rounded skeletal **test** enclosing the soft parts. The mouth opens through the center of the lower surface of the test, and the anus opens through a small hole in the upper surface. Radiating upward from mouth to anus are five rows of minute holes through which the numerous tube feet project. These five rows correspond to the under surface of the five arms in sea stars. If we imagine the arms of a sea star bending upward to meet, and if, at the same time, we fill in the angles between them by elevating and reducing the size of the disk so that the sieve plate lies near the ends of the rows of tube feet, we can see how the spherical sea urchin is similar to the five-rayed sea star.

Sea urchins have the same type of water vascular system as

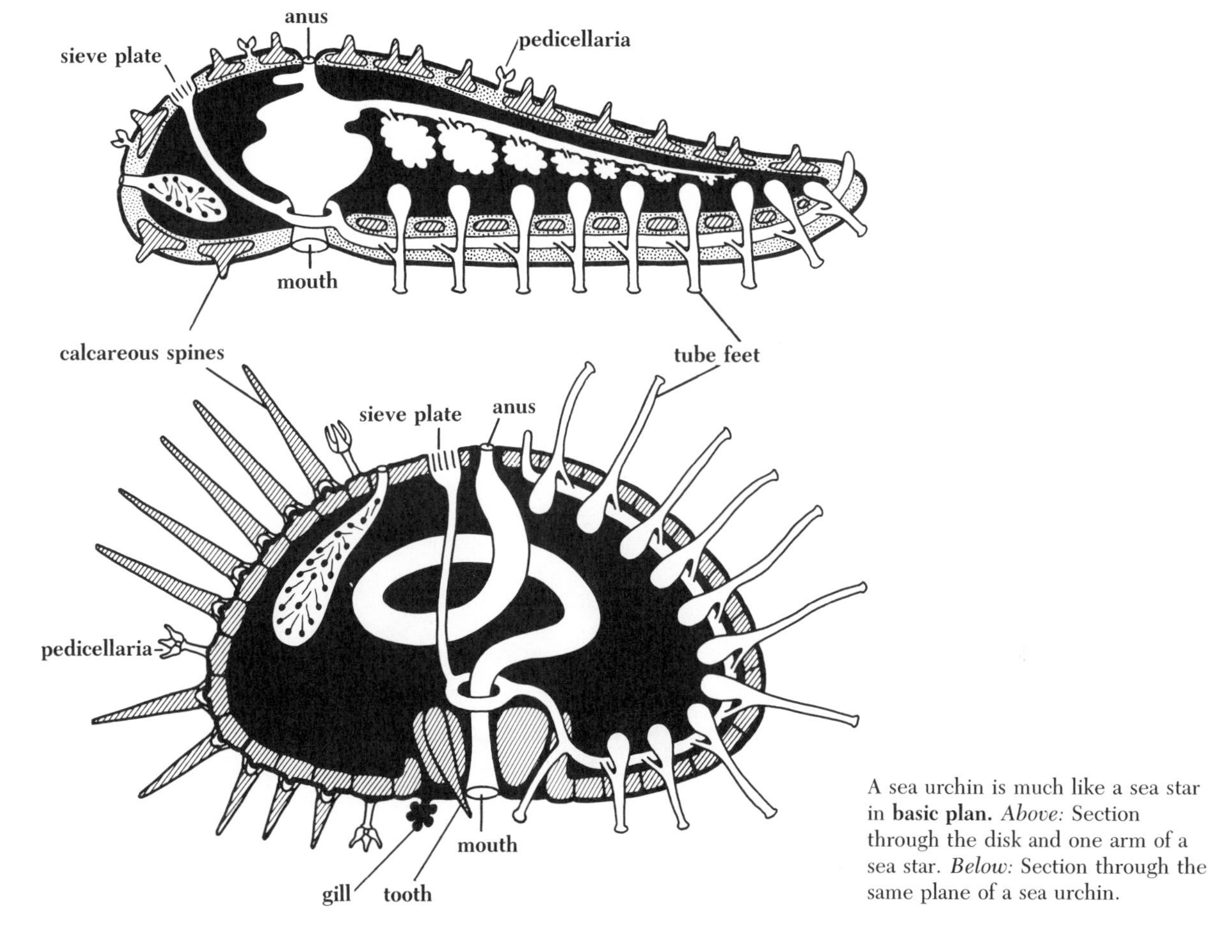

A sea urchin is much like a sea star in **basic plan.** *Above:* Section through the disk and one arm of a sea star. *Below:* Section through the same plane of a sea urchin.

was described for sea stars, but the tube feet are more slender and longer and project beyond the spines. The large pedicellarias have three jaws rather than two and are usually on long stalks. The digestive tract is longer than that of sea stars, as sea urchins feed mostly on plant material that is held in the gut while being digested. There are no digestive glands. Around the mouth is an elaborate set of five **teeth,** arranged radially and worked by a complex set of muscles and calcareous plates ("Aristotle's lantern") to cut the food into small pieces. Other systems are much like those described for sea stars. The gonads open from five pores near the anus, and eggs and sperms are shed into the sea where fertilization and embryonic and larval development take place. Metamorphosis from a bilaterally symmetrical free-swimming, feeding **pluteus larva** into a tiny globular sea urchin is similar to that in sea stars.

Sea urchin gonads nearly fill the test of the animals. They are a delicacy in many parts of the world; typically they are served fresh with a little lemon juice. Puerto Montt, Chile. (R. Buchsbaum)

Early development of a sea urchin is like that of most other animals. As the embryo divides, the cells become progressively smaller, until by the blastula stage, at the time of hatching from the egg membrane, individual cells are barely distinguishable under the light microscope. Actual size of unfertilized egg, 0.1 mm. (D. P. Wilson)

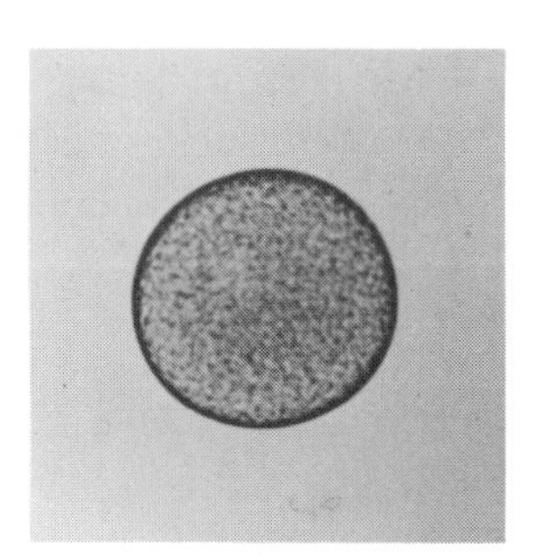

1. Unfertilized egg (ovum).

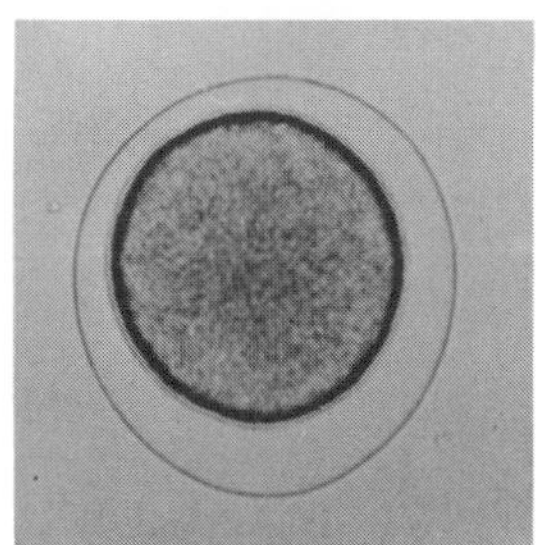

2. Fertilized egg (zygote).

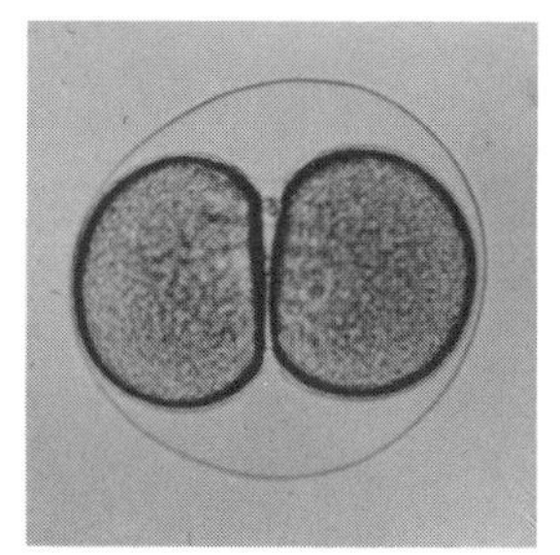

3. Two-cell stage.

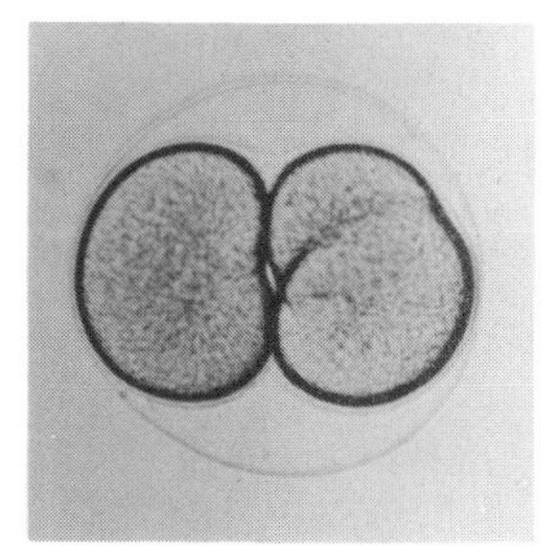

4. Further division.

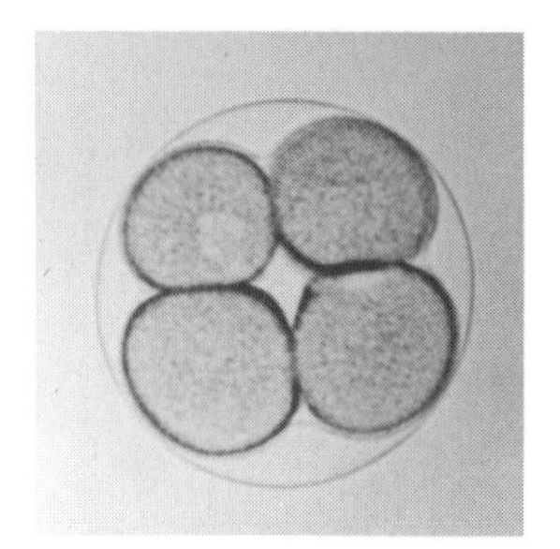

5. Four-cell stage.

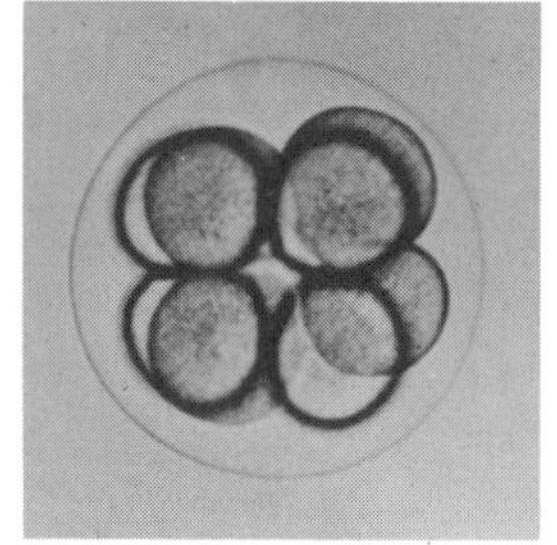

6. Eight-cell stage.

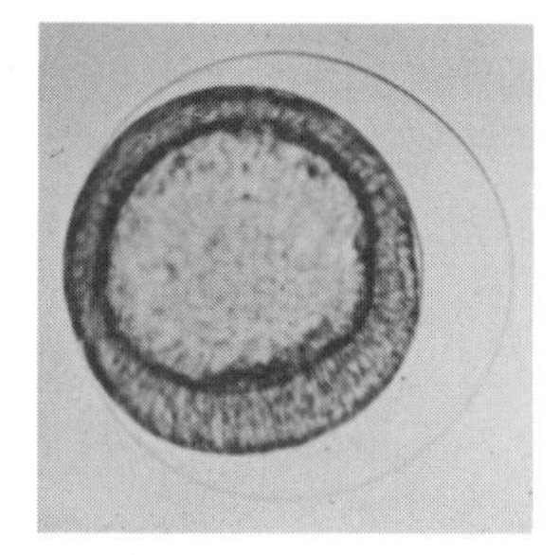

7. Blastula.

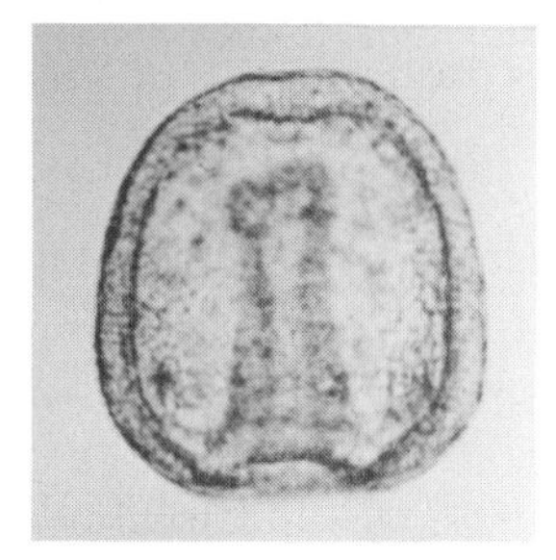

8. Gastrula.

Sand dollars are like sea urchins but are flattened and are covered with short, fine spines with which they move and burrow in the sand. Large, flattened tube feet protrude from the five double rows of holes on the upper surface and serve as gills. Smaller, mucus-covered tube feet, on both surfaces of the test, collect small organic particles and pass them to ciliated grooves that sweep them to the mouth to be eaten. *Above:* Living sand dollars. *Below:* Dried tests of two sand dollars arranged in positions corresponding to the pair above. The left ones show the upper surface; the right ones, the lower surface with the mouth opening in the center and the anus near the margin. *Dendraster*. Monterey Bay, California. (R. Buchsbaum)

Other groups of echinoderms—brittle stars and basket stars, sea cucumbers, sea lilies and feather stars, and sea daisies—differ from sea stars about as much as do sea urchins. They have characteristic spiny or leathery skins, calcareous internal plates, an extensive body coelom, a water vascular system, secondary radial symmetry usually based on a pentagonal body plan, and bilateral larvas that undergo marked metamorphosis. These groups are illustrated by photographs and drawings.

Brittle stars have long, flexible arms that often break off and are then regenerated, giving the animals their common name. The arms bear small tube feet that are used to catch organic particles and small animal prey. Placed together in a restricted space, brittle stars twine their arms about one another. *Ophiopholis.* Maine. (R. Buchsbaum)

Serpent stars is another common name for brittle stars because of the agile, snakelike movements of the arms by which these echinoderms crawl over the ocean floor, as shown by these photos of the same animal taken in rapid succession. *Ophiothrix.* Monterey Bay, California. (R. Buchsbaum)

Basket star is a type of brittle star with branching arms. It can roll into a small ball or open to spread the arms over a large area and capture small planktonic food organisms and organic particles. *Gorgonocephalus.* Friday Harbor, Washington. (R. Buchsbaum)

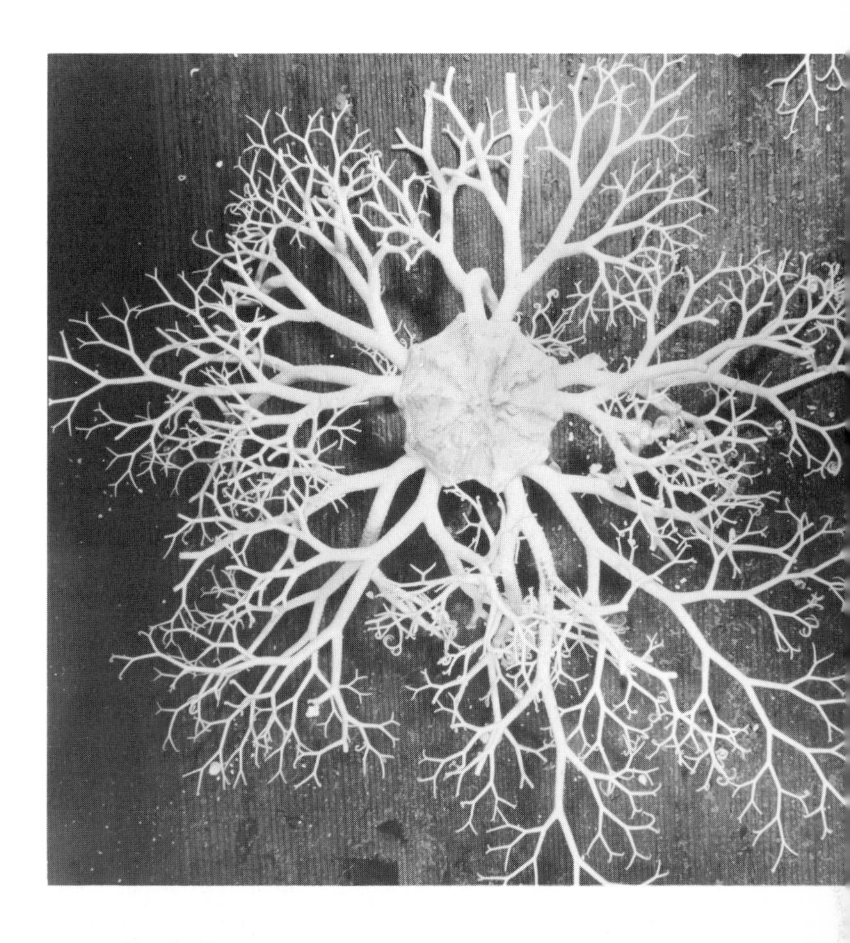

Below: **Developmental stages of a brittle star.** *Left:* The bilaterally symmetrical pluteus larva with long arms used for feeding and swimming in the plankton. *Center:* At metamorphosis, the larval arms shorten and the tiny juvenile brittle star emerges from the side of the larva. *Right:* The secondarily radially symmetrical juvenile begins life on the bottom. Plymouth, England. (D. P. Wilson)

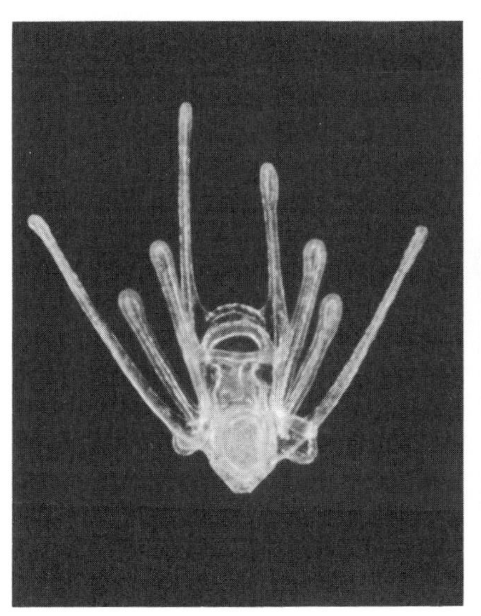

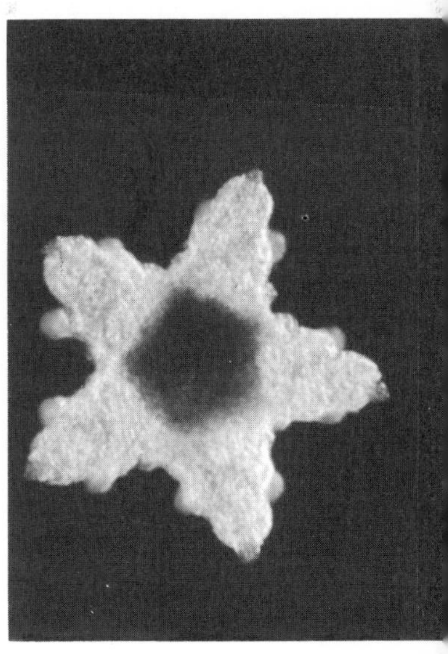

Sea cucumbers are fleshy echinoderms without a hard test or spines. The body wall contains toxins that discourage predators; some sea cucumbers also throw out their viscera or discharge special slime threads when disturbed. Nevertheless, when dried, the body wall and muscles of many sea cucumbers are a prized ingredient in Oriental soups. The mouth of sea cucumbers is at one end of the elongated body and the anus at the other. Around the mouth are feeding tentacles, which are enlarged tube feet. Five rows of smaller tube feet usually run the length of the body. This animal is a deposit feeder: the tentacles pass sand and mud into the mouth, and the small amount of organic material contained is digested. In areas where these kinds of sea cucumbers are abundant, most of the surface sand and mud passes through their guts several times each year. *Holothuria*. Plymouth, England. (R. Buchsbaum)

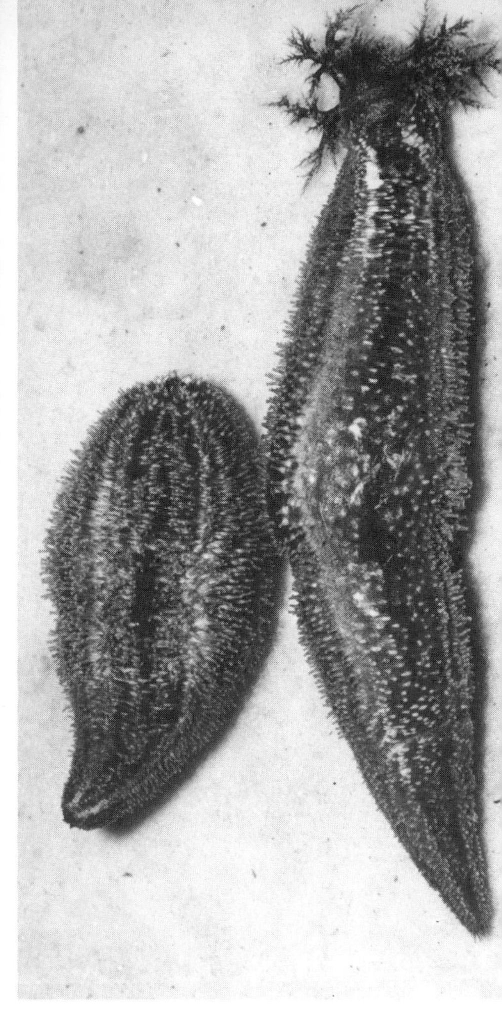

The tentacles of suspension-feeding sea cucumbers are large and branching. When held upright into the currents, they collect fine food particles and pass them to the mouth. Smaller tube feet along the body assist in holding the animal in place. When not feeding, or when stressed, the animal withdraws the delicate feeding tentacles. *Left:* Tentacles withdrawn. *Right:* Tentacles extended. *Thyonella*. Panacea, N.W. Florida. (R. Buchsbaum)

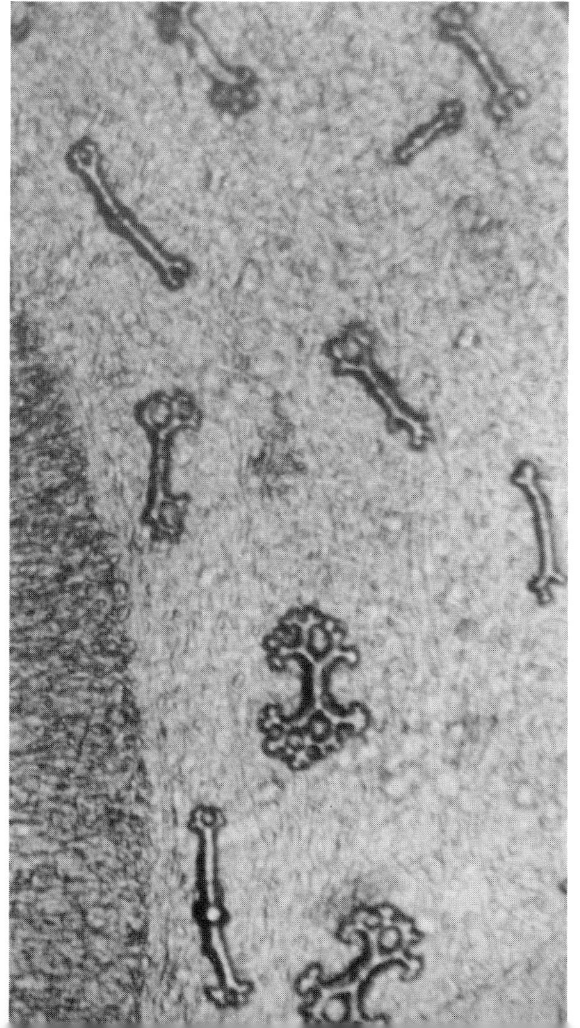

Wormlike sea cucumber common in shallow tropical waters. This animal does not have small tube feet along its body wall, but instead uses its tentacles to pull its long body slowly along, while at the same time the tentacles collect organic particles from the bottom. *Opheodesoma.* Oahu, Hawaii. (A. Reed)

Minute ossicles, rather than large plates and spines, are embedded in the soft body wall of sea cucumbers. The ossicles can be seen when thin pieces of tissue are examined with a microscope. Nearly all skeletal components of echinoderms, even large sea urchin spines, are organized at the microscopic level in the same way as these ossicles, with a highly ordered arrangement of holes and calcium carbonate struts. (R. Buchsbaum)

Sea lily, a stalked crinoid brought up from deep water by a dredge. Only a small part of the stalk, which was attached to a rock, is shown. When undisturbed, the animal spreads the branched arms widely; small mucus-coated tube feet snare organic particles from the water and pass them into ciliated grooves that sweep them into the central mouth. (A. H. Clark)

Feather star, an unstalked crinoid that can move along the bottom or even swim short distances by waving its arms up and down. Juveniles have a stalk and are attached to the bottom like sea lilies, but they soon break free. When feeding they usually perch on rocky outcrops and feed in the same way as sea lilies. *Antedon.* Banyuls, Mediterranean coast of France. (R. Buchsbaum)

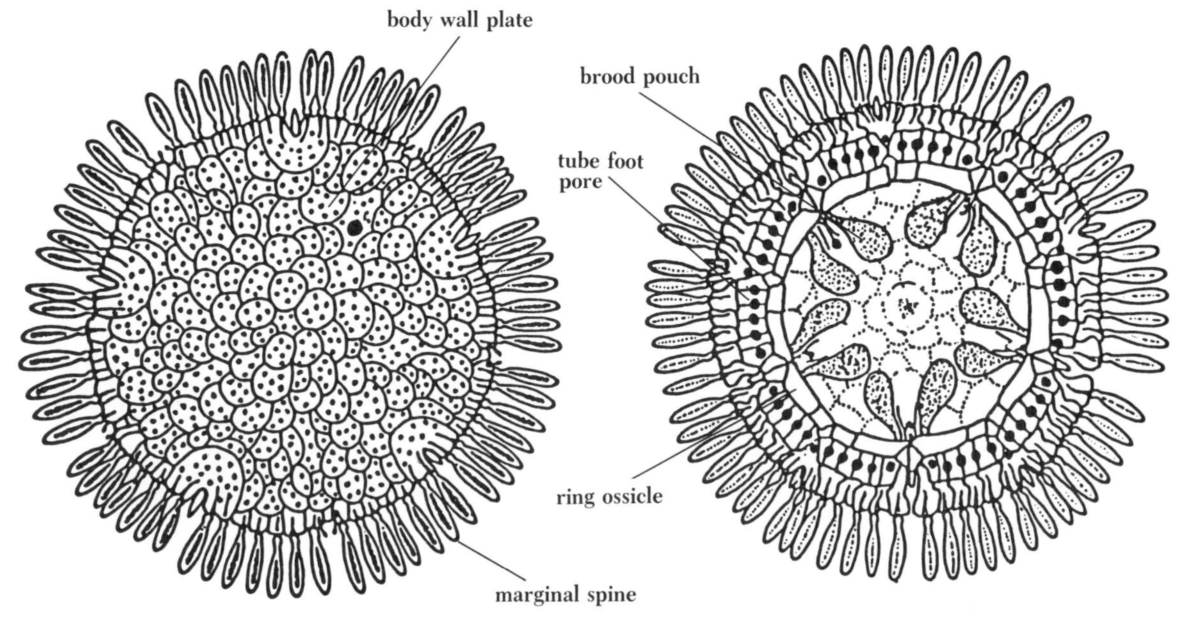

Sea daisies are the most recently discovered (1986) group of echinoderms. They are disk-shaped animals, less than 1 cm in diameter, that occur on sunken wood collected from the deep sea. Though superficially resembling cnidarian medusas, they have pentamerous symmetry, calcareous spines and internal plates, and a water vascular system. The tube feet are arranged in a ring around the edge of the disk, and there is no mouth or internal gut; the animals apparently absorb nutrients through the thin membrane covering the lower surface. Embryos within 5 pairs of brood pouches develop directly into juveniles similar to the adults. The drawing above is of the skeletal system of one species, *Xyloplax medusiformis*, first collected from over 1,000 m depth off the coast of New Zealand. A second species has been found off the Bahamas. (From Baker, Rowe, and Clark)

Chapter 22

Chordate Beginnings

The only major group we have not yet considered is our own phylum **CHORDATA,** composed almost entirely of animals *with* backbones and therefore no proper subject for this book. There are, however, a few chordates that have no vertebral column. These invertebrate chordates—lancelets and tunicates—are mostly inconspicuous animals, seldom seen, or at least not usually noticed, though lancelets are so abundant on the seacoast of China that during certain months they are collected by the ton for human consumption. Tunicates of several kinds are also eaten in some parts of the world, but these animals are more likely to be regarded as a nuisance when they foul ships and docks. None of the invertebrate chordates is of much economic importance, and they are described here in some detail chiefly because they share with the vertebrates certain distinctive characters found nowhere else in the animal kingdom, and so help to link the vertebrates with the invertebrates.

Lancelets

Lancelets are members of the chordate subphylum **CEPHALOCHORDATA.** These small, laterally compressed, semitransparent animals live on shallow sandy bottoms in tropical and warm-temperate marine waters around the world. A lancelet (or amphioxus, "sharp at both ends") can swim about by fishlike undulations of the body, but spends most of the time buried in the sand with only the anterior end protruding into the overlying water. In this position it feeds by drawing into the mouth a steady current of water, from which it strains suspended microscopic organisms.

Opposite: **Lancelet** lies quietly buried in sand during the day, with only its head end out, its body mostly concealed from predatory fishes. After sunset it may rise from the sand and swim about. *Branchiostoma.* Florida. (R. Buchsbaum)

The most distinctive chordate character, and the one from which the name of the phylum is derived, is the **notochord,** a turgid rod that extends the length of the body. It serves as a firm

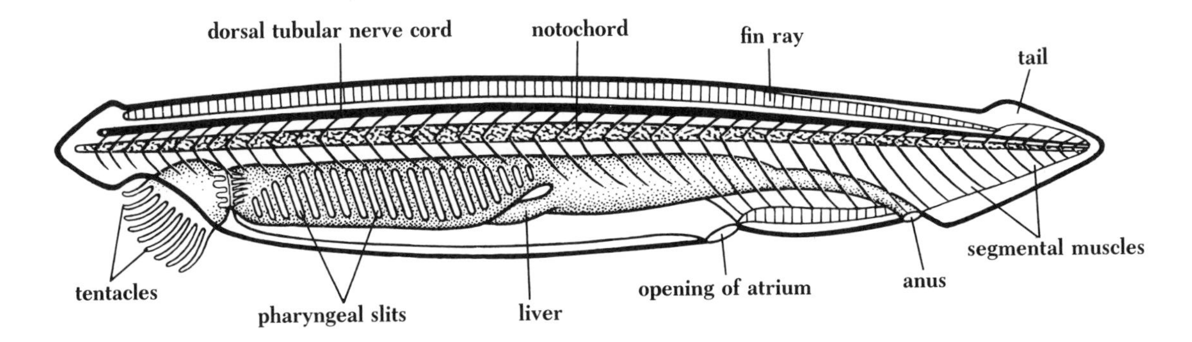

Diagram of a lancelet.

but flexible axis and so permits powerful side-to-side undulatory movements of the whole body that propel the animal through the water with a speed unattainable by flabby animals such as flatworms and annelids.

The strong, swift swimming movements of aquatic chordates, made possible by a flexible internal skeletal axis, were probably a major factor in the early success of the group. Besides great muscularity (as shown in the cross-section of the lancelet), chordates are characterized by the prolongation of the body beyond the anus as a *tail*, a region that contains little else but the skeletal axis, nerves, and muscles. In aquatic chordates it is specialized for swimming, and in terrestrial types it aids in other kinds of locomotion.

The notochord develops from the roof of the primary gut and is present in the embryos of all chordates including humans. It persists as the functional skeletal axis in the adults of certain primitive fishes (for example, lampreys and sturgeons), and these have only the beginnings of a backbone, a series of cartilaginous elements that appear on each side of the notochord. In all other vertebrates a backbone composed of a series of separate cartilaginous or bony vertebras forms around the notochord during development; and in adults the backbone largely or entirely replaces the notochord as the mechanical axis of the body. Portions of the notochord may remain in the cushioning disks between the vertebras.

A second chordate character is the **dorsal tubular nerve cord.** In all other invertebrates that have been described, the principal nerve cord is ventral or lateral in position; but in lancelets and other chordates it lies between the notochord and the dorsal body wall, and it is hollow. From the cord go a pair of nerves to each of the segmentally arranged bundles of muscles.

The third important chordate character is the structure of the pharynx, which is perforated by pairs of slitlike openings. The **pharyngeal slits,** which have mainly a respiratory function in fishes, serve chiefly in straining food from the water in lancelets. The pharynx is lined with cilia that beat steadily, producing a current of water than enters the mouth and passes out through the pharyngeal slits, leaving behind the suspended particles. The slits do not open directly to the exterior as in fishes, but into a chamber, the **atrium,** which surrounds the pharynx and opens to the exterior by a pore some distance anterior to the anus. The atrium is lined with ectoderm and is formed, in the embryo, by the outgrowth of two folds of skin, one on each side, which finally fuse in the midline. The walls of the pharynx, perforated from top to bottom by the pharyngeal slits, are supported by rods that run in the walls bounding the slits.

Adult lancelets have about 180 pairs of pharyngeal slits. This large and indefinite number is probably a primitive condition; modern fishes have a small and definite number; for example, sharks commonly have 6 pairs. Land vertebrates, which breathe by lungs, have no pharyngeal slits in the adult stage, but their embryos have pharyngeal pouches, and slits may make a fleeting appearance during embryonic development, or functional slits may be present in a larval stage, as in frog tadpoles. In the pharyngeal pouches of human embryos, slits break through only rarely and are easily closed by surgery shortly after birth.

Many of the structures associated with the **feeding mechanism** of lancelets are peculiar to these animals and not necessarily primitive chordate characters. At the anterior end of the animal is a funnel-like hood fringed with a row of stiffened tentacles that act as a sieve, keeping out large particles. The mouth lies at the back of the hood and is bounded by a rim of small tentacles that further screen the particles in the feeding current. During feeding, mucus is continuously secreted in a *ciliated groove* in the floor of the pharynx, and a sheet of mucus is moved by cilia upward along the walls of the pharynx on each side. Suspended organisms and particles that enter the pharynx in the water current are trapped in this mucus. The food-laden mucus collects in another ciliated groove in the roof of the pharynx and is moved backward to the intestine. A mode of filter feeding like that of lancelets, with mucus sheets secreted by a ventral ciliated groove, is found also in the larvas of certain primitive fishes and is part of the evidence for ancestral ties between invertebrate chordates and vertebrates. In these fishes, the ciliated groove later develops into the thyroid gland.

In the intestine the food and mucus are digested and almost completely absorbed. From the anterior end of the intestine is given off a hollow, glandular digestive cecum, or liver, which extends forward along the right side of the pharynx and can be seen in the cross-section. Since it arises in the same way as the vertebrate liver, by an outpocketing of the digestive tract, the two organs are thought to be homologous. Further, the blood leaving the capillaries of the intestine of lancelets is not returned to the general circulation until it has passed through the capillaries of the liver. Such a path for the blood is found nowhere else except in vertebrates, and it furnishes further evidence that lancelets are descended from the same primitive stock that gave rise to the backboned animals.

The **circulatory system** is a closed one; and there is no heart, the blood being pumped by contractile vessels, including a large ventral one. In this connection it is interesting to note that in all vertebrate embryos the blood is first pumped by a simple, pulsating tube that only later becomes bent and constricted to form the heart. The blood receives oxygen as it flows through vessels in the tissue between the pharyngeal slits, which are in close contact with the steady current of water passing through the

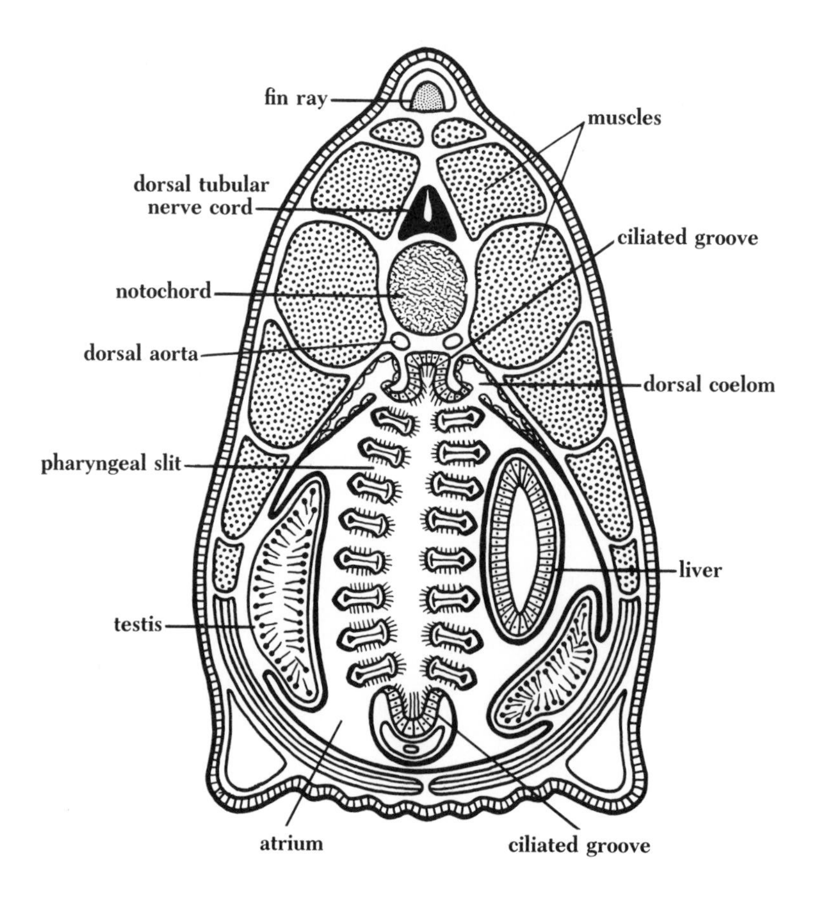

Cross-section through a lancelet. The pharyngeal slits are set obliquely along the pharynx, so that several appear in a single cross-section. Likewise, the chevron-shaped segmental muscles overlap, so that several muscle blocks (shown stippled) are cut by the section.

pharynx. After passing through the pharyngeal vessels, the blood flows into two vessels, the dorsal aortas, which unite behind the pharynx into a single vessel that supplies the intestine. Wastes are extracted by **excretory organs** that lie against the dorsal wall of the pharynx, a pair to every other pair of pharyngeal slits. Although these excretory organs superficially resemble those of annelids and several other miscellaneous phyla, they are more like certain cells in vertebrate kidneys. The **coelom** of lancelets has been partly crowded out by the expansion of the atrium, and in the pharyngeal region it is represented only by a number of small cavities with complex connections. The sexes are separate, and the **reproductive organs** are paired, segmentally repeated sacs that lie along the sides of the body and push the atrial walls inward so that in cross-section, they appear to lie in the cavity of the atrium. When mature, the gonads burst through the atrial wall; the gametes are released into the atrium and exit through the atrial opening.

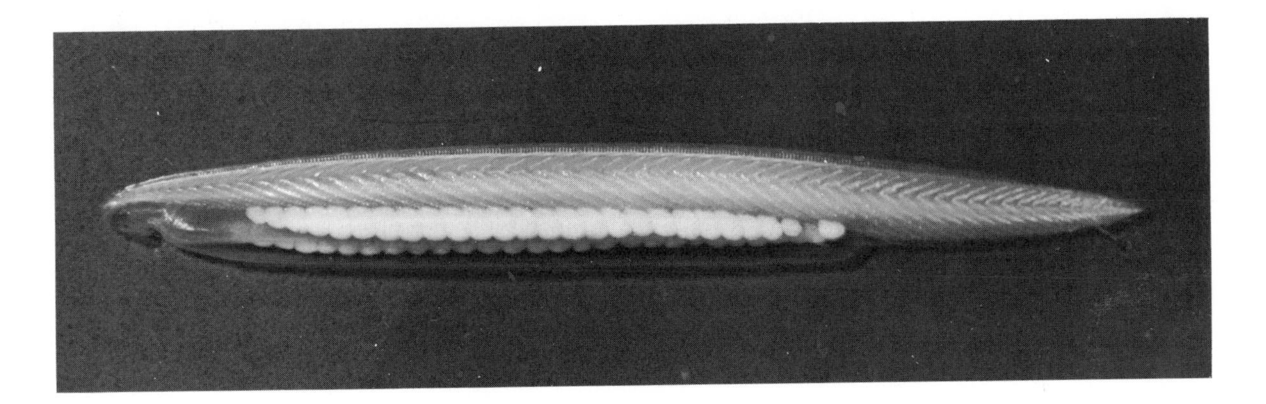

Two rows of gonads and chevron pattern of the muscles show clearly in this lancelet, *Branchiostoma*. Length, 5 cm. (R. Buchsbaum)

There are no paired eyes or other well-developed sense organs, though there is a large pigment spot at the anterior tip of the nerve cord and a row of smaller pigment spots along the lower edge of the cord. The nerve cord does not expand at the anterior end into a brain but tapers to a point. The apparent simplicity of the sense organs and central nervous system is probably not a primitive condition but is more likely a reduction of the head region associated with the sedentary habits of lancelets.

In spite of the various reductions and specializations associated with their particular way of life, lancelets are probably the only living forms to which we can look for a concrete idea of what the primitive chordates that gave rise to the vertebrates might have been like.

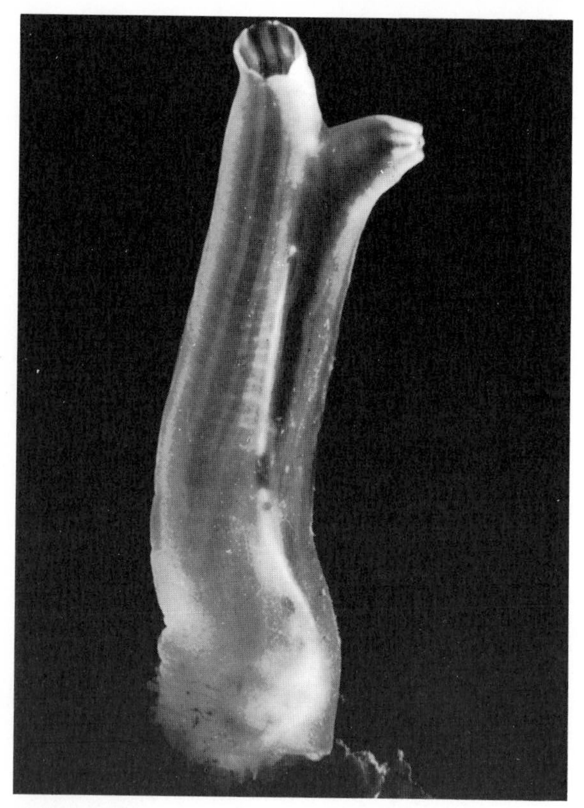

Tube tunicate, *Ciona intestinalis* (3/4 natural size), has a semitransparent tunic through which can be seen the muscles of the body wall. Water enters the upper (mouth) opening and leaves by way of the lower (atrial) opening. Plymouth, England. (D. P. Wilson)

Cluster of tunicates, *Dendrodoa grossularia,* consists of separate individuals growing close together. Shown here about natural size, this small red tunicate grows in berry-like bunches on rocks near low-water mark. Plymouth, England. (D. P. Wilson)

Tunicates

Tunicates are so named because they are covered by a tough, often translucent **tunic** made of proteins, carbohydrates, and, mostly, seawater. One of the common components is cellulose, a carbohydrate that we usually associate with plants but has been found also in the connective tissue of humans and other mammals, especially older individuals. Some tunicates are pelagic; but most forms are sessile, growing permanently attached to rocks or seaweeds. The simplest kind looks like an upright sac with two openings: one at the top, and one somewhat lower down on one side. When the animal is disturbed, the body wall contracts suddenly, and the water contained within the body is forced out in two jets—hence the common name "sea squirt."

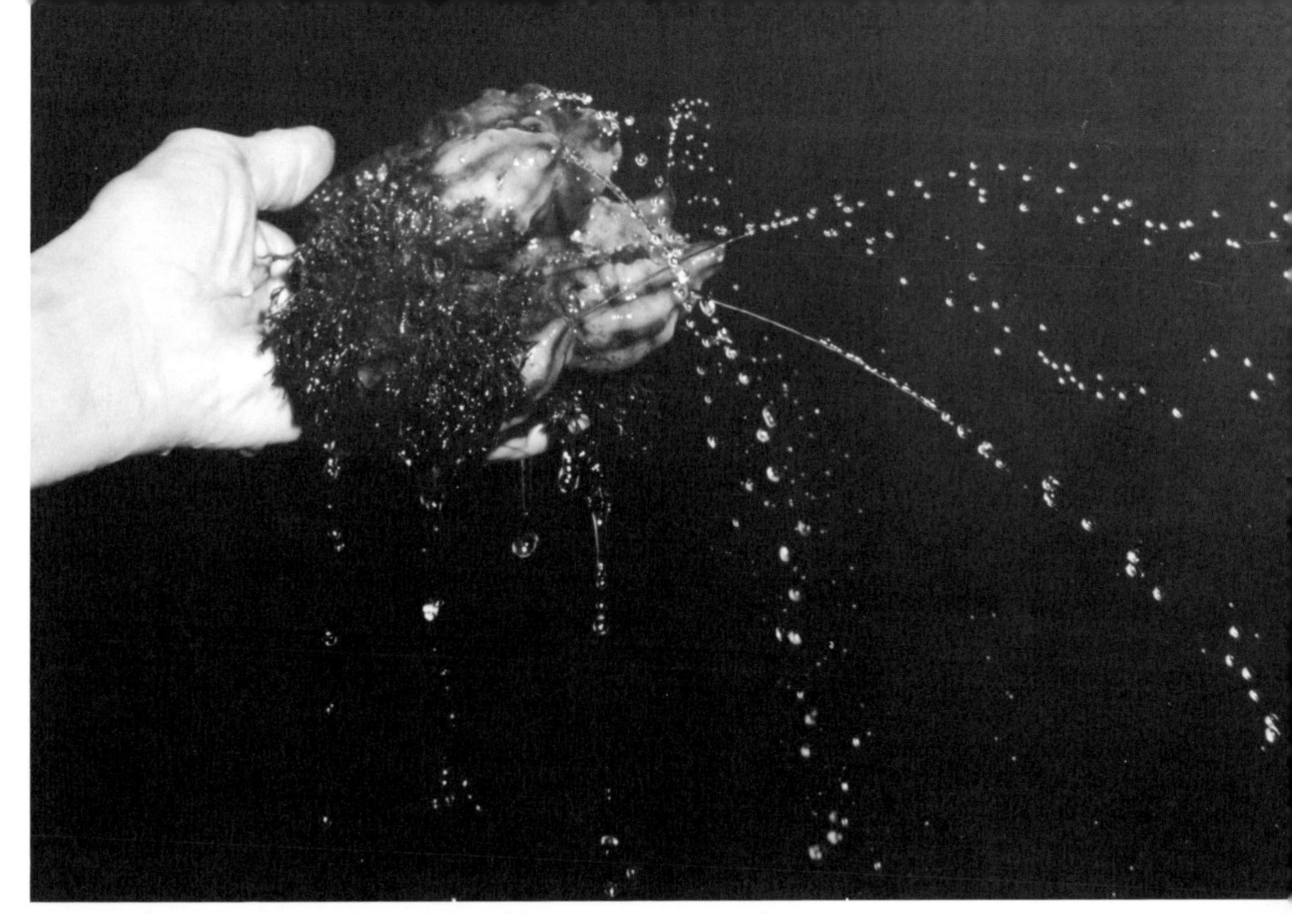

"**Sea squirts**" is a common name for tunicates, and here several specimens of *Styela* are demonstrating why. Disturbed by their human captor, they are forcefully ejecting water from both mouth and atrial openings. Panacea, N.W. Florida. (R. Buchsbaum)

The interior of the body is occupied for the most part by a large saclike pharynx perforated by many rows of **pharyngeal slits.** Cilia around the edges of the slits create a current that draws water into the mouth at the top, through the pharyngeal slits, and out into the **atrium,** a cavity surrounding the pharynx. Food particles are trapped by mucus sheets that are secreted in a ciliated groove along the ventral wall of the pharynx, and water leaves by an opening from the atrium, all much as in lancelets.

The atrial opening also serves as an exit for feces and gametes, since the anus and the gonads of both sexes (tunicates are hermaphroditic) open into the atrium. The circulatory system consists of unlined vessels and sinuses that supply blood to the tunic as well as all the internal organs. The heart is remarkable in that it drives the blood in one direction during several minutes and then reverses direction.

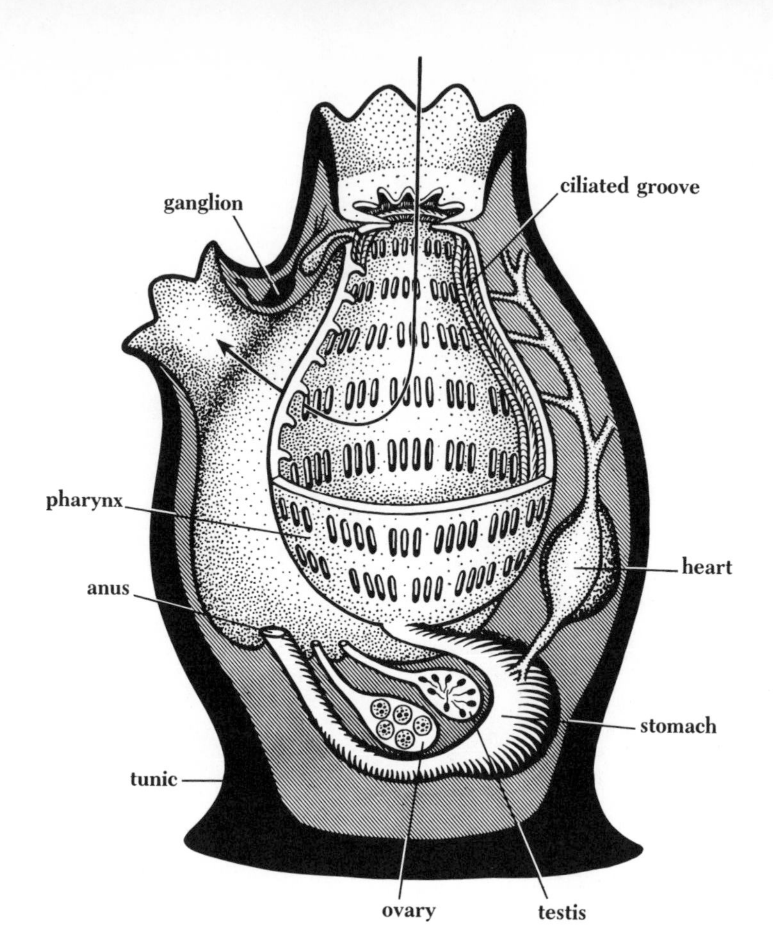

Tunicate. Arrow indicates direction of water flow: into the mouth, through the pharyngeal slits, and out the opening of the atrium. (Combined from various sources.)

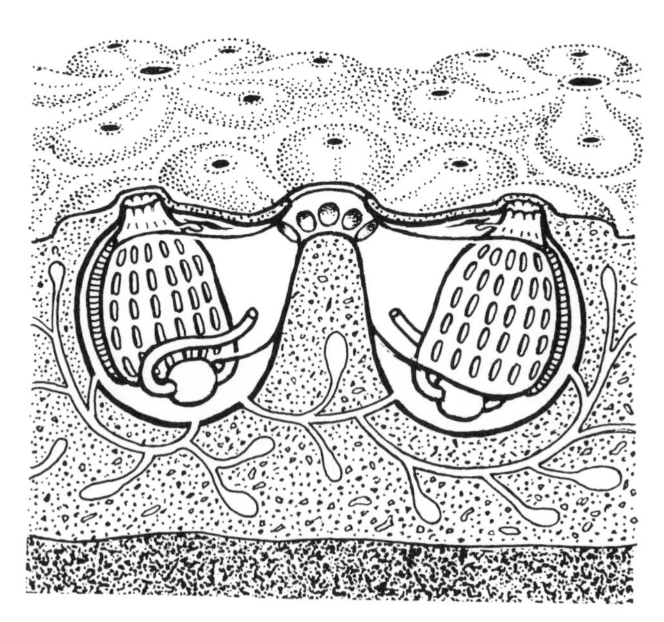

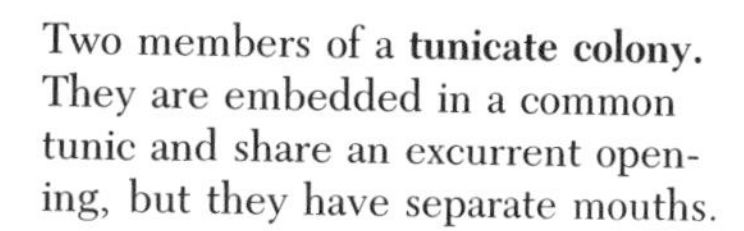

Two members of a **tunicate colony.** They are embedded in a common tunic and share an excurrent opening, but they have separate mouths.

Many tunicates proliferate by budding. The budded individuals remain together as a colony, sometimes embedded in a common tunic. Colonies of sessile tunicates grow as encrusting masses over the surface of rocks, shells, or seaweeds. The members may be arranged in small groups, each such cluster looking like a flower with the separate mouths of the members at the tips of the "petals." At the center of the cluster is a common opening for the exit of water. Colonies of pelagic tunicates are sometimes extremely abundant and important in the ecology of the oceans.

Tunicate colony, *Botryllus schlosseri.* Each star-shaped pattern represents a group of several colony members with their individual mouths at the outer points of the star and their common excurrent opening at its center. Helgoland, North Sea. (R. Buchsbaum)

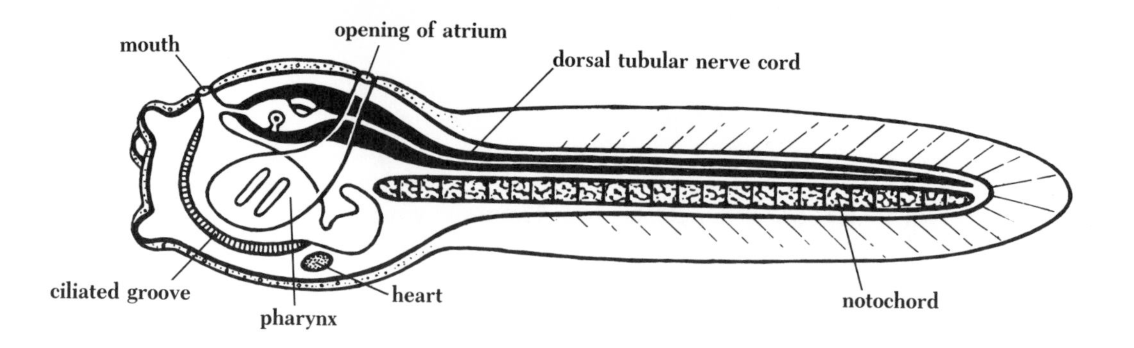

Free-swimming **larva of a tunicate** shows chordate characters. (Combined from various sources.)

Except for the pharyngeal slits there is very little about such unresponsive animals as bottom-living tunicates to suggest any reason for including them in the same phylum with fishes or mammals. But the development of a tunicate tells another story. The **larva** is a free-swimming animal that reminds one of a tadpole. It has a large tail that contains, besides muscles, a well-developed **notochord,** which places this group clearly among the chordates as the subphylum **UROCHORDATA** ("tail-chord"). Other chordate characters include a **dorsal tubular nerve cord** and **pharyngeal slits.** When the larva finally settles down on a rock, it resorbs the tail, notochord, and most of the larval nervous system, and its whole structure becomes reorganized into the sessile, adult form. The adult has no trace of a notochord, and the central nervous system is represented only by a ganglion in the dorsal region of the body between the two openings. Here we have another striking example of how animal relationships are revealed by study of the young stages, even though the similarities of the adult forms are quite obscure.

Acorn Worms

Acorn worms, members of the phylum **HEMICHORDATA,** are soft, elongate animals that burrow in the sand or mud of seashores. At the anterior end of the body is a muscular *proboscis* joined by a narrow stalk to a short, wide *collar* and followed by a long *trunk.* The proboscis and collar are used in burrowing and feeding. The proboscis is forced through the sand, with the collar following. Distension of the collar firmly anchors the anterior end of the worm, so that contraction of the muscles in the trunk region draws the trunk forward.

The mouth is in the middle of the ventral surface, at the base of the proboscis and concealed by the edge of the collar. As the animal burrows, sand or mud passes into the mouth, through the

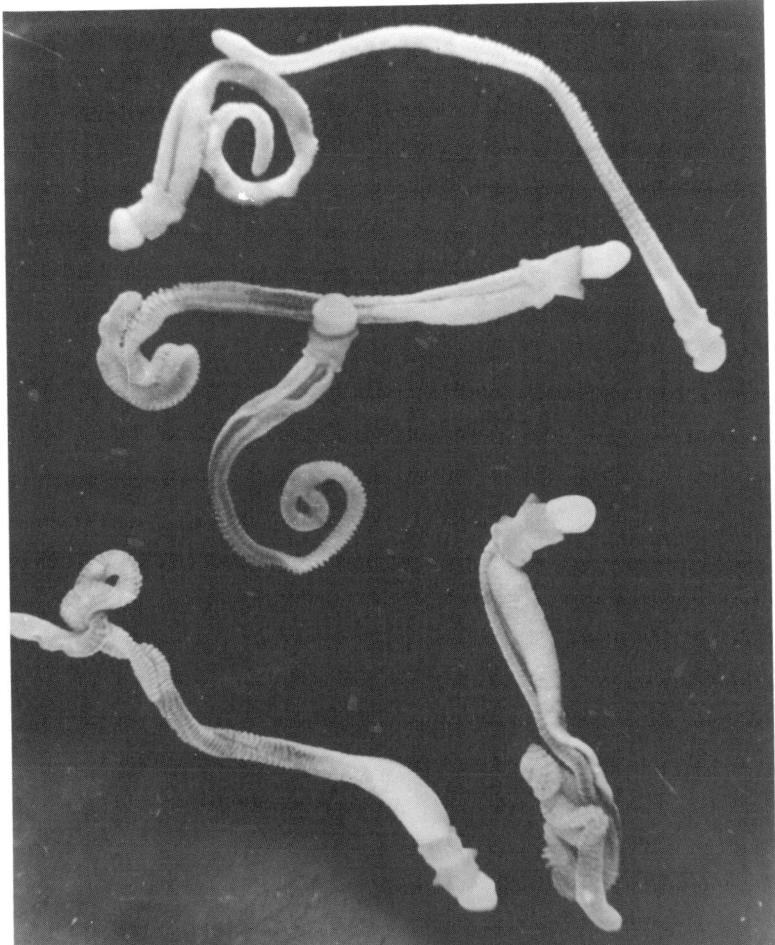

Acorn worms are sluggish burrowers on sandy or muddy bottoms in shallow marine waters. Unlike the muscular chordates, the soft bodies of acorn worms are delicate and tend to fragment if brought up in a dredge. The worms shown here *(Ptychodera bahamensis)* were collected intact by diving to a sandy bottom about 6 meters deep and gently scooping them up by hand. Bermuda. (R. Buchsbaum)

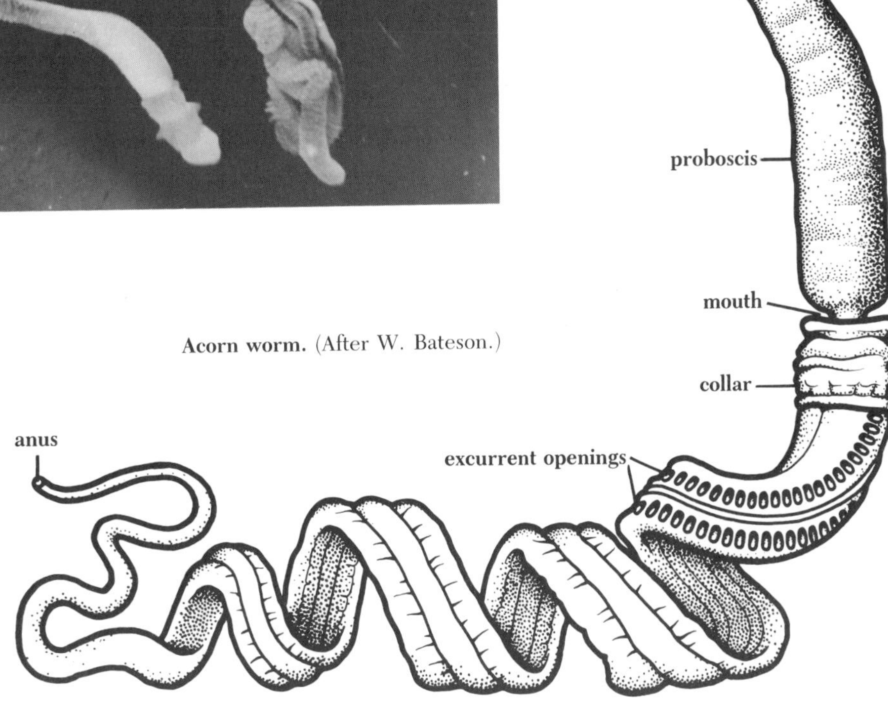

Acorn worm. (After W. Bateson.)

digestive tract, and out the anus at the posterior tip. Organic materials present in the sand or mud are digested. Acorn worms may also feed by collecting organic particles that stick to mucus on the proboscis. Cilia sweep the food-laden mucus over the surface of the proboscis and into the mouth. Water taken in passes out through **pharyngeal slits** and finally exits through openings in the dorsal wall of the anterior part of the trunk. These pharyngeal slits are a character shared only with the chordates.

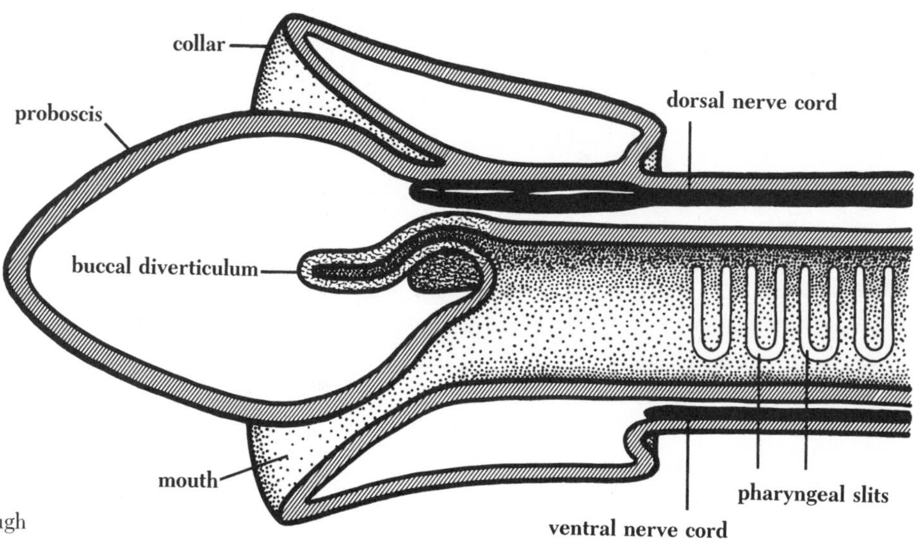

Longitudinal section through **anterior end of acorn worm.** (Combined from various sources)

In almost all respects the nervous system of acorn worms is among the simplest and least centralized of any group of animals having an organ-system level of organization. It consists of a network extending under the whole of the surface ectoderm and in the trunk region is concentrated along the mid-dorsal and mid-ventral lines of the body as two nerve cords. Only the **dorsal cord** extends into the collar, where it is especially thick, and in some species of acorn worm it is hollow, suggesting a resemblance to the dorsal tubular nerve cord of chordates.

A short rodlike growth (buccal diverticulum) of the anterior end of the digestive tract into the base of the proboscis is composed of vacuolated cells like those in the notochord of lancelets and tunicate larvas, but that it corresponds to the notochord of chordates seems unlikely.

Aside from their interest as animals that appear to have branched from some early prechordate stock, the acorn worms, through their larvas, furnish one of the few real clues that link the chordates with any other phylum. The early larvas of acorn worms look so surprisingly like the larvas of certain echinoderms that one can be mistaken for the other. Moreover, various features of the development of larvas from both phyla are similar, such as the manner of formation of the coelom. The later development of the larvas is very different, however, and the adults display little in common.

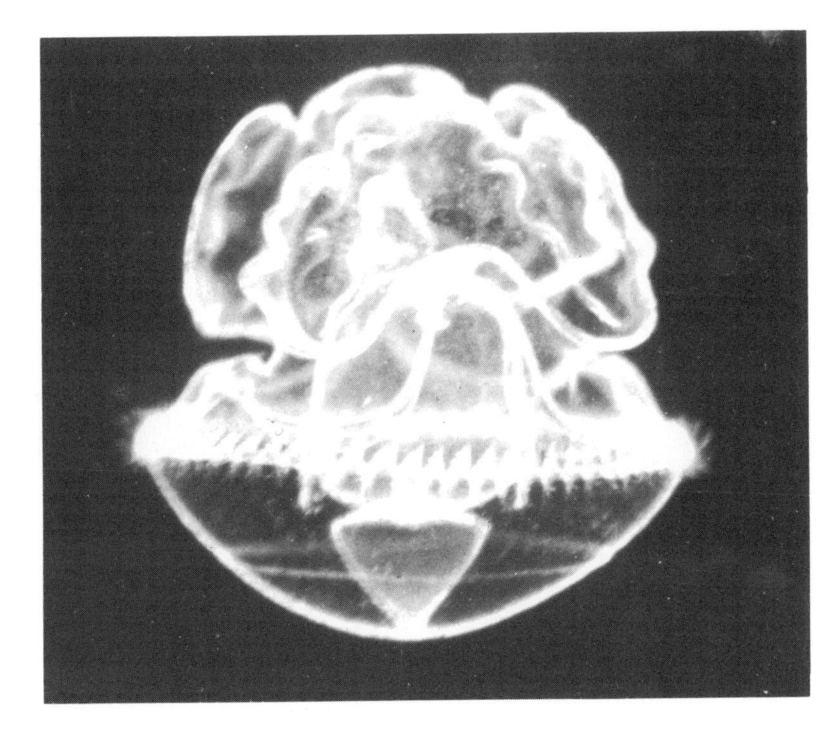

Free-swimming **larva of an acorn worm.** Plymouth, England. (D. P. Wilson)

Members of the phylum Chordata typically have, at some time in their life history, a notochord, dorsal tubular nerve cord, pharyngeal slits, and a tail. But in most other respects they fall into three groups—vertebrates, lancelets, and tunicates—so different that each is designated a subphylum. Lancelets show all these chordate characters as adults. Tunicates and most vertebrates lose one or more at an early stage. Acorn worms, in the related phylum Hemichordata, have only the pharyngeal slits and dorsal tubular nerve cord. It is these similarities (and others not detailed here) that link the hemichordates and invertebrate chordates with the vertebrates, including humans, and that must also have been present in the early invertebrate stock that gave rise to our phylum.

Chapter 23

Records of the Invertebrate Past

Although animals have left no birth certificates, marriage contracts, tombstone inscriptions, or written documents, upon which students of human history depend so much, there are abundant records of the invertebrate past—not just of the past few thousand years but of some 600 million years or more.

Any evidence of the materials or rocks of the earth's surface that gives some idea of the size, shape, or structure of the whole or any part of a plant or animal that lived in bygone times is called a **fossil.** The name is derived from the Latin verb "to dig" because it was, at one time, applied to almost anything of interest that was dug up. Petroleum is an indication of past life, but it cannot be considered a fossil because by itself it gives no idea of the character of the organisms that were responsible for its formation. Nor would we classify as a fossil any empty snail shell that turned up while we were digging about in the garden, for although the shell shows something of the character of the living organism, animal remains become fossils only when they also show changes of a sort that come with age. There is no exact age at which animal remains become fossils; most, though not all, are of species now extinct.

Fossils can be formed in a variety of ways. Most have undergone considerable changes since the death of the organism. Horny coverings, such as the chitinous exoskeletons of arthropods, leave thin **films of carbon** between layers of sedimentary rock. Calcareous shells and other skeletal structures can be slowly dissolved by water percolating through the ground to be gradually **replaced by minerals,** such as calcite, silicon dioxide, or iron sulfide. Or the shells may be completely dissolved away, leaving only a cavity, on the wall of which is a **mold** of the exter-

Opposite: **Fossils** are concrete records of the invertebrate past. These fossil crinoids (echinoderms) are from the early Carboniferous period, when crinoids were a dominant group at the peak of their diversity and abundance. Their modern descendants are relatively few and play a far more modest part in the marine economy. Indiana. (R. Buchsbaum)

Paleontology is the study of the life of past geological periods as known from fossil remains. *Above, left:* A class in invertebrate zoology is examining a much-studied paleontological site at Scientists' Cliffs, Maryland. *Below, left:* A closeup of one portion of the cliff shows mostly fragments of fossil clam shells from the Tertiary period. (R. Buchsbaum)

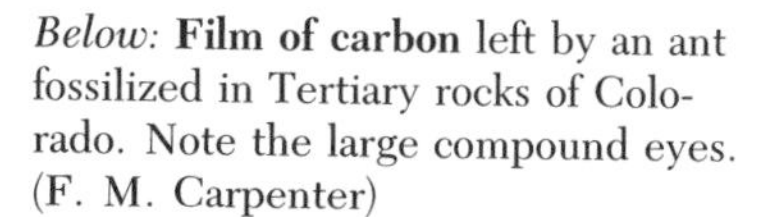

Below: **Film of carbon** left by an ant fossilized in Tertiary rocks of Colorado. Note the large compound eyes. (F. M. Carpenter)

nal surface of the shell. The cavity later may be filled with some other rocky material, forming a **cast** of the original fossil.

Some of the rarest and most beautiful fossils are of animals entombed within **amber,** fossilized resin from coniferous trees. Such fossils are found in considerable numbers in the Baltic region of Europe, which in mid-Tertiary times (about 38 million years ago) was covered with a dense coniferous forest. Drops of gummy resin, dripping from the trees, trapped spiders and mites scurrying across the forest floor or insects on the wing. The sticky resin, on fossilization, became hard amber; and the arthropods within left exquisite casts outlined by delicate carbon films. In exceptional cases, parts of the animals were permeated with resin and internal structures were preserved within the amber; even some of the internal parasites can be identified.

Amber is the fossilized resin of coniferous trees. This piece reveals two termites that lived in the Middle Tertiary (about 38 million years ago) but look as if they had died only yesterday. *Reticulotermes antiquus.* Baltic. (P. S. Tice)

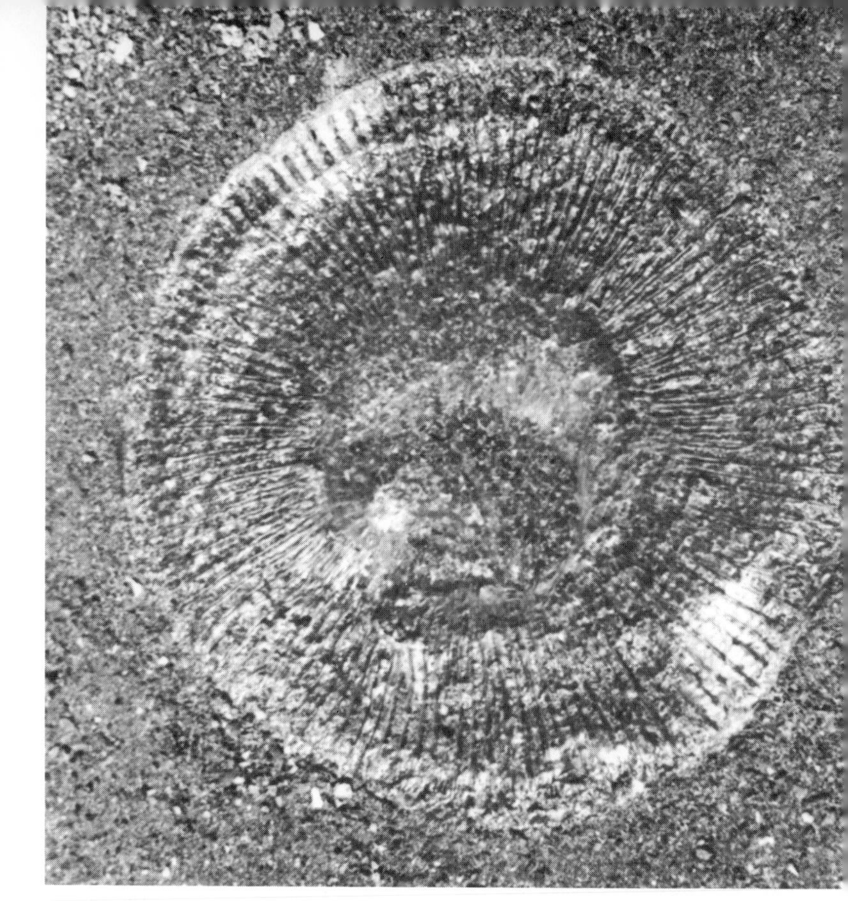

Future fossil? This large hydrozoan medusa (*Aequorea,* 20 cm diameter), which was cast ashore and dried on a hard surface, shows clearly its radiating gastrovascular canals. Though it looks like a promising candidate for fossilization, it will most likely be washed away in the next tide. Few soft-bodied animals meet with all of the conditions successively necessary for fossilization. Gulf of California. (R. Buchsbaum)

As we would expect, the vast majority of fossils are from animals with hard parts: calcareous and siliceous shells and spicules, or chitinous coverings and jaws. But even soft animals such as medusas may leave **impressions** on soft mud which, if soon covered by a layer of fine sediment, can be preserved when the mud hardens into rock. In a similar way, activities of animals can form **trace fossils** that are indirect evidence of their existence. *Tracks or trails* impressed on muddy or sandy bottoms sometimes indicate the kind of appendage that left the record, especially when compared with those left by living animals. *Burrows* in mud or sand and *borings* in rock or wood also can be identified by comparison with those left by living organisms, or with those in which fossil shells are left in the cavity. Certain irregular, straplike, or pelletlike fossils are interpreted as fossilized feces because they are similar to feces of living animals. In some cases these *coprolites,* as they are called, have revealed what extinct animals ate.

To be fossilized, an animal usually needs to be buried at the time of death or very shortly thereafter; otherwise the body is likely to be eaten by scavengers or decomposed by bacteria and

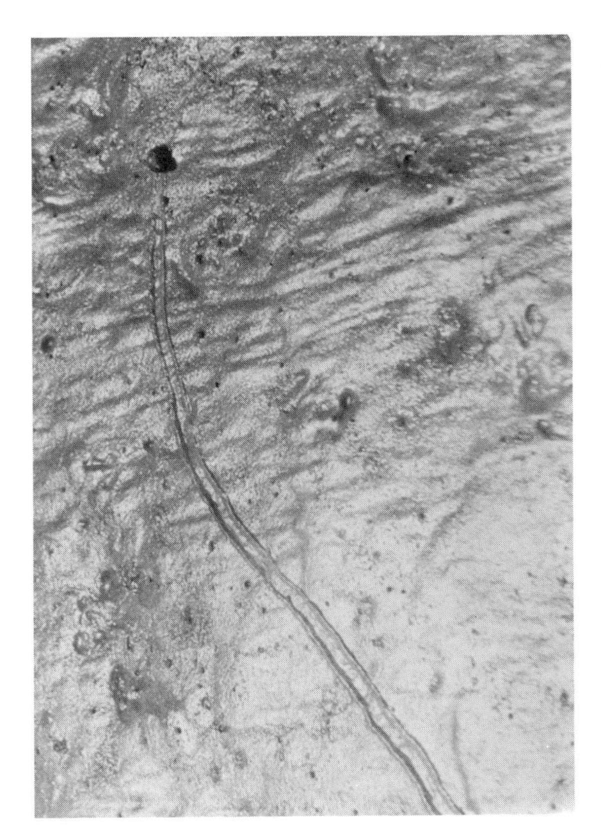

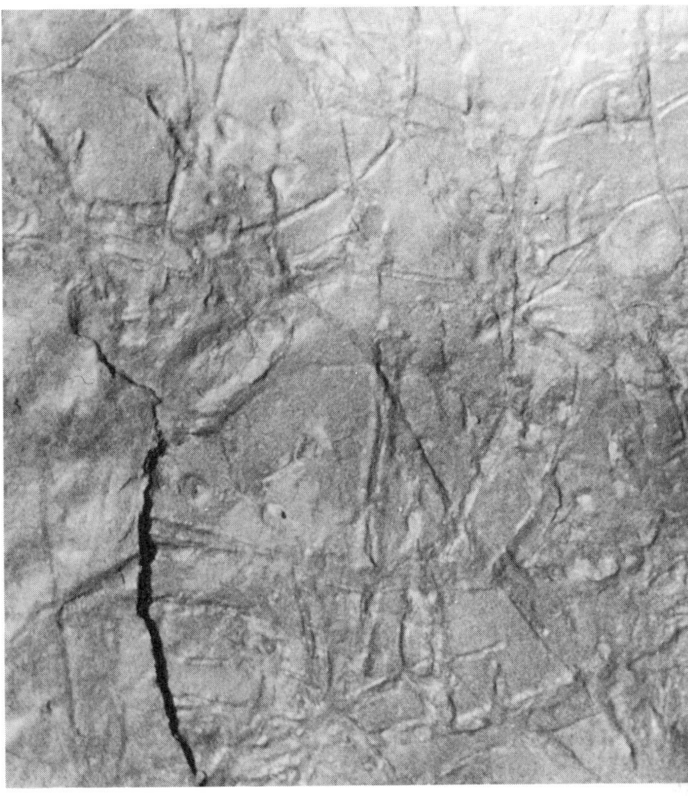

Trace fossils are indirect evidence of the past activities of animals. Compare the tracks of a living snail *(left)* with the fossil snail tracks on a Cambrian rock *(right)*. (M. Fenton)

fungi. **Fossils formed on land** are relatively rare, for even when the bodies of land animals are covered by wind-blown soil or sand, or sink into muddy bottoms of shallow ponds, the potential fossils are usually removed later by erosion. Exceptional conditions, such as prevail when animals are trapped under thick layers of volcanic ash or under muds formed during floods, provide most of the fossils that contribute to our sketchy picture of ancient terrestrial life. **Fossils of marine invertebrates,** however, are abundant because constantly shifting marine sediments quickly bury the dead bodies of many of these animals. Later accumulations of sediments further bury the fossils, sometimes many meters deep, where they can remain safely for many millions of years. When geologic processes eventually lift the marine sediments above sea level and they are cut by stream erosion or by faults, a legible record is presented, with fossils of the earliest animals preserved in the lowermost layers and those of the most recent forms in the uppermost layers. In most places we can examine only those fossil-bearing rocks that lie close to the surface, and have no access to the many fossils locked in the underlying rocks. Fortunately, however, uplifting and further erosion expose many older rocks. For example, the walls of the Grand Canyon, where the Colorado River has cut a cross-section a mile deep through the earth's crust, provide a record hundreds of millions of years old.

The systematic study of fossils in successive layers of rock reveals not only that animals have been present on earth for at least 600 million years, but that *the deeper we go into the rocks the less and less familiar are the fossils that we find.* Fossils a mere million years old are of animals much like living forms. Those several million years old are more different, and those still older must be assigned to orders, and even whole classes and phyla, of animals that no longer exist. In other words, the fossil record furnishes direct evidence that animals were not always as they are today. Modern forms, most of which do not occur as fossils, therefore must be descended from the earlier animals like those whose remains we find in the rocks. In some cases the record is complete enough to trace the changes, or *evolution,* of forms from one layer of rock to the next.

Strangely enough, the science of ancient life, or **paleontology,** has been developed not by biologists, who have been busy enough studying living organisms, but by geologists, who have found that rocks of different ages are characterized by distinctive assemblages of plant and animal fossils. Certain of these fossils, which are worldwide in distribution, distinctive, and restricted to limited periods of geologic time, serve as *index fossils* by which rocks can be recognized and dated no matter where they occur.

Geologic time has been divided into a number of *eons* and *eras,* the end of each marking the time of some significant geologic event such as changes in sedimentary rocks, glaciation, continental breakup, and continental collisions. Less profound changes in the face of the earth, especially as indicated by changes in fossil remains, form the basis for dividing the later eras into a number of *periods,* and these are often further divided. The major time units, their estimated age and duration, together with some of the invertebrate groups characteristic of each, are summarized in the accompanying table.

The first eon, the **Hadean** ("hellish"), marks the origin of the earth and the differentiation of the core and mantle through extensive volcanism and extrusion of molten igneous rock; no life was present. This eon lasted three-quarters of a billion years; it ended when the surface of the earth had cooled enough for water to accumulate, which in turn allowed sediments to form. Extensive volcanism continued into the second eon, the **Archean** ("primitive"), when the first sedimentary rocks were formed.

Table of Geologic Time

Eons (Millions of years ago)	Eras (Millions of years ago)	Periods (Duration in millions of years)	Principal evolutionary events among invertebrates
Phanerozoic	Cenozoic 64	Quaternary 2	Modern species; insects and gastropods most numerous species on land and in sea.
		Tertiary 62	Recovery of most groups after Cretaceous extinction; most modern genera and families established.
	Mesozoic 225	Cretaceous 71	Continuous development, then gradual decline of many groups until abrupt extinction of remaining ammonoids and many species of other groups, possibly from results of impact of a giant meteorite.
		Jurassic 57	Ammonoids abundant; most modern orders well established and flourishing.
		Triassic 33	Slow and erratic recovery of groups surviving the Permian extinction.
	Paleozoic 650	Permian 55	Decline of most groups of animals with extinction of trilobites, possibly from major cooling period; dragonflies, beetles, and bugs present.
		Carboniferous 65	Crinoids and blastoids peak and begin slow decline; winged insects (mayflies, grasshoppers, cockroaches) appear.
		Devonian 50	Continued development of many forms on land and in the sea; brachiopods and eurypterids peak; arachnids and wingless insects present on land.
		Silurian 35	Extensive tabulate coral reefs; graptolites and trilobites begin to decline; millipedes appear on land.
		Ordovician 70	Most major classes of animals present. Trilobites, echinoderms, graptolites, and nautiloids near their peak.
		Cambrian 70	Animals with hard skeletons first appear. Most phyla of animals present; archeocyathids, trilobites, and brachiopods numerous.
		Ediacarian 80	First animal fossils, not clearly recognizable as members of modern phyla but resembling cnidarians, worms, and arthropods.
Paleophytic 2,000		1,350	No animal fossils but sea floor covered with algal mats (stromatolites); oxygen atmosphere begins to develop; earliest nucleated cells with meiotic sex.
Proterophytic 2,600		600	Development of bacterium-like forms, some filamentous and photosynthetic; first evidence of free oxygen.
Archean 3,750		1,150	Earliest known sedimentary rocks, some with microfossils of bacterium-like cells; probable time when life arose.
Hadean 4,500		750	Formation of earth; differentiation of earth's rocks; radiometric clocks set.

Chemical evolution led to the first forms of life during this time and some of the sedimentary rocks contain tiny *microfossils* of bacterium-like organisms, some of which accumulated in such quantities as to form laminated mounds called *stromatolites*. In the following eon, the **PROTEROPHYTIC** ("first plants"), diverse kinds of bacterium-like cells appeared. Bands of oxidized sediments that appear in the rocks at this time indicate that some of these cells have evolved the ability to produce oxygen through photosynthesis. By the next eon, the **PALEOPHYTIC** ("ancient plants"), thick mats of *photosynthetic cyanobacteria* (blue-green algas) covered the shallow sea floor and formed massive stromatolites. Larger algalike cells with distinct nuclei also appeared during this time. These almost certainly developed the ability to undergo meiosis and fertilization, or *sexual recombination*. The cyanobacteria and algas continued to produce oxygen, which accumulated in the water and atmosphere until there was enough free oxygen to support animal life.

The first animal fossils appeared about 650 million years ago in the **PHANEROZOIC** ("visible animals"). This eon is divided into three eras, each of which is divided into several periods. The first era, the **Paleozoic** ("ancient animals"), began with the *Ediacarian* period. Rocks of this period contain fossils of animals that appear similar to jellyfishes, sea pens, and various worms and arthropods, but cannot be assigned with certainty to any known phyla. By 570 million years ago, with the beginning of the *Cambrian* period, we find representatives of most phyla living today (except for some that we would not expect to fossilize). Why so

Cambrian scene, a reconstruction of a marine bottom of this period. Swimming and crawling everywhere are several sizes of trilobites, extinct arthropods. Also conspicuous in this scene are tubular sponges and branching gorgonians (cnidarians), but fossils show that most phyla of animals were already present. Based on Burgess Shale, British Columbia.

Above: **Silurian scene,** probably brackish or freshwater, with a large and several small eurypterids, extinct chelicerate arthropods.

Below: **Devonian scene** features several fishes, a large nautiloid (cephalopod mollusc) at the left, and a diverse marine community of invertebrates. At the lower right are siliceous sponges covered with protruberances; a photo of a fossil of one of these appears later in this chapter.

Carboniferous scene. Between two nautiloid cephalopods at the center and right are several brachiopods, one large spiny one and some smaller clamlike types. In the left foreground are large cylindrical corals and, in the background, tall lobed sponges. (All reconstructions on this and previous two pages by Carnegie Museum, Pittsburgh, Pennsylvania. Photos, R. Buchsbaum.)

many different groups of animals appeared so "suddenly" in the Cambrian period has been a great puzzle, still not completely explained, but their appearance seems to have been related to the slow accumulation of oxygen in this atmosphere, and the end of extensive glaciation. Once established, many groups diversified, and by the end of the *Ordovician,* the third period of the Paleozoic, many invertebrate groups were already at their peak of abundance, and vertebrates were on the scene. During the following 200 million years plants and animals moved onto land and fluctuated in diversity on land and in the sea. The close of the Paleozoic, at the end of the *Permian* period, was marked by the extinction of most of the species of most animal groups. Trilobites disappeared altogether. This time of mass extinction, the greatest that has ever occurred, seems to have been related to the collision and fusion of most of the continental land masses, decreasing areas of shallow shorelines and leading to worldwide cooling and glaciation over millions of years.

Many of the groups that survived the Permian extinction recovered during the following era, the **Mesozoic** ("middle animals"), and certain large, shelled cephalopods reached their climax in diversity and abundance during this time. Crabs, predaceous snails, bony fishes, and large predaceous swimming reptiles are diversified in the sea during this time, while insects and various reptiles, including dinosaurs, thrived on land. A gradual decline in diversity of species during the *Cretaceous* period of the Mesozoic ended when many species of many groups, including all species of ammonoid cephalopods (and dinosaurs), abruptly became extinct, perhaps when a large meteorite collided with the earth 64 million years ago. The survivors of the Cretaceous extinction recovered early in the next era, the **Cenozoic** ("new animals"), and developed into the modern groups living on earth today.

The above description of events that occurred over the past 4.5 billion years is only a general glimpse of what is known of our planet's history from the fossil record. Each period, or even parts of periods, can be described in much greater detail as different species appear and become extinct. Rather than focus on periods of time, however, we will examine the fossil history of the main invertebrate groups treated in this book, and then touch on some forms that apparently left no descendants.

One would not expect to find many fossils of **protozoans** because most are small and delicate. However, a number of kinds, including flagellate, ameboid, and ciliate protozoans, produce distinctive skeletons that are preserved in the rocks as microfossils. The best record of ancient protozoan life is provided by some of the ameboid protozoans, especially the *radiolarians* with their delicate siliceous skeletons, and the *foraminiferans* with their calcareous shells. Good fossils of both groups are first found in Cambrian rocks. Their abundance in number of species and individuals subsequently fluctuated over time. Because the minute calcareous shells of foraminiferans are easily preserved and have distinctive shapes and markings, and because different species have been abundant and widespread during discrete periods of time, they are extremely valuable as index fossils. Even small pieces of rock from different parts of the world can be compared to see if they were formed from rocks far below the surface in borings brought up by drills. For this reason, paleontologists specializing in foraminiferans often work for oil companies and direct well-drilling operations to rock layers with foraminiferan fossils known to be associated with oil.

Foraminiferans, *Nummulites,* shells of which are about 15 to 20 mm in diameter, flourished on the sea bottom in the early Tertiary and formed great beds of limestone, now exposed in the Alps and northern Africa. The Egyptian pyramids of Gizeh, near Cairo, are built of nummulitic limestone.
(R. Buchsbaum)

The fossil record of **sponges** is not an abundant one, but calcareous and siliceous spicules, impressions of soft spongin fibers, and even whole sponges similar to those living today have been found in Cambrian rocks. Glass sponges of the past, like contemporary ones, seem to have been associated with deep water. Unlike any modern sponges, some kinds of sponges were important reef builders in shallow Paleozoic and Mesozoic seas; they may have been replaced in recent seas, at least in part, by modern forms of corals.

Fossils nearly 700 million years old, found in Ediacarian rocks of Australia as well as in other parts of the world, provide some of the earliest known animal records. Among these fossils are some that appear to have been formed in the sand of shallow seas as impressions of animals that resembled **cnidarians**—jellyfishes and sea pens—but how they were related to later cnidarians, or if they were, remains obscure. Fossils of undisputed jellyfishes and other cnidarians that lack hard skeletons are rare in later rocks.

A variety of different kinds of **corals,** with massive calcareous skeletons, have left an undisputed fossil record extending back to the Ordovician. *Rugose corals* (horn corals) and *tabulate cor-*

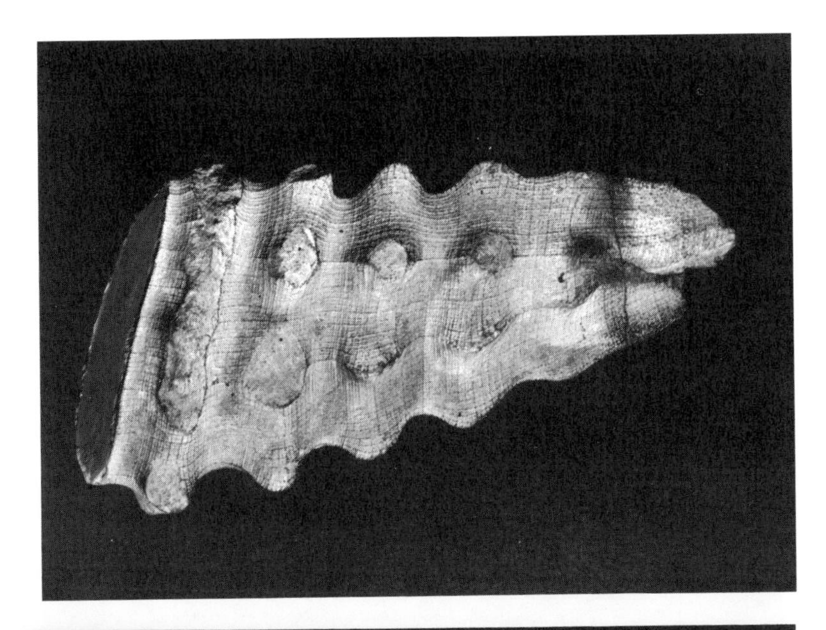

Siliceous sponge, *Hydnoceras,* lived on the bottom of a Devonian sea in what is now New York State. Length, about 15 cm. See reconstruction in Devonian scene earlier in this chapter. (R. Buchsbaum)

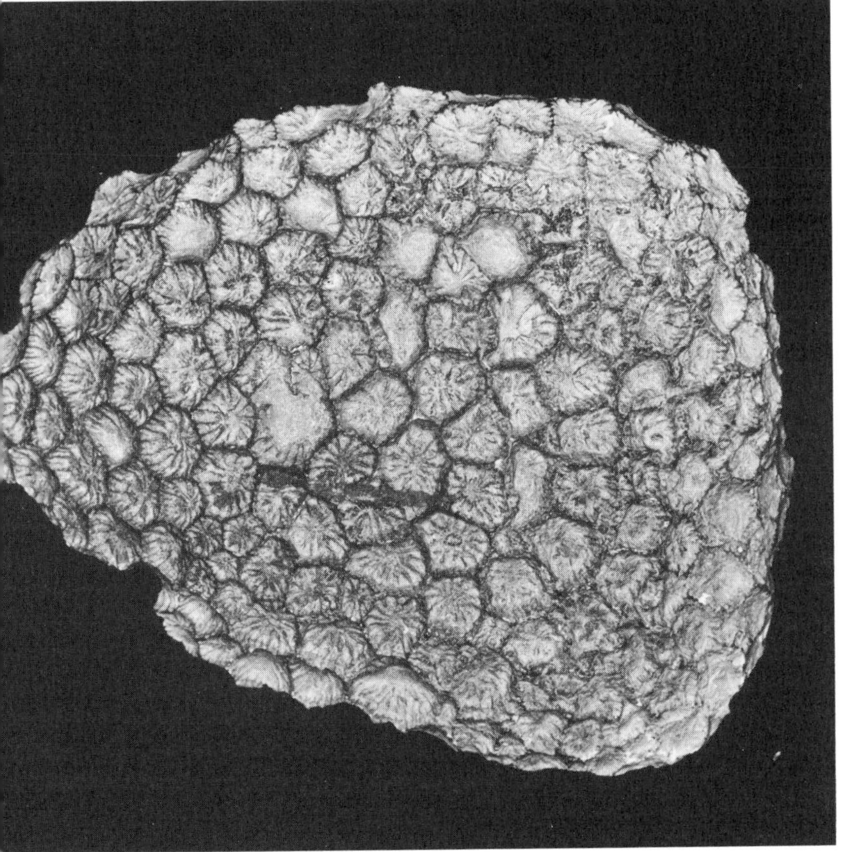

Paleozoic coral, *Columnaria alveolata,* from the Ordovician, this colony about 7 cm across. Called "tetracorals" because their skeletal partitions were in multiples of four, these cnidarians became extinct at the end of the Paleozoic and were replaced by modern stony corals with partitions in multiples of six. (R. Buchsbaum)

als, now all extinct, formed massive reefs in the Paleozoic. Some fossils of rugose corals show traces of growth lines similar to those seen in modern coral skeletons. Counts of the rings, like counts of tree rings, suggest that individual corals lived for well over 100 years. Moreover, the rings seem to have daily and monthly components, so day and month length can be estimated. In the Devonian period there were apparently 13 lunar months and about 400 days in each year, supporting conclusions from astronomical calculations that the earth rotated faster in these ancient times than it does now. The *stony corals* (scleractinian corals) that today form such extensive reefs in shallow, tropical waters did not appear until the early Mesozoic, when they replaced the earlier types of corals.

Most wormlike invertebrates have left a very poor fossil record, and this is largely a record of trails, tracks, tubes, and burrows; it is usually impossible to identify the animals responsible for such trace fossils. **Flatworms** have left no recognizable remains,

Ediacarian fossil, *Spriggina floundersi,* shown as an external mold (*left*) and a latex cast (*right*), is strikingly segmented and resembles a modern polychete, *Laetmonice producta,* shown in ventral view (*center*). However, the "head shield" at the top of the fossil is unlike the head of any polychete and is reminiscent of an arthropod head. Ediacarian fossils, dated at more than 600 million years, are among the oldest known metazoan fossils. The fossil shown here is about 4 cm long, a little less than half as long as the polychete. (M. F. Glaessner)

Brachiopod shells, when very numerous, form solid beds of limestone called "lamp-shell coquina." *Above:* A piece of Ordovician coquina containing shells of *Dalmanella.* (M. Fenton) *Below:* A common fossil type of winglike brachiopod shell, *Spirifer.* (R. Buchsbaum)

but a fossil **nemertean** has been found in rocks of Jurassic age in Germany. A fossil parasitic **roundworm** has been found in the fossilized leg of a Paleozoic scorpion, and others have been found in Cenozoic fossils of insects and vertebrates.

Other worms have very old fossil records. **Arrow worms, peanut worms,** and worms of some other small phyla have left fossils in Cambrian rocks. Fossils of several annelid-like animals occur in Ediacarian rocks, and by the Cambrian, several types of **polychete annelids,** with well-developed bristles, were present. The small, toothed jaws of polychetes have formed particularly fine fossils and are found in many rocks from the Ordovician on.

Bryozoans appear first in the Ordovician, and have been abundant from that time to the present; nearly 20,000 species are known, of which over 15,000 are extinct. Because the small individual cases of bryozoans are often decorated with pits and spines and leave distinctive fossils, many fossil bryozoans are used by paleontologists (as fossil foraminiferans are) for comparing rocks from different areas. Most of the Paleozoic groups of bryozoans nearly disappeared at the end of the Permian, and most modern bryozoans belong to a group that did not appear until the Jurassic.

We are accustomed to think of **brachiopods** as belonging to a small and unimportant phylum, which in this book has been described with the bryozoans as one of the "lesser lights." But when the animal kingdom is viewed in terms of the past as well as the present, brachiopods take a prominent place, and they have one of the most beautifully preserved of all fossil records. Over 30,000 species of brachiopods lived in the past; only about 260 species are alive today. Not only do brachiopods have hard shells that lend themselves readily to preservation, but the animals live in shallow seas where the chance of fossilization is greatest. Moreover, they often live in aggregations, and their remains can be so abundant as to form most of the rock in which they occur. The group was well established at the beginning of the Cambrian, and species of most orders were abundant in the Ordovician. One genus, *Lingula,* has been present on earth nearly unchanged since the Cambrian, surviving one period of extinction after another. Many other groups were not so fortunate; most species of most groups of brachiopods disappeared during the Permian extinction, and although a few remained abundant and important into the Mesozoic, the decline has generally continued to the present. The decline of brachiopods in the Mesozoic is associated with a dramatic rise of burrowing bivalves, which may have displaced (or simply replaced) the brachiopods that reigned supreme for so long.

Fossil scallop shell, still embedded in the rock at Scientists' Cliffs, Maryland. Bivalved molluscs and other organisms with calcareous shells or skeletons have naturally left a far more complete fossil record than have soft-bodied organisms. (R. Buchsbaum)

Like brachiopods, **molluscs** have external shells that provide for an excellent, unbroken fossil record from the Cambrian to the present. Most of the classes of molluscs living today, plus at least one extinct class, were already present in the Cambrian. *Gastropods* have increased in number of species and individuals from that time to the present, and are probably close to their peak of development now, comprising 90,000 of the 110,000 species of living molluscs. *Bivalves* also increased in diversity and abundance, especially in the Mesozoic and Cenozoic, after their burrowing foot and siphons were developed; they are now near their peak of development. In contrast, *cephalopods* peaked in the Mesozoic and are now a shadow of their past glory. Most modern cephalopods, such as octopuses and squids, have small internal shells or none at all. Similarly, the first cephalopods may not have been shelled; they are known only by fossilized teeth from the early Cambrian. But most known fossil cephalopods had strong external shells like those of the few relict species of *Nautilus* living today. The early *nautiloids* of the Cambrian had small, slightly curved, cap-shaped shells. By the Ordovician there were many sizes and shapes of nautiloid shells; they reached over 5 meters in length, and were straight, curved, or, as in the modern nautilus, tightly coiled. The various shapes probably reflect different habits, with the tightly coiled shell providing the best buoyancy control for active swimmers. Nautiloids peaked in the mid-Paleozoic and declined as they were replaced by another great group of shelled cephalopods, the *ammonoids.*

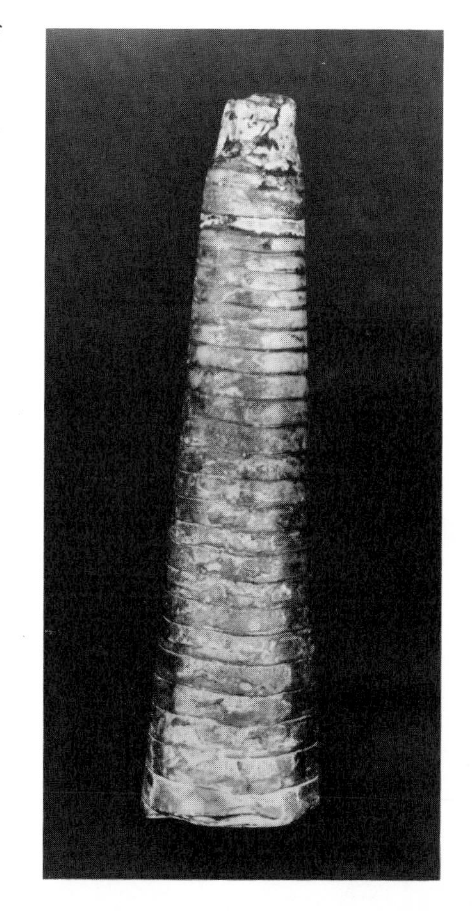

Shelled cephalopods. *Right:* The straight chambered shell of *Orthoceras*, shaped like a cone (about 18 cm long). Some shells of the straight type approached lengths of 6 m (20 ft). *Below, left:* Curved shell of *Jolietoceras* (about 18 cm long). *Below, right:* Coiled shell of an ammonite, *Coeloceras* (about 6 cm across). Certain coiled ammonites of the Cretaceous had shells about 2 meters (almost 7 ft) in diameter—the largest shelled invertebrates of all time. (R. Buchsbaum)

Ammonoids, with their thin coiled shells strengthened by elaborate suture lines between the chambers, were among the dominant animals of Mesozoic seas before declining and finally becoming extinct at the end of the Cretaceous. Their decline, at least in part, was probably related to the rise of shell-crushing predators, such as crabs, bony fishes, and swimming reptiles that could prey on ammonoids, themselves the previously unchallenged major marine predators. Like nautiloids, some ammonoids had partly uncoiled, straight, or even twisted shells; these probably were specialized to float passively or live a sedentary life on the sea floor. The third major group of extinct cephalopods, the *belemnites,* with an internal gas-filled chambered shell somewhat like that of modern cuttlefishes, were probably agile predators comparable to modern squids. They also peaked in the Mesozoic before becoming extinct in the mid-Cenozoic.

The chitinous and often calcareous exoskeletons of **arthropods** are excellent for leaving fossilized molds, carbon films, and impressions. The resulting fossil record attests to the importance and abundance of arthropods throughout the history of animal life. The earliest animal fossils in the Ediacarian rocks contain forms that may have been arthropods, and three of the four major subphyla of arthropods, as well as a marine onychophoran, were well established in the Cambrian.

The most numerous early arthropods were **trilobites,** which constitute over half of all known Cambrian fossils. The name trilobite means "three-lobed" and refers to the fact that the dorsal surface of the body is divided, by two longitudinal furrows, into three lobes. The body is also divided transversely into three regions: the head, the thorax, and the abdomen. The head bears a pair of compound eyes. The exquisite calcite lenses of the eyes of some trilobites are among the most optically perfect lenses found in any animal, living or extinct; it is unknown, however, how they worked, or what they saw of the ancient world. Also on the head were a pair of antennas and four pairs of similar, jointed, two-branched appendages. The outer branch, which is flattened and has a row of bristles along its posterior edge, probably served for respiration and swimming. The inner branch apparently was used for walking and gathering food, and inwardly directed teeth on the basal part of each limb could macerate food, as do some of the limbs of modern horseshoe-crabs, arachnids, and crustaceans. In contrast to the diversified appendages of nearly all other arthropods, similar appendages occur on all the segments of the trilobite body, all apparently used simultaneously for a variety of functions. Most trilobites are from 1 to 5 centimeters long, but some forms reached a length of over 50

Trilobites are extinct aquatic arthropods. These two beautifully preserved fossils of *Neolenus* from mid-Cambrian rocks clearly show the tri-lobed body. Notice the antennas on the one at the left. Burgess Shale, British Columbia. (C. E. Resser)

centimeters. They all lived in the sea, where they probably fed on various seaweeds, sedentary animals, and organic debris. Most of the early trilobites were extinct by the end of the Cambrian, but the survivors flourished and diversified in the Ordovician only to enter a long decline ending in complete extinction in the Permian. The decline may be related to the rise of cephalopods in the Ordovician, and fishes in the Devonian, both of which probably fed on trilobites.

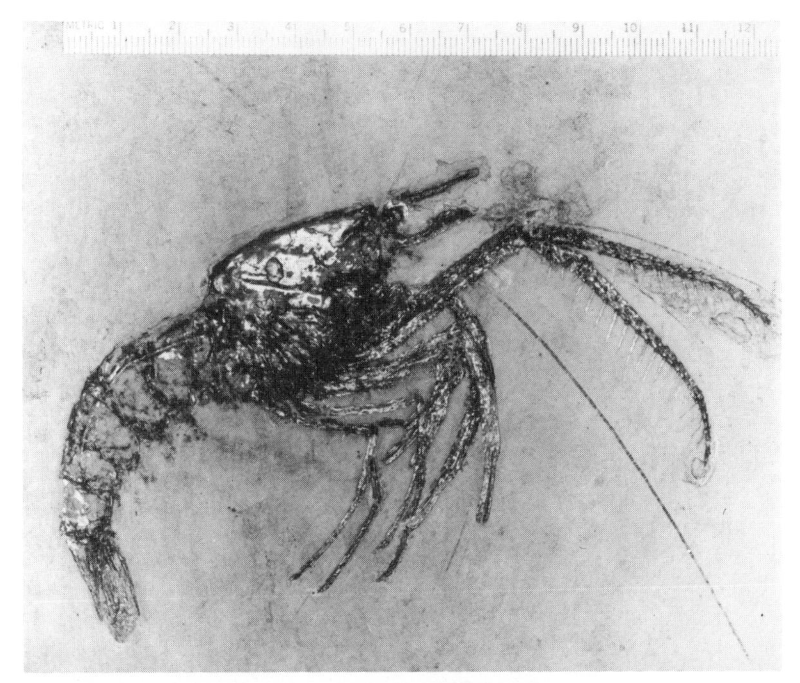

Fossil crustacean. This one, *Penaeus speciosus,* preserved in fine detail, can be determined to be quite modern in structure and has been assigned to a contemporary genus. W. Germany. (R. Buchsbaum)

Ancient merostome, member of a marine chelicerate group from which modern horseshoe-crabs, *Limulus polyphemus* and a few related species, are the only living descendants. This fossil is of *Prestwitchia* from the late Carboniferous. Length, about 5 cm, including tail spine. (R. Buchsbaum)

Eurypterid, an extinct aquatic chelicerate arthropod. *Eurypterus,* from the Silurian, shows some small appendages and two large ones. The long segmented abdomen has no appendages and ends in a spine. Length, about 15 cm. (R. Buchsbaum)

Besides trilobites, **crustaceans** inhabited Cambrian seas. Tiny *ostracods,* similar to those living today, are abundant in the fossil record from the Cambrian onward and, like foraminiferans, are valuable to paleontologists for comparing rocks from different areas. *Shrimplike animals* that presumably gave rise to most modern crustaceans were also present in the Cambrian, but the familiar lobsters and crabs, as well as copepods and most kinds of barnacles, did not appear until the Mesozoic.

Chelicerates also appeared in the Cambrian. Among these were **merostomes** like the modern horseshoe-crabs such as *Limulus,* which survive today as remnants of ancient seas. Other aquatic chelicerates were the spectacular **eurypterids,** some of which exceeded 2 meters in length, the largest size known for any arthropod. They probably swam near the bottoms of brackish estuaries and freshwater streams because their fossils are rarely found with those of known marine animals. Eurypterids appeared in the Ordovician, flourished into the Devonian, and then gradually declined to final extinction in the Permian. The other main chelicerate group, the **arachnids,** first appeared as apparently aquatic *scorpions* in the Silurian, while fossils of terrestrial arachnids, including *spiders* and *mites* similar to those of today, are found in later Paleozoic rocks.

In spite of the slim chances of fossilization on land, there is a fair record of most groups of **myriapods and insects.** *Millipedes* were present in the Silurian, presumably feeding on the first land plants; *centipedes* and *springtails* were present in the Devonian; *grasshoppers* and *cockroaches* were part of the Carboniferous scene; while *dragonflies, bugs,* and *beetles* appeared by the Permian. Fossils of other familiar groups of insects, such as *earwigs, termites, flies, bees,* and *ants* have been found in Mesozoic rocks, but there is no record of *butterflies* until the Cenozoic, after flowering plants developed.

Largest insect that ever lived had a wingspread of about 70 cm. It belonged to a group (extinct since the end of the Triassic period) that probably gave rise to modern dragonflies. This reconstruction is of a specimen found in Permian rocks in Kansas. (Field Museum of Natural History)

Fossil dragonfly, *Libellulium longialatum* of the Jurassic, is an exceptional specimen showing the details of the fine veins of the wings. West Germany. (R. Buchsbaum)

The characteristic calcareous ossicles, plates, and spines of **echinoderms** have provided a rich and complex fossil record. More than 20 classes of echinoderms were present in Paleozoic seas; only 6 classes survive today. And many of the extinct classes have little in common with the familiar living ones. Among the earliest echinoderms were small globular forms, called *helicoplacoids,* which were covered with small ossicles arranged spirally around the body; they probably burrowed in soft sediments. Other bizarrely shaped asymmetrical forms, such as the *homalozoans,* apparently lay on top of soft sediments and held their tube feet up into the water to collect food particles, in some cases holding the tube feet aloft on a single arm. *Edrioasteroids* had five rows of tube feet radiating away from the mouth on the upper surface of cylindrical bodies; the lower surface attached to rocks and the bodies of other animals, such as trilobites. *Eocrinoids* were flattened to globular animals that attached to the sea floor by a long stalk and extended their feeding tube feet upwards along variable numbers of long rows of ossicles, called brachioles. In the Ordovician, most other classes of echinoderms, including those living today, appeared. Most were sessile forms that attached to the bottom by long stalks; in *cystoids* and *blas-*

Helicoplacoid, a member of an early class of echinoderms, now extinct and quite unlike any modern form. *Helicoplacus,* about 17 mm across. (R. Buchsbaum)

Cystoid, *Pleurocystis,* from the Ordovician. Note the broken bases of two brachioles and the long stalk by which this sessile echinoderm attached, as some modern crinoids do. Cystoids became extinct at the end of the Paleozoic. Body about 3 cm long. (R. Buchsbaum)

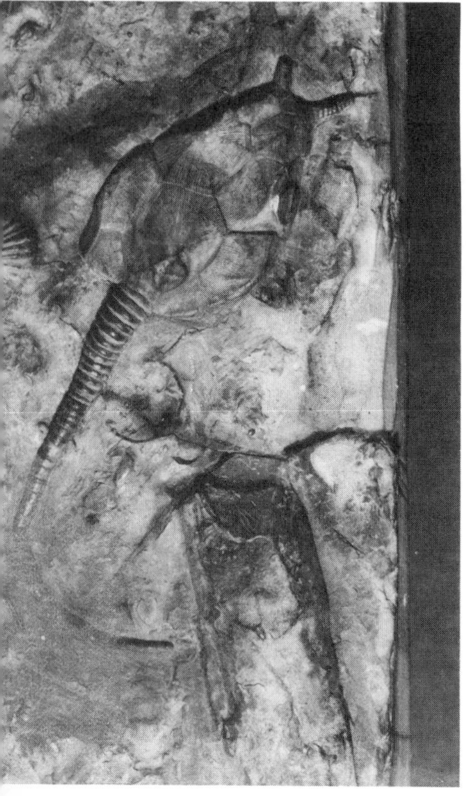

Edrioasteroids, small five-rayed echinoderms of an extinct class, here attached to a trilobite. Such fossils that reveal hints about the interactions of different kinds of animals are particularly intriguing. (R. Buchsbaum)

toids the tube feet were on five sets of brachioles, as in the earlier eocrinoids; *crinoids* bore their tube feet on five sets of branching arms, as they do today. These sessile echinoderms were dominant members of shallow Paleozoic seas, and they must have formed spectacular "forests" gently swaying in the ancient currents. They peaked in abundance and diversity in the Carboniferous, then entered a long decline that ended with the extinction of most echinoderms in the Permian.

Only one group of crinoids survived the Permian extinction, and these are present today as sea lilies and feather stars. The other main classes of living echinoderms—*sea stars*, *brittle stars*, *sea urchins*, and *sea cucumbers*—left a poor fossil record. They were all present in the Ordovician, however, and they expanded in diversity and abundance after the Permian extinction so that they are dominant members of our seas today.

The fossil record of the echinoderms reveals many kinds of animals completely unknown today; it is only by their characteristic calcite skeletons that they can be recognized as echinoderms. Fossils of some other extinct animals are so unlike any living today that they have been considered to represent entirely extinct phyla. **Archeocyathids,** for example, left numerous small cup-shaped fossils in Cambrian rocks; sometimes they were so

Sea star, *Petraster,* from the Ordovician. Length of each arm, about 4 cm. (R. Buchsbaum)

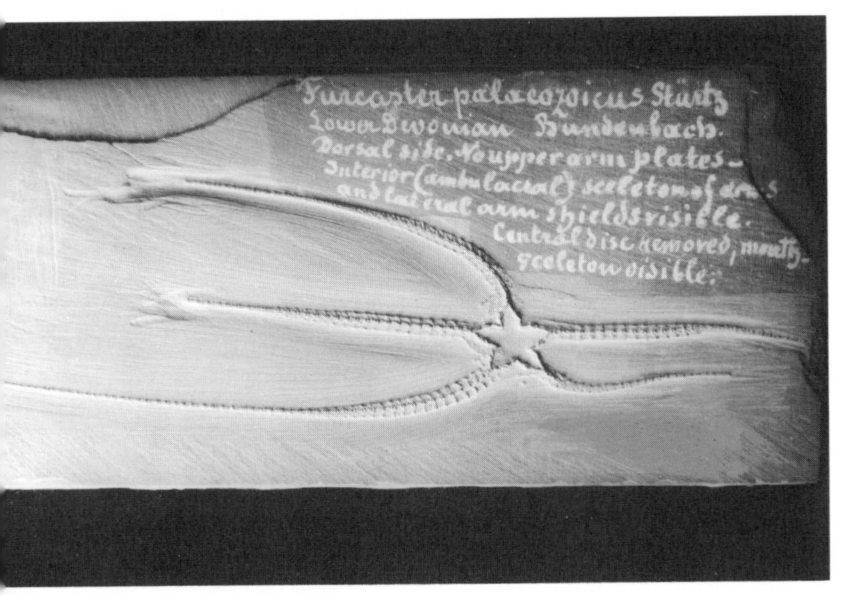

Brittle star, *Furcaster,* from the Devonian. Longest arm, about 9 cm. (R. Buchsbaum)

Sea urchin, *Cidaris,* a test without spines, from the Jurassic. About 5 cm across. (R. Buchsbaum)

abundant that they formed extensive reefs, but none apparently survived after the Cambrian. **Graptolites** survived longer and left an extensive fossil record through the first half of the Paleozoic. They apparently were colonial animals, similar to colonial hydroids, bryozoans, and hemichordates living today, and formed long rows of tiny toothed cups, each probably holding a feeding subindividual. Many were attached to small floatlike structures and are presumed to have drifted on the surface of Paleozoic seas, perhaps as siphonophores do today. There are many other fossils, sometimes abundant, but usually rare, of animals unknown today. Among the most spectacular are those of *Tullimonstrum* ("Tully monsters," named after Francis Tully, an amateur fossil collector, who discovered them in 1958), found by the thousands in iron concretions exposed after areas near Chicago were strip-mined for coal. These fossils are impressions of delicate wormlike animals, about 10 centimeters long, with a segmented body ending in a paddlelike tail. A transverse bar, ap-

Graptolites are carbonized fossils left by extinct colonial forms. *Diplograptus,* from the Ordovician, has stems, each with a double row of tiny cups presumably occupied by the individual members of the colony. About 10 cm across. (R. Buchsbaum)

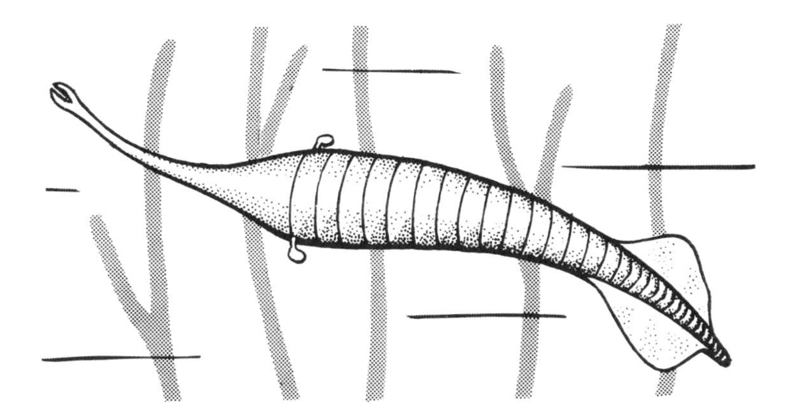

Tully monster, as it might have appeared in life, a reconstruction from many fossils. (Modified after E. S. Richardson Jr.)

parently bearing an eye at each end, crossed the front half of the body, and there was a long flexible anterior proboscis ending in a formidable claw equipped with sharp teeth. These animals were presumably open-ocean pelagic predators, like arrow worms of today, and were numerous in Carboniferous seas, but rarely came into nearshore environments where their bodies might be fossilized. This lucky fossil find makes one wonder how many other creatures thrived in earlier times without leaving any fossil evidence that they ever lived.

Chapter 24

Invertebrate Relationships

Opposite: **A phylogenetic tree** showing the authors' current view of how the major phyla of animals are related to each other. The two main branches—the annelid-arthropod and echinoderm-chordate lines—could also be viewed as separate trees with independent origins.

Everyone enjoys the unraveling of a good mystery, but no one likes to read on from clue to clue, until the earliest and most important events seem about to be disclosed, only to find that the rest of the pages in the book are missing. Just this kind of exasperating situation confronts us when we try to relate different phyla of animals to one another in an orderly scheme. Anyone can see that honey bees are much like bumble bees, that bees resemble flies more than they do spiders, and that spiders are more like lobsters than like clams. But when we attempt to relate groups, especially phyla, which, by definition, are groups of animals with fundamentally different body plans, there is little we can say with certainty. The different groups of arthropods are clearly allied to each other as well as to annelids; but how arthropods are related to each other, or to such utterly different animals as sea stars or vertebrates, remains quite a mystery.

The **fossil record,** which might be likened to our mystery book, provides many examples of species that are clearly descended one from another, but it is of practically no use in relating the phyla to each other. For, as we "turn the pages," digging deeper and deeper in the rocks and expecting to find intermediate forms linking different phyla, we instead continue to find fossils of animals that are readily identifiable as members of phyla living today—but few intermediates. In the earliest rocks for which we have good animal fossils, those of the Cambrian period, all the important animal phyla are already represented. Still older rocks contain very few animal fossils, and most of those are difficult to assign even to the phyla represented in the Cambrian. Thus, while the fossil record tells us a great deal about what the early forms of most phyla looked like, and the order in which species, or even genera, families, and orders appeared, it has little to say about whether the different phyla are related to each other, and if they are, the sequence in which they appeared. Despite this, the situation is by no means hopeless. Good detectives have been known to reconstruct, in detail, the events leading up to a crime to which there were no witnesses. And biologists have been able to find some definite clues to events that happened considerably more than 500 million years ago.

The most important kind of evidence is that based on **comparative morphology,** the study of the structures and body plans of the various groups living today. Similar structures that are judged to have formed from a common ancestral structure are

Analogous structures, the result of convergent evolution, are the eyes, fins, and streamlined shape of squids and fishes. (R. Buchsbaum)

said to be *homologous*. The greater the number of homologous structures that are shared by two groups, the closer their relationship is assumed to be. Sometimes one or both homologous structures of two related groups have diverged from their original function and have changed many of their attributes, for example, the foot of a snail compared with that of a clam or a squid. In other cases, structures in two unrelated groups are quite similar because they have converged to meet the demands of similar functions, for example, the eye of a squid compared with that of a fish; such structures are said to be *analogous*. The challenge to the biologist, of course, is to distinguish homologous structures, which contain information about ancestral relationships, from analogous structures, which contain information about similar function.

The adult structures of some related groups are so highly modified in adaptation to their different ways of life that homologous structures are difficult to recognize. This is the case, for example, with barnacles and crabs, or sea squirts and fishes. Yet the early embryonic stages in such cases can be almost identical, indicating that the groups are related. Early stages of development tend to be more conservative than later ones, and often retain evidence of similarities that are later lost. Thus the study of the **comparative embryology** of animals has often revealed relationships that were otherwise not suspected. As with evidence from comparative anatomy, however, that from comparative embryology must be interpreted with care. There may be limits to the kinds of configurations that embryos can adopt under similar conditions. Young embryos are mostly spherical, and larvas that float at the surface in marine waters tend to retain this shape for some time; they swim and feed with bands of beating cilia on their outer surfaces. It may take careful study of such relatively simple forms to tell which ones are closely related and which are not. Conversely, some species that have almost identical structure as adults pass through markedly different sequences during their development. For example, many marine snails have larvas that swim and feed in the sea before settling on rocks and metamorphosing into juvenile snails, while species that are morphologically nearly identical have no such larvas, and their juveniles simply hatch out of egg capsules and crawl away.

Comparisons of other features among animal groups also may reveal relationships. Using **comparative biochemistry and molecular biology,** relationships can be inferred on the basis of similarity or lack thereof between complex molecules or metabolic pathways. However, distinguishing between homologous and analogous biochemical characters can be even trickier than distinguishing between structures or developmental pathways.

Many biochemical characters seem easily changeable with relatively few mutations, and therefore are more likely to converge or diverge, depending on the level of natural selection. Nevertheless, detailed similarities between homologous proteins, or sequences of nucleotides in nucleic acids, may be particularly valuable for indicating relationships, especially when comparing several species that are already known to be related. On the other hand, when comparing phyla that are only distantly related, if at all, biochemical or molecular comparisons may be particularly uncertain.

Careful studies of comparative anatomy, embryology, and molecular biology, as well as consideration of the fossil record, have enabled biologists to construct **phylogenetic trees** that show how different groups are thought to be related, and in particular, which might have evolved from which. Considering the inconclusive nature of much of the evidence, it is clear that any particular "tree" must be considered only as a highly speculative hypothesis. However, as such, trees can be modified and improved as more information is accumulated and analyzed so that they come to reflect a more and more accurate hypothesis of events that happened long ago.

One version of a "tree" summarizing much of the information currently available about relationships between the major phyla of animals, as seen by the authors of this book, is presented at the beginning of this chapter; it undoubtedly will be modified as we acquire additional information. Nevertheless, it can serve as a framework for what seems to us to be a plausible account of the evolution of the main phyla of animals.

It is highly probable that multicellular animals arose from one or more types of single-celled ancestral forms, similar to some of the modern protozoans. Exactly how this happened, of course, we cannot now determine, but some pretty good guesses can be made. In many protozoans and single-celled algas, individual cells remain together after division to form colonies of cells. In some, such as colonial collar flagellates, all the cells are similar and relatively independent of each other, but in others, such as colonies of *Volvox*, there is some differentiation of cells for different functions; in particular, only some cells become gametes. Using such colonies as models, we can propose that further differentiation of similar colonies in the past led to the formation of simple multicellular animals. For example, if in addition to gametes, other cells became specialized to form protective epithelia, feeding surfaces, and skeletal elements, the colony would be considered to be a simple multicellular animal.

In many ways, *sponges* seem closest today to the kind of animals just proposed. They have a variety of different types of cells, but these are hardly developed into tissues, and they could be considered as elaborate colonies with a **cellular level of organization.** Their unique body plan (without mouth or digestive tract, but with feeding collar cells similar to collar flagellates) sets them apart from all other phyla and suggests that they evolved independently from some type of colonial protozoan. And although sponges are probably not ancestral to any other phyla, they provide a model of how colonial protozoans may have evolved into multicellular animals.

Most other multicellular animals have a well-defined digestive cavity with a mouth to allow for the entry of food. Such an organization includes at least an outer epithelial layer covering the animal and an inner epithelial layer forming the gut. Animals so organized may have evolved from colonial protozoans shaped like balls of cells, such as *Volvox.* If a portion of the colony became specialized for ciliar swimming and protection, while the opposite side became specialized for feeding, there would be the beginning of differentiation into a multicellular organism. Most animals today achieve a two-layered organization during their embryonic development by first forming a single-layered ball of cells, the blastula, then proceeding by various means to form a two-layered gastrula. Because this course of development is so widespread among different phyla, it is often considered to be a fundamental feature reflecting the original course of early animal evolution. On the other hand, the gastrula stage is a necessary intermediate in the development of a gut. Moreover, gastrulas develop in many different ways. Therefore, their widespread occurrence does not necessarily mean that most or all animal phyla with a gut evolved from a single gastrula-like ancestor, and gastrulation may be a convergent process.

Among the phyla living today, *cnidarians* are the most prominent group of essentially two-layered animals constructed at a **tissue level of organization.** There is an outer ciliated epithelium specialized for locomotion, protection, and sensation, and an inner epithelium specialized for digestion. There are relatively few cells within the mesoglea between the outer and inner epithelia. Cnidarians are radially symmetrical and use unique stinging cells to capture prey. They cannot be readily related to any other phylum. Perhaps, like sponges, they arose independently of the other phyla, or they may be viewed as an early offshoot of some of the earliest multicellular animals.

Except for sponges, cnidarians, and a few other small phyla, most animals have **bilateral symmetry.** Even those that appear radial, such as sea stars, are clearly derived from bilateral ancestors. Moreover, many of the earliest fossils are bilaterally symmetrical, so it is not clear whether these animals were derived from radial ancestral forms resembling some sort of cnidarian, or from ancestral forms that were originally bilateral. Planulas of cnidarians are similar in form and behavior to some small flatworms, swimming with one end forward, and some biologists have proposed that planulas may represent the bridge between radial cnidarians and bilateral animals like flatworms. On the other hand, some protozoans, including multinucleate flagellates and ciliates, are wormlike and swim with one end forward. Ancestral forms of such protozoans could have given rise directly to bilateral animals.

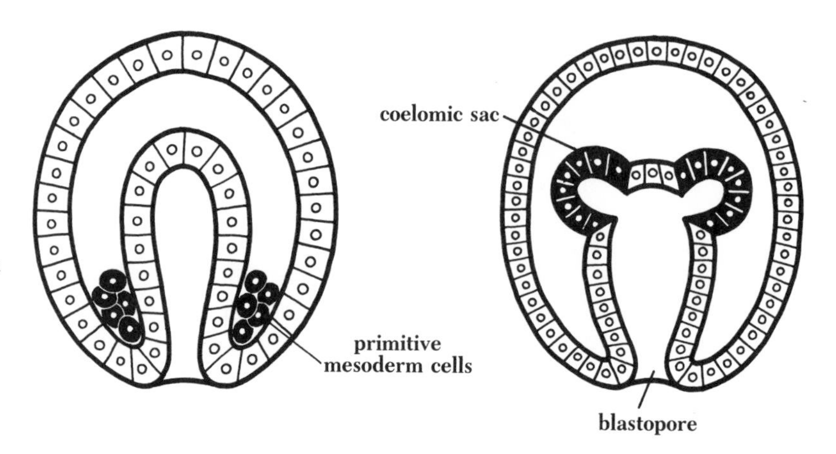

Origin of the mesoderm in annelid-arthropod and echinoderm-chordate lines. *Left:* A mollusc gastrula showing the primitive mesoderm cells that were set aside early in development. *Right:* An echinoderm gastrula showing the mesoderm being budded off as pouches from the sides of the primitive endoderm.

Most animals other than cnidarians have well-developed **organ systems** derived at least in part from the **mesoderm** located between the outer epithelium (ectoderm) and the inner epithelium (endoderm). Mesoderm generally forms during embryonic development in one of two main ways. In flatworms, nemerteans, molluscs, annelids, and arthropods it usually originates from a small number of cells, the "primitive mesoderm cells," which are set aside early in development even before the blastula forms. On the other hand, in echinoderms, hemichordates, and chordates, the mesoderm forms much later as outpocketings of the endoderm in the gastrula. There are other developmental differences between these two groups of animals. In most members of the first group, early cleavage has a distinct spiral pattern; the mouth forms from the original opening present in the gastrula; and coelomic cavities arise from splits in the mesoderm. In most members of the second group, early cleavage follows a radial pattern; the mouth never forms from the original opening in the gastrula but always at a later stage from a new opening; and coelomic cavities are derived from the mesodermal outpockets of the gut. For these reasons and others, two major lines of animal

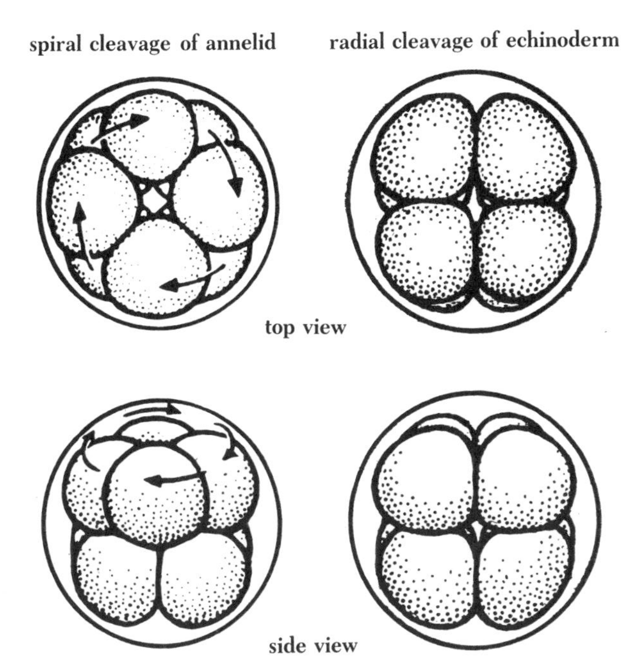

Cleavage patterns of early embryos with small eggs are typically either radial or spiral. *Left:* **Spiral cleavage** in an annelid with the upper tier of cells twisted 45° out of line with the lower tier. *Right:* **Radial cleavage** in an echinoderm with one tier of cells directly on top of the other.

evolution are recognized: the **annelid-arthropod line** and the **echinoderm-chordate line.** These are indicated in the tree at the beginning of this chapter as two major branches. However, it is not known whether they formed independently from separate protozoan ancestors or diverged shortly after the ancestral animals appeared.

Within the annelid-arthropod line, the *flatworms* appear relatively simple and are considered most similar to the ancestral forms. Flatworms make substantial use of the mesoderm for organ systems, but do not have internal body cavities apart from the digestive cavity. Nor do they have an anus. *Nemerteans* are similar to flatworms but have a digestive tract with both mouth and anus as well as a simple circulatory system. A much more complex organization is found in *molluscs* with their well-developed and integrated organ systems. Although the adult body plans of molluscs and *annelids* are very different (annelids are segmented while molluscs are not), they are unmistakably related as shown by their embryology. The early embryos of many marine molluscs and annelids are nearly identical, cell for cell, and they hatch as free-swimming **trochophore larvas.** A trochophore larva, with a characteristic double ciliated band around the equator for feeding and swimming and an apical sensory tuft of flagella, is found not only in many molluscs and annelids but also in members of some other phyla, and thus serves to link together many of the groups in the annelid-arthropod line.

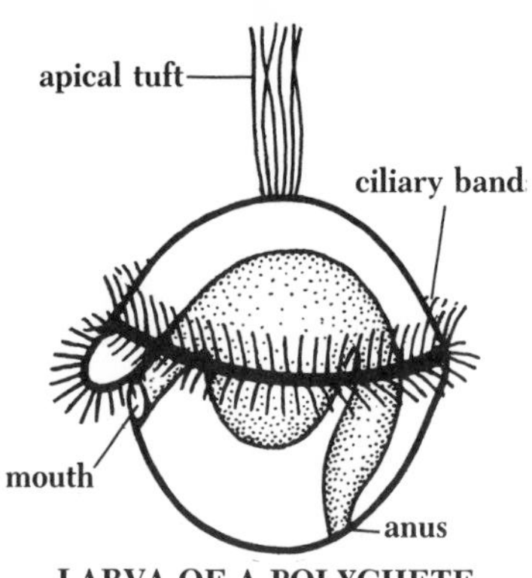

LARVA OF A POLYCHETE

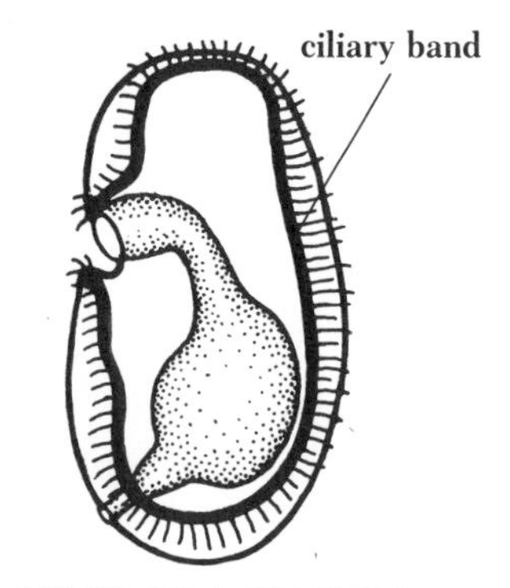

LARVA OF A SEA STAR

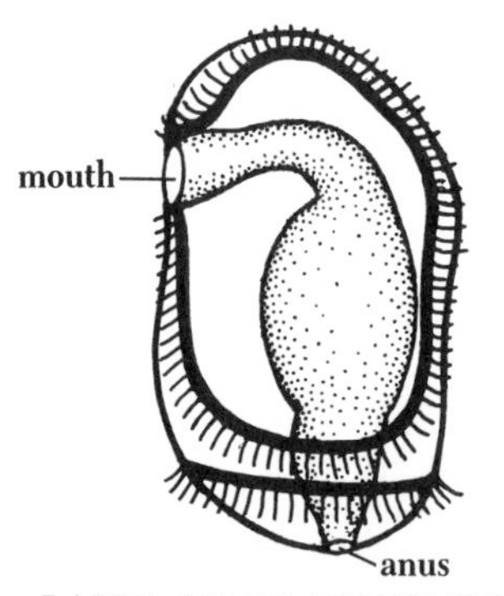

LARVA OF AN ACORN WORM

Early larval stages may reveal relationships between groups in which the adults appear unrelated. Trochophores of polychete annelids and molluscs are remarkably similar but strikingly different from the auricularias of echinoderms and the tornarias of acorn worms.

None of the *arthropods* have trochophore larvas, but early development in many is similar to that in annelids that also lack trochophore larvas, such as earthworms. Moreover, the adult body plans of the various arthropod groups are so similar in so many respects to that of annelids that there is little doubt that they all have a common segmented ancestor. Thus annelids and arthropods are linked mainly through their similar adult body plans, while annelids are linked to molluscs and other groups in the lineage mainly through their similar early development and larval types.

At first glance, *echinoderms* such as sea stars and sea urchins would appear to have very little in common with *chordates* such as fishes and ourselves. Linkage is made primarily through another group of animals in the echinoderm-chordate line: *hemichordates*, or acorn worms. The early development and larvas of echinoderms and hemichordates are strikingly similar, so much so that when hemichordate larvas were first collected in the plankton they were described as echinoderm larvas. Because these larvas all live in the surface waters of the ocean, feeding on microscopic organisms, their similarities could be due to convergent evolution. On the other hand, the larvas of molluscs and annelids also live in the same waters and feed on the same microscopic organisms; yet they are not at all like echinoderm or hemichordate larvas. Thus the striking developmental and larval similarities in echinoderms and hemichordates, like those in molluscs and annelids, seem best explained as due to common ancestry. And, like molluscs and annelids, echinoderms and hemichordates are linked mainly by their embryological and larval similarities.

No chordate has a larva like the planktonic larvas of echinoderms and hemichordates, but many aspects of early development, including cleavage pattern and formation of the mouth, mesoderm, and coelom, are similar in all three groups, and different from those of the annelid-arthropod line. Moreover, many adult features of acorn worms are similar to those of chordates. In particular, the pharyngeal gill slits in acorn worms are nearly identical to those in lancelets and are clearly homologous to those in tunicates and aquatic vertebrates (and even to the gill pouches formed during embryonic development of terrestrial vertebrates such as ourselves). Other adult features in hemichordates, including a dorsal tubular nerve cord, further link hemichordates with chordates, and like the link between annelids and arthropods, that between hemichordates and chordates is based mainly on adult characters. The connection between echinoderms and chordates, like that between molluscs and arthropods, therefore appears rather circuitous: echinoderms and chordates

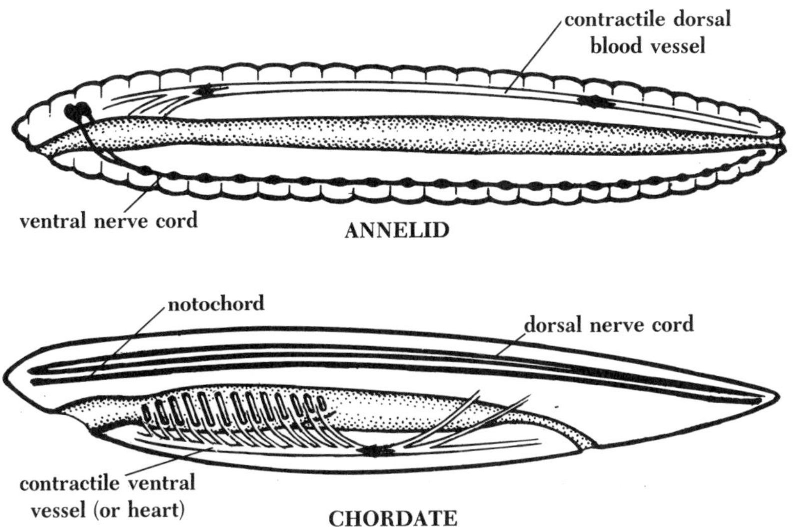

Broad differences in body plan, as shown diagrammatically here, characterize members of the annelid-arthropod and the echinoderm-chordate lines. In an annelid the nerve cord is ventral, a contractile dorsal vessel drives blood anteriorly, and the anus usually opens at the posterior tip of the body. In a chordate the nerve cord is dorsal, a contractile ventral vessel or heart drives blood anteriorly, and a tail typically extends beyond the anal opening. Although both have segmental characters, these differ between the two lines both in their manner of development and in final form.

are linked mainly by their common ties to hemichordates, while molluscs and arthropods are linked mainly because of their common ties to annelids.

In some of the earlier schemes of animal evolution proposed in the nineteenth century, animals were pictured as ascending, on a vertical ladder, directly from amebas to humans, with the other animals placed on intermediate rungs according to the level of complexity they were judged to display. However, as more was learned about how groups differ and are related, it was recognized that they do not form a continuous series from simplest to most complex. All groups have their own long evolutionary histories; the fossil record shows that nearly all have been on earth for over 500 million years. A branching "tree," like that shown at the beginning of this chapter, or perhaps a "forest" of separate trees, not only fits the facts more closely than a ladder, but with the many branches and twigs, reflects the exquisite and endless variety found among animals without backbones.

Chapter 25

Further Knowledge

This chapter is for the inquiring student who wants to know about the sources of the materials presented in earlier chapters as well as how to gain access to the large and ever-growing volume of knowledge about invertebrates.

How new facts are discovered, who performs the observations and experiments, where such work is done, and how it is made available to everyone interested in learning about it are questions that cannot be completely answered within the scope of an introductory book. Here it is possible only to allude briefly to these questions and to erect a few signposts that indicate the general direction in which students may set out to learn what they wish to know.

Opposite: **Knowledge about invertebrates** usually begins with reading what is already known about some question in invertebrate biology. Then, one might write to or talk with other biologists to learn more about where the animals occur and about methods for working with them. The next step is usually to go to wherever the animals live and find them. In these photos from Puget Sound, Washington, a dredge that has been dragged along the bottom is brought up and emptied on deck to reveal the marine animals it has collected. Selected animals will be taken to the laboratory for study and the published results will add further knowledge about invertebrates. (R. Buchsbaum)

Most of the facts, principles, and problems that have been presented in this book are based on the work of scientists who may be broadly called **zoologists**—biologists whose research involves protozoans or animals. But each modern zoologist is a specialist who works within a limited field and is usually labeled according to the nature of the problems that he or she studies. **Ecologists** are concerned with the relationships between organisms and their environment and must necessarily consider all the different species in a given habitat, plants as well as animals. Some are *population or community ecologists,* who examine the interactions of organisms or the effects of environmental changes on groups of organisms. Others are *physiological ecologists,* who focus primarily on how the physical or chemical environment influences the functioning of individual organisms. The latter can also be considered as **physiologists,** zoologists who are concerned with how animals carry out their life activities such as digestion, respiration, excretion, reproduction, and so on. Physiologists must invariably specialize in one or more limited areas, but often study an organism at many different levels simultaneously; for example, a *neurophysiologist* might examine biochemical and electrical changes in individual nerve cells and attempt to correlate these with the behavior of the whole animal. Or a **behavioral biologist** might attempt to relate an animal's behavior to its ecology or to the changing problems it faces in the course of its development. **Developmental biologists** study the regulation of growth and morphogenesis ("taking form") in embryos and in regenerating animals. **Cell and molecular biologists** are concerned with elucidating structure and function at the cellular, subcellular, or molecular level. **Geneticists** study the mechanisms of heredity, and most modern geneticists are allied to one of the fields already mentioned, identifying themselves as *population geneticists, developmental geneticists*, or *molecular geneticists*.

Another way of classifying biologists, besides those mentioned in the text, is by the *habitats* in which they work. **Marine biologists** are seen here working on a rocky intertidal shelf of the central California coast, using a one-meter-square grid to make a quantitative record of the animals and plants present. Marine biologists study shallow subtidal or open-ocean habitats by scuba diving. They collect animals from deeper waters by trawling with nets, and from deep bottoms by dredging (as seen in the chapter heading) or by setting traps. Some marine biologists descend to great depths to observe and collect from pressurized submersibles. (T. O'Leary)

Between high and low tidemarks is the intertidal zone, where marine invertebrates are accessible to any marine biologist who is willing to dodge the waves when necessary and to get up before dawn, or work until dark, if that is when the low tide happens to occur. A typical tidal range is a couple of meters or so, but in a few places the water level rises and falls as much as 18 m (about 60 ft) between tides. These photos are of high and low tides on the coast of Roscoff, France, where the Station Biologique of the University of Paris provides laboratories for marine biologists. (R. Buchsbaum)

Freshwater biologists work in habitats that are far more variable than the oceans and that also undergo greater changes from season to season. Much of their work focuses on how various animals cope with the special ecological and physiological problems of freshwater environments. *Left:* A crustacean biologist is using a net to strain small animals from the surface water at a marshy spot on the shore of Lake Erie. *Right:* A specialist in lake biology (limnology) is holding a small dredge that can be lowered to sample mud from a lake bottom. (R. Buchsbaum)

Terrestrial biologists work with animals in a limited spectrum of phyla—many arthropods (mostly insects), some molluscs, soil nematodes and annelids, vertebrate chordates, and a few others. *Above:* Students are using nets to collect flying insects in an open field in Michigan. Later they will find a greater variety of terrestrial invertebrates by searching among the branches of shrubs and trees, in the grass, under logs and stones, in rotting wood, among ground litter, and in the soil. *Below:* A biologist working at an inland tropical field station, the Barro Colorado Island Biological Laboratory in Panama, administered by the Smithsonian Institution. (R. Buchsbaum)

Other zoologists focus their interests around particular groups of protozoans or animals, rather than on a kind of problem. Specialists on certain groups are called *protozoologists, malacologists, entomologists,* and so on. This is necessarily true of **taxon-**

omists, who describe, name, and classify animals. Because the number of species is so great, a taxonomist can become expert on no more than one or a few small groups. However, modern taxonomists need to master many different techniques, from electron microscopy to biochemistry to sophisticated genetic and statistical analysis. **Morphologists** describe the structure of the animals and, again, usually limit themselves either to one group of organisms or to a comparative study of a particular kind of structure as it occurs throughout the animal kingdom. Many morphologists delve into aspects of physics and engineering in order to understand the forms and functions of biological materials and structures. **Paleontologists** study fossils, and **evolutionary biologists** ideally should have broad backgrounds in paleontology, as well as in comparative morphology and development, and in genetics. Among the various fields of research, the newest areas often tend necessarily to be more descriptive, although they may involve complex laboratory techniques, while older areas that have accumulated a good descriptive background may become largely experimental; but significant biological questions demand both descriptive and experimental approaches.

Invertebrate zoologists of all types are represented on the faculties of academic institutions—colleges and universities—where they devote themselves both to teaching and to basic research in their area of expertise, which may require special laboratory or field facilities. Zoologists in private or government research institutions are mostly concerned with laboratory or field problems of economic or health importance. Some zoologists work in zoos or aquariums, and the essential work of describing and classifying animals goes on most actively in museums, some of which send out field expeditions to collect and study animals all over the world. Another area of employment for zoologists is in private consulting firms that work under contract to government or industry to evaluate the environmental consequences of public or commercial projects, such as construction or oil-drilling.

Zoologists who wish to study animals that cannot be obtained in their own locality often have them shipped in by special arrangements with their colleagues in other localities or may buy them from collectors or from the various biological supply houses, whose facilities are available to anyone who wishes to buy living or preserved animals of a great variety. In addition, there are special **field stations** maintained all over the world to provide laboratory facilities to scientific investigators at places where there is ready access to a rich and varied fauna. Some of these stations offer classes for students and opportunities to see an abundance of invertebrates of many kinds.

Above: The Laboratory of the Marine Biological Association of the United Kingdom, Plymouth, England. (R. Buchsbaum)

Marine Biological Laboratory, Woods Hole, Cape Cod, Massachusetts. (Mar. Biol. Lab.)

Friday Harbor Laboratories of the University of Washington, San Juan Island, Puget Sound (R. Buchsbaum)

Marine stations with laboratory and library facilities, on seacoasts around the world, provide opportunities for students to attend classes and for investigators to do research. Some also maintain aquariums open to the public.

Above: Hopkins Marine Station of Stanford University, Pacific Grove, California. (R. Buchsbaum)

Long Marine Laboratory of the University of California at Santa Cruz. (R. MacTavish)

Knowledge that is gained by scientific investigation is made available in several ways. Every year various scientific groups hold meetings at which members present their original research in short talks, often illustrated with slides or motion pictures. Or they may prepare exhibits of their work. At such meetings they discuss the progress of their research and exchange ideas with other scientists investigating related problems. But by far the most important medium of communication between scientists is publication of papers in scientific journals. There are many hundreds of such journals in biology alone, and in most of them there are papers on some phase of the biology of invertebrates. A small sample of the journals most accessible to and most useful for the student interested in invertebrates is listed in the Bibliography. They can be found in the libraries of large academic and research institutions.

Zoologists read regularly in order to learn as much as possible about what is known in the fields of their investigations, as well as to ascertain that they are not trying to find out something that has already been discovered. A major part of the reading consists of papers in scientific journals because the most complete and reliable sources of information are the published accounts written by the investigators who did the original research. Most scientific papers are primarily reports of original investigations, but almost all of them begin by summarizing what has already been accomplished on the immediate problem under consideration. Scientific papers are the units of which the body of scientific literature is built. The majority of modern scientific articles are difficult to understand without considerable background.

The scientific literature, the whole of published writings of original scientific research, is so voluminous and is increasing at such a rate that scientists find it difficult to scan all the journals to find pertinent papers or to read in their entirety all the papers covering even a limited area of biology. To make their task easier, biologists collectively support, write, and edit a special journal, ***Biological Abstracts***, which consists of short summaries or abstracts of currently published papers, all classified according to subject matter and indexed by authors and subject. There are also present in biology libraries a number of special indexes for finding references to scientific work by subject matter, by author, or by the group of animals concerned, such as the ***Zoological Record.***

Important aids to scientific workers in keeping up with the broad front of advancing zoological knowledge outside the limited field in which they read most intensively are **reviews**—long summarizing articles that synthesize and critically evaluate the evidence bearing on some problem. Such reviews, written by leading scholars in various fields, coordinate and compact the large and often unwieldy literature (sometimes more than a thousand papers) dealing with a single problem. Some journals are devoted entirely to reviews, and in some fields reviews are published annually in a special volume. Since most reviews document the sources in the scientific literature on which the authors have based each of their statements and have extensive bibliographies, they can serve as good starting points from which students can find their way about in the literature of some field that attracts their interest.

Other useful publications on invertebrates are specialized **books** and **monographs** that are devoted to a single subject or animal group. Broader but often just as intensive treatments are the advanced **treatises** written by one or by a number of cooperating specialists and often consisting of many volumes. Treatises may be organized group by group, or they may proceed from topic to topic, using various invertebrates where they most appropriately demonstrate principles or illustrate diversity. Good accounts of the various invertebrate groups can be found in some of the better encyclopedias.

Finally there are **textbooks,** advanced and elementary, in which the authors have selected what they regard as the essential and better-established aspects of the subject for presentation to students. Many provide bibliographies of sources in the scientific literature to which the reader may refer for a more detailed treatment of the topic. *Advanced textbooks* are usually less detailed than treatises, but are organized in the same ways. They hold accessible answers to many of the questions about animals that are asked by students who already have a professional interest in invertebrates.

Elementary textbooks such as ***Animals Without Backbones*** are intended to introduce the invertebrates to a broader readership and are not complete, systematic treatments of all of the various topics or groups that might be of interest to a professional invertebrate zoologist. Certain animals may be emphasized or omitted according to whether the authors choose them to illustrate certain facts or principles.

Few of the statements in textbooks (or in many books, treatises, and encyclopedias) are individually documented, and this anonymity may give some students the erroneous impression that the materials presented are based entirely on the personal experiences and investigations of the authors. On the contrary, the great majority of these materials are based on the scientific literature.

The Bibliography that follows includes some of the kinds of publications in which information about invertebrates can be found—some suitable for the beginning student, others more technical. It is meant only as a small sampling and is much biased toward English-language and especially American sources. But the references contained in these works will lead the interested seeker to a vast literature in many languages and treating localities around the world.

Students interested in pursuing a particular problem should look first in an encyclopedia, an advanced textbook, or a treatise. If the treatment in these is not satisfactory, students should scan the bibliographies of such publications. Each pertinent reference will lead to other sources in the literature. These may reveal that there is a thoroughly worked-out answer to the question, that there are conflicting views on the subject, or that it has not yet even been investigated and is an area wide open for the student's own original research.

Selected Bibliography

Journals Containing Popular Articles on Invertebrates

Audubon. National Audubon Society.
BBC Wildlife.
Endeavour.
Natural History. American Museum of Natural History, New York.
Oceans. Oceanic Society.
Pacific Discovery. California Academy of Sciences.
Scientific American.

Journals Containing Technical Articles on Invertebrates

American Scientist. Sigma Xi.
American Zoologist. American Society of Zoologists.
Animal Behaviour. Association for the Study of Animal Behaviour.
Biological Bulletin. Marine Biological Laboratory, Woods Hole, Mass.
Bioscience. American Institute of Biological Sciences.
Bulletin of Marine Science.
Ecology. Ecological Society of America.
Journal of Experimental Marine Biology and Ecology.
Journal of Marine Biological Association, U.K.
Journal of Morphology. Wistar Institute.
Journal of Parasitology. American Society of Parasitologists.
Marine Biology.
Marine Ecology Progress Series.
Ophelia. Marine Biological Laboratory, Helsingor, Denmark.
Proceedings of the Royal Society of London.
Veliger. California Malacozoological Society.

Journals Containing Review Articles on Invertebrates

Advances in Marine Biology.
Advances in Parasitology.
Annual Reviews of Ecology and Systematics.
Biological Reviews of the Cambridge Philosophical Society.
Quarterly Review of Biology.

Systematic and General References on Invertebrates

Alexander, R. McN. 1979. *The Invertebrates*. Cambridge: Cambridge Unversity Press.
Barnes, R. D. 1987. *Invertebrate Zoology*. 5th ed. Philadelphia: Saunders College.
Encyclopaedia Britannica. 1980. 30 vols. 15th ed. Chicago: Encyclopaedia Brittanica.
Grassé, P.-P., dir. 1949–82. *Traité de Zoologie*. Vols. 1–17. Paris: Masson.

Grassé, P.-P., R. A. Poisson, and O. Tuzet. 1961. *Zoologie I: Invertébrés*. Paris: Masson.

Hyman, L. H. 1940–67. *The Invertebrates*. Vols. 1–6. New York: McGraw-Hill.

Kaestner, A. 1967–70. *Invertebrate Zoology*. Vols. 1–3. Trans. and adapted by H. W. Levi and L. R. Levi. New York: John Wiley & Sons.

Lutz, P.E. 1986. *Invertebrate Zoology*. Reading, Mass.: Addison-Wesley.

McGraw-Hill Encyclopedia of Science and Technology. 1982. 5th edition. New York: McGraw-Hill.

Margulis, L., and K. V. Schwartz. 1982. *Five Kingdoms: An Illustrated Guide to the Phyla of Life on Earth*. San Francisco: W. H. Freeman.

Marshall, A. J., and W. D. Williams, eds. 1972. *Textbook of Zoology: Invertebrates*. 7th ed. New York: American Elsevier.

Meglitsch, P. A., 1972. *Invertebrate Zoology*. 2d ed. New York: Oxford University Press.

Parker, S. B., ed. 1982. *Synopsis and Classification of Living Organisms*. Vols. 1 and 2. New York: McGraw-Hill.

Pearse, V., J. Pearse, M. Buchsbaum, and R. Buchsbaum. 1987. *Living Invertebrates*. Palo Alto: Blackwell Scientific Publications; and Pacific Grove, Calif.: Boxwood Press

Pechenik, J. A. 1985. *Biology of the Invertebrates*. Boston: Willard Grant Press.

Russell-Hunter, W. D. 1979. *A Life of Invertebrates*. New York: Macmillan.

References on Special Aspects of Invertebrate Biology

Barrington, E. J. W. 1979. *Invertebrate Structure and Function*, 2d ed. New York: John Wiley & Sons.

Bullock, T. H., and G. A. Horridge. 1965. *Structure and Function in the Nervous Systems of Invertebrates*. Vols. 1 and 2. San Francisco: W. H. Freeman.

Calow, P. 1981. *Invertebrate Biology: A Functional Approach*. London: Croom Helm.

Chandler, A. C., and C. P. Read. 1961. *Introduction to Parasitology*. 10th ed. New York: John Wiley & Sons.

Clarkson, E. N. K. 1979. *Invertebrate Paleontology and Evolution*. London: George Allen & Unwin.

Corning, W. C., J. A. Dyal, and A. O. D. Willows, eds. 1973–75. *Invertebrate Learning*. Vols. 1–3. New York: Plenum Press.

Fairbridge, R. W., and D. Jablonski. 1979. *The Encyclopedia of Paleontology*. Stroudsburg, Pa.: Dowden, Hutchinson, & Ross.

Florkin, M., and B. T. Scheer, eds. 1967–74. *Chemical Zoology*. Vols. 1–8. New York: Academic Press.

Giese, A. C., and J. S. Pearse, eds. 1974–79. *Reproduction of Marine Invertebrates*. Vols. 1–5, other volumes to follow. New York: Academic Press.

Halstead, B. W. 1978. *Poisonous and Venomous Marine Animals of the World*. 2d ed. Princeton, N.J.: Darwin Press.

Moore, R. D., ed. 1953–83. *Treatise on Invertebrate Paleontology*. Parts A–W, other volumes to follow. Geological Society of America and University of Kansas Press.

Noble, E. R., and G. A. Noble. 1982. *Parasitology: The Biology of Animal Parasites*. 5th ed. Philadelphia: Lea & Febiger.

Schmidt, G. D., and L. S. Roberts, 1985. *Foundations of Parasitology*. 3d ed. St. Louis: C. V. Mosby.

Wainwright, S. A., W. D. Biggs, J. D. Currey, and J. M. Gosline. 1982. *Mechanical Design in Organisms*. 2d ed. Princeton, N.J.: Princeton University Press.

Books on Natural History and Identification of Invertebrates

Buchsbaum, R., and L. J. Milne. 1962. *The Lower Animals: Living Invertebrates of the World.* Garden City, N.Y.: Doubleday.

Gosner, K. L. 1971. *Guide to Identification of Marine and Estuarine Invertebrates: Cape Hatteras to the Bay of Fundy.* New York: John Wiley & Sons.

Grzimek, B. 1974–75. *Grzimek's Animal Life Encyclopedia.* Vols. 1–3. New York: Van Nostrand & Reinhold.

Kozloff, E. N. 1983. *Seashore Life of the Northern Pacific Coast.* Seattle: University of Washington Press.

MacGinitie, G. E., and N. MacGinitie. 1968. *Natural History of Marine Animals.* 2d ed. New York: McGraw-Hill.

Meinkoth, N. A. 1981. *Audubon Society Field Guide to North American Seashore Creatures.* New York: Alfred A. Knopf.

Milne, L., and M. Milne. 1972. *Invertebrates of North America.* New York: Doubleday.

Morris, R. H., D. P. Abbott, and E. C. Haderlie. 1980. *Intertidal Invertebrates of California.* Stanford, Cal.: Stanford University Press.

Ricketts, E. F., J. Calvin, and J.W. Hedgpeth. 1985. *Between Pacific Tides.* 5th ed. Revised by D.W. Phillips. Stanford, Calif.: Stanford University Press.

Smith, R. I., and J. T. Carlton., eds. 1975. *Light's Manual: Intertidal Invertebrates of the Central California Coast.* 3d ed. Berkeley: University of California Press.

Books on Marine Biology

Carefoot, T. 1977. *Pacific Seashores: A Guide to Intertidal Ecology.* Seattle: University of Washington Press.

Hardy, A. 1956. *The Open Sea: The World of Plankton.* London: Collins.

McConnaughey, B. H. 1978. *Introduction to Marine Biology.* 3d ed. Saint Louis: C. V. Mosby.

Nybakken, J. W. 1982. *Marine Biology: An Ecological Approach.* New York: Harper & Row.

Stephenson, T. A., and A. Stephenson. 1972. *Life between Tidemarks on Rocky Shores.* San Francisco: W. H. Freeman.

Sumich, J. L. 1980. *An Introduction to the Biology of Marine Life.* 2d ed. Dubuque, Iowa: William C. Brown.

Books on Freshwater Invertebrate Biology

Edmondson, W. T., ed. 1959. *Fresh-Water Biology.* 2d ed. New York: John Wiley & Sons.

Klots, E. B. 1966. *The New Field Book of Freshwater Life.* New York: G. P. Putnam Sons.

Needham, J. G., and P. R. Needham. 1962. *A Guide to the Study of Freshwater Biology.* 5th ed. San Francisco: Holden-Day.

Pennak, R. W. 1978. *Freshwater Invertebrates of the United States.* 2d ed. New York: John Wiley & Sons.

Books and Monographs on Various Invertebrate Groups

Protozoans:

Grell, K. G. 1973. *Protozoology.* New York: Springer-Verlag.

Sleigh, M. A. 1973. *The Biology of Protozoa.* London: Edward Arnold.

Porifera:

Bergquist, P. R. 1978. *Sponges.* London: Hutchinson.

Cnidaria (Coelenterata):

Muscatine, L., and H. M. Lenhoff, eds. 1974. *Coelenterate Biology: Reviews and New Perspectives.* New York: Academic Press.

Platyhelminthes:

Arme, C., and P. W. Pappas, eds. 1984. *Biology of the Eucestoda.* Vols. 1 and 2. New York: Academic Press.

Chandebois, R. 1976. *Histogenesis and Morphogenesis in Planarian Regeneration.* Monographs in Developmental Biology, vol. 11. Basel: S. Karger.

Smyth, J. D., and D. W. Halton. 1983. *The Physiology of Trematodes.* 2d ed. Cambridge: Cambridge University Press.

Nemertea:

Gibson, R. 1973. *Nemerteans.* London: Hutchinson University Library.

Nematoda:

Croll, N. A., and B. E. Matthews. 1977. *Biology of Nematodes.* New York: John Wiley & Sons.

Maggenti, A. 1981. *General Nematology.* New York: Springer-Verlag.

Poinar, G. O. 1983. *The Natural History of Nematodes.* Englewood Cliffs, N. J.: Prentice-Hall.

Bryozoa:

Larwood, G. P. and C. Nielsen, eds. 1981. *Recent and Fossil Bryozoa.* Fredensborg, Denmark: Olsen and Olsen.

Ryland, J. S. 1970. *Bryozoans.* London: Hutchinson University Library.

Brachiopoda:

Rudwick, M. J. S. 1970. *Living and Fossil Brachiopods.* London: Hutchinson University Library.

Mollusca:

Abbott, R. T., 1974. *American Seashells: The Marine Mollusca of the Atlantic and Pacific Coasts of North America.* New York: Van Nostrand Reinhold.

Boyle, P.R. 1981. *Mollusca and Man.* Institute of Biology's Studies in Biology no. 134. London: Edward Arnold.

Lane, F. W. 1957. *Kingdom of the Octopus.* London: Jarrolds.

Morton, J. E. 1979. *Molluscs.* 5th ed. London: Hutchinson University Library.

Purchon, R. D. 1977. *The Biology of the Mollusca.* 2d ed. Pergamon Press.

Rehder, H. A. 1981. *Audubon Society Field Guide to North American Seashells.* New York: Alfred A. Knopf.

Solem, G. A. 1974. *The Shell Makers: Introducing Molluscs.* New York: John Wiley & Sons.

Wilbur, K. M., ed. 1983–84. *The Mollusca.* Vols. 1–7. New York: Academic Press.

Yonge, C. M., and T. E. Thompson. 1976. *Living Marine Molluscs.* London: Collins.

Annelida:

Dales, R. P. 1967. *Annelids.* London: Hutchinson University Library.

Edwards, C. A., and J. R. Lofty. 1977. *Biology of Earthworms.* 2d ed. New York: John Wiley & Sons.

Mann, K. H. 1962. *Leeches (Hirudinea): Their Structure, Physiology, Ecology, and Embryology.* Elmsford, N. Y.: Pergamon Press.

Arthropoda:

Bliss, D. E., ed. 1982–85. *The Biology of Crustacea.* Vols. 1–10. New York: Academic Press.

Borror, D. J., D. M. DeLong, and C. A. Triplehorn. 1976. *An Introduction to the Study of Insects*. 4th ed. New York: Holt, Rinehart, & Winston.

Bristowe, W. S. 1958. *The World of Spiders*. London: Collins.

Cloudsley-Thompson, J. L. 1958. *Spiders, Scorpions, Centipedes, and Mites*. New York: Pergamon Press.

Daly, H. V., J. T. Doyen, and P. Ehrlich. 1978. *Introduction to Insect Biology and Diversity*. New York: McGraw-Hill.

Foelix, R. F. 1982. *Biology of Spiders*. Cambridge, Mass.: Harvard University Press.

Green, J. 1961. *A Biology of Crustacea*. Chicago: Quadrangle Books.

Insects of Australia. 1970 (Suppl. 1974). Department of Entomology, Commonwealth Scientific and Industrial Research Organization, Canberra. Carlton, Aust.: Melbourne University Press.

Klots, A. B., and E. B. Klots. 1959. *Living Insects of the World*. Garden City, N. Y.: Doubleday.

Linsenmaier, W. 1972. *Insects of the World*. New York: McGraw-Hill.

McLaughlin, P. A. 1980. *Comparative Morphology of Recent Crustacea*. San Francisco: W. H. Freeman.

Milne, L., and M. Milne. 1980. *Audubon Society Field Guide to North American Insects and Spiders*. New York: Alfred A. Knopf.

Richards, O. W., and R. G. Davies. 1977. *Imms' General Textbook of Entomology*. 10th ed. Vols. 1 and 2. London: Chapman & Hall.

Wenner, A. M. 1971. *The Bee Language Controversy*. Boulder, Colo.: Educational Programs Improvement Corp.

Wigglesworth, V. B. 1964. *The Life of Insects*. London: Weidenfeld & Nicolson.

Wilson, E. O. 1971. *The Insect Societies*. Cambridge, Mass.: Harvard University Press.

Sipuncula **and** ***Echiura:***

Stephen, A. C., and S. J. Edmonds. 1972. *The Phyla Sipuncula and Echiura*. London: British Museum (Natural History).

Echinodermata:

Clark, A.M. 1968. *Starfishes and their Relations*. Brit. Mus. (Natl. Hist.), London.

Nichols, D. 1969. *Echinoderms*. London: Hutchinson University Library.

Hemichordata **and** ***Chordata:***

Barrington, E. J. W. 1965. *The Biology of Hemichordata and Protochordata*. San Francisco: W. H. Freeman.

Appendix: Classification

Systems of classification, whether arrived at by consensus of many biologists or by a single highly opinionated author, are somewhat arbitrary, based on incomplete information, and in need of constant rethinking and changes. Consequently, they vary from one source to another. Even the spelling of group names (above the level of family) varies from one scheme to another.

The scheme of classification used throughout this text is summarized below. It was selected for its familiarity and relatively wide acceptance. Classification of protozoans and other protists is in such flux that it seems more useful simply to mention some of them by their traditional names and to group them into broad categories. Classes are listed only for some of the major animal phyla. Certain small or rare phyla of living invertebrates, which were not described in the text, are treated briefly here. We have omitted all extinct groups, except for a few mentioned in the chapter on fossils.

Classification of Organisms
(Numbers refer to text pages)

Kingdom **Archetista:** acellular, saprobic. Viruses, etc.

Kingdom **Monera:** no definite nuclei; autotrophic and/or saprobic. Bacteria, cyanobacteria (blue-green "algas"), spirochetes. (5)

Kingdom **Protista (Protoctista):** nucleated cells, as single cells or colonies; photosynthetic, saprobic, and/or ingestive. (Only forms commonly considered "protozoans" are listed here.) (5)

- **Flagellated protozoans:** euglenoids, volvox, dinoflagellates, collar flagellates, trypanosomes, trichomonads, opalinids, etc. (42–50)
- **Ameboid protozoans:** amebas, foraminiferans, heliozoans, radiolarians, etc. (20–30; 51–56)
- **Spore-forming protozoans:** gregarines, coccidians and hemosporidians, cnidosporidians, etc. (57–60)
- **Ciliated protozoans:** holotrichs, spirotrichs, peritrichs, suctorians. (31–39; 60–67)

Kingdom **Fungi:** nucleated cells, often organized into tissues and organs sometime during the life history; mainly saprobic. Slime molds, yeasts, molds, rusts, and mushrooms.

Kingdom **Plantae (Metaphyta):** nucleated cells usually organized into tissues and organs; mainly photosynthetic. Red, brown, and green algas; liverworts and mosses; ferns; gymnosperms and angiosperms.

Kingdom **Animalia (Metazoa):** nucleated cells usually organized into tissues and organs: mainly ingestive.

Group 1: Acoelomate Phyla

Phylum **Archaeocyatha:** archeocyathids (extinct) (529)

Phylum **Porifera:** sponges (68–85)

Phylum **Placozoa:** Only one species, *Trichoplax adhaerens.* Dorsoventrally flattened and differentiated: ciliated dorsally and ventrally. No permanent anterior-posterior polarity. Single layer of cells sandwiching a core of mesenchyme. Asexual and possibly sexual reproduction. Diameter, 1 to 5 mm. So far found only in marine aquariums containing rocks and organisms from coral reefs.

Phylum **Cnidaria (Coelenterata):** cnidarians (86–147)
- Class **Hydrozoa:** hydromedusas, hydroids, hydrocorals
- Class **Scyphozoa:** jellyfishes
- Class **Anthozoa:** sea anemones, corals, sea fans, etc.

Phylum **Ctenophora:** comb jellies (148–53)

Phyla **Dicyemida** and **Orthonectida:** mesozoans. Microscopic, flattened, ciliated parasites consisting only of a small group of outer cells enclosing one or more reproductive cells. Complicated life history of both asexual and sexual phases. Dicyemid mesozoans common in excretory organs of octopuses. Orthonectid mesozoans found in internal spaces and tissues of turbellarians, nemerteans, polychetes, gastropods and bivalves, and brittle stars.

Phylum **Platyhelminthes:** flatworms (154–203)
- Class **Turbellaria:** turbellarians
- Class **Monogenea:** monogeneans
- Class **Trematoda:** flukes
- Class **Cestoidea:** tapeworms

Phylum **Gnathostomulida:** Elongate ciliated worms, mostly microscopic. Hermaphroditic and without an anus or body cavity, they resemble turbellarian flatworms, but differ in their feeding apparatus of tiny hard jaws and in other features. Found in poorly oxygenated, sandy marine sediments, rich in microorganisms and organic material on which gnathostomulids feed.

Phylum **Nemertea:** proboscis worms (204–11)

Group 2: Pseudocoelomate Phyla

The first six to eight phyla may be grouped as a single phylum, Aschelminthes.

Phylum **Nematoda:** roundworms (212–27)
Phylum **Nematomorpha:** horsehair worms (228–30)
Phylum **Rotifera (Rotatoria):** rotifers (232–37)
Phylum **Gastrotricha:** gastrotrichs (237)
Phylum **Kinorhyncha (Echinodera):** microscopic with a spiny cuticle and a thick thatch of spines on the head that anchors the segmented body in the shallow marine muds where kinorhynchs are mostly found. Feed on microorganisms and organic particles with a sucking pharynx like that of nematodes or gastrotrichs.
Phylum **Loricifera:** newest phylum of animals (described 1983). Microscopic members with spiny heads resemble kinorhynchs, but have unsegmented body encased in a vase-shaped cuticle (lorica). Found in clean coarse marine sand or gravel.
Phylum **Priapula:** cylindrical fleshy worms that burrow in marine sediments; ridged, spiny body from less than a millimeter to several centimeters long. As in sipunculans, the anterior part of the body can be alternately everted or inverted into the trunk, as the priapulan burrows. Spines around the mouth and in the pharynx are everted to capture prey.
Phylum **Acanthocephala:** spiny-headed worms (230–31)
Phylum **Kamptozoa (Entoprocta):** small hydroid-like animals, solitary or colonial, mostly marine. A stem supports the bowl-shaped body with a circle of ciliated feeding tentacles around the rim. The digestive tract is U-shaped, and both mouth and anus open within the circle of tentacles. Asexual reproduction by budding. Ciliated free-swimming larva.

Group 3: Phyla with Coelom Formed by Splitting of Mesoderm Bands; Mouth Develops from Blastopore (Protostomes)

Phylum **Mollusca:** molluscs (248–89)
- Class **Aplacophora:** solenogasters
- Class **Monoplacophora:** *Neopilina*
- Class **Polyplacophora:** chitons
- Class **Gastropoda:** limpets, abalones, snails, slugs
- Class **Scaphopoda:** tooth or tusk shells
- Class **Bivalvia (Pelecypoda, Lamellibranchia):** clams, shipworms, mussels, scallops, oysters.
- Class **Cephalopoda:** squids, octopuses, nautiluses

Phylum **Annelida:** segmented worms (290–317)
Class **Polychaeta:** polychetes
Class **Oligochaeta:** earthworms and other oligochetes
Class **Hirudinea:** leeches
Phylum **Echiura:** spoonworms (461–62)
Phylum **Sipuncula:** peanut worms (463)
Phylum **Pogonophora:** beardworms have a "beard" of tentacles at the front end of an exceedingly long slender body, encased in a secreted tube. They live embedded in soft marine sediments and apparently take all nourishment through the body surface or from symbiotic bacteria within the body, as there is no mouth or digestive tract at any stage. Giant pogonophores called vestimentiferans, some over 1.5 m long, found around vents of warm oxygen-poor water on the deep ocean floor.
Phylum **Tardigrada:** waterbears (461)
Phylum **Pentastomida:** tongueworms. Parasites in lungs or air passages of reptiles, a few in birds and mammals. Up to 16 cm long, they attach to host tissue with 2 pairs of hooks and feed on blood and tissue fluids. Eggs eaten by an intermediate host hatch into tiny 4- or 6-legged larvas. Humans may ingest eggs or larvas in contaminated water or food, and may suffer irritating symptoms, but pentastomids do not mature in human hosts. Sometimes classified as crustaceans.
Phylum **Onychophora:** peripatuses (456–60)
Phylum **Arthropoda:** arthropods
Subphylum **Trilobita:** trilobites (extinct) (511, 512, 522, 523, 528)
Subphylum **Crustacea:** crustaceans (318–51)
Class **Remipedia:** cave swimmers
Class **Cephalocarida:** cephalocarids
Class **Branchiopoda:** brine-shrimps, fairy-shrimps, water-fleas, tadpole shrimps, clam shrimps
Class **Malacostraca:** shrimps, lobsters, crabs, isopods, amphipods, euphausids, mysids, mantis shrimps, etc.
Class **Maxillopoda:** copepods, barnacles, fishlice, etc.
Class **Ostracoda:** seed shrimps
Subphylum **Chelicerata:** chelicerates (352–71)
Class **Arachnida:** spiders, scorpions, mites, etc.
Class **Merostomata:** horseshoe-crabs
Class **Pycnogonida:** sea-spiders
Subphylum **Uniramia:** uniramians
Superclass **Myriapoda** (449–55)
Class **Chilopoda:** centipedes
Class **Diplopoda:** millipedes
Classes **Symphyla** and **Pauropoda:** tiny, blind soil myriapods

Superclass **Insecta (Hexapoda):** insects (372–448)
Classes **Protura, Collembola, Diplura,** and **Thysanura:** wingless insects
Class **Pterygota:** winged insects

Group 4: Lophophorate Phyla

Phylum **Bryozoa:** moss animals (238–43)
Phylum **Brachiopoda:** lampshells (243–47)
Phylum **Phoronida:** Phoronids are marine worms that live in a secreted tube usually embedded vertically in shallow sediments. Visible above the sediment surface is the lophophore of ciliated tentacles, which is horseshoe-shaped and spirally coiled at the ends. Mouth, anus, and excretory pores open close together at the upper exposed end. Adults are sedentary; ciliated larvas, free-swimming.

Group 5: Phyla with Coelom Formed from Outpocketing of Mesoderm Pouches; Mouth Does Not Develop from Blastopore (Deuterostomes)

Phylum **Chaetognatha:** arrow worms (247)
Phylum **Echinodermata:** echinoderms (464–89)
Class **Crinoidea:** feather stars and sea lilies
Class **Asteroidea:** sea stars (starfishes)
Class **Ophiuroidea:** brittle stars or serpent stars
Class **Echinoidea:** sea urchins, sand dollars, etc.
Class **Holothuroidea:** sea cucumbers
Class **Concentricycloidea:** sea daisies.
Phylum **Hemichordata:** hemichordates (500–503)
Class **Pterobranchiata:** pterobranchs
Class **Enteropneusta:** acorn worms
Phylum **Graptolithina:** graptolites (extinct). Sometimes considered related to hemichordates, sometimes assigned to cnidarians. (530)
Phylum **Chordata:** chordates (490–500)
Subphylum **Urochordata (Tunicata):** tunicates, salps, etc.
Subphylum **Cephalochordata (Acrania):** lancelets
Subphylum **Vertebrata (Craniata):** fishes, amphibians, reptiles, birds, mammals

Index

A

B

C

D

E

F

G

H

I

J

O

P

T

U

V

W

Y

Z